Withdrawn
University of Waterloo

D0999765

Handbook of Industrial Cell Culture

Handbook of Industrial Cell Culture

Mammalian, Microbial, and Plant Cells

Edited by

Victor A. Vinci, PhD

Eli Lilly and Company,
Indianapolis, IN

and

Sarad R. Parekh, PhD

Dow AgroSciences,
Indianapolis, IN

Humana Press
Totowa, New Jersey

© 2003 Humana Press Inc.
999 Riverview Drive, Suite 208
Totowa, New Jersey 07512
humanapress.com

All rights reserved. No part of this book may be reproduced, stored in a retrieval system, or transmitted in any form or by any means, electronic, mechanical, photocopying, microfilming, recording, or otherwise without written permission from the Publisher.

All authored papers, comments, opinions, conclusions, or recommendations are those of the author(s), and do not necessarily reflect the views of the publisher.

This publication is printed on acid-free paper. ∞
ANSI Z39.48-1984 (American Standards Institute)
Permanence of Paper for Printed Library Materials.

Production Editor: Mark J. Breaugh.

Cover illustration: (Abstraction of) Maternal inheritance of plastid transgenes. *See* Fig. 3 on page 265.

Cover design: Patricia Cleary.

For additional copies, pricing for bulk purchases, and/or information about other Humana titles, contact Humana at the above address or at any of the following numbers: Tel.: 973-256-1699; Fax: 973-256-8341; E-mail: humana@humanapr.com

Photocopy Authorization Policy:
Authorization to photocopy items for internal or personal use, or the internal or personal use of specific clients, is granted by Humana Press, provided that the base fee of US $20.00 per copy, is paid directly to the Copyright Clearance Center at 222 Rosewood Drive, Danvers, MA 01923. For those organizations that have been granted a photocopy license from the CCC, a separate system of payment has been arranged and is acceptable to Humana Press Inc. The fee code for users of the Transactional Reporting Service is: [0-58829-032-8/03 $20.00].

Printed in the United States of America. 10 9 8 7 6 5 4 3 2 1

Library of Congress Cataloging in Publication Data

Handbook of industrial cell culture : mammalian, microbial, and plant cells / edited by Victor A. Vinci and Sarad R. Parekh.
     p. cm.
  Includes bibliographical references and index.
  ISBN 1-58829-032-8 (alk. paper); 1-59259-346-1 (ebook)
    1. Cell culture--Handbooks, manuals, etc. 2. Biotechnology--Handbooks, manuals, etc.
I. Parekh, Sarad R., 1953– II. Vinci, Victor.

TP248.25.C44 H36 2003
660.6--dc21
                                         2002068470

Preface

The essential and determining feature of an industrial bioprocess is the culturing of cells that yield a desired product. Mammalian, microbial, and plant cells are traditionally used for the manufacture of products derived directly or semisynthetically from cellular metabolites. These cells are increasingly used as the cellular machinery to express recombinant proteins of considerable economic and therapeutic value. The choice of cell culture type determines the degree of success in obtaining a clinically useful product, as well as in achieving an economical process, by facilitating acceptable yield and purity. Each of the major classes of industrially relevant cultures is manipulated by a variety of means, selected for desired phenotypes, and is exploited either in bioreactors or in the field by functionally similar approaches.

The knowledge of how best to achieve this utility has its roots in empirical learning that reaches back many thousands of years. Much later, Pasteur, Koch, and others dramatically advanced our knowledge of the underlying cellular nature of bioprocesses during classical studies in the 19th century using modern scientific methods. Following World War II, the advent of modern industrial production methods, inspired by the discovery and isolation of penicillin, brought the first boom in natural product biotechnology. More recently, the dramatic acceleration in identifying protein biopharmaceutical candidates, as well as the current rebirth in natural product discovery, have been driven by molecular genetics. Likewise, plant cell culture and engineered crops have already impacted agriculture and are poised to revolutionize biotechnology.

The progression of transgenic animal and plant methodologies from laboratory to industrial scale production has resulted in the most recent, and perhaps most dramatic, step in using cells to make products. Supporting the production of novel therapeutics in mammalian, microbial, and plant cells is an impressive array of new methodologies from the fields of molecular genetics, proteomics, genomics, analytical biochemistry, and screening. For an industrial bioprocess, manipulation and propagation of cells in order to elicit expression of a product is followed by the recovery, analysis, and identification of these products. The methodology for successfully developing a commercial process is functionally similar across the spectrum of cell types.

Handbook of Industrial Cell Culture: Mammalian, Microbial, and Plant Cells attempts to link these common approaches, while also delineating those specific aspects of cell types, to give the reader not only an overview of the best current practices, but also of today's evolving technologies, with examples of both their practical applications and their future potential. Many scientists currently in the field find their careers transitioning across work with mammalian, microbial, and plant bioprocesses; thus they are very much in need of a book linking these disciplines in a single format. Moreover, the next generation of scientists and engineers will interface across these disciplines and likely see even more dramatic enhancements in technology. Our hope is that this Handbook will prove especially useful not only to those involved in biotechnology as a broad discipline, but also assist experienced practitioners in perfecting the special art of industrial cell culture.

Victor A. Vinci
Sarad R. Parekh

Contents

Contributors

RICHARD H. BALTZ • *Cubist Pharmaceuticals, Lexington, MA*

GERALD W. BECKER • *Proteomics, Eli Lilly and Company, Indianapolis, IN*

THOMAS BLACK • *Bioprocess Research and Development, Eli Lilly and Company, Indianapolis, IN*

STEVE DEL CARDAYRE • *Maxygen, Redwood City, CA*

NEAL C. CONNORS • *Biocatalysis & Fermentation Development, Merck Research Laboratories, Rahway, NJ*

LAWRENCE M. GELBERT • *Functional Genomics, Lilly Research Laboratories, Eli Lilly and Company, Indianapolis, IN*

JOHN E. HALE • *Proteomics, Eli Lilly and Company, Indianapolis, IN*

HANSJÖRG HAUSER • *Gesellschaft für Biotechnologische Forschung (GBF), Braunschweig, Germany*

JOSEPH J. HEIJNEN • *Laboratory for Biotechnology, Delft University of Technology, Delft, The Netherlands*

MATTHEW D. HILTON • *Bioprocess Fermentation Development, Eli Lilly and Company, Indianapolis, IN*

PONSAMUEL JAYAKUMAR • *Plant Trait Discovery Research, Dow AgroSciences, Indianapolis, IN*

NIGEL JENKINS • *Bioprocess Research and Development, Eli Lilly and Company, Indianapolis, IN*

MICHAEL D. KNIERMAN • *Proteomics, Eli Lilly and Company, Indianapolis, IN*

IGNACIO M. LARRINUA • *Molecular Biology and Traits, Dow AgroSciences, Indianapolis, IN*

DONALD J. MERLO • *Molecular Biology and Traits, Dow AgroSciences, Indianapolis, IN*

JAMES R. MOLDENHAUER • *Central Culture Preservation, Eli Lilly and Company, Indianapolis, IN*

MAURICE M. MOLONEY • *Plant Biotechnology, The University of Calgary, Calgary, Canada*

PETER P. MUELLER • *Gesellschaft für Biotechnologische Forschung (GBF), Braunschweig, Germany*

SARAD R. PAREKH, *Supply Research and Development, Dow AgroSciences, Indianapolis, IN*

JEFFREY S. PATRICK • *Analytical Department, Eli Lilly and Company, Indianapolis, IN*

JOSEPH F. PETOLINO • *Plant Transformation and Gene Expression, Dow AgroSciences, Indianapolis, IN*

KEITH POWELL • *Maxygen, Redwood City, CA*

DIVAKAR RAMAKRISHNAN • *Bioprocess Research and Development, Eli Lilly and Company, Indianapolis, IN*

SAM REDDY • *Molecular Biology and Traits, Dow AgroSciences, Indianapolis, IN*

JEAN L. ROBERTS • *Plant Transformation and Gene Expression, Dow AgroSciences, Indianapolis, IN*

STEVEN ROSE • *Bioprocess Research and Development, Eli Lilly and Company, Indianapolis, IN*

MAX O. RUEGGER • *Molecular Biology and Traits, Dow AgroSciences, Indianapolis, IN*

PAVEL SHIYANOV • *Proteomics, Eli Lilly and Company, Indianapolis, IN*

VIPULA K. SHUKLA • *Molecular Biology and Traits, Dow AgroSciences, Indianapolis, IN*

JEFFREY M. STAUB • *Plant Genetics Division, Monsanto Company, St. Louis, MO*

YUEJIN SUN • *Molecular Biology and Traits, Dow AgroSciences, Indianapolis, IN*

JACQUELINE UNSINGER • *Gesellschaft für Biotechnologische Forschung (GBF), Braunschweig, Germany*

WILLIAM H. VELANDER • *Department of Chemical Engineering, Virginia Polytechnic Institute, Blacksburg, VA*

KEVIN E. VAN COTT • *Department of Chemical Engineering, Virginia Polytechnic Institute, Blacksburg VA*

WALTER M. VAN GULIK • *Laboratory for Biotechnology, Delft University of Technology, Delft, The Netherlands*

WOUTER A. VAN WINDEN • *Laboratory for Biotechnology, Delft University of Technology, Delft, The Netherlands*

VICTOR A. VINCI • *Bioprocess Commercial Development, Eli Lilly and Company, Indianapolis, IN*

DAGMAR WIRTH • *Gesellschaft für Biotechnologische Forschung (GBF), Braunschweig, Germany*

I

MAMMALIAN CELL CULTURE

1

Analysis and Manipulation of Recombinant Glycoproteins Manufactured in Mammalian Cell Culture

Nigel Jenkins

1. Introduction

Mammalian cell culture has progressed in recent years from a lab-scale enterprise, mainly used for in vitro testing of chemical agents, to a major source of therapeutic and prophylactic agents. Industrial-scale cell culture had its origins in vaccine production from adherent fibroblasts in roller bottles; however, the dominant technology employed today uses suspension-adapted Chinese Hamster ovary (CHO) or hybridoma cells in large-scale (up to 12,000 L) tanks.

Since microbial cells are easier to manipulate and have lower production costs than mammalian cells, they are the method of choice for peptides and simple proteins such as insulin and growth hormone. However, many therapeutic proteins require complex post-translational modifications such as glycosylation, gamma-carboxylation, and site-specific proteolysis, and mammalian cells are uniquely equipped to perform these operations.

This chapter explores the methods and interpretation used to optimize the most extensive post-translational protein modification: glycosylation.

Many animal-cell proteins are modified by the covalent additions of oligosaccharides. These additions often result in significant mass increases and can cover a large portion of their surfaces. There are two common classes of oligosaccharides: N- and O-linked (Fig. 1). The oligosaccharides (also called glycans) usually contain high levels of sialic acids, mannose, galactose, N-acetylglucosamine (GalNAc) and fucose residues. N-linked oligosaccharides are linked through an N-glycosidic bond between a GlcNAc residue and an asparagine residue (1), and fall into two broad categories: high mannose and complex types. In most cases, the core unit is composed of a pentasaccharide core consisting of three branched mannose residues sequentially attached to two GlcNAc residues via an amide linkage to asparagine. This common core structure is formed from the same precursor (dolichol-linked oligosaccharide), which is responsible for the glycan transfer to nascent peptides, and is then further processed to form a wide variety of structures. In high and oligo-mannose glycans found in yeast and insect cells no further glycan processing is observed. In proteins containing and the complex glycans typically found in mammals and higher plants, further processing takes place

From: *Handbook of Industrial Cell Culture: Mammalian, Microbial, and Plant Cells*
Edited by: V. A. Vinci and S. R. Parekh © Humana Press Inc., Totowa, NJ

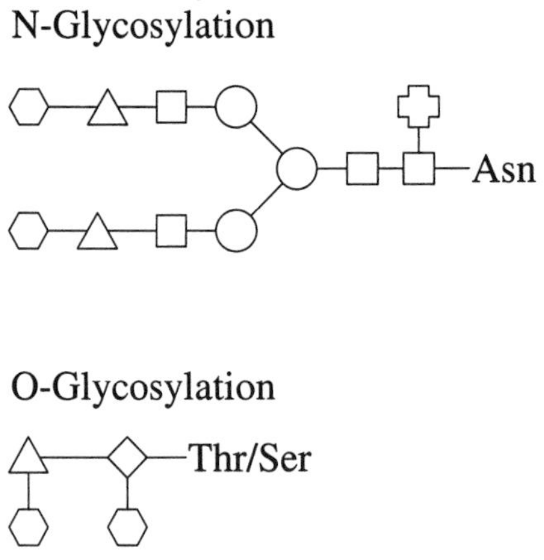

N-Glycosylation

O-Glycosylation

Fig. 1. Carbohydrate structures associated with N- and O-glycosylation.

in *cis*, medial, and trans components of the Golgi apparatus. This processing adds GlcNAc, fucose, galactose, and sialic acids to the core structure.

In addition to the bi-antennary structures containing two arms for each glycan, complex triantennary and tetra-antennary structures are also found on some glycoproteins such as erythropoetin (EPO) and blood-clotting factors. The O-linked oligosaccharide is linked through an O-glycosidic bond between the GlcNAc residue and either serine or threonine, and is not processed by the same enzymes as N-glycans. Multiple O-glycosylation sites are found in mucin-type proteins. Individual glycosylation sites may contain a mixture of oligosaccharides (microheterogeneity). This is believed to be caused by competition between glycosyltransferase enzymes that add nucleotide-monosaccharide units to the glycoprotein as it travels through the Golgi compartments.

Translation of both cytoplasmic and membrane proteins is initiated on the processed mRNA in the cytoplasm. During and following protein translation in the endoplasmic reticulum (ER), a number of modifications can occur such as proteolytic cleavage, protein folding, and glycosylation. There is evidence that certain chaperone proteins recognize different glycosylation modifications and act as a quality-control mechanism for ER-Golgi and intra-Golgi translocations *(2)*. The presence of cellular proteases and exoglycosidases in the culture media—particularly those released by dead and dying cells—can often result in product degradation toward the end of culture.

Glycosylation is believed to have several biological roles, including protein stability (resistance to proteolysis and aggregation, improved protein folding), clearance in vivo, antigenicity, and in some cases, biological activity. Both mannose and asialoglycoprotein (ASG) receptors in the liver remove glycoproteins with terminal mannose or galactose residues from the bloodstream *(3)* and influence the plasma half-life of a glycoprotein in vivo. In the case of immunoglobulins, effector functions such as Fc-receptor binding and complement activation can be influenced by the glycosylation state of the Fc region *(4)*.

2. Glycoprotein Analysis

Increased availability and reliability of carbohydrate isolation and detection methods have led to major advances in our understanding of protein glycosylation and how it is controlled. In general, the level of carbohydrate analysis expected by the regulatory authorities increases as the glycoprotein drug passes from early- (IND) to late-phase (BLA) clinical trials. A common theme is a requirement to demonstrate the influences of oligosaccharide content on the glycoprotein's biological efficacy in humans.

Basic information on the presence or absence of oligosaccharide (macroheterogeneity) can be obtained by using polyacrylamide gel electrophoresis (PAGE), Western blotting using carbohydrate-specific lectins, or fluorophore-labeled carbohydrates *(5)*. Another technique used for the rapid analysis of glycoproteins is capillary electrophoresis (CE), involving the separation of glycoforms by electrophoretic flow through a narrow bore capillary based on net charge, mol wt, and micellar association. Recent improvements on the basic CE technique include microbore capillary isoelectric focusing (cIEF).

Common methods for the analysis of microheterogeneity include lectin-affinity chromatography, exoglycosidase digestion, mass spectrometry (MS), high-performance anion-exchange chromatography, and nuclear magnetic resonance (NMR). Lectins are carbohydrate-binding proteins, which are specific for oligosaccharide structures, and when these are covalently immobilized in silica, individual oligosaccharides or glycopeptides can be separated from a heterogeneous population. Lectin-affinity HPLC analysis can provide a basic fingerprint for most glycoproteins, although several lectin columns would be needed to complete the structural analysis *(6)*.

The availability of larger quantities (>1 g) of biopharmaceuticals allows analytical methods to reach higher accuracy but with low sensitivity. NMR is a powerful technique that allows complete structural information to be recovered with the disadvantage of requiring large amounts of protein. The interpretation of 1H NMR spectra requires knowledge of uniquely known structures such as the H-2 and H-3 of mannose, H-3 of sialic acid, the methyl protons of N-acetyl groups and specific glycan linkages. In contrast, the most sensitive approach to glycosylation analysis involves various forms of mass spectrometry. Fast-atom bombardment (FAB-MS) has the highest mass accuracy of the spectrometric methods, yet it is also expensive and requires a large amount of sample. Two other recently developed methods of ionization are particularly suitable for glycoprotein and glycan analysis: matrix-assisted laser desorption/ionization (MALDI-MS) and electrospray ionization (ES-MS). Both are relatively rapid techniques used to study both the primary (peptide) structure and post-translation modifications of recombinant proteins. For MALDI-MS, the sample is mixed with excess matrix that strongly absorbs the laser light (e.g., sinapinic acid). When the mixture is subjected to laser light, it is vaporized into ions which can be measured by a time-of-flight (TOF) analyzer, with an effective mass range of approx 0.5–200 kDa. Several laboratories have successfully used this technique to elucidate mol wt and structural details of oligosaccharides in combination with an exoglycosidase array sequencing *(7,8)*. ES-MS detectors are often coupled to high-performance liquid chromatography (HPLC), allowing in-line analysis of glycopeptides generated by the controlled digest of glycoproteins. This method allows complete structural analysis of glycoproteins with a very high mass accuracy.

3. Cell Type, Metabolism, and Environment

The cell line or tissue in which a glycoprotein is produced has dramatic effects on the oligosaccharide attached to the glycoprotein. This is caused by the differences in relative activities and groups of glycosyltransferases and nucleotide-sugar donors that differ between species and tissue type. The N-linked carbohydrate populations associated with both Asn25 and Asn97 glycosylation sites of interferon IFN-γ have been characterized by MALDI-MS in combination with exoglycosidase array sequencing *(9–11)*. Recombinant IFN-γ produced by CHO cells showed that N-glycans were predominantly of the complex bi- and triantennary type at both N-glycosylation sites. In the same recombinant glycoprotein produced by baculovirus-infected Sf9 insect cells, glycans were mainly tri-mannosyl core structures, with no direct evidence of post-ER glycan-processing events other than core fucosylation and de-mannosylation. Transgenic mouse-derived IFN-γ exhibited considerable site-specific variation in N-glycan structures with Asn25-linked carbohydrates of the complex, core fucosylated type and Asn97-linked carbohydrates were mainly of the oligomannose type, with smaller proportions of hybrid and complex N-glycans *(12)*. These data indicate that ER to *cis*-Golgi transport is a predominant rate-limiting step in both expression systems. However, both sialyltransferase activity and CMP-NeuAc substrate were found to be absent in uninfected or baculovirus-infected Sf9, Sf21, and Ea4 insect cells. These data demonstrate the profound influence of host-cell type and protein structure on the N-glycosylation of recombinant proteins, and the considerable challenges faced in re-engineering nonvertebrate cells to reproduce mammalian-type glycosylation.

4. Effects of Chemical Supplements on Protein Glycosylation

Changes in the cell culture environment can also lead to changes in the oligosaccharide structures observed *(13)*. Some of the conditions studied include the age of the batch culture, glucose limitation *(14)*, oxygen starvation *(15)*, intracellular ammonium ion accumulation *(16)*, and pH excursions *(17)*. In general, these batch and fed-batch conditions have led to poorer glycosylation profiles of recombinant glycoproteins compared to those produced in perfusion or chemostat cultures *(18)*. This is most probably caused by a combination of nutrient depletion and waste metabolite accumulation in batch and fed-batch systems.

The control of protein glycosylation is a major goal of the biotechnology industry, since glycosylation is the most extensive post-translation modification made by animal cells. Protein structure, cell type, and cell-culture environment all affect the N-glycosylation machinery and contribute to glycoprotein heterogeneity. Exogenous chemicals that are not essential to maintain cell growth in defined media are sometimes added to improve recombinant protein production. Sodium butyrate has been shown to unbind the chromatin structure, thus making it more accessible to RNA polymerase and promoting mRNA replication *(19)*. As transfected genetic material will integrate at the most accessible loci, it has been shown that these genes are most likely to be transcribed during sodium butyrate treatment. It has also been shown that sodium butyrate can increase the expression of certain chaperone-like proteins (i.e., GRP78 and 94) by preventing histone phosphorylation and increasing hyperacetylation. Sodium butyrate can also cause a dilation of the ER and influence cell-cycle kinetics by stabilizing a population in the G1 phase. Lamotte et al. *(20)* also studied the effects of butyrate on the glycosylation pathway of CHO cells, and found a positive correlation between both IFN-γ production and N-glycan complexity with increasing doses of sodium butyrate.

Dichloroacetate (DCA) is known to inhibit cell growth, glutamine catabolism, and the production of pyruvate and alanine. Little is known about the effect of this chemical on glycosylation, but the literature shows that recombinant antibody yields can be increased by up to 60% *(21)*. N-acetyl mannosamine, a sialic acid precursor capable of entering the cell, has been shown to be effective in increasing the intracellular sialic acid pool and sialic acid content of recombinant glycoproteins produced by CHO cells, but its high cost may prohibit its utility in large-scale *(22)*.

Lipids play an essential role in the glycosylation of many proteins. They act as sugar carriers, and the dolichol donor is used for the block transfer of oligosaccharide units from the cytosol to the ER. Both complex lipid supplements and dolichol derivatives have been shown to improve glycosylation at Asn 97 of γ-IFN *(13,23)*. Cholesterol can also be an important growth supplement for CHO cells, and is required for optimal NS0 cell growth. Thus, cholesterol, phospholipids (i.e., phosphatidyl choline), and fatty acids (linoleic and oleic acids) are usually included in serum-free formulations.

The following section provides a detailed description of two examples from our laboratories using transfection of the 2,6-sialyltransferase (2,6-ST) gene to correct an inherent defect in CHO cells and produce recombinant glycoproteins with modified properties.

5. Transfection of 2,6-ST to Produce a Universal Host CHO Line and its Effects on Recombinant IFN-γ

CHO are widely used to produce glycosylated recombinant proteins, and are equipped with a glycosylation machinery very similar to the human equivalent *(24)*. A notable difference concerns sialylation: N-linked glycans of human origin carry terminal sialic acid residues in both α2,3- and α2,6-linkages, whereas only α2,3-terminal sialic acids are found in glycoproteins from CHO and baby hamster kidney (BHK) cells. Indeed, these cell lines lack a functional copy of the gene that codes for 2,6-ST (EC 2. 4. 99. 1). Sialic acid residues confer important properties to glycoproteins because of their negative charge and the terminal position. Specific receptors, such as the ASG receptor, have evolved in the liver and in macrophages to bind and internalize glycoproteins devoid of terminal sialic acid molecules *(4,25,26)*. Moreover, the binding of various members of the sialoadhesin family *(27)* and CD22 ligand *(28)* to terminal sialic acid residues is specific for either the α2,3 or α2,6 linkage.

Our group and others have demonstrated that the sialylation defect of CHO cells can be corrected by transfecting the 2,6-ST cDNA. Glycoproteins produced by such CHO cells display both α2,6- and α2,3-linked terminal sialic acid residues, similar to human glycoproteins. Here, we have established a CHO cell line that stably expresses α2,6-ST, providing a universal host (UH) for further transfections of human genes. Several relevant parameters of the UH cell line were studied, demonstrating that the α2,6-ST transgene was stably integrated into the CHO cell genome, that transgene expression was stable in the absence of selective pressure, that the recombinant sialyltransferase was correctly localized in the Golgi, and finally, that the bioreactor growth parameters of the UH were comparable to those of the parental cell line. A second step consisted in the stable transfection into the UH of cDNAs for human glycoproteins of therapeutic interest such as IFN-γ *(29)*.

5.1. Materials and Methods

The availability of DNA sequences that code for glycosyltransferases *(30)* has raised the possibility to genetically alter the glycosylation capabilities of a cell line. In particular, the cDNA for the rat liver α2,6-ST has been transiently or stably transfected into BHK *(18)* or CHO cells *(30,24)*, and the sialylation of a target recombinant protein has been confirmed to be in both α2,3- and α2,6-linkages. The plasmids pSfiSV-2,6-ST, pSfiSVneo, pSfiSVdhfr and pCISfiT: a new vector for in vitro amplification *(24)* carrying the human CMV promoter were built from plasmid pCI (Promega, Madison, WI). First, one *Sfi*I site was inserted into the unique *Not*I site and the resulting plasmid (pCISfi) was further modified by introduction into the filled-in *Bam*HI site of the oligonucleotides containing the 41-basepair (bp) stretch corresponding to nucleotides +189 to +229 of the human gastrin gene, reported to act as a transcription terminator *(29)*. The resulting vector was named pCISfiT, and contained the *Sfi*I site necessary for multimerization at the end of the multiple cloning site, plus a transcription terminator sequence downstream of the polyadenylation signal, to improve transcription efficiency from multimerized transcription units. Plasmid pCISfiT-IFN-γ contained the 1-kb open reading frame (ORF) coding for human IFN-γ, excised from plasmid pSVL-IFN-γ (obtained from Glaxo Smith-Kline) by cleavage with *Bam*HI, filling in with the Klenow fragment of T4-DNA polymerase and cleavage with *Xho*I.

The fragment was inserted between the *Xho*I and *Sma*I sites of the vector pCISfiT. DNA concatenamers for stable transfections were prepared according to the in vitro amplification method *(31)* by mixing the *Sfi*I-flanked expression units and the selectable marker unit at the following ratios: α2,6-ST transfection into CHO Dux B11 cells: pSfiSV-2,6-ST / pSfiSVneo = 10:1; and IFN-γ into the UH: pCISfiT-IFN-γ / pSfiSVdhfr = 20:1. Ligation products were confirmed to extend between 50 and at least 200 kb by pulsed-field electrophoresis extracted sequentially with phenol/chloroform, according to standard procedures and used for lipofection-mediated transfections.

For Southern blot analysis, CHO genomic DNA was extracted from cells by a salting-out procedure, and completely digested with *Sfi*I. 10 μg of digested DNA were separated on a 0.8% DNA gel, and adequate amounts of the *Sfi*I-cleaved plasmids pSfiSV-2,6-ST and pCISfiT-IFN-γ were loaded along with the samples, to allow estimation of band intensities. DNA was then blotted onto a HybondN+ nylon membrane (Amersham Pharmacia Biotech, Uppsala, Sweden) and hybridized to the ^{32}P-labeled, 607-bp *Bgl*II fragment of the rat α2,6-ST cDNA, according to standard techniques. The filter was then stripped and re-hybridized to the ^{32}P-labeled, 411-bp *Ssp*I fragment of the human IFN-γ cDNA. Adjacent cassettes of the α2,6-ST cDNA integrated in tandem into the genome were revealed as a single 3.5-kb band.

5.2. Results

Approximately ten copies of the 2,6-ST expression unit were detected in both in the UH and clone 54 (IFN-γ producing) cell lines, and were also maintained in clone 54 cells grown in 2 μM methotrexate (MTX). A similar analysis was performed on UH cells grown without G418 for 2 mo, and a comparable number of integrated copies was obtained indicating the stability of 2,6-ST expression in the absence of selection agent. Conversely, less than 10 copies of the IFN-γ cassette were integrated into the genome of clone 54, and as expected a dramatic increase in IFN-γ copy number was observed in clone 54 cells grown in 2 μM MTX amplification agent.

For transfection, 5×10^5 CHO-DxB11 cells were plated onto 60-mm dishes and 10 μg of DNA concatenamers, were mixed with 2 m*M* ethanolic solution of DOGS (Transfectam®, Gibco-LTI). Two days after transfection, cells were exposed to selective medium: either complete medium supplemented with 900 μg/mL G418 for neomycin resistance, or MEM alpha medium without nucleosides and deoxynucleosides for dhfr selection. Resistant colonies were obtained and individually grown in 96-well plates.

Amplification in MTX was performed by exposing individual clones to increasing concentrations ranging from 10 nM to 2 μ*M*. Cells were exposed to each MTX concentration for at least 3 wk; resistant cells were pooled and expanded. Control UH cells were maintained in the above medium, but containing 500 μg/mL G418 for neomycin selection. Clone 54 cells, deriving from the stable transfection of the UH with IFN-γ cDNA, and CHO clone 43 cells, stably producing IFN-γ were grown in MEM alpha medium without nucleosides and deoxynucleosides, and 2 μ*M* MTX, the medium for clone 54 only also contained 500 μg/mL G418.

Flow cytometry using fluorescently labeled probes is a powerful method of searching for rare clones expressing a particular phenotype from a mixed population of cells. We used fluorescein-5-isothiocyanate (FITC)-labeled lectin from *Sambucus nigra* (SNA) that recognizes the Gal-α2,6-sialic acid epitope to isolate cells expressing the recombinant 2,6-ST enzyme transfected into CHO cells. The technique can be used in two modes, either to correct the sialylation defect in cells that are already making a recombinant protein of interest (in this case IFN-γ) or to correct the defect in the parental cell line, which can subsequently be used as a "universal host" for expressing recombinant glycoproteins with both α2,3- and α2,6-sialylation. It could also be used to pick clones that produce large amounts of the recombinant protein of interest (using a fluorescent antibody) and simultaneously check for authentic product quality (sialylated species using a different fluorophore covalently linked to SNA lectin).

In this experiment, the presence of α2,6-sialic acid linkages was examined in isolated clones by flow cytometry analysis of cells incubated with fluorescein-5-isothiocyante (FITC)-labeled SNA lectin *(24,29)*. 2,6-ST enzyme activity was again assayed in cell lysates from each clone by a lectin-based microtiter plate assay *(32)*.

CHO Dux B11 cells were stably transfected with DNA concatenamers composed of an average of 10 expression units of rat 2,6-ST per neomycin resistance expression unit, using the in vitro amplification method. Twenty-seven G418-resistant colonies were screened by exploiting the specific binding of α2,6-sialic acid residues on plasma-membrane glycoproteins to SNA lectin, using flow cytometry analysis. Fourteen (52%) colonies displayed a positive binding of between 75 and 100% of each cell population. α2,6-ST enzyme activity was assayed on these colonies by the specific solid-phase assay *(32)*, and ranged between 12 and 68 IU of α2,6-ST/mg protein. Figure 2 shows the correlation between flow cytometry positivity and α2,6-ST activity, and the clone that displays the highest indices of both parameters was chosen for further characterization and named the UH. The stability of expression of α2,6-ST in the UH was tested by growing the cells in the presence or in the absence of G418 selective pressure for 2 mo, and binding to SNA lectin was again analyzed by flow cytometry. Identical flow cytometry profiles were obtained in both cases, indicating a stable expression of the 2,6-ST enzyme in the UH.

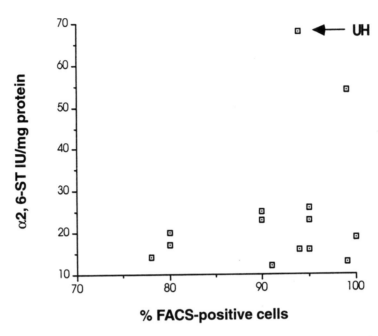

Fig. 2. α2,6-ST expression in individual CHO Dux B11 clones. Correlation between positive FITC-SNA-lectin binding by flow cytometry analysis and α2,6-ST enzyme activity in clones of CHO Dux B11 cells stably transfected with α2,6-ST cDNA. The arrow indicates the clone chosen as universal host (UH).

In order to demonstrate that the α2,6-ST introduced into UH cells can act on the glycans of recombinant glycoproteins, the cDNA for human IFN-γ (placed under the hCMV promoter) was transfected into the UH, using *dhfr* as the second selectable marker. Conditioned media from 200 individual clones were analyzed by an IFN-γ-specific enzyme-linked immunosorbent assay (ELISA), and displayed up to 1 pg IFN-γ/cell.d. To reach higher production levels, five of these clones were exposed to gradually increasing MTX concentrations, up to 2 μ*M*. A steady increase in IFN-γ levels was observed, reaching a maximum of 37 pg/cell (18 mg/L) per 24 h in clone 54. To test for expression stability, clone 54 cells were grown in the presence or in the absence of MTX for 2 mo and assayed for IFN-γ production and for α2,6-ST activity. IFN-γ levels were completely abolished following removal of MTX, and full 2,6-ST activity was maintained as determined by flow cytometry analysis, demonstrating that 2,6-ST expression was not affected by MTX treatment. The fact that no variation in copy number was observed after 2 mo in culture without selective pressure indicates that the 2,6-ST concatenamer was stably integrated into the UH genome, and that no significant recombination events occurred during the observed time span. The copy number of the ST transgene did not change even after transfection with IFN-γ and MTX treatment, indicating that the two transgenes had integrated independently.

Naturally occurring sialyltransferases have been localized to the Golgi by immunofluorescence *(33)* and immunoelectron microscopy *(34)*. We utilized a rabbit antiserum to rat 2,6-ST (a gift from Prof. E. Berger, Zurich) and tetramethylrhodamine B isothiocyanate (TRITC)-conjugated polyclonal secondary antibody against rabbit Ig from DAKO to localize this enzyme. A trans-Golgi marker, NBD-C_6-ceramide from Molecular Probes, was used to confirm the organelle location of 2,6-ST. Cells were grown for 2 d on glass cover slips, fixed in 2% paraformaldehyde, washed in phosphate-buffered saline (PBS), saline, and fatty acid-free bovine serum albumin (BSA). For α2,6-ST detection, cells were permeabilized with 0.1% Triton X-100 for 10 min and incubated with the rat α2,6-ST antiserum for 1 h followed by the secondary antibody for 30 min. Negative control experiments were also performed, in which the primary antibody was omitted. For Golgi staining, the permeabilization step was omitted and cells were incubated with 5 µ*M* NBD-C_6-ceramide for 60 h. Cover slips were mounted with a phenylenediamine-mounting medium and analyzed by the fluorescence photomicroscope. The 2,6-ST antiserum strongly labeled vesicles in limited areas of the cytoplasm near the nucleus, consistent with a localization in the trans-Golgi apparatus, as demonstrated by the identical distribution of NBD-C_6-ceramide. The intracellular distribution of the recombinant 2,6-ST in the UH line was similar to the one observed for the natural enzyme in the human HepG2 cell line, known to express α2,6-ST.

The UH cell line was grown in a stirred-tank bioreactor to verify that the integration of the exogenous gene coding for the α2,6-ST enzyme did not compromise the growth and metabolism of the cell line in suspension culture. The culture conditions have been described elsewhere *(35)*, but medium was supplemented here with adenosine, cytidine, guanosine, uridine, 2'-deoxyadenosine, 2'-deoxycytidine, 2'-deoxyguanosine, and 2'-deoxyuridine to achieve a final concentration of 1 mg/L. The cell-growth phase took place over a period of 50 h, and cell viability remained higher than 95% during this period, with a maximal viable cell concentration of 1. 2 3 10^6 cells/mL. The maximal cell-specific growth rate (0. 04 h^{-1}) was high as compared with other CHO cell lines, but the uptake of nutrients (glucose and glutamine) and the production of toxic substances (lactate and ammonium) were similar to those observed with the parental cell line.

For IFN-γ purification, an IFN-γ affinity column was generated by coupling 20B8 monoclonal anti-IFN-γ antibody (Lonza Biologics) to CNBr-activated Sepharose. The conformation of sialic acids in IFN-γ purified from either UH cells (clone 54) or CHO Dux B11 cells (clone 43) was determined in duplicate by reverse-phase HPLC utilizing 2,3-sialic acid specific sialidase from Newcastle Disease Virus and the non-discriminating sialidase from *Arthrobacter ureafaciens*. The total amount of sialic acid was determined spectrophotometrically, following acid-catalyzed hydrolysis of duplicate samples *(36)*. The sialic acid conformation on IFN-γ purified from clone 54 (2,6-ST$^+$) was 40.4% sialic acid in α2,6-linkage and 59.6% in α2,3-linkage. Conversely, IFN-γ from clone 43 (control, nontransfected) exclusively displayed α2,3-linkages, as expected. The sialic acid content was similar in both IFN-γ preparations (1.16 and 1.12 moles sialic acid per mole of protein, respectively). The two IFN-γ preparations were therefore comparable in the of extent of sialylation, and differ only in the presence of α-2,6-linked sialic acids in the protein produced in 2,6-ST+ (clone 54) cells. This indicates that the stable expression of 2,6-ST did not hamper the sialylation capability of the cells, and that the added sialyltransferase competes with the endogenous one for the pool of donor CMP-sialic acid under these culture conditions.

In a pharmacokinetic study, female Sprague-Dawley rats (160–270 g) were anesthetized with Hypnorn, and IFN-γ was administered by intravenous (iv) infusion (0.3 mg/kg). The tail was clipped, and blood (100 µL) was then collected via the tail vein at 3,

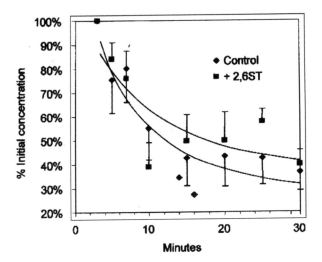

Fig. 3. Clearance of IFN-γ glycoforms in rats. IFN-γ from the universal host (■) and from parental CHO (◆) cell lines were injected into rats and serum concentration was determined at the defined times.

5, 7, 10, 15, 20, 25, 30, and 35 min post-administration. Blood samples were centrifuged, plasma IFN-γ levels were determined by ELISA, and data was processed using the S-Plus statistical package from MathSoft Inc. (Fig. 3). The difference of log-AUC (Area Under Concentration curve) between the two groups was statistically significant ($p = 0.013$), and indicates that the recombinant IFN-γ from UH cells persists longer in the blood compartment as compared to the same protein made in control cells. The clearance values obtained for the two groups were highly statistically significant ($p < 0.001$).

To test whether the IFN-γ produced in the UH line as well as in normal CHO cells was biologically active, upregulation of the endogenous expression of major histocompatibility complex (MHC) class II *(37)* in human colon carcinoma LoVo cells was assessed by flow cytometry analysis. 100 ng/mL each of IFN-γ purified from either clone 54 or clone 43 elicited an upregulation in MHC class II of 3.00 ± 0.61-fold and 3.52 ± 0.52-fold, respectively. In comparison, 100 ng/mL of purified, standard human IFN-γ at 10 IU/ng produced a 4.50 ± 0.49-fold increase in MHC class II expression in the same cell line (means ± SE for two measurements, run in quadruplicate). The observed differences between the two samples and between each sample and the standard were not statistically significant, as determined by analysis of variance (ANOVA) analysis, indicating that the biological activity of all three IFN-γ preparations was similar.

5.3. Conclusion

Our group and several others have successfully transfected mammalian cells with the genes for various glycosyltransferases. Here, we have established and fully characterized a CHO universal host line with improved glycosylation capabilities, thus providing a relevant tool for the production of recombinant proteins for human therapy. Such a cell line is able to produce glycoproteins carrying both α2,6- and α2,3-linked terminal sialic acids, as found in human glycoproteins but not in products from normal CHO cells. The in vitro amplification transfection method *(31)* was used. In keeping with our previous results *(24)*, a high proportion (52%) of positive clones was obtained. The average number of expression cassettes per DNA concatenamer was increased in the present work from 5 to 10, in an attempt to enhance the expression of α2,6-ST, and clones with high α2,6-ST activity were obtained. The best of these clones, named "universal host," was studied to determine its suitability as a cellular host for the production of recombinant proteins.

In theory, an alternative strategy to the two-step process requiring preparation of the UH line first, followed by transfection of the target glycoprotein, could consist of the co-expression of both the α2,6-ST and IFN-γ in one step, by transfection of DNA concatenamers, including both expression cassettes and the selectable marker gene. However, preliminary transfections of CHO Dux B11 cells with IFN-γ DNA concatenamers, including the neomycin resistance gene as a selectable marker, had failed to yield stable clones displaying satisfactory production levels. The use of the *dhfr* selectable marker gene coupled to MTX treatment was therefore determined by the mandatory. The two-step strategy discussed here was eventually selected, to allow more flexibility in individually modulating the expression levels of the two genes. Indeed, continued selective pressure was not needed to maintain satisfactory α2,6-ST activity in the UH cell line. Conversely, high-level production of IFN-γ in MTX-amplified cells proved unstable, as is often the case in the CHO/dhfr expression system *(38)*.

Recently, TNK-tPA, a variant of wild-type tissue plasminogen activator, was produced in CHO cells that over-expressed α-2,3-ST *(39)*. The purified glycoprotein displayed increased sialylation levels, as compared to the control protein produced in wild-type CHO cells; this imparted improved pharmacokinetic properties to the protein following iv injection into rabbits. The prolonged permanence of TNK-tPA in circulation was attributed to decreased recognition of the highly sialylated product by the ASG receptor. The differences in pharmacokinetics observed for IFN-γ are caused by subtle changes in the stereochemistry of the glycan structure. Although further work is required to investigate other possible alterations in biological activity resulting from the presence of α2,6-sialic acid residues, this is already a positive indication of the effects of the linkage alteration we have introduced in the glycosylation properties of CHO cells.

In conclusion, we have established a CHO UH cell line with improved sialylation capabilities. The UH line can be further engineered to produce therapeutically relevant glycoproteins, whose sialylation more closely resembles the pattern found in human products. Correctly sialylated and biologically active IFN-γ was produced by the UH line, and displayed prolonged pharmacokinetics.

6. Sialylation of Human IgG-Fc Carbohydrate by Transfecting Rat α2,6-sialyltransferase

The IgG molecule is composed of four structural and functional regions: two identical antigen-binding Fab regions linked through a flexible hinge to an Fc region that has binding/activation sites for effector ligands, e.g., C1q and Fcγ receptors *(40,41)*. The Fc region is comprised of part of the hinge with inter-heavy-chain disulfide bridges, two unpaired glycosylated C_H2 domains, and two noncovalently paired C_H3 domains. Glycosylation of the C_H2 domain at Asn-297 is essential to activation of effector functions *(4)*, and together with structure at or proximal to the N-terminus of the C_H2 domain, generates recognition sites for leukocyte Fcγ receptors and complement *(25)*. Secreted IgG is a heterogeneous mixture of glycoforms that exhibit variable addition of the sugar residues fucose, galactose, sialic acid, and bisecting N-acetylglucosamine. The efficiency of effector function activation can vary between the glycoforms—e.g., galactosylation enhances recognition of IgG by complement or by human Fcγ receptors.

6.1. Materials and Methods

In an initial study, a recombinant IgG3 immunoglobulin against the hapten nitroiodophenacetyl (NIP) with Phe-243 replaced by Ala (FA243) was expressed in a CHO-K1 parental cell line. The resulting IgG-Fc-linked carbohydrate was significantly α2,3-sialylated (53% of glycans), as indicated by normal and reverse-phase HPLC analyses (42) in contrast to wild-type immunoglobulins, which are poorly (<5%) sialylated. We transfected the rat 2,6-ST gene into this FA243 cell line, to investigate the biological properties of this new IgG3 *(43)*.

The vector used, pCISfiT containing an insert encoding for the rat 2,6-ST gene, was a gift from Dr. Monaco. A *Kpn*1/*Xho*1 digested fragment of pCISfiT containing the gene (1.7 Kb) was ligated into the single *Kpn*1/*Xho*1 digested site of the pBK-CMV expression vector (Stratagene, UK) and transformed into E. coli (TG1) competent cells by calcium shock. Plasmid DNA (pBK-CMV-α2,6-ST) was isolated from a number of individual kanamycin-resistant colonies, and shown by restriction with *Kpn*I/*Xho*I to contain the expression vector and the 1.7-Kb insert encoding the α2,6-ST gene. The CHO-K1 cell line FA243 was transfected with pBK-CMV-2,6-ST by spheroplast fusion. This cell line had previously been transfected with the genes encoding for FA243 IgG3 and was selected on the basis of its resistance to mycophenolic acid *(42)*. A 24-well plate was seeded with the transfected cells and left for 48 h prior to selection with the antibiotic G418. The adherent cell lines were incubated in RPMI-1640 medium containing G418 (0.6 m*M*). After 2 wk in selection medium, transfected colonies were visible in all the wells when limiting dilution was used to clone the transfectants. FA243 CHO-K1 cells that had not been transfected with the pBK-CMV-α2,6-ST vector construct did not form colonies. Once the cloned cells reached confluence, they were checked for maintained IgG$_3$ production by ELISA.

Clones that maintained levels of IgG production were assayed for 2,6-ST activity using the SNA-ELISA method *(32)*. The final FA243-ST clone was selected on the basis of optimal α2,6-ST activity (60.8 ± 6.6 mU/mg [*n* = 4]) relative to the FA243 cell line (14.1 ± 2.9 mU/mg [*n* = 4]), and amplified in roller-bottle culture in order to produce IgG3 antibody. Cells were grown to a final cell density of 2.5 3 10^6 cells/mL in 2-L roller bottles. The supernatant was collected, and IgG3 antibodies were purified from cell supernatant on a column of NIP-Sepharose 4B and eluted with 0.5 m*M* NIP. Sodium dodecyl sulfate (SDS) PAGE was used to check the purity of the IgG3 proteins.

To release oligosaccharides from IgG3 antibodies, 3 nmoles of purified antibody was incubated with protoeglycanase F and the reducing ends of the released N-linked oligosaccharides were reductively aminated with the fluorophore 2-aminoacridone using sodium cyanoborohydride. Labeled glycans were analyzed by normal-phase HPLC using a GlycoSep N column (Oxford Glycosciences), and detected fluorescence using excitation and emission wavelengths of 428 and 525 nm. The sialic acid content and the nature of its linkage to the oligosaccharide were evaluated as described previously in this chapter.

Table 1
Sialic Acid Composition and Linkage of Oligosaccharides Derived
from Anti-NIP IgG3 Wild-Type and Mutant Antibodies

Sialic Acid Linkage	Wild-type	Std. Dev.	FA243	Std. Dev.	FA243-ST+	Std. Dev.
% α2,3	2.4	0.1	56.7	0.4	35.4	3.5
% α2,6	0.0	0.0	1.9	0.6	30.1	3.0
Total	2.4	0.1	58.6	0.4	65.5	0.4

The nmoles sialic acid per nmole of IgG were determined in each case from 3 nmoles IgG, so that a theoretical yield of 100% corresponds to 6 nmoles sialic acid. Data are presented for three replicate determinations. The α2,3 and total sialic acids were determined following release with sialidase from Newcastle disease virus and *A. ureafaciens*, respectively, and the α2,6 values determined by subtraction of α2,3 from the total value.

6.2. Results

All samples displayed a heterogeneous range of complex biantennary oligosaccharide structures with a single N-glycosylation site on each heavy chain at Asn-297 within the C_H2 domain of IgG, but the oligosaccharide profiles varied between the IgG3 antibodies. The glycans released from the wild-type IgG3 showed low levels of galactosylation as compared to those observed for the FA243 and FA243-ST IgG3 antibodies. Similarly, there was a much higher degree of sialylation (mono-sialyl + di-sialyl) for glycans from FA243 IgG3 and FA243-ST IgG3 than for the wild-type, which had minimal levels of sialylation. Thus, galactosylation and sialylation are inhibited for the nascent oligosaccharide chains of the wild-type IgG relative to its FA243 replacement. IgG-Fc-linked glycans were sialylated (60% of glycans) such that the ratio of α2,6-α2,3-linked sialic acids was almost equal (Table 1).

The FA243 replacement in IgG3 permits significant α2,6-sialylation, comparable to the physiological level found for human polyclonal IgG, which has 27% ± 7% of oligosaccharide moieties that are α2,6 sialylated. One explanation for these findings is that sialylation is inhibited by virtue of extensive noncovalent interactions of the C_H2 domain with the nascent oligosaccharide chain *(44)* interactions, which are reduced in the case of the replacement FA243. We suggest that for wild-type IgG3 expressed in CHO-K1 cells the extensive noncovalent interactions strongly inhibit galactosylation and sialylation, which is overcome to some extent in the case of the FA243 replacement, but still sufficient to exert the predominant limiting influence on the extent of sialylation. Thus, sialylation is controlled primarily by the protein structure local to the carbohydrate, and the two sialyltransferases compete to sialylate the nascent oligosaccharide.

Glycoproteins other than IgG have only minimal noncovalent interactions between the carbohydrate and the protein. An increased level of total sialylation (95%) has been reported *(45)* for recombinant human trace β-protein expressed by BHK cells relative to the native protein (60%) isolated from human cerebrospinal fluid (CSF). Thus, it is reasonable to conclude that various factors can limit α2,3-ST and α2,6-ST activity associated with these different glycoprotein substrates.

It has previously been shown that outer-arm galactosylation enhances recognition of IgG by FcγRI, C1q, and complement *(46)*. We hypothesized that remodeling the glycans on the IgG3 would modulate its biological activities. The ability of the wild-type, NA297, FA243, and FA243-ST IgG3 antibodies to trigger lysis mediated through human complement was determined. Sheep red blood cells (RBC) were NIP-derivatized and tested with various antibody preparations, followed by either Guinea pig or human serum to quantify complement-mediated cell lysis. The FA243 IgG3 required a 1.8-fold increased antibody concentration to yield 50% lysis compared to the wild-type IgG3. However, FA243-ST+ IgG3 demonstrated a capacity to trigger complement-mediated cell lysis similar to that of wild-type IgG3, indicating that the presence of the $\alpha2,6$ sialylation linkage compensates for the effect of the replacement FA243. A similar trend was obtained using guinea pig complement, so that FA243-ST required only a 1.4-fold increased IgG concentration to yield 50% lysis relative to the wild-type IgG3, compared to 1.9-fold for the FA243 and sevenfold for NA297 IgG3.

Superoxide anion was measured as lucigenin-enhanced chemiluminescence, using U937 cells treated with IFN-γ, and IgG3-sensitized NIP-derivatized human RBC in the presence of 0.25 mM lucigenin. This assay measures the ability to trigger a respiratory burst mounted through FcγRI expressed on IFN-γ stimulated U937 cells. FA243 IgG3 exhibited a reduced ability to trigger superoxide relative to the wild-type IgG3 (~ 70% of the maximal response). However, the FA243-ST IgG3 triggered a maximal response that was 1.3-fold greater than that observed for the wild-type protein. Thus, $\alpha2,6$ sialylation contributes to recognition of IgG by human FcγRI.

Binding of antigen-complexed IgG3 to K562 cells via FcγRII receptors was determined by rosette formation *(47)* using human red cells were derivatized with NIP and sensitized with anti-NIP IgG3. The aglycosylated IgG3 antibody NA297 showed a greatly reduced capacity to rosette, confirming the requirement for carbohydrate to support recognition of IgG by human FcγRII. FA243 IgG3 exhibited a twofold reduced capacity to rosette, relative to the wild-type IgG3. However, FA243-ST IgG3 showed a capacity to rosette that was similar to the wild-type IgG3, indicating that $\alpha2,6$ sialylation contributes to recognition of IgG by human FcγRII.

6.3. Conclusions

These results suggest that the nature of the sialylation linkage can influence recognition by C1q/complement, FcγRI and FcγRII. A plausible explanation is that the $\alpha2,6$-sialic acid linkage can affect the structural dynamics of the oligosaccharide moiety, particularly the primary and secondary GlcNAc residues, which have a predominant influence on the recognition of IgG by its effector ligands *(48)*. The oligosaccharide moieties may additionally affect the conformation of the lower hinge-binding site for effector ligands on the IgG protein, stabilizing conformations that are more tightly bound by the effector ligands.

7. Conclusion

This chapter review detailed examines the multiple sources of glycan heterogeneity that can be found in both natural and recombinant glycoproteins. Advances in carbohydrate analysis have facilitated detailed, site-specific determinations of each glycoform, and have paved the way to find methods by which glycans can be controlled and manipulated. We have reviewed several chemical approaches used to improve protein glycosylation, and have detailed two of our own case studies (using recombinant IFN-γ and IgG3), in which genetic manipulation of sialyltransferase in the host CHO cell line resulted in significant changes to the properties of each molecule. A similar genetic strategy has also been used to incorporate bisecting GlcNAc residues into recombinant antibodies, which proved useful for enhancing antibody-dependent cytotoxicity *(49)* using the enzyme Glycosyltransferase III.

In the future, these chemical and genetic methods will also be assessed against the recently introduced technique of post-fermentation manipulation of recombinant glycoproteins—using cheap, microbial-derived sialyltransferases and sugar-nucleotide precursors invented by J.C. Paulsen and commercialized by Neose Technologies. Although these alternatives pose their own challenges in scale ability and enzyme elimination in the final product, time will reveal whether they provide more cost-effective methods of manipulating recombinant glycoproteins.

References

1. Kornfeld, R. and Kornfeld, S. (1985) Assembly of asparagine-linked oligosaccharides. *Annu. Rev. Biochem.* **54**, 631–664.
2. Fagioli, C. and Sitia, R. (2001) Glycoprotein quality control in the endoplasmic reticulum. Mannose trimming by endoplasmic reticulum mannosidase I times the proteasomal degradation of unassembled immunoglobulin subunits. *J. Biol. Chem.* **276**, 12,885–12,892.
3. Bianucci, A. M. and Chiellini, F. (2000) A 3D model for the human hepatic asialoglycoprotein receptor (ASGP-R). *J. Biomol. Struct. Dyn.* **3**, 435–451.
4. Jefferis, R., Lund, J., and Pound, J. D. (1998) IgG-Fc mediated effector functions: molecular definition of interaction sites for effector ligands and the role of glycosylation. *Immunol. Rev.* **163**, 59–76.
5. Jackson, P. (1996) The analysis of fluorophore-labeled carbohydrates by polyacrylamide gel electrophoresis. *Mol. Biotechnol.* **5**, 101–123.
6. Lee, K. B., Loganathan, D., Merchant, Z. M., and Linhardt, R. J. (1990) Carbohydrate analysis of glycoproteins. A review. *Appl. Biochem. Biotechnol.* **23**, 53–80.
7. Sutton, C. W., O'Neill, J. A., and Cottrell, J. S. (1994) Site-specific characterization of glycoprotein carbohydrates by exoglycosidase digestion and laser desorption mass spectrometry. *Anal. Biochem.* **218**, 34–46.
8. James, D. C., Goldman, M. H., Hoare, M., Jenkins, N., Oliver, R. W. A., Green, B. N., et al. (1996) Post-translational processing of recombinant human interferon-gamma in animal expression systems. *Protein Sci.* **5**, 331–340.
9. Hooker, A. D., Green, N. H., Baines, A. J., Bull, A. T., Jenkins, N., Strange, P. G., et al. (1999) Constraints on the transport and glycosylation of recombinant IFN-gamma in Chinese hamster ovary and insect cells. *Biotechnol. Bioeng.* **63**, 559–572.
10. James, D. C., Freedman, R. B., Hoare, M., Ogonah, O. W., Rooney, B. C., Larionov, O. A., et al. (1995) N-glycosylation of recombinant human interferon-γ produced in different animal expression systems. *Bio-technology* **13**, 592–596.
11. Ogonah, O. W., Freedman, R. B., Jenkins, N., Patel, K., and Rooney, B. C. (1996) Isolation and characterization of an insect-cell line able to perform complex N-linked glycosylation on recombinant proteins. *Bio-technology* **14**, 197–202.
12. Kemp, P. A., Jenkins, N., Clark, A. J., and Freedman, R. B. (1996) The glycosylation of human recombinant alpha-1-antitrypsin expressed in transgenic mice. *Biochem. Soc. Trans.* **24**, 339.
13. Jenkins, N., Parekh, R. B., and James, D. C. (1996) Getting the glycosylation right—implications for the biotechnology industry. *Nat. Biotechnol.* **14**, 975–981.

14. Hayter, P. M., Curling, E. M. A., Baines, A. J., Jenkins, N., Salmon, I., Strange, P. G., et al. (1991) Chinese hamster ovary cell growth and interferon production kinetics in stirred batch culture. *Applied Microbio. Biotechnol.* **34**, 559–564.

15. Regoeczi, E., Kay, J. M., Chindemi, P. A., Zaimi, O., Suyama, K. L.(1991) Transferrin glycosylation in hypoxia. *Biochem. Cell Biol.* **69**, 239–244.

16. Yang, M. and Butler, M. (2000). Effect of ammonia on the glycosylation of human recombinant erythropoietin in culture. *Biotechnol. Prog.* **16**, 751–759.

17. Borys, M. C., Linzer, D. I., and Papoutsakis, E. T. (1993) Culture pH affects expression rates and glycosylation of recombinant mouse placental lactogen proteins by Chinese hamster ovary (CHO) cells. *Bio-technology* **6**, 720–724.

18. Grabenhorst, E., Schlenke, P., Pohl, S., Nimtz, M., and Conradt, H. S. (1999) Genetic engineering of recombinant glycoproteins and the glycosylation pathway in mammalian host cells. *Glycoconj. J.* **16**, 81–97.

19. Sowa, Y. and Sakai, T. (2000) Butyrate as a model for "gene-regulating chemoprevention and chemotherapy." *Biofactors* **12**, 283–287.

20. Lamotte, D., Eon-Duval, A., Acerbis, G., Distefano, G., Monaco, L., Soria, M., et al. (1997) Controlling the glycosylation of recombinant proteins expressed in animal cells by genetic and physiological engineering, in *Animal Cell Technology* (Carrondo, M. J. T., ed.) Kluwer Academic Publishers, The Netherlands, pp. 761–765.

21. Murray, K. and Dickson, A. J. (1997) Dichloroacetate inhibits glutamine oxidation by decreasing pyruvate availability for transamination. *Metabolism* **46**, 268–472.

22. Gu, X. and Wang, D. I. (1998) Improvement of interferon-gamma sialylation in Chinese hamster ovary cell culture by feeding of N-acetylmannosamine. *Biotechnol. Bioeng.* **58**, 642–648.

23. Green, N. H., Hooker, A. D., James, D. C., Baines, A. J., Strange, P. G., Jenkins, N., et al. (1997) Control of interferon-gamma glycosylation by the addition of defined lipid supplements to batch cultures of recombinant chinese hamster ovary cells, in *Animal Cell Technology, Basic & Applied Aspects*, Vol 8. (Funatsu, A., ed.) Kluwer Academic Publishers, The Netherlands, pp. 339–345.

24. Monaco, L., Marc, A., Eon-Duval, A., Acerbis, G., Distefano, G., Lamotte, D., et al. (1996) Genetic engineering of $\alpha2,6$-sialyltransferase in recombinant CHO cells and its effects on the sialylation of recombinant interferon-γ. *Cytotechnology* **22**, 197–203.

25. Lund, J., Tanaka, T., Takahashi, N., Sarmay, G., Arata, Y., and Jefferis, R. (1990) A protein structural change in aglycosylated IgG3 correlates with loss of huFcγRI and huFcγRII binding and/or activation. *Mol. Immunol.* **27**, 1145–1153.

26. Sarmay, G., Lund, J., Rozsnyay, Z., Gergely, J., and Jefferis, R. (1992) Mapping and comparison of the interaction sites on the Fc region of IgG responsible for triggering antibody dependent cellular cytotoxicity (ADCC) through different types of human Fc receptor. *Mol. Immunol.* **29**, 633–639.

27. Morgan, A., Jones, N. D., Nesbitt, A. M., Chaplin, L., Bodmer, M. W., and Emtage, J. S. (1995) The N-terminal end of the C_H2 domain of chimeric human IgG1 anti-HLA-DR is necessary for C1q, FcγRI and FcγRII binding. *Immunology* **86**, 319–324.

28. Crocker, P. R., Hartnell, A., Munday, J., and Nath, D. (1997) The potential role of sialoadhesin as a macrophage recognition molecule in health and disease. *Glycoconj. J.* **14**, 601–609.

29. Bragonzi, A., Distefano, G., Buckberry, L. D., Acerbis, G., Foglieni, C., Lamotte, D., et al. (2000) A new Chinese hamster ovary cell line expressing alpha-2,6-sialyltransferase used as universal host for the production of human-like sialylated recombinant glycoproteins. *Biochim. Biophys. Acta* **1474**, 273–282.

30. Lee, E. U., Roth, J., and Paulsen, J. C. (1989) Alteration of terminal glycosylation sequences on N-linked oligosaccharides of chinese hamster ovary cells by expression of β-galactoside $\gamma2,6$-sialyltransferase. *J. Biol. Chem.* **264**, 13,848–13,855.

31. Monaco, L., Tagliabue, R., Soria, M. R., and Uhlen, M. (1994) An in vitro amplification approach for the expression of recombinant proteins in mammalian cells. *Biotechnol. Appl. Biochem.* **20**, 157–171.
32. Mattox, S., Walrath, K., Ceiler, D., Smith, D. F., and Cummings, R. D. (1992) A solid-phase assay for the activity of CMPNeuAc-Gal β1,4GlcNAc-R α2,6-sialyltransferase. *Anal. Biochem.* **206**, 430–436.
33. Lipsky, N. G. and Pagano, R. E. (1985). A vital stain for the Golgi apparatus. *Science* **228**, 745–747.
34. Marks, D. L., Wu, K., Paul, P., Kamisaka, Y., Watanabe, R., and Pagano, R. E. (1999) Oligomerization and topology of the Golgi membrane protein glucosylceramide synthase. *J. Biol. Chem.* **274**, 451–456.
35. Hayter, P. M., Curling, E. M., Gould, M. L., Baines, A. J., Jenkins, N., Salmon, I., et al. (1993) The effect of dilution rate on CHO cell physiology and recombinant interferon-g production in glucose-limited chemostat cultures. *Biotechnol. Bioeng.* **42**, 1077–1085.
36. Fukuda, M. and Kobata, A. (1994) Glycobiology, A Practical Approach, IRL Press, Oxford, UK.
37. Waldburger, J. M., Masternak, K., Muhlethaler-Mottet, A., Villard, J., Peretti, M, Landmann, S., et al. (2000) Lessons from the bare lymphocyte syndrome: molecular mechanisms regulating MHC class II expression. *Immunol. Rev.* **178**, 148–165.
38. Kaufman, R. J., Wasley, L. C., Spiliotes, A. J., Gossels, S. D., Latt, S. A., Larsen, G. R., et al. (1985) Coamplification and coexpression of human tissue-type plasminogen activator and murine dihydrofolate reductase sequences in Chinese hamster ovary cells. *Mol. Cell Biol.* **5**, 1750–1759.
39. Weikert, S., Papac, D., Briggs, J., Cowfer, D., Tom, S., Gawlitzek, M., et al. (1999) Engineering chinese hamster ovary cells to maximize sialic acid content of recombinant glycoproteins. *Nat. Biotechnol.* **17**, 1116–1121.
40. Shields, R. L., Namenuk, A. K., Hong, K., Meng, Y. G., Rae, J., Briggs, J., et al. (2000) High resolution mapping of the binding site on human IgG1 for FcγRI, FcγRII, FcγRIII and FcRn and design of IgG1 variants with improved binding to FcγR. *J. Biol. Chem.* **276**, 6591–6604.
41. Sondermann, P., Huber, R., Oosthuizen, V., and Jacob, U. (2000) A structural basis for immune complex recognition: the 3.2Å crystal structure of the human IgG1 Fc-fragment-FcγRIII complex. *Nature* **406**, 267–273.
42. Lund, J., Takahashi, N., Popplewell, A., Goodall, M., Pound, J. D., Tyler, R., King D. J., and Jefferis, R. (2000) Expression and characterization of truncated forms of humanized L243 IgG1: architectural features can influence synthesis of its oligosaccharide chains and affect superoxide production triggered through human Fcγ receptor I. *Eur. J. Biochem.* **267**, 7246–7256.
43. Jassal, R., Jenkins, N., Charlwood, J., Camilleri, P., Jefferis, R., and Lund, J. (2001) Sialylation of human IgG-Fc carbohydrate by transfected rat alpha2,6-sialyltransferase. *Biochem. Biophys. Res. Commun.* **286**, 243–249.
44. Deisenhofer, J. (1981) Crystallographic refinement and atomic models of a human Fc fragment and its complex with fragment B of protein A from Staphylococcus aureus at 2.9- and 2.8Å resolution. *Biochemistry* **20**, 2361–2370.
45. Hoffmann, A., Nimtz, M., Wurster, U., and Conradt, H. S. (1994) Carbohydrate structures of β-trace protein from human cerebrospinal fluid- evidence for brain-type N-glycosylation. *J. Neurochem.* **63**, 2185–2196.
46. Tsuchiya, N., Endo, T., Matsuta, K., Yoshinoya, S., Aikawa, T., Kosuge, E., et al. (1989) Effects of galactose depletion from oligosaccharide chains on immunological activities of human IgG. *J. Rheumatol.* **16**, 285–290.
47. Lund, J., Takahashi, N., Pound, J. D., Nakagawa, H., Goodall, M., and Jefferis, R. (1995). Oligosaccharide-protein interactions in IgG can modulate recognition by Fcγ receptors. *FASEB J.* **9**, 115–119.

48. Mimura, Y., Church, S., Ghirlando, R., Ashton, P. R., Dong, S., Goodall, M., et al. (2000). The influence of glycosylation on the thermal stability and effector function expression of human IgG1-Fc: properties of a series of human glycoforms. *Molec. Immunol.* **37**, 697–706.

49. Umana, P., Jean-Mairet, J., Moudry, R., Amstutz, H., and Bailey, J. E. (1999). Engineered glycoforms of an antineuroblastoma IgG1 with optimized antibody-dependent cellular cytotoxic activity. *Nat. Biotechnol.* **17**, 176–180.

2

Genetic Approaches to Recombinant Protein Production in Mammalian Cells

Peter P. Mueller, Dagmar Wirth, Jacqueline Unsinger, and Hansjörg Hauser

1. Control of Gene Expression

Genetically engineered mammalian cells play an essential role in many processes, from basic research to high-throughput screening and pharmaceutical protein production. In basic research, mammalian cells serve to study gene function and mechanisms of regulation. Important health-related applications include drug screening and production of secreted, pharmaceutically active proteins. The reason that mammalian cells are preferred is the close relationship to cells and their products in the human body. In particular, mammalian cells have the unique capability to authentically process, fold, and modify secreted human proteins. The resulting products are free of microbial contaminants, thereby minimizing the risk of immunogenic and inflammatory responses, respectively. In addition, human-like modifications extend the in vivo lifetime of therapeutic proteins. This translates into therapeutic products that are safe and highly active.

However, compared to the alternative cell-culture systems, protein production in mammalian cells is time-consuming and expensive. Multiple factors affect recombinant protein production in mammalian cells, such as the specific cell line used and the expression vector, chromosomal integration site, and copy number of the integrated recombinant gene, selection procedures, cell-culture conditions, and medium employed (1).

A standard recombinant protein production strategy is to first choose a suitable producer-cell line. Then, the DNA carrying the gene of interest and a selection marker gene are transfected into the host cells. Only a small minority of the transfected cells will integrate the recombinant DNA into the genome. Cells that express the selection marker grow and form clones under appropriate conditions, whereas untransfected cells are eliminated. However, there is little correlation between selection-marker gene expression and the expression level of the gene of interest. The productivity of newly isolated clones is unpredictable, variable, and usually modest. Since the optimal producer clones are generally not those with the highest initial productivity, a large number of clones must be screened and characterized further. Limitations may occur at various levels within a producer-cell clone, including transcriptional silencing, inefficient or aberrant mRNA processing, instability of the recombinant mRNA, low transla-

From: *Handbook of Industrial Cell Culture: Mammalian, Microbial, and Plant Cells*
Edited by: V. A. Vinci and S. R. Parekh © Humana Press Inc., Totowa, NJ

tional efficiency, and bottlenecks in post-translational modifications and in secretion. In addition, mammalian cells are fragile, they grow slowly, and require complex media and sophisticated fermentation setups for the production process. Because of the limited maximal cell densities achieved, the overall production is one or more orders of magnitude below values achieved with bacterial, fungal, or insect cell systems. For these reasons, sophisticated vector design and elaborate producer-cell adaptation and optimization procedures are used to improve process productivity.

Interestingly, in their native environment highly specialized cells represent the most efficient mammalian expression systems known in the body, such as ß-globin synthesis in erythroblasts or immunoglobulin secretion by B-cells. Understanding the principles of gene expression in such cells could lead to recombinant cell lines with superior production properties.

1.1. Vector Design

Generally, mammalian expression vectors are circular shuttle plasmid vectors (Fig. 1). Such plasmids usually carry a mammalian gene-expression cassette and an optional mammalian selection-marker gene. Alternatively, a mammalian selection marker can be provided on a second plasmid that is co-transfected with the expression construct. Since standard transfection protocols result in the integration of multiple copies of the transfected DNA into the mammalian genome, the frequent co-integration of the expression construct and at least one copy of a drug-resistant gene results in the selection of mostly producer cells in the presence of the drug. To facilitate vector DNA preparation, an origin of replication and a bacterial selection marker gene allow for the propagation and selection of these plasmids in *E. coli*.

In most cases, the expression of the gene of interest is driven by strong constitutive viral or cellular promoters, such as the cytomegalovirus (CMV) promoter or the translation elongation factor EF2α gene promoter. Mammalian promoters are generally extremely complex, but contain two basic parts, enhancer sequences and a minimal promoter sequence (Fig. 1). Enhancers often contain repeated sequences that are binding sites for transcriptional activators, whose position relative to the transcription start site can be surprisingly variable, and can even function downstream of the transcribed region. The minimal promoter sequence binds factors of the basic RNA polymerase II-dependent transcriptional machinery and determines the transcriptional start site. In practice, natural promoters are much more complex, and the interactions and contributions of the various transcriptional elements have rarely been investigated in detail. However, the composite nature of mammalian promoters allows construction or adaptation for special purposes. In particular, for regulated gene expression, enhancer elements of the CMV promoter can be replaced by binding sites for recombinant or regulated transcription activators such as binding sites for the tetracycline-regulated transactivator. In addition, cell-specific enhancers can even be added downstream of the transcription unit to increase expression in the cell line of interest. Immediately downstream of the promoter, multiple unique restriction endonuclease recognition sequences (multiple cloning site) allow the insertion of the reading frame of interest.

Most mammalian genes contain introns. For recombinant gene expression, the much shorter minimal protein coding sequences that are derived from reverse transcription of

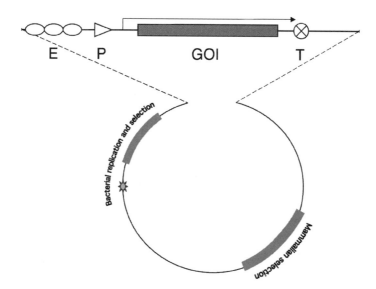

Fig. 1. Typical components of a basic expression vector. An enhancer (E) consisting of multiple transcriptional activator binding sites is a major determinant of the transcriptional activity. The minimal promoter element (P) frequently contains a conserved sequence element (TATA box) that constitutes a binding site for the RNA polymerase II transcription factor TBP and determines the transcription initiation site (arrow). Transcription of the reading frame of the gene of interest (GOI) is terminated by an mRNA precursor cleavage site and poly(A) addition signal (T) that defines the 3' end of the mRNA. Optionally, the vector may carry a second transcription unit encoding a mammalian selection gene. A bacterial origin of replication and an antibiotic resistance selection gene facilitate DNA manipulations, cloning, and plasmid DNA preparation in *E. coli*.

mature intronless mRNA (cDNA) are often used. However, introns and splice sites are also recognition sites for proteins that facilitate the mRNA export from the nucleus. Therefore, some expression constructs that contain a natural or artificial intron are more efficiently expressed than an intronless version of the same gene. Further improvement of gene expression can sometimes be achieved by replacing unfavorable secondary RNA structures and upstream initiation codons in the 5' noncoding region with a favorable unstructured 5' mRNA with an optimal consensus initiation sequence *(2)*, or by removing mRNA destabilizing sequences downstream of the protein coding region that are present in many relevant cytokine mRNAs. To avoid the generation of unstable transcripts, the protein-coding region must be followed by a cleavage and poly(A) tail addition site. This appears to be a somewhat less critical choice than promoter selection, and there are various efficient poly(A) addition sites available that are currently used.

1.2. Regulated Gene Expression

Regulated promoters can be used to restrict recombinant gene expression of toxic or growth-inhibitory products to the final production phase, in which further cell growth is no longer essential. Many such regulatory systems are available that respond to specific medium additives or to growth conditions. One major goal is to achieve efficient expression under inducing conditions while keeping basal expression low in the noninduced state. Presently, the most popular system for this purpose consists of two elements, a chimeric transactivator protein and a DNA construct containing an inducible promoter with multiple transactivator binding sites *(3)*. The transactivator tTA comprises the DNA-binding domain of the prokaryotic tetracycline repressor protein fused to a eukaryotic transcriptional-activating domain. In the original construct, DNA-binding of the tTA transactivator is prevented in the presence of tetracycline. When tetracycline is omitted, the tTA protein binds to the cognate promoter and efficiently activates transcription. The system has been developed to such a high degree of sophistication that it is the method of choice for regulated expression. In some cell lines, very high transcription rates have been achieved, as well as ranges of expression spanning five orders of magnitude *(3)*. In addition to the tetracycline-repressed transactivator, an inverse system that is tetracycline-inducible has been developed *(4)*. The level of the transactivator protein present in the cell plays a major role in the transcriptional activity of the tetracycline-dependent promoter. A number of such cell lines expressing tetracycline-regulated transactivators as well as the tTA-dependent promoter system are commercially available (Tet-On™/Tet-Off™; Clontech).

Another development of the tTA system are autoregulatory expression systems in which both, the gene of interest and the transactivator are expressed from the tTA-dependent promoter. Low basal levels of transactivator gene expression allow self-amplified activation upon withdrawal of tetracycline. In addition, tTA overexpression can negatively affect the cell growth rate or even reduce cell viability. An advantage of autoregulation is that tTA-expression and its potentially toxic side effects are restricted to the time of induction. Furthermore, the autoregulatory cassettes can be expressed in a tissue-independent manner *(5)*. Because of the necessary minimal basal expression level in the repressed state, autoregulatory expression units cannot be applied for expression of highly toxic proteins, which are lethal even if expressed in minimal amounts. Autoregulatory expression cassettes based on modified bidirectional tTA-dependent promoters have been established *(6–8)*.

A related streptogramin-regulated system is available, which can be used as an alternative to the tetracycline system or in tandem with the same cells to regulate the expression of two different products independently *(9)*. Another type of regulated expression system that is widely used is steroid hormone-receptor-responsive transactivators. Hormone-dependent transcription activators have been constructed by fusing hormone-binding domains of steroid receptors to DNA-binding domains of unrelated heterologous proteins *(10)*. DNA-binding domains from the yeast *Saccharomyces cerevisiae* Gal4 protein have been used in combination with the hormone-binding domain of the mammalian estrogen receptor, which can be induced by estradiol *(11)*. Analogously, a progesterone-receptor fusion protein that can be activated by RU 486 at concentrations much lower than those required for antiprogesterone activity *(12,13)* and a system based on the heterodimeric insect ecdyson receptor can be induced by the synthetic ecdysteroid compound known as ponasterone A *(14)*. In the absence of hormone, the receptors form an inactive complex with heat-shock proteins. Ligand addition induces DNA-binding and transcriptional activation of the target promoters.

2. Dominant Role of the Chromosomal State in Gene Expression

2.1. Influence of Position and Gene Copy Number on Number on Transcription Levels

Transcription is the first step of recombinant gene expression, and has remained a focal point of interest. Most current applications make use of cell lines with recombinant genes stably integrated into the genome of a host cell under the control of a strong cellular or viral promoter *(15)*. Some of these viral promoters, including the CMV promoter *(16)*, can be used in different cell lines for high-level protein expression, but their transcriptional activity varies depending on the cellular levels of the relevant transcription factors, on the copy number of the integrated DNA, and on the integration site.

DNA in the cell nucleus is organized in chromatin that determines the transcriptional capacity of a particular chromosomal region to a large degree. Chromatin consists of DNA that is wrapped around histones, forming the so-called nucleosomes. Chromatin can condense into higher-order structures, resulting in tightly packed DNA-histone complexes that prevent access to transcription factors, producing inactive genes. The accessibility of chromatin to transcription factors is determined to a large degree by its acetylation status. Histone acetylation leads to a net increase in negative charges and is believed to loosen the histone-DNA interactions, thereby facilitating access to activating transcription factors. The acetylation status is determined by the antagonistic activities of histone acetyltransferases (acetylases) and histone deacetylases. Sequence-specific DNA-binding transcriptional activating proteins frequently recruit histone acetylases, whereas transcriptional repressors and silencers recruit histone deacetylases. The effects of these enzymes can be localized to a few nucleosomes, and can be specific for a given promoter. However, the acetylation status can spread from so-called locus control regions (LCR) over an entire chromatin domain and can affect multiple genes simultaneously. Further spreading is believed to be prevented by specialized DNA regions, such as insulator sequences *(17)* or AT-rich nuclear scaffold/matrix attachment regions, (S/MAR) elements *(18)* (*compare* Table 1).

Several strategies to maximize gene expression at the transcriptional level aim to increase histone acetylation. Expression vectors have been constructed that carry S/MAR elements to specifically protect the insert from negative influences of the surrounding chromatin at the integration site *(18)*. Alternatively, cell-culture medium additions of deacetylase inhibitors such as butyrate can increase recombinant gene expression by enhancing the overall level of histone acetylation *(24)*.

Aside from the chromosomal integration site, the copy number of the inserted gene can have a significant influence on the expression characteristics. On average, higher gene copy numbers correlate with higher expression levels. However, when individual clones are compared, this correlation breaks down, because of the drastic differences in the expression level of individual clones. Also, there appear to be cellular mechanisms that recognize repeated sequences and preferentially inactivate them transcriptionally in the absence of selection pressure, resulting in a gradual decrease in specific productivity. Thus, in some applications, stable expression from single or low gene copy numbers is preferred. Alternatively, recombinant protein overexpression in the most widely used Chinese Hamster Ovary cells (CHO) is achieved by gene amplification. To obtain such cell lines, an amplification marker—usually the dihydrofolate reductase gene

Table 1
DNA Elements That Can Affect Promoter Activity

Epigenetic element	Description	Reference
Enhancer	Increases transcription of nearby promoters	*(19)*
Silencer	Reduces transcription of nearby promoters	*(20)*
Scaffold/matrix attached region	Anchoring of DNA to the nuclear scaffold; structural boundaries of chromatin	*(18,21)*
Insulator	Boundary element, blocks enhancer and silencer action	*(17, 22)*
Locus control region (LCR)	Dominant activating sequence that confers position and copy number-dependent expression on a linked gene	*(23)*

(DHFR)—is cotransfected together with the gene of interest. Resistant cell clones that first have low expression levels of both the gene of interest and DHFR are successively selected for increasing resistance to the amplification drug (methotrexate). This is often caused by an increase in the copy number (amplification) of a chromosomal domain containing the amplification marker gene and the gene of interest *(25,26)*. The expression level of both heterologous genes increases with the domain copy number present in the genome. This procedure to isolate high copy number clones is extremely time-consuming. The initial expression level is not meaningful, and amplifiable clones cannot be identified in any other way. Therefore, a large number of clones must be handled initially, and the entire procedure to obtain a final producer strain may involve a period of several months to years. Because of the unstable expression characteristics, the cells must be cultivated under constant selection pressure. However, the presence of toxic substances in the final production process is generally omitted because of difficulties with the disposal of large volumes of toxic waste. Therefore, gene amplification using methotrexate is preferentially used in limited batch processes. Selection marker genes such as glutamine synthetase are used for continuous production processes by using nontoxic glutamine-free selection medium instead of methionine sulphoximine-containing medium. Depending on the particular product, final expression levels in CHO cells in the range of 10–100 pg/cell/d are generally considered satisfactory, although higher titers have been reported *(27)*.

A major drawback of isolating stable, inducible, or constitutive high-level producer-cell lines is usually a time-consuming selection procedure. In order to facilitate isolation of high-producing cell clones with desirable characteristics, a number of alternative techniques have been developed.

2.2. Controlled and Targeted Chromosomal Integration Expression Strategies

Although producer cells with multiple copies of the transgene often yield large amounts of the protein, the genetic instability of these multimeric cassettes requires continuous growth in the presence of the respective selection drug. Selected cell clones carrying a single copy of the transgene can overcome the problem of instability without a loss in productivity. However, high screening efforts usually must be made in order to isolate appropriate clones. The novel strategies discussed in the following section provide a means to overcome this problem.

Generally, transgenes expressed from ectopic chromosomal sites are subjected to position-dependent expression phenomena. Such epigenetic features usually arise through local modifications by chromatin structure that generally result from interactions between the transgene and chromatin at the site of integration. In some circumstances, this process can be bidirectional, so that the transgene changes expression from the endogenous gene at the site of integration. Numerous experiments using transgenic animals have confirmed the complexity of this epigenetic influence on gene expression, and have shown that some genomic sites develop very complex interactions and others are essentially neutral, allowing a transgene to be expressed according to patterns seen at the natural chromosomal locus. In recent years, a number of chromosomal elements have been identified that are capable of influencing the expression of a transduced gene (Table 1). However, currently no efficient protocols exist to use these elements for efficient protection of a transduced gene from influences mediated by the nature of the integration site.

Nevertheless, many examples from practical applications show that single-copy integration of a transgene can lead to high and stable expression. This proves that favorable chromosomal sites exist in the genome of currently used production cell lines. However, to identify such high-expressing cell clones, a substantial effort is required. The major recurring task for every gene of interest is the time-consuming screening for appropriate levels of expression and stability followed by validation procedures.

Consequently, in recent years, efforts have been made in order to reuse one of the rare sites on the chromosome that allow high and stable gene expression from a single integrated gene copy. Classically, the specific targeting of a transgene to a predetermined chromosomal site can be achieved by homologous recombination. However, this method requires a detailed characterization of the bordering fragments and long stretches (~5 kb) of flanking DNA must be available to construct the recombination vector. In addition, the frequency of recombination also varies with the accessibility of the integration site, and appropriate selection strategies are required. Although this technique is routinely applied in murine ES cells for the establishment of transgenic mice, the method is hardly feasible for transformed cell lines used for biotechnological purposes. This is because in the latter cells, homologous recombination is masked by the more frequently occurring illegitimate recombination events that result in a random integration of the respective transgene *(28)*.

New perspectives have been reached by application of the sequence-specific recombinases Cre or Flp in mammalian cell lines *(29,30)*. These enzymes are derived from bacteriophage P1 and yeast, respectively, and mediate site-specific recombination between short recombination targets, so-called loxP (for Cre) and FRT sites (for Flp). These recombination targets consist of two inverted 13-basepair (bp) repeats flanking an 8-bp spacer element. In the FRT sites, a third inverted repeat is present. The recombinases bind to the inverted repeat elements and perform cleavage and religation of the spacer region. Elements flanked by two correspondingly orientated loxP or FRT sites are efficiently excised by the respective recombinase (Fig. 2A). Generally, these enzymes are also capable of catalyzing the backward reaction, which corresponds to a site-specific integration into a loxP or FRT tagged chromosomal

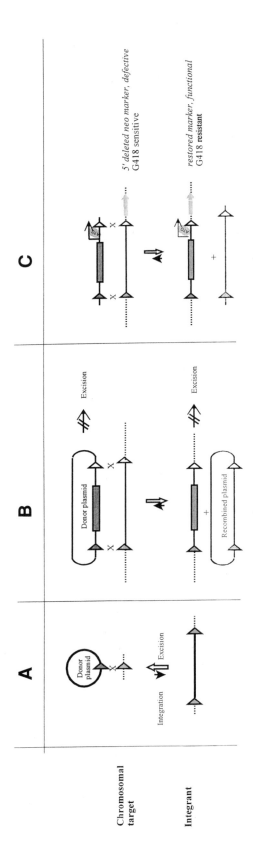

Fig. 2. Strategies for targeted integration. (**A**) Principle of Flp- or Cre-mediated integration. A donor plasmid tagged with a recombinase recognition target sequence can be specifically integrated into a correspondingly tagged chromosomal target site. However, the backward reaction is much more efficient. (**B**) Preventing excision by use of non-interacting recombination sites. A chromosomal locus is tagged with an expression cassette flanked by a set of non-interacting FRT sites (as symbolized by the black-and-white triangle). Transfection of a plasmid with corresponding FRT sites in presence of Flp recombinase results in the exchange of the intervening sequence. Excision is circumvented (*35,37*). (**C**) Advanced targeted integration strategy. The efficiency can be further improved by combining the double FRT principle as shown in (**B**), with the recombination-mediated activation of a selection marker (*see* refs. *39,41*).

integration site. However, in this simple design, this method has not found a broad application. This is because the excision reaction is much more favored over specific integration both for thermodynamic and kinetic reasons: as soon as the integrated state is formed, it will be readily excised again in the presence of residual amounts of the recombinase. As a result, the intermediate integrated state cannot be fixed. Certain strategies restrict the action of the recombinase, either by expressing it in a timely restricted fashion *(31,32)* and/or by using (hormone-) activatable fusions *(33,34)*, yet could not significantly improve this application.

To overcome these limitations, a sophisticated version of this site-specific recombination strategy has been developed. This advanced strategy is based on sets of two heterospecific recombination target sites (spacer mutants). These target sites can recombine efficiently with homologous counterparts, but do not recombine heterospecifically. Such mutants have been developed for the Flp/FRT system *(35)*. Corresponding mutants for loxP have been developed, yet have not been shown to provide similar specificity *(36)*. Thus, an expression cassette flanked with a set of heterospecific mutants cannot be excised, and is stable even in the presence of the recombinase (Fig. 2B). However, such a cassette can be precisely exchanged for a novel cassette flanked with corresponding recombination target sites. Through application of a negative selection strategy (e.g., thymidine kinase) to eliminate cells which have not undergone recombination, cells with successful targeted exchange can be isolated *(37,38)*.

The efficiency of this approach can be further improved by combining the heterospecific FRT sites with a highly effective selection procedure for correct cassette replacement *(39)*. Thereby, the efficiency of targeted exchange is 90–100%, which reduces screening to a minimum. This strategy is based on an initiation codon-deficient drug-resistance marker, which can only be complemented by site-specific in-frame integration of the initiator codon (Fig. 2C). Only cells that have recombined precisely become drug-resistant. Most importantly, this technique has shown that the prediction of expression levels of a targeted transgene is feasible as long as the basic structure of the expression cassette is maintained *(40,41)*.

In summary, these advanced recombination-based techniques should now allow for a more rapid production of a new product. Integration sites with the desired expression properties are first screened using a reporter cassette flanked by heterospecific recombination target sites. Selected integration sites of choice are then efficiently targeted with the expression cassette of choice through a standardized process. The cultivation procedures developed for one such cell line could be applied to diverse products. Since all parameters except for the product gene are conserved, this would also greatly facilitate the approval of novel producer-cell lines by the regulatory authorities.

3. Co-Expression of Multiple Genes

Current applications in mammalian cell culture increasingly require the defined co-expression of different genes *(42)*. Four typical applications are described here: (i) Co-expression of a selectable marker together with the protein of interest, which is routinely used to establish stable cell lines. For industrial applications, overexpression of a recombinant protein in Chinese Hamster Ovary (CHO) cells is achieved by gene amplification *(see* **Subheading 2.1.**). (ii) Defined but unequal expression. Some applications require that the protein of interest is expressed at a much higher level than the selectable marker. If a defined relationship of co-expression of both genes can be adjusted, cells that are resistant to the respective selective drug will produce the protein of interest at levels which cannot be below a critical threshold required for survival in the presence of the respective drug concentrations. (iii) Equivalent expression: expression of genes at a one-to-one ratio. Subunits of heteromultimeric protein complexes need to be synthesized in equal amounts. Examples are cytokine receptors, antibodies, or other di- or trimeric proteins. An additional application involves the co-expression at similar amounts of several enzymes that form a metabolic pathway. (iv) Quantitation of a protein for which no satisfactory assay method is available. If a strict coupling of expression of a reporter gene and the gene of interest is achieved [compare (ii)], a calibration for reporter and product level can indirectly determine the quantity of the protein of interest.

In nature, multiple levels of regulatory mechanisms ensure stoichiometric expression of genes in mammalian cells. Transcription, post-transcriptional processing, mRNA transport, stability, and translational efficiency adjust the correct levels of the synthesized proteins. A simulation of these events for genetic manipulation of cells would imply coordinated engineering efforts of all crucial steps. With the current techniques, this type of co-expression is nearly impossible to achieve. The currently applied methods for co-expression are summarized in Fig. 3.

3.1. IRES-Mediated Initiation of Translation

Although reinitiation of ribosomes is generally inefficient, internal ribosomal entry sites (IRES)-mediated internal initiation can be as efficient, or in rare cases even more efficient than cap-dependent initiation. Therefore, IRES elements are the method of choice to co-express multiple reading frames *(43)*.

IRES have been identified in viral and cellular eukaryotic mRNAs. Most currently well defined IRES elements are present in the 5' untranslated regions (5' UTR) of mRNAs and range from about 200–1300 nucleotides.

The type of IRES element used and its efficiency are of major importance for applications in expression vectors. The currently used classification of IRES elements is based on the picornaviral IRES classification *(44,45)*, which relies on the position of the initiation codon relative to the IRES element and additional downstream sequence requirements. Type I IRES elements include the Polio virus IRES element *(46)* and all cellular IRES elements characterized thus far. Type I IRES elements can be located at a variable distance from the downstream reading frame, and can be as far as 50–100 nucleotides upstream of the initiation codon. Type II IRES elements are characterized by the strict position requirements for the initiation codon at the 3' boundary of the IRES *(47)*. Encephalomyocarditis virus (EMCV) *(48)* is regarded as a prototype for class II elements. In type III IRES elements, the initiation codon position requirements are similar to type II IRES. However, additional sequences that are important for IRES function are located in the downstream coding sequence. The prototype of type III IRES is represented by the Hepatitis C virus (HCV) IRES.

A highly efficient cellular type I IRES element has been identified that shows a higher relative activity in various cell lines than picornaviral IRES elements (49). Cellular elements were discovered much later than the viral IRES, and presumably for this reason they are currently not used in such a broad range of applications as the viral

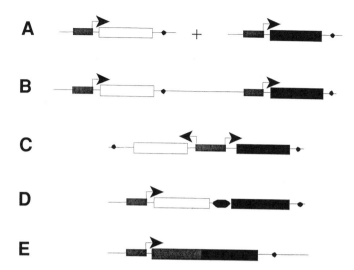

Fig. 3. Strategies for construction of polyvalent vectors (adapted from ref. *42*). Cassettes for the expression of two different genes (open and black boxes) under the control of unidirectional or bidirectional promoters (striped boxes) are shown. Arrows symbolize transcription starts, and filled circles depict polyadenylation sites. The filled oval symbolizes an IRES element. Vector sequences are drawn as black lines. A simple way to obtain co-expression of two proteins is to transfect cells with two independent constructs (**A**) or by introducing a single vector harboring two discrete expression cassettes (**B**). The first approach is often limited by the inefficiency and unpredictability of cotransfection. Another method relies on bidirectional promoters (**C**). Although this method is currently not in common use, it opens a new level of co-expression technology. The methods depicted from A to C suffer from the fact that even if the same promoter strength is given, the two transcripts may significantly differ because of variations in processing, half-life time, and translational efficiency of the mRNA. Di- and polycistronic mRNAs can be constructed by using internal ribosomal entry sites (IRES) that allow the co-expression from a single mRNA (**D**). In this way, variable expression ratios resulting from unpredictable transcription efficiencies of separate expression cassettes are circumvented. Finally, for special applications, fusion proteins provide the strictest coupling of two protein functions (**E**). However, for various reasons, this is often not possible.

ones. The use of various IRES elements facilitates the construction of multicistronic vectors with more than two reading frames by avoiding the potential for genetically unstable repeated insertion of identical IRES sequences.

Artificial IRES elements have been developed. A 9-nt segment of a cellular mRNA functions as an IRES, and when present in linked multiple copies, efficiently directs internal initiation *(50)*.

3.2. Polycistronic Vectors

Many commercially available dicistronic expression cassettes harbor a selectable marker or a reporter gene 3' of an IRES element and a multiple cloning site (MCS) for the insertion of the gene of interest. These cassettes are transcriptionally initiated by a promoter/enhancer for cap-dependent expression of the first cistron, followed by an

IRES element to permit translation of the second cistron and also provide a polyadenylation site at the 3' end. The insertion of genes into such vectors is straightforward but it is difficult to replace the second cistron by other genes needed.

Vectors or vector systems that allow the construction of di-, tri-, or multicistronic expression cassettes have been developed. Translation of mRNA that encodes the reading frames occurs upon cap-dependent initiation for the first cistron, and translation reinitiation of the second and all other downstream cistrons mediated by IRES elements. In most cases, IRES elements from Polio virus or EMCV are used *(51–57)*.

Compared to monocistronic mRNAs, dicistronic mRNAs are considerably longer and more complex. To achieve optimal expression, care must be taken to avoid the unintentional inclusion of special regulatory sequences such as polyadenylation signals or mRNA destabilizing elements. The expression level depends not only on the particular IRES element used, but also on both the sequence and on the order of the reading frames.

IRES-mediated translation efficiency is highly context-dependent, requiring experimental confirmation of the expression characteristics of each individual vector construct *(58)*. A systematic study of dicistronic vectors expressing two unrelated luciferase genes from Renilla and firefly, respectively, showed that the presence of the firefly luciferase in the first cistron has a surprisingly drastic inhibitory effect on internal initiation. This effect is independent of the promoter and IRES element used, and is observed in several cell lines *(59)*. Therefore the nature of the cistrons and their positioning is of key importance for their expression.

Overall, the efficiency of IRES elements can vary considerably, depending on the particular construct, on the experimental setup, or on the host cell used, and in some cases is even dependent on the physiological status of the cells. Despite the unquestionable advantages of multicistronic expression vectors, each individual construct must be tested, and if necessary, expression levels must be optimized to achieve satisfactory results.

3.3. Bidirectional Promoters

Coordinate expression of two transcription units can be obtained by using bidirectional promoters. Although for viral bidirectional promoters (e.g., SV40 and adenovirus) expression of the divergent transcription units is temporarily controlled in the course of infection mammalian bidirectional promoters allow simultaneous transcription in both directions. Most of the known mammalian bidirectional promoters are TATA box-deficient, and mediate low-level transcription of housekeeping genes. Generally, they are asymmetric, and one direction is preferentially transcribed.

The general potential of artificial bidirectional promoters for simultaneous expression of two genes has been recently shown for the artificial promoter P_{bi-1} *(60)*. This promoter is derived from a unidirectional tetracycline promoter *(3)*. The promoter is regulated by a tetracycline-dependent transactivator *(see* **Subheading 1.2.***)*.

The potential of the bidirectional promoter can be substantially extended if it is combined with polycistronic expression cassettes. Although simple cloning vectors are not yet available, the advantages of such a vector design are obvious: two coordinately transcribed mRNAs encoding several cistrons can be achieved. Thereby, certain problems or limitations arising from IRES-mediated translation of certain genes can be bypassed.

Applications of IRES elements and bidirectional promoters as outlined here are not restricted to DNA vectors. The expression cassettes described may be used to construct viral vectors and by this method, gene-transfer recombinant cells or organisms can be obtained with much higher efficiency.

4. Viral Vectors

One alternative to the transfection of DNA is the use of viral vector systems. Apart from an efficient transduction of different cell lines, viral systems offer high-level transient or stable gene expression, respectively.

A wide variety of viral vectors are used to deliver recombinant genes into mammalian cells, either in vitro (in established immortalized cell culture) for basic and applied research and particularly in vivo (in animals) and ex vivo (in primary cell cultures) for gene-therapy purposes. Viruses have developed highly efficient strategies to enter host cells and to alter the physiology of the host in order to achieve optimal conditions for viral gene expression.

In general, these viral properties can be used for the expression of a desired recombinant gene. For this purpose, certain viral sequences are replaced by the gene of interest. To allow packaging, the deleted viral proteins are offered in trans. Therefore, so-called packaging or helper cell lines have been constructed that express the missing viral proteins. The transfer of the recombinant viral vector into a helper-cell line results in the packaging of the vector into infectious viral particles. The particles are released from the cell, and can be used for infection. This strategy restricts the amplification of the virus to the packaging cells. Once the virus has entered the target cell, viral amplification is not possible because of the lack of viral proteins.

Since the virus species differ within their specific properties, the selection of the viral system depends on the experimental requirements. One of the major selection criteria is the requirement of transient or stable expression of the transgene.

4.1. Viral Vectors for Transient Expression

Viral vectors are used for transient gene expression in mammalian cells as an alternative to the cumbersome and time-consuming selection procedure of stable cell lines, because they are more efficient than nonviral transfection strategies for rapid and efficient but limited production of small quantities of recombinant product protein. Alphaviruses and adenoviruses, the most commonly used systems for transient protein production, will be discussed in the following section.

4.1.1. Alphaviruses

Alphavirus-derived vectors have been developed for protein production *(61)*, since they have a broad host range and can even efficiently infect nondividing cells. Semliki Forest virus (SFV), Sindbis virus (SIN), and several pathogenic encephalitis-producing viruses are all members of the alpha virus genus *(62)*, which replicates in a large number of animal hosts ranging from mosquitoes to avian and mammalian species *(63)*.

The typical alphavirus consists of an enveloped nucleocapsid containing a single-stranded, positive-polarity RNA genome of approx 12,000 nucleotides that is capped at the 5' end and polyadenylated at the 3' end. The RNA consists of two expression units with an internal promoter that mediates transcription of a subgenomic mRNA. The viral structural genes encoded by the subgenomic transcription unit can be replaced by the gene of interest. The alphaviral genomic mRNA encodes its own replicase. The replicase mediates the replication of the plus-strand genome into full-length minus strands, which efficiently produce both new genomic and subgenomic RNAs that encode for viral proteins. Natural alphavirus gene expression and replication results in cell lysis caused by the avalanche-like amplification of the genomic mRNA. Replication of alphavirus-based systems is extremely efficient, leading to approx 10^5 new virions per cell. In this manner, the host's own translational machinery is used.

This replication efficiency has been utilized for the development of RNA- and DNA-based expression systems for the production of recombinant proteins in eukaryotic cells. *(61,62,64–68)*.

Expression vectors have been developed from SFV, SIN, and Venezuelan Equine Encephalitis virus (VEE). Since the alphaviral genome consists of a positive single-stranded RNA, the alphaviral RNA vector can be delivered directly either as naked RNA, as naked DNA, or as viral particles (Fig. 4). In all cases, high-level expression of the heterologous proteins is obtained. For virus production, the alphaviral recombinant RNA is cotransfected with a helper vector. High viral titers of approx 10^9–10^{10} per mL can be produced.

Since the large-scale production of RNA in vitro is expensive, a strategy where DNA as vector is used have been developed (Fig. 4c) *(69–71)*. In this system, the full-length recombinant alphaviral cDNA under control of an eukaryotic promoter is delivered directly to the cell by conventional DNA transfection. In the cell nucleus, the complete transferred unit is transcribed into RNA and transported to the cytoplasm. Translation of the RNA results in the production of the viral replicase, which initiates the replication of the entire molecule (*see* ref. *65*).

Alphavirus vectors have been used to express proteins for many different purposes, including large-scale production, protein characterization, and functional studies. In addition, alphavirus vectors can also be used for gene delivery in vivo with potential applications in vaccination and gene therapy *(72)*. One disadvantage is that for every production cycle of the gene of interest, DNA or viruses must be newly transferred.

4.1.2. Adenoviruses

Adenoviral transfer systems allow easy production and concentration to high viral titers (up to 10^{12} colony-forming units [CFU]/mL), a high level of transgene expression and a broad host range. Various cell types, including nondividing cells, can be infected. However, because of the lack of stable integration of the DNA, expression is only transient in proliferating cells, whereas in nonproliferating cells the genome persists as an episome and continues to express for a longer period of time.

Adenoviruses are nonenveloped viruses with a double-stranded DNA genome of 36 kb. Viral replication occurs within the nucleus of the cell, without integration into the host DNA. Immediately after infection, the early viral genes (E1 to E4) are expressed, producing polypeptides that are important for regulation of gene expression, replication, and the inhibition of cellular apoptosis. Activation of the late genes results in expression of polypeptides required for encapsidation of the virus. At the end of the replication cycle, viruses are released by cell lysis (*see* ref. *73*).

Of many strains of adenoviruses, only strains 5 and 2 from the subgroup C *(74)* have been predominantly used to make vectors.

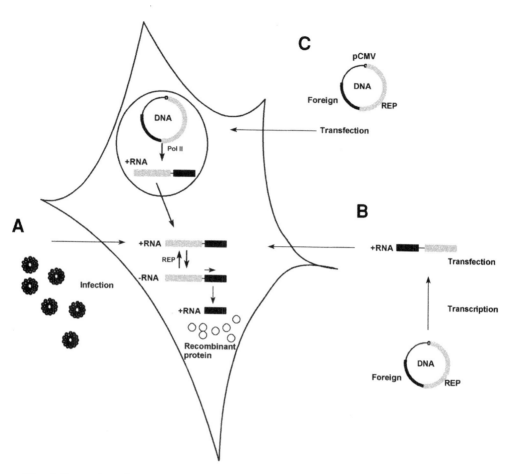

Fig. 4. Strategies of recombinant protein production using alphaviral vectors (adapted from ref. *72*). (**A**) Recombinant gene transferred by infection. The virus infects the cell by receptor-mediated endocytosis and the recombinant RNA that serves as mRNA is released and translated into the viral replicase. The replicase produces new full-length RNAs via a negative-strand template, which in turn serves as a template for the translation of the recombinant gene. (**B**) Plasmid DNA can be used for in vitro transcription of the viral mRNA, which is then transported directly to the cells. (**C**) An alternative is the transduction of a plasmid vector containing the viral replicase (REP) and the gene of interest. After transcription, recombinant viral RNA is transferred to the cytoplasm and amplified by the viral replicase and the gene product is translated.

Within the first-generation adenoviral vector, the E1 and E3 genes important for the activation of the viral early genes and for modulation of the immune response are replaced by the gene of interest. This leads to an insertion capacity of up to 8 kb. The E1/E3 replication-defective virus can be propagated in a cell line that complements E1, such as the human embryonic kidney-cell line 293 (Fig. 5). The transgene can be cloned into an adenoviral vector by recombination of a shuttle plasmid that contains the gene of interest and a second plasmid containing essentially the entire adenoviral genome in a circular form *(75)*, or by using a cosmid system.

More recently, defective adenovirus vectors were constructed in which all viral coding regions were removed, leaving only the inverted terminal repeats (ITR), the

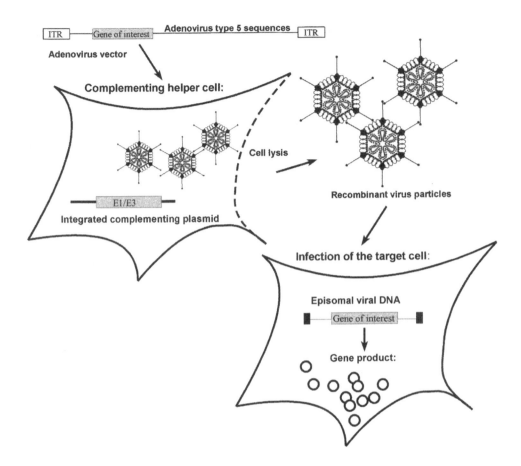

Fig. 5. Recombinant adenoviral vector production. The recombinant adenoviral vector is transferred to the complementing helper cell by conventional DNA transfection. In the course of viral amplification, the helper cells (often human embryonic kidney cell line 293) are lysed and the recombinant virus is harvested. After infection of a specific target cell, the linear DNA is transported to the nucleus and associated with the nuclear matrix. Because of the lack of chromosomal integration, expression of the transgene is only transient and decreases with the proliferation of the target cells.

transgene and the psi packaging sequences in these "gutless" vectors. Approximately 28 kb can be inserted. Such adenoviral vectors were successfully utilized for the expression of full-length genes such as the dystrophin gene and the cystic fibrosis transmembrane conductance regulator (CFTR) gene *(76,77)*. Since the production of gutless vectors depends on the co-expression of a helpervirus, strategies must be employed to separate both the helper and the recombinant virus carrying the transgene.

4.2. Viral Vectors for Stable Expression

A cell line that stably expresses a particular protein of interest is required in order to study the function of a given gene product or to continuously produce huge amounts of a biologically active protein. For this purpose, the retroviral transfer is a powerful tool.

4.2.1. Retroviruses

Retroviral vectors are effective transfer systems whenever stable introduction of foreign genes into target cells is needed. Recombinant retroviruses are efficiently used for the analysis of cDNA libraries *(78)*. Expression cloning is a powerful tool to isolate a cDNA of interest. Alternative techniques are based on transient expression of cDNA libraries in CHO cells which has obvious limitation in searches for proteins with various functions in specialized cell types, since the function of some proteins only become obvious during long-term expression. In contrast, retroviral gene transfer offers long term expression in a wide range of target cells, since the delivered transgene stably integrates into the host genome. This characteristic is also useful for fast and efficient establishment of target cells for high-throughput screening, and for validation of specific protein functions.

Retroviruses are RNA viruses that reverse-transcribe their diploid positive-stranded RNA genome into a double-stranded viral DNA, which is then stably inserted into the host DNA. Members of this class of RNA viruses are the murine leukemia viruses (MuLV) and the lentiviruses, which are extensively used for virus vector engineering. The advantage of the retroviral transfer system is its property of stable integration into the host genome, which leads to a stable expression of the gene of interest. Additionally, retroviruses possess the ability to infect a broad range of different cell types. Viral titers of up to 10^7 infectious particle per mL can be produced for efficient in vitro gene transfer. Moreover, single integration events can be easily established by using an appropriate ratio of virus particles and cells to be infected—multiplicity of infection (MOI). However, MLV-based vectors can only infect proliferating cells, because they need the breakdown of the nuclear membrane to deliver the preintegration complex into the cell nucleus.

The retroviral vectors currently mostly used for gene transfer are derived from the Moloney murine leukemia virus (MoMulV) *(see ref. 79)*. These viruses possess a relatively simple genome consisting of three structural genes—gag, pol, and env—responsible for replication, encapsidation, and infection. They are flanked by the viral long-terminal repeats (LTR), which are responsible for expression of the viral genome. By deleting the structural genes, but maintaining the *cis*-acting elements (LTR and packaging signal Psi) a space of 8 kb can be replaced by the gene of interest. For the production of virus, the recombinant vector is transduced into a packaging cell line that provides the viral proteins in trans. The recombinant viral RNA is packaged into viral particles and released from the helper cell. Packaging cell lines that yield titers of greater than 10^7 infectious viral particles per mL have been developed *(80)* (Fig. 6).

Because of the design of retroviral vectors and packaging cell lines, the formation of replication-competent retroviruses (RCRs) is avoided, thereby making this transfer applicable for various purposes including gene therapy *(81,82)*.

Typically, a selectable marker is co-inserted to isolate stably infected cells. A wide variety of vector constructions have been described that differ in size, orientation of the gene, and in the promoters used. In many retroviral vectors, the gene of interest is controlled by the viral LTR (LTR-based vectors). Additional internal promoters are used to express a second gene (internal promoter vectors). However, these vectors often suffer from reduced expression levels as a consequence of promoter interference *(83)*. An alternative to internal promoters offer the IRES leading to the co-expression of two

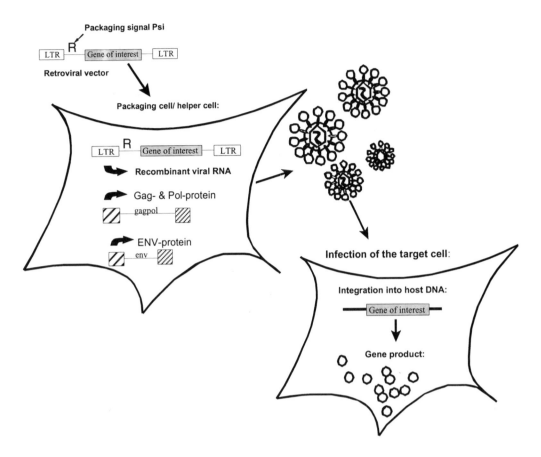

Fig. 6. Recombinant retroviral vector production. The recombinant retroviral DNA vector carrying the gene of interest is transferred into a helper cell that stably expresses the viral proteins. The transcribed genomic viral RNA is packaged into viral particles, which are released from the cell by a budding mechanism. After infection of the target cells the viral RNA genome is reverse-transcribed into DNA and integrated into the genome of the cell, leading to a stable expression of the protein.

genes from the LTR (*see* Chapter 3). A second possibility to circumvent promoter interferences is the use of retroviral self-inactivating vectors (SIN-vectors). In these vectors, the viral enhancer elements of the LTRs are deleted in infected cells *(84)*. SIN vectors are often used for transduction of regulated expression cassettes in order to reduce activating influences on the regulation properties of the transgene.

4.2.2. Stable Expression from Episomal Vectors

Some efforts have been made to construct vectors that replicate episomally in higher eukaryotic cells concomitant with the cell cycle. Although the expression of a stable integrated gene of interest is always influenced by the site of chromosomal integration, episomal expression offers a more predictable expression behavior *(85)*. A number of DNA viruses, such as SV40, BPV, or Epstein-Barr virus (EBV) replicate episomally in mammalian cells. This process depends on both viral transacting factors and accessory activities recruited from the host-cell replication machinery. The viruses acquire centromer function by associating with the host chromosomes. Responsible for the episomal maintenance is a viral protein (EBNA-1 for EBV; large T-antigen for SV-40) that serves as replication and transcriptional enhancer.

EBV, as a herpesvirus with a double-stranded DNA genome (165 kb) is engineered to express large DNA fragments or entire authentic genes, including the native regulatory sequences of these genes for efficient gene expression in target cells *(86)*. The properties of EBNA-1 and oriP were successfully combined with the high and efficient adenoviral transfer system in order to extend the stability of expression of adenovirally transduced DNA. Therefore, the efficient adenoviral transfer system is used to deliver an EBV-based episome. After transduction, the episome is excised, circularized and thereby activated by the sequence-specific recombinase system Cre-loxP *(87)*. This technique extends expression time in proliferating cells; a loss of 2–5 % of vector copies occurs per cell division *(88,89)*

Another approach to achieve stable expression upon adenoviral gene transfer is the construction of chimeric adeno-retroviral vectors. The adenoviral part enables the efficient infection of proliferating and nonproliferating cells, whereas the retroviral part enhances the integration frequency of recombinant chimeric vectors into the genome of the target cells *(90,91)*. The adenoviral transfer of both the retroviral helper genes and a recombinant retroviral vector allows effective establishment of virus producers *in situ*. Recent developments could show that efficient integration does not require the retroviral helper genes *(91)*. Adeno-retroviral vectors free of any retroviral helper function are potent vectors for transduction and integration of regulatable expression cassettes (Unsinger, J., unpublished).

5. Metabolic Engineering of Mammalian Cells

The currently available cell lines for biotechnological use do not meet all criteria for optimal production of the protein of interest. The limitations may be essential for the quality of the protein or for the production process, yet the latter in turn often influences product quality. For this chapter, we have selected examples of metabolic engineering to demonstrate the problems and the principles of the current approaches.

5.1. Glycosylation Engineering

Post-translational processes were recognized as metabolic bottlenecks and potential targets to improve the quality of the products. This modification of a product protein includes various potentially rate-limiting interactions with proteins in the cytoplasm, the endoplasmatic reticulum (ER), and in the compartments of the Golgi apparatus. In many studies of mammalian cells the protein productivity and quality were improved by augmenting the post-translational capabilities. The major reason for using mammalian cell cultures for pharmaceutical protein production is the human-like post-translational modifications. Aside from protein processing, protein glycosylation is of central importance. Complex oligosaccharide side chains are covalently attached to newly synthesized peptide chains. The majority of secreted protein is glycosylated. Glycosylation can affect protein solubility, activity, and in vivo lifetime. The immune system recognizes nonhuman oligosaccharides as foreign, causing inflammatory and immune responses. For biopharmaceutical manufacture, the absence of sialic (neuraminic) acid on the termini of complex carbohydrate structures—which results in more rapid clearance, as does the presence of high-mannose oligosaccharides—is of key importance.

N-linked oligosaccharides are transferred initially as a preformed core oligosaccharide from a dolichol lipid to the amide nitrogen of asparagine. In contrast, O-linked oligosaccharides are built starting with the attachment of N-acetylgalactosamine to the hydroxyl side chain of a serine or threonine residue. These initial steps are followed by a series of trimming or addition reactions in the ER and in the Golgi compartments. Glycoproteins are built up in a complex cascade of sequential, enzyme-catalyzed reactions and intracompartmental transport processes, often with multiple enzymes acting on common substrates to yield alternative oligosaccharide products *(92)*. The resulting glycosylation pattern of a protein is heterogeneous, and is highly dependent on multiple parameters such as primary amino acid sequence and protein conformation.

The glycosylation pattern can be partially altered by changing the culture medium composition, or by adding precursors, intermediates, or inhibitors of glycosylation *(93)*. Many of these reagents are toxic, and some alter glycosylation patterns inefficiently. For optimal glycosylation of therapeutic proteins, the choice of cell line plays a key role *(94)*. The human cell lines such as the 293 HEK line may be preferred, but may not be ideal. Producer-cell lines most widely used for pharmaceutical protein production, e.g., CHO and BHK cells, glycosylate recombinant proteins in a human-compatible way.

Recombinant-DNA technology for the metabolic engineering of glycosylation of mammalian cells aims to extend the host's oligosaccharide-biosynthesis capabilities by introducing genes that encode heterologous carbohydrate-synthesis enzymes. Glycosylation engineering strategies predominantly consider enzymes to be responsible for the terminal steps of complex oligosaccharide biosynthesis *(95–98)*. The following strategies are followed for this type of engineering: (i) glycosylation activities can be increased based either on gene activation or by expression of recombinant glycosylation genes; (ii) conversely, anti-sense RNA, ribozymes, or and other methods can be used to block undesirable glycosylation. Several glycosylation enzymes have been cloned and are available for metabolic engineering. Apart from analytical approaches, genetic engineering of cell lines aim to achieve more human-like glycosylation patterns and to improve the degree of terminal sialylation. Thus, CHO cells do not express $\alpha2,6$-sialyltransferase (2,6 ST) *(99)*. Recombinant expression of a 2,6-ST gene results in a more human-like glycosylation profile, including both, $\alpha2,3$- and $\alpha2,6$-linked sialic acids *(96)*. Similarly, the glycosylation pattern in BHK cells has been successfully rendered more human-like by expression of $\alpha2,6$-sialyltransferase and 1,3-fucosyltransferase-III genes.

Our understanding of the potential of glycosylation is incomplete. Although some general concepts have been clarified, important questions remain concerning activity and other relevant in vivo characteristics of individual glycoforms of a particular protein. Secreted protein products are a mixture of a number of different yet related isoforms. For clarification, these are purified and analyzed separately. These data serve as a basis for a rational engineering approach. Genes for enzymes involved in the important steps were identified, and are used for directed cell engineering. This process is intended to achieve pharmaceutical products with optimal in vivo characteristics and to understand the involvement of glycoslyation in protein secretion and in the function of certain glycosylated forms of proteins.

5.2. Controlled Proliferation Technology

The control of cell growth and division is of fundamental importance for multicellular organisms. In the first developmental phase, cell proliferation is essential for growth of the organism. After terminal differentiation, the major growth phase is completed. Then, control of proliferation becomes the dominant aspect of the genetic stability of higher organisms. Nevertheless, the growth-arrested cells produce and secrete proteins continuously during their lifetime. This type of regulation is preferable in standard biotechnological production processes. However, for transformation of these processes, permanently proliferating animal cells are used. An ideal production process would include proliferation control to first allow cells to rapidly grow to high cell densities, followed by a proliferation-inhibited production phase in which the cells devote their metabolic capability to the formation of a product. Higher productivity of proliferation-inhibited cells was indeed observed with antibody-producing hybridoma cells *(100–102)*.

The first attempt to control cell proliferation by genetic engineering of the BHK cell line was based on an estrogen-regulated interferon-responsive factor 1 (IRF-1)-estrogen-receptor fusion *(103–107)*. IRF-1 is a DNA-binding transcription activator that accumulates in cells in response to interferons and has antiviral, antiproliferative, and anti-tumor activities *(108–112)*. Its function as a proliferation inhibitor relies on the induction of downstream genes *(104,109)*. Recombinant protein production can be strongly enhanced in IRF-1-arrested cells, by placing the transcription of the corresponding gene under the control of an IRF-1-responsive promoter *(113)*. In addition, expression of estrogen-responsive IRF-1 in a dicistronic configuration concomitant with the selection marker stabilizes the growth control in a way that allows several cycles of growth and growth-arrested states to be performed with the same culture *(107,114)*. Conditions to control cell growth and improve recombinant gene expression were extensively studied in BHK cells *(115,117)*. It was shown that the glycoslyation properties of the cells are not altered during the growth-inhibition phase *(118)*.

This genetic growth-control system can be used to facilitate the handling of helper cell lines in co-culture for cell and tissue engineering.To demonstrate this, the growth-control system was established in a human stroma feeder cell line. Transient estrogen exposure was sufficient to reduce cell growth for more than 1 wk. The proliferation-controlled cells could therefore be co-cultured with hematopoietic stem cells under conditions that are identical to those used with conventionally irradiation-arrested stroma cells. Analysis of the proliferation-controlled stroma cells revealed that the relevant characteristics of the parental L88/5 cells have been preserved in the regulated cell clone *(119)*.

The cyclin-dependent-kinase (cdk) inhibitors *p21* and *p27*, and the tumor-suppressor gene *p53* were used to reversibly arrest CHO cells at the G1-phase of the cell cycle *(120)*. Overexpression of these genes leads to a prevention of S-phase entry *(120)*. In the G1-phase of the cell cycle, cells check their physiological state and can arrest to repair genetic defects. In CHO cells *p21*, *p27*, and *p53* were expressed under tetracycline control. Transient expression of *p27* leads to growth arrest and gives rise to higher specific protein productivity *(120)*. In contrast, continuous overexpression of *p53* leads to rapid cell death. Because the permanent expression of *p21* is not possible, tetracycline-regulated expression technology, in which the reporter SEAP and *p21* are co-expressed, was used. With this construct, growth arrest for several weeks and a significantly higher specific productivity was achieved when compared to control cells *(120)*. In other examples for controlled proliferation technology based on a *p27*-encoded expression unit, CHO cells were blocked in proliferation and reporter-gene expression was stimulated *(121)*.

5.3. Apoptosis Engineering

Apoptosis plays a fundamental role in multicellular life. In biotechnological production processes, apoptosis is an undesirable phenomenon, Processes are often limited by rapid cell death in the decline phase of a culture. Some commercially important production cell lines are sensitive to apoptosis, particularly hybridoma and myeloma cell lines *(122,123)*. In other cell lines, proliferation in response to nutrient limitations or genotoxic stress is blocked and no apoptosis is found. This allows the cells to replenish their metabolic precursors or repair DNA damage *(124)*.

Apart from the elimination of nutrient deprivation by feeding strategies, chemical-medium additives are used to block apoptosis pathways. Suppression of apoptosis in cell-culture processes is applied to engineer the cells using antiapoptotic genes. Mammalian cells were shown to be successfully protected against stress-induced apoptosis by overexpression survival genes from the bcl-2 *(125)*.

The level of protection varies between different cell types and cell lines and, in most cases, Bcl-2 overexpression cannot prevent cell death, but it can extend cellular lifetime and lead to increased production *(102,125–127)*. Attempts to complement the action of Bcl-2 using concomitant overexpression of the anti-apoptotic genes bag-1, bcl-x_L or the adenoviral E1B-19K gene were encouraging and showed that the protective effects of individual genes were equal or even additive when, for example, Bag-1 and Bcl-2 were co-expressed *(128–131)*. More recent strategies to prevent apoptosis concern the inhibition of caspases. These are downstream actors of the programmed cell death cascade. Using peptide inhibitors and the expression of genes encoding dominant negative caspases, effective inhibition of apoptosis could be observed *(132)*.

References

1. Hauser, H. and Wagner, R., eds. (1997) *Mammalian Cell Biotechnology in Protein Production*, W. DeGruyter, New York, pp. 1–190.
2. Kozak M. (1999) Initiation of translation in prokaryotes and eukaryotes. *Gene* **234**, 187–208.
3. Gossen, M. and Bujard, H. (1992) Tight control of gene expression in mammalian cells by tetracycline-responsive promoters. *Proc. Natl. Acad. Sci. USA* **89**, 5547–5551.
4. Gossen, M., Freundlieb, S., Bender, G., Müller, G., Hillen, W., and Bujard, H. (1995) Transcriptional activation by tetracyclines in mammalian cells. *Science* **268**, 1766–1769.
5. Shockett, P. E., Difilippantonio, M., Hellman, N., and Schatz, D. G. (1995) A modified tetracycline-regulated system provides autoregulatory, inducible gene expression in cultured cells and transgenic mice. *Proc. Natl. Acad. Sci. USA* **92**, 6522–6526.
6. A-Mohammadi, S. and Hawkins, R. E. (1998) Efficient transgene regulation from a single tetracycline-controlled positive feedback regulatory system. *Gene Ther.* **5**, 76–84.

7. Strathdee, C. A., McLeod, M. R., and Hall, J. R. (1999) Efficient control of tetracycline-responsive gene expression from an autoregulated bi-directional expression vector. *Gene* **229**, 21–29.

8. Unsinger, J., Kröger, A., Hauser, H., and Wirth, D. (2001) Retroviral vectors for transduction of an autoregulated bidirectional expression cassette. *Mol. Ther.* **4**, 484–489.

9. Fussenegger, M., Morris, R. P., Fux, C., Rimann, M., von Stockar, B., Thompson, C. J., et al. (2000) Streptogramin-based gene regulation systems for mammalian cells. *Nat. Biotechnol.* **18**, 1203–1208.

10. Eilers, M., Picard, D., Yamamoto, K. R., and Bishop, J. (1989) Chimaeras of myc oncoprotein and steroid receptors cause hormone-dependent transformation of cells. *Nature* **340**, 66–68.

11. Braselmann, S., Graninger, P., and Busslinger, M. (1993) A selective transcriptional induction system for mammalian cells based on Gal4-estrogen receptor fusion proteins. *Proc. Natl. Acad. Sci. USA* **90**, 1657–1661.

12. Wang, Y., O'Malley, Jr., B. W., Tsai, S. Y., and O'Malley, B. W. (1994) *Proc. Natl. Acad. Sci. USA* **91**, 8180–1884.

13. Burcin, M. M., O'Malley, B. W., and Tsai, S. Y. (1998) A regulatory system for target gene expression. *Front. Biosci.* **3**, 1–7.

14. No, D. Yao, T. P., and Evans, R. M. (1996) Ecdysone-inducible gene expression in mammalian cells and transgenic mice. *Proc. Natl. Acad. Sci. USA* **93**, 3346–3351.

15. Kaufman, R. (1990) Vectors used for expression in mammalian cells. *Methods Enzymol.* **185**, 487–512.

16. Boshart, M., Weber, F., Jahn, G., Dorsch-Häsler, K., Fleckenstein, B., and Schaffner, W. (1985) A very strong enhancer is located upstream of an immediate early gene of human cytomegalovirus. *Cell* **41**, 521–530.

17. Bell, A. C. and Felsenfeld, G. (2001) Gene regulation: insulators and boundaries: versatile regulatory elements in the eukaryotic genome. *Science* **291**, 447–450.

18. Bode, J., Benham, C., Knopp, A., and Mielke, C. (2000) Transcriptional augmentation: modulation of gene expression by Scaffold/matrix attached regions (S/MAR elements). *Crit. Rev. Eukaryot. Gene Expr.* **10**, 73–90.

19. Fiering, S., Whitelaw, E., and Martin, D. I. K. (2000) To be or not to be active: the stochastic nature of enhancer action. *Bioessays* **22**, 381–387.

20. Ogbourne, S. and Antalis, T. M. (1998) Transcriptional control and the role of silencers in transcriptional regulation in eukaryotes. *Biochem. J.* **331**, 1–14.

21. Bode, J., Schlake, M., Ríos-Ramírez, M., Mielke, C., Stengert, M., Kay, V., et al. (1995) Scaffold/matrix-attached regions: structural properties creating transcriptioally active loci, in *Structural and Functional Organization of the Nuclear Matrix (International Review of Cytology)*. (Berezney, R. and Jeon, K. W., eds.), Academic Press, San Diego, CA, pp. 389–453.

22. Bell, A. C. and Felsenfeld, G. (1999) Stopped at the border: boundaries and insulators. *Curr. Opin. Genet. Dev.* **9**, 191–198.

23. Li, Q. L., Harju, S., and Peterson, K. R. (1999) Locus control regions – coming of age at a decade plus. *Trends Genet.* **15**, 403–408.

24. Dorner, A. J., Wasley, L. C., and Kaufman, R. J. (1989) Increased synthesis of secreted proteins induces expression of glucose- regulated proteins in butyrate-treated Chinese hamster ovary cells. *J. Biol. Chem.* **264**, 20,602–20,607.

25. Kaufman, R. J., Wasley, L. C., Spiliotes, A. J., Gossels, S. D., Latt, S. A., Larsen, G. R., et al. (1985) Coamplification and coexpression of human tissue-type plasminogen activator and murine dihydrofolate reductase sequences in Chinese hamster ovary cells. *Mol. Cell. Biol.* **5**, 1750–1759.

26. Looney, J. E. and Hamlin, J. L. (1987) Isolation of the amplified dihydrofolate reductase domain from methotrexate-resistant Chinese hamster ovary cells. *Mol. Cell. Biol.* **7**, 569–577.

27. Brown, M, E., Renner, G., Field R, P., and Hassell, T. (1992) Process development for the production of recombinant antibodies using the glutamine-synthetase (GS) system. *Cytotechnology* **9**, 231–236.

28. Cappechi, M. R. (1990) Gene targeting: how efficient can you get? *Nature* **348**, 109.

29. Kilby, N. J., Snaith, M. R., and Murray, J. A. H. (1993) Site-specific recombinases: tools for genome engineering. *Trends Genet.* **9**, 413–421.

30. Baer, A. and Bode, J. (2001) Coping with kinetic and thermodynamic barriers: RMCE, an efficient strategy for the targeted integration of transgenes. *Curr. Opin. Biotechnol.* **12**, 473–480.

31. Klehr-Wirth, D., Kuhnert, F., and Hauser, H. (1997) Generation of mammalian cells with conditional expression of cre recombinase. *Technical Tips online*, T40067., URL: http://tto.trends.com.

32. Kühn, R., Rajewski, K., and Müller, W. (1995) Inducible gene targeting in mice. *Science* **269**, 1427.

33. Metzger, D., Clifford, J., Chiba, H., and Chambon, P. (1995) Conditonal site-specific recombination in mamalian cells using a ligand-dependent chimeric Cre recombinase. *Proc. Natl. Acad. Sci. USA* **92**, 6991–6995.

34. Kellendonk, D., Tronche, F., Monaghan, A. P., Angrand, P. O., Stewart, F., and Schütz, G. (1996) Regulation of cre recombinase activity by the synthetic steroid RU486. *Nucleic Acids Res.* **24**, 1404–1411.

35. Schlake, T. and Bode, J. (1994) Use of mutated Flp recognition target (FRT) sites for the exchange of expression cassettes at defined chromosomal loci. *Biochemistry* **33**, 12,746–12,751.

36. Lee, G., Kim, S., Lee, G. K. M., and Park, J. (2000) An engineered lox sequence containing part of a long terminal repeat of HIV-1 permits cre recombinase-mediated excision. *Biochem. Cell Biol.* **78**, 653–658.

37. Seibler, J. and Bode, J. (1997) Double-reciprocal crossover mediated by Flp recombinase. A concept and an assay. *Biochemistry* **36**, 1740–1747.

38. Seibler, J., Schübeler, D., Fiering, S., Groudine, M., and Bode, J. (1998) DNA cassette exchange in ES cells mediated by Flp recombinase: an efficient strategy for repeated modification of tagged loci by marker-free constructs. *Biochemistry* **37**, 6229–6234.

39. Verhoeyen, E., Hauser, H., and Wirth, D. (1998) Efficient targeting of retrovirally FRT-tagged chromosomal loci. Techn. Tips online T01515 URL: http://tto.trends.com.

40. Schübeler, D., Maass, K., and Bode, J. (1998) Retargeting of retroviral integration sites for the pedictable expression of transgenes and the analysis of cis-acting sequences. *Biochemistry* **37**, 11,907–11,914.

41. Verhoeyen, E., Hauser, H., and Wirth, D. (2001) Evaluation of retroviral vector design in defined chromosomal loci by Flp-mediated cassette replacement. *Hum. Gene Ther.* **12**, 933–944.

42. Müller, P., Oumard, A., Wirth, D., Kröger, A., and Hauser, H. (2001) Polyvalent vectors for coexpression of multiple genes, in *Plasmids for Therapy and Vaccination* (Schleef, M., ed.), Wiley-VCH, Weinheim, pp. 119–136.

43. Martinez-Salas, E. (1999) Internal ribosome entry site biology and its use in expression vectors. *Curr. Opin. Biotechnol.* **10**, 458–464.

44. Jackson, R. J. and Kaminski, A. (1995) Internal initiation of translation in eukaryotes: the picornavirus paradigm and beyond. *RNA* **1**, 985–1000.

45. Sachs, A. B., Sarnow, P., and Hentze, M. W. (1997) Starting at the beginning, middle, and end: translation initiation in eukaryotes. *Cell* **89**, 831–838.

46. Pelletier, J. and Sonenberg, N. (1988) Internal initiation of translation of eukaryotic mRNA directed by a sequence derived from poliovirus RNA. *Nature* **334**, 320–325.

47. Pestova, T. V., Hellen, C. U. T., and Shatsky, I. N. (1996) Canonical eukaryotic initiation factors determine initiation of translation by internal ribosome entry. *Mol. Cell. Biol.* **16**, 6859–6869.

48. Jang, S. K., Krausslich, H. G., Nicklin, M. J., Duke, G. M., Palmenberg, A. C., and Wimmer, E. (1988) A segment of the 5' nontranslated region of encephalomyocarditis virus RNA directs internal entry of ribosomes during in vitro translation. *J. Virol.* **62**, 2636–2643.

49. Oumard, A., Hennecke, M., Hauser, H., and Nourbakhsh, M. (2000) Translation of NRF mRNA is mediated by highly efficient internal ribosome entry. *Mol. Cell. Biol.* **20**, 2755–2759.

50. Chappell, S. A., Edelman, G. M., and Mauro, V. P. (2000) A 9-nt segment of a cellular mRNA can function as an internal ribosome entry site (IRES) and when present in linked multiple copies greatly enhances IRES activity. *Proc. Natl. Acad. Sci. USA* **97**, 1536–1541.

51. Dirks, W., Schaper, F., Kirchhoff, S., Morelle, C., and Hauser, H. (1994) A multifunctional vector family for gene expression in mammalian cells. *Gene* **149**, 387–388.

52. Dirks, W., Wirth, M., and Hauser, H. (1993) Dicistronic units for gene expression in mammalian cells. *Gene* **128**, 247–249.

53. Zitvogel, L., Tahara, H., Cai, Q., Storkus, W. J., Muller, G., Wolf, S. F., et al. (1994) Construction and Characterization of retroviral vectors expressing biologically active human interleukin-12. *Human Gene Ther.* **5**, 1493–1506.

54. Fussenegger, M., Mazur, X., and Bailey J. E. (1998) pTRIDENT, a novel vector family for tricistronic gene expression in mammalian cells. *Biotechnol. Bioeng.* **51**, 1–10.

55. Schirmbeck, R., von Kampen, J., Metzger, K., Wild, J., Grüner, B., Schleef, M., et al. (1999) DNA-based vaccination with polycistronic expression plasmids. *Methods in Molecular Medicine* **29**, 313–322.

56. Kwissa, M., Unsinger, J., Schirmbeck, R., Hauser, H., and Reimann, J. (2000) Polyvalent DNA vaccines with bidirectional promoters. *J. Mol. Med.* **78**, 495–506.

57. Mielke, C., Tümmler, M, Schübeler, D., von Hoegen, I., and Hauser, H, (2000) Stabilized, long-term expression of heteromeric proteins from tricistronic mRNA. *Gene* **254**, 1–8.

58. Attal, J., Theron, M. C., and Houdebine, L. M. (1999) The optimal use of IRES (internal ribosome entry site) in expression vectors, *Genet. Anal.* **15**, 161–165.

59. Hennecke, M., van Kampen, J., Metzger, K., Schirmbeck, R., Reimann, J., and Hauser, H. (2001) The strength of IRES-driven translation of bicistronic expression vectors depends on the composition of the mRNA. *Nucleic Acids Res.* **29**, 3327–3334.

60. Baron, U., Freundlieb, S., Gossen, M., and Bujard, H. (1995) Co-regulation of two gene activities by tetracycline via a bidirectional promoter. *Nucleic Acids Res.* **23**, 3605–3606.

61. Liljestrom, P. (1994) Alphavirus expression systems. *Curr. Opin. Biotechnol.* **5**, 495–500.

62. Johnston, R. E. and Peters, C. J. (1996) Alphaviruses, in *Fields Virology* (Fields B. N., Knipe D. M., Howley P. M., eds.), Lippincott-Raven: Philadelphia, PA, pp. 842–898.

63. Strauss, J. H. and Strauss, E. G. (1994) The alphavirus: gene expression, replication and evolution. *Microbiol. Rev.* **58**, 491–562.

64. Pushko, P., Parker, M., Ludwig, G. V., Davis, N. L., Johnston, R. E., and Smith, J. L. (1997) Replicon helper systems from attenuated Venezuelan equine encephalitis virus: expression of heterologous genes in vitro and immunization against heterologous heterologous genes in vitro and immunization against heterologous pathogens in vivo. *Virology* **39**, 389–401.

65. Berglund, P. M., Sjoberg, M., Garoff, H., Atkins, G. J., Sheahan, B. J., and Liljestrom, P. (1993) Semiliki Forest virus expression systems: production of conditionally infectious recombinant particles. *Bio-technology* **11**, 916–920.

66. Schlesinger, S. (1993) Alphaviruses-vectors for the expression of heterologous genes. *Trends Biotechnol.* **11**, 18–22.

67. Schlesinger, S. (1995) RNA viruses as vectors for the expression of heterologous proteins. *Mol. Biotechnol.* **3**, 155–165.

68. Wahlfors, J. J., Zullo, S. A., Nelson, D. M., and Morgan, R. A. (2000) Evaluation of recombinant alphaviruses as vectors in gene therapy. *Gene Ther.* **7**, 472–480.

69. Herweijer, H., Latendresse, J. S., Williams, P., Zhang, G., Danko, J., Schlesinger, S., et al. (1995) A plasmid-based self amplifying [Sindbis] virus vector. *Hum. Gene Ther.* **6**, 1161–1167.

70. Dubensky, T.W., Driver, D. A., Polo, J. M., Belli, B. A., Latham, E. M., Ibnaez, C. E., et al. (1996) Sindbis virus DNA-based expression vectors: utility for in vitro and in vivo gene transfer. *J. Virol.* **70**, 508–519.

71. Perri, S., Driver, D. A., Gardner, J. P., Sherrill, S., Belli, B. A., Dubensky, T. W., et al. (2000) Replicon vectors derived from Sindbis Virus and Semliki forest virus that establish persistent replication in host cells. *J. Virol.* **74**, 9802–9807.

72. Tubulekas, I., Berglund, P., Fleeton, M., and Liljeström, P. (1997) Alphavirus expression vectors and their use as recombinant vaccines: a minireview. *Gene* **190**, 191–195.

73. Horwitz, M. S. (1996) Adenoviruses, in *Fields Virology* (Fields, B. N., Knipe, D. M., and Howley, P. M., eds.), Lippincott-Raven: Philadelphia, PA, pp. 2149–2171

74. Romano, G., Micheli, P., Pacilio, C., and Giordano, A. (2000) Latest developments in gene transfer technology: achievments, perspectives, and controversies over therapeutic applications. *Stem Cells* **18**, 19–39.

75. Hitt, M., Bett, A. J., Prevec, L., and Graham, F. L. (1994) Construction and propagation of human adenovirus vectors, in *Cell Biology: A Laboratory Handbook.* Academic Press, San Diego, CA, pp. 479–490.

76. Kochanek, S., Clemens, P. R., Mitani, K., Chen, H. H., Chan, S., and Caskey, C. T. (1996) A new adenoviral vector: replacement of all viral coding sequences with 28 kb of DNA independently expression both full length dystrophin and ß-galactosidase. *Proc. Natl. Acad. Sci. USA* **93**, 5731–5736.

77. Fisher, K. J., Choi, H., Burda, J., Chen, S. J., and Wilson, J. M. (1996) Recombinant adenovirus deleted of all viral genes for gene therapy of cystic fibrosis. *J. Virol.* **217**, 11–22.

78. Kitamura, T., Onishi, M., Kinoshita, S., Shibuya, A., Miyajima, A., and Nolan, G. P. (1995) Efficient screening of retroviral cDNA expression libraries. *Proc. Natl. Acad. Sci. USA* **92**, 9146–9150.

79. Walther, W. and Stein, U. (2000) Viral vectors for gene transfer: a review of their use in the treatment of human diseases. *Drugs* **60**, 249–271.

80. Cosset, F. L., Takeuchi, Y., Battini, J. L., Weiss, R. A., and Collins, M. K. (1995) High titer packaging cells producing recombinant retroviruses resistant to human serum. *J. Virol.* **69**, 7430–7436.

81. Chong, H. and Vile, R. C. (1996) Replication-competent retrovirus produced by a "split-function" third generation amphotropic packaging cell line. *Gene Ther.* **3**, 624–629.

82. Palù, G., Parolin, C., Takeuchi, Y., and Pizzato, M. (2000) Progress with retroviral gene vectors. *Rev. Med. Virol.* **10**, 185–202.

83. Emerman, M., and Temin, H. M. (1986) Comparison of a promoter suppression in avian and murine retrovirus vectors. *Nucleic Acids Res.* **14**, 9381–9396.

84. Yu, S. F., von Ruden, T., Kantoff, P. W., Garber, C., Seiberg, M., Ruther, U., et al. (1995) Self-inactivating retroviral vectors designed for transfer of whole genes into mammalian cells. *Bio-technology* **13**, 389–392.

85. Bode, J., Fetzer, C. P., Nehlsen, K., Scinteie, M., Hinrichsen, B., Baiker, A., et al. (2001) The hitchhiking principle: Optimization episomal vectors for the use in gene therapy and biotechnology. *Gene Ther. Mol. Biol.* **6**, 33–46.

86. Van Craenenbroeck, K., Vanhoenacker, P., Duchau, H., and Haegeman, G. (2000) Molecular integrity and usefulness of episomal expression vectors derived form BK and Epstein-Barr virus. *Gene* **253**, 293–301.

87. Tan, B. T., Wu, L., and Berk, A. (1999) An Adenovirus–Epstein–Barr virus hybrid vector that stably transforms cultured cells with high efficiency. *J. Virol.* **73**, 7582–7589.

88. Leblois, H., Roche, C., Falco, D., Orsini, P., Yeh, P., and Perricaudet, M. (2000) Stable transduction of actively dividing cells via a novel adenoviral/episomal vector. *Molecular Therapy* **1**, 314–322.

89. Krougliak, V. A., Krougliak, N., and Eisensmith, R. C. (2000) Stabilization of transgenes delivered by recombinant adenovirus vectors through extrachromosomal replication. *J. Gene. Med.* **3**, 51–58.

90. Feng, M., Jackson, W. H., Goldman, C. K., Rancourt, C., Wang, M., Dusing, S. K., et al. (1997) Stable in vivo gene transduction via a novel adenoviral/retroviral chimeric vector. *Nat. Biotechnol.* **15**, 866–870.

91. Zheng, C., Baum, B. J., Iadarola, M. J., and O'Connell, B. C. (2000) Genomic integration and gene expression by a modified adenoviral vector. *Nat. Biotechnol.* **18**, 176–180.

92. Umaña, P. and Bailey, J. E. (1997) A mathematical model of N–linked glycoform biosynthesis. *Biotechnol. Bioeng.* **55**, 890–908.

93. Minch, S. L., Kallio, P. T., and Bailey, J. E. (1995) Tissue plasminogen activator coexpressed in Chinese hamster ovary cells with alpha(2,6)–sialyltransferase contains NeuAc alpha(2,6) Gal beta(1,4)Glc–N–AcR linkages. *Biotechnol. Prog.* **11**, 348–351.

94. Elbein, A. (1991) Glycosidase inhibitors: inhibitors of N–linked oligosaccharide processing. *FASEB J.* **5**, 3055–3063.

95. Grabenhorst, E., Costa, J., and Conradt, H. S. (1997) in *Animal Cell Technology* (Carrondo, M. J. T., Griffiths, B., and Moreira, J. L. P., eds.), Kluwer Academic Publishers, Dordrecht, The Netherlands, pp. 481–487.

96. Grabenhorst, E., Hoffmann, A., Nimtz, M., Zettlmeissl, G., and Conradt, H. S. (1994) Construction of stable BHK-21 cells coexpressing human secretory glycoproteins and human Gal(beta 1-4)GlcNAc–R alpha 2,6-sialyltransferase alpha 2,6-linked NeuAc is preferentially attached to the Gal(beta 1-4)GlcNAc(beta 1-2)Man(alpha 1-3)–branch of diantennary oligosaccharides from secreted recombinant beta–trace protein. *Eur. J. Biochem.* **232**, 718–725.

97. Sburlati, A., Umaña, P., Prati, E. G. P., and Bailey, J. E. (1998) Synthesis of bisected glycoforms of recombinant IFN–beta by overexpression of beta-1,4-N-acetylglucosaminyltransferase III in Chinese hamster ovary cells. *Biotechnol. Prog.* **14**, 189–192.

98. Li, E., Gibson, R., and Kornfeld, S. (1980) Structure of an unusual complex-type oligosaccharide isolated from Chinese hamster ovary cells. *Arch. Biochem. Biophys.* **199**, 393–399.

99. Costa, J., Grabenhorst, E., Nimtz, M., and Conradt, H. S. (1996) Stable expression of the Golgi form and secretory variants of human fucosyltransferase III from BHK-21 cells. Purification and characterization of an engineered truncated form from the culture medium. *J. Biol. Chem.* **272**, 11,613–11,621.

100. Suzuki, E. and Ollis, D. F. (1990) Enhanced antibody production at slowed growth rates: experimental demonstration and a simple structured model. *Biotechnol. Prog.* **6**, 231–236.

101. Franek, F. and Dolnikova, J. (1991) Hybridoma growth and monoclonal antibody production in iron-rich protein-free medium: effect of nutrient concentration. *Cytotechnology* **7**, 33–38.

102. Al-Rubeai, M., Emery, A. N., Chalder, S., and Jan, D. C. (1992) Specific monoclonal antibody productivity and the cell cycle–comparisons of batch, continuous and perfusion cultures. *Cytotechnology* **9**, 85–97.

103. Kirchhoff, S., Schaper, F., and Hauser, H. (1993) Interferon regulatory factor 1 (IRF-1) mediates cell growth inhibition by transactivation of downstream target genes. *Nucl. Acids Res.* **21**, 2881–2889.

104. Kirchhoff, S., Koromilas, A., Schaper, F., Grashoff, M., Sonenberg, N., and Hauser, H. (1995) IRF-1 induced cell growth inhibition and interferon induction requires the activity of the protein kinase PKR. *Oncogene* **11**, 439–445.

105. Köster, M., Kirchhoff, S., Schaper, F., and Hauser, H. (1995) Proliferation control of mammalian cells by the tumor suppressor IRF–1. *Cytotechnology* **18**, 67–75.

106. Köster, M., Kirchhoff, S., Schaper, F., and Hauser, H. (1995) Proliferation control of mammalian cells by the tumor suppressor IRF-1, in *Animal Cell Technology: Developments Towards the 21st Century* (Beuvery, C., Griffiths, B., and Zeijlemaker, W. P., eds.) Kluwer Academic Publishers, Dordrecht, The Netherlands, pp. 33–44

107. Carvalhal, A. V., Moreira, J. L., Müller, P. P., Hauser, H., and Carrondo, M. J. T. (1998) Cell growth inhibition by the IRF-1 system, in *New Developments and New Applications in Animal Cell Technology* (Merten, O.-W., Perrin, P, and Griffiths, B., eds.), Kluwer Academic Publishers, pp. 215–217.

108. Schaper, F., Kirchhoff, S., Posern, G., Köster, M. Oumard, A., Sharf, R., et al. (1998) Functional domains of Interferon Regulatory Factor 1 (IRF-1). *Biochemical J.* **335**, 147–157.

109. Kirchhoff, S., Wilhelm, D., Angel, P., and Hauser, H. (1999) NFκB activation is required for IRF-1 mediated IFN-β. *Eur. J. Biochem.* **261**, 546–554.

110. Kirchhoff, S., Oumard, A., Nourbakhsh, M., Levi, B.-Z., and Hauser, H. (2000) Interplay between repressing and activating domains defines the transcriptional activity of IRF–1. *Eur. J. Biochem.* **267**, 6753–6761.

111. Müller, P. P., Carvalhal, A. V., Moreira, J. L., Geserick, C., Schroeder, K., Carrondo, M. J. T., et al. (1999) Development of an IRF-1-based proliferation control system, in *Cell Engineering 1* (Al-Rubeai, M., Betenbaugh, M., Hauser, H., Jenkins, N., MacDonald, C., Merten, A.-W., eds.), Kluwer Academic Publishers, Dordrecht, The Netherlands, pp. 220–238.

112. Kröger, A., Köster, M., Schroeder, K., Hauser, H., and Mueller, P. P. (2002) Activities of IRF-1. *J. Interferon and Cytokine Res.* **22**, 5–14.

113. Geserick, C., Bonarius, H. P. J., Kongerslev, L., Hauser, H., and Müller, P. P. (2000) Enhanced productivity during controlled proliferation of BHK cells in continuously perfused bioreactors. *Biotech. Bioeng.* **69**, 266–274.

114. Müller, P. P., Kirchhoff, S., and Hauser, H. (1998) Sustained expression in proliferation controlled BHK–21 cells, in New developments and new applications in animal cell technology (Merten, O.-W., Perrin, P, and Griffiths, B., eds.), Kluwer Academic Publishers, Dordrecht, The Netherlands, pp. 209–214.

115. Kirchhoff, S., Kröger, A., Cruz, H., Tümmler, M., Schaper, F., Köster, M., et al. (1996) Regulation of cell growth by IRF-1 in BHK-21 cells. *Cytotechnology* **22**, 147–156.

116. Geserick, C., Schroeder, K., Bonarius, H., Kongerslev, L., Schlenke, P., Hauser, H., et al. (1999) Recombinant pharmaceutical protein overexpression in an IRF-1 proliferation controlled production system, in *Animal Cell Technology: Products from Cells, Cells as Products* (Bernard, A., Griffiths, B., Noé, W., and Wurm, F. eds.), Kluwer Academic Publishers, The Netherlands, pp. 3–10.

117. Carvalhal, A. V., Moreira, J. L., Cruz, H., Mueller, P. P., Hauser, H., and Carrondo, M. J. T. (2000) Manipulation of culture conditions for BHK cell growth inhibition by IRF-1 activation. *Cytotechnology* **32**, 135–145.

118. Müller, P. P., Grabenhorst, E., Conradt, H. S., and Hauser, H. (1999) Recombinant glycoprotein product quality in proliferation controlled BHK-21 cells. *Biotechnol. Bioeng.* **65**, 529–536.

119. Schroeder, K., Koschmieder, S., Ottmann, O. G., Hoelzer, D., Hauser, H., and Mueller, P. P. (2002) Genetic proliferation control facilitates the handling of a human stromal feeder cell line during cocultivation with hematopoietic progenitor cells. *Biotechnol. Bioeng.* **78**, 346–352.

120. Fussenegger, M., Schlatter, S., Dätwyler, D., Mazur, X., and Bailey, J. E. (1998) Controlled proliferation by multigene metabolic engineering enhances the productivity of Chinese hamster ovary cells. *Nat. Biotechnol.* **16**, 468–472.

121. Fussenegger, M. and Bailey, J. E. (1998) Molecular regulation of cell-cycle progression and apoptosis in mammalian cells: implication for biotechnology. *Biotechnol. Prog.* **14**, 807–833.

122. Mercille, S. and Massie, B. (1999) Apoptosis-resistant E1B-19K-expressing NS/0 myeloma cells exhibit increased viability and chimeric antibody productivity under perfusion culture conditions. *Biotechnol. Bioeng.* **63**, 529–543.

123. Cotter, T. G. and Al-Rubeai, M (1995) Cell death (apoptosis) in cell culture systems. *Trends Biotechnol.* **13**, 150–155.

124. Perreault, J. and Lemieux, R. (1993) Essential role of optimal protein synthesis in preventing the apoptotic death of cultured B cell hybridomas. *Cytotechnology* **13**, 99–105.

125. Singh, R. P., Emery, A. N., and Al-Rubeai, M. (1996) Enhancement of survivability of mammalian cells by overexpression of the apoptosis-suppressor gene bcl-2. *Biotechnol. Bioeng.* **52**, 166–175.

126. Chung, J. D., Sinskey, A. J., and Stephanopoulos, G. (1998) Growth factor and bcl-2 mediated survival during abortive proliferation of hybridoma cell line. *Biotechnol. Bioeng.* **57**, 164–171.

127. Simpson, N. H., Milner, A. E., and Al-Rubeai, M. (1997) Prevention of hybridoma cell death by bcl-2 during suboptimal culture conditions. *Biotechnol. Bioeng.* **54**, 1–16.

128. Huang, D. C. S., Cory, S., and Strasser, A. (1997) Bcl-2, Bcl-XL and adenovirus protein E1B19kD are functionally equivalent in their ability to inhibit cell death. *Oncogene* **14**, 405–414.

129. Terada, S., et al. (1997) Anti-apoptotic genes, bag-1 and bcl-2, enabled hybridoma cells to survive under treatment for arresting cell cycle. *Cytotechnology* **25**, 17–23.

130. Zanghi, J. A., Fussenegger, M., and Bailey, J. E. (1999) Serum protects protein–free competent Chinese hanster ovary cells against apoptosis induced by nutrient deprivation in batch culture. *Biotechnol. Bioeng.* **5**, 573–582.

131. Mastrangelo, A. J., Hardwick, M. J. Zou, S., and Betenbaugh, M. J. (2000) Part 2. Overexpression of bcl-2 family members enhances survival of mammalian cells in response to various culture insults. *Biotechnol. Bioeng.* **67**, 555–564.

132. Vives, J., Juanola, S., Gabernet, C., Prats, E., Cairó, J. J., Cornudella, L., et al. (2001) Genetic strategies for apoptosis protection and hybridoma cells based on overexpression of cellular and viral proteins, in *Animal Cell Technology: From Target to Market* (Lindner-Olsson, E., et al., eds.), Kluwer Academic Publishers, Dordrecht, The Netherlands, pp. 230–233.

3

Protein Expression Using Transgenic Animals

William H. Velander and Kevin E. van Cott

1. Introduction

In general, the need for more economical, abundant, safe, and efficacious supplies of therapeutic proteins has motivated research and development into the use of transgenic animals as bioreactors since about 1987. These proteins have traditionally been derived from human plasma where supply and safety issues have inspired the development of recombinant versions from mammalian cells. However, a lack of development and production capacity for recombinant mammalian cell culture has provided added impetus to use transgenic livestock as sources of these proteins *(1)*. The use of transgenic animals as bioreactors for human wild-type or genetically engineered variants of human proteins is one of the most advanced examples of recombinant biology because it must achieve the biosynthesis of a protein both in a temporal and tissue-specific manner without harming the host animal. The preferred tissue for expression of the recombinant protein is one that naturally produces and exports high concentrations of protein to enable easy harvesting. In particular, the expression of recombinant proteins into the milk of transgenic rabbits and livestock has been a central focus of the transgenic animal bioreactor theme, although expression in urine and blood has been also studied *(2a,b)*.

Research experience from the last decade has helped to delineate the molecular biology needed to obtain tissue-specific and temporal control of expression as well as control over the level of expression. Progress also has been made toward understanding the species and tissue-specific limitations of producing biologically active, complex mammalian proteins at the highest rates of biosynthesis possible in transgenic animals. As a result of this progress, in 1997 the United States Food and Drug Administration (USFDA) established a Points to Consider document for production of therapeutics by transgenic animals *(3)*. These guidelines parallel those concerning safety and efficacy associated with therapeutics produced from cultured mammalian cells. These issues revolve around the maintenance of genotypic and phenotypic stability as well as the specific pathogen-free pedigree of the transgenic animal lineages used for production. Cell storage banks for semen, embryos, and other tissues that can regenerate a perpetual supply of the production animals are also addressed. Although no therapeutic products made by transgenic livestock have yet been approved by the USFDA, several

From: *Handbook of Industrial Cell Culture: Mammalian, Microbial, and Plant Cells*
Edited by: V. A. Vinci and S. R. Parekh © Humana Press Inc., Totowa, NJ

products made in the milk of sheep *(4)*, goats *(5)*, and rabbits *(6)* have advanced to a phase III stage of human clinical trials. Thus, the initial issues of safety and dose-dependent response have been favorably established for these particular examples.

The combination of high expression levels that are naturally exported in volumes that are easily harvested gives milk the potential for dramatically better balance of economics and abundance than production from mammalian cell culture. This is largely because of the fact that process purification costs are logarithmically decreased as the concentration of the target protein is increased. Transgenic livestock can typically make complex mammalian proteins at over 200-fold higher concentration per time in milk because of the high epithelial-cell density of the mammary gland, at a rate of synthesis that is similar to the higher-producing cell types used in recombinant mammalian cell culture. In addition, the ability to rapidly and cost-effectively add more animals to an already established milking herd circumvents scale-up lag times associated with adding new mammalian cell-culture production facilities. Although the initial production of founder transgenic animals can be limiting and expensive, a combination of techniques used for gene delivery in vitro with embryonic-cell cloning has been developed for livestock to improve the efficiency of making founder transgenic animal lineages.

Post-translational processing leads to the biochemical alteration of a specific amino acid within the polypeptide backbone, and this intracellular derivatization is termed a post-translational modification (PTM) *(7)*. Examples of these PTMs include γ-carboxylation of glutamic acid, N-linked glycosylation at asparagines (Asn), O-linked glycosylation at serine or threonine (Ser, Thr), sulfation of tyrosine, phosphorylation of serine, and proteolytic processing. Many of these modifications are required for secretion and biological activity of human proteins. Most human therapeutic proteins have complex PTMs, and therefore cannot be made in genetically engineered bacteria or yeast because of the primitive post-translational (PT) processing that is present in these microorganisms *(8)*. More advanced cells types, such as those found in plants and insects, are also unable to make fully functional recombinant forms of many human proteins because they lack the appropriate enzymes to make the necessary PTMs of the recombinant protein *(9)*. Even mammalian cells can vary in their capability to post-translationally process proteins *(10)*. Therefore, the choice of a particular animal for protein production at large scale is then typically evaluated on the optimal combination of volume needed to satisfy clinical demand and also on the PT processing capability of the host animal for that protein.

2. Techniques for the Production of Transgenic Animals

Transgenic animals can be made using several different techniques that introduce the foreign gene (here termed "transgene") into cells at an early embryonic stage. The transgene that dictates the primary structure of protein can consist of cDNA, genomic, or hybrid combinations of cDNA, and genomic sequences called minigenes. In general, these DNA constructions can be made using classical gene-cloning techniques, and are usually excised from the cloning vector, purified, reconstituted in low ionic-strength buffers, and introduced as linearized instead of circular DNA sequences. The natural biochemistry of chromosomal maintenance is believed to be the key mechanism for the insertion of foreign DNA. The goal of most DNA delivery techniques is to be compatible or to encourage chromosomal integration events that produce animals with the properties of Mendelian inheritance with respect to the transgene. Outbreeding with nontransgenic animals enables genotypic stability to be more easily studied, such as detecting transgene insertions on multiple chromosomes as well as unstable insertions that result in whole or partial transgene deletion from generation to generation.

2.1. Microinjection as a Technique for Producing Transgenic Animals

There are several techniques designed to deliver the transgene and ultimately produce a viable transgenic embryo. The simplest technique is the introduction of DNA into the nucleus of a one- or two-celled embryo by microinjection, and then surgically transfer the embryos to a hormonally synchronized recipient surrogate mother *(11)*. Microinjection into the pronucleus of a newly fertilized embryo is most often used in mice, rabbits, pigs, goats, and sheep. The pregnancy frequency resulting from surgically transferred, microinjected embryos is typically less than 50%, and about 10–30% of the animals born in mice, rabbits, and pigs are transgenic. Pronuclear microinjection is most efficient in mice, rabbits and pigs, which have short gestation periods and five or more offspring per pregnancy. For example, the pig has a gestation of less than 4 mo and typically has up to12 piglets in a litter. Thus, there is a good chance of making transgenic piglets in every litter. In contrast to the pig, microinjection is very inefficient for species such as the cow, which typically have only one healthy offspring per pregnancy and have gestation periods of 10 mo *(12)*.

Common to all gene delivery techniques is the need to have an ample source of embryos and then to identify the most viable transgenic embryos after gene delivery. The greatest inefficiency of gene transfer is related to physical damage as well as developmental asynchrony and retardation that are created from manipulation in vitro. After gene delivery, the embryo is cultured for a short time to determine viability from the physical damage caused by the injection needle. For example, about 50% of the embryos survive microinjection until the two-cell stage *(13)*. However, at this stage it is not possible to reliably detect which embryos are transgenic without destroying the embryo *(14)*. Further culture to past the morula stage of development (about 8 cells) decreases embryonic viability. Therefore, one- to four-cell-stage embryos are most often transferred into a recipient surrogate mother that is hormonally synchronized to become pregnant without knowledge of which embryos are transgenic. Because the viability of manipulated embryos is lower than that of unmanipulated embryos, about threefold more embryos are transferred into the reproductive tract than the number that would normally occur in vivo during the natural fertilization and uterine impregnation process. The chief inefficiency of microinjection and other embryonic gene transfer techniques is the number of viable embryos that must be surgically harvested from synchronized, and naturally or artificially inseminated donor animals. In addition, it is estimated that the transgene integration event occurs at a two-cell or later stage in about 30–60% of all embryos *(14)*. This results in a nonuniform distribution between cells of the integrated transgene known as mosaicism. Mosaicism can mask the true potential expression-level of the transgene and decrease the chance of Mendelian transmission of the transgene to offspring. For example, if only half of the mammary cells have the integrated transgene, then the expression level will be lower than if the mammary gland were totally uniform in its transgene distribution.

2.2. Animal Cloning Techniques Used to Produce Transgenic Animals

With respect to the production of transgenic animals for use as bioreactors, the development of animal cloning techniques has been partly inspired by the low transgenesis efficiencies *(11)*. This is further exacerbated in animals such as cows and sheep that have a small number of offspring per pregnancy. In addition, the desire to knock-out genes that code for certain endogenous enzymes associated with post-translational processing and then replace them with genes that code for enzymes that would exact a more efficient and perhaps more humanized PTM upon the target recombinant protein. The applications for animal cloning also extend to general livestock improvement *(15)* and to humanizing the tissue of pigs for xenograft transplantation *(16)*. Examples of replacement genes include those that code for propeptide cleavage enzymes and certain glycan processing enzymes.

All cloning methods used to achieve transgenesis rely on the establishment of primary cell lines from embryos, fetal, or somatic tissue with nuclei that have retained the potential for subsequent progeneration of an animal *(17)*. Most cloning methods first screen for transgenesis from tens of thousands of primary cell lineages derived from embryonic or fetal cells that have been transfected in culture. These primary culture methods typically yield cell passages of about thirty to forty generations before senescence occurs. Secondary cloning is frequently done from transgenic fetuses to establish a developmental synchrony, and thus more viable fetuses and born animals as well as a more stable transgenic genotype from the first clonal lineage. This second-pass transgenic progenitor cell line is then expanded to provide multiple clones that can be stored frozen. Electroporation and/or calcium phosphate precipitation is most often used to introduce DNA into the primary embryonic cell lines used in cloning. However, a transgenic primary cell line can be established from an existing transgenic animal made first by microinjection *(18a,b)*. Retroviral transfection can also be achieved with reasonable efficiency on cultured cells, but it poses a potential risk because of the incorporation of viral DNA elements *(11)*. This method involves limitations in the size of the transgene that can be packaged into virion particles. Once the transgenic primary cell lineage is established, the detection and characterization of the transgene can be easily done in vitro from a sample of the cultured lineage.

Nuclear transfer from genetically stable and outgrown transgenic cells into enucleated oocytes is then performed to establish a transgenic embryo *(17)*. Thus, the transfer of nuclei taken from these established transgenic cell lineages into enucleated oocytes results in 100% transgenic frequency at the embryonic level. The abundance of recipient oocytes harvested from slaughterhouse cattle or pigs makes cloning efficient from a donor-cell perspective. Viable embryos are then selected after further culture, and are usually transferred at the morula to blastocyst stage into a hormonally synchronized, recipient surrogate mother. It is important to note that the animals produced by cloning will be uniformly transgenic, and the resulting genotype will not be complicated by transgene mosaicism. In summary, cloning strategies have been combined with DNA delivery techniques so that virtually 100% of all born animals are stable transgenic lineages.

2.3. Disease Transmission Issues

In an analogous manner to that of classical mammalian cell-culture methods, transgenic livestock represent a mammalian cell source that can be maintained with a specific pathogen-free (SPF) pedigree in the same context as any other mammalian-cell source for therapeutic proteins. A summary criterion for maintaining any mammalian-cell source with a SPF-pedigree is that the source must be both reliable and safe from known zoonotically transmissible diseases. In the case of transgenic livestock, a SPF-pedigree can be maintained by simple restricted access procedures already used in many high-volume commercial swine facilities. The main concern is contamination of the SPF herd and facilities with animal diseases carried by visitors who have recently visited ordinary farms. Thus, a post-farm-visitor quarantine, shower-in only facilities, and post-shower uniformed employees are recommended operating procedures for maintaining indoor SPF-conditions. In the case of sheep and cattle, an immunologically stronger and disease resistant animal may result from outside pasturing, and pigs have been domesticated for continuous indoor housing. An indefinitely stable and renewable supply of these SPF-animals can be insured by frozen storage of embryo and semen. Embryos and semen provide additional barriers to the transmission of disease, as low levels of infectivity of most pathogens occur in these cell types.

The presence of pathogens that integrate into the genome of the host mammalian cell is also of concern in transgenic animals as in cultured mammalian cells (3,19). For example, the presence of endogenous retroviruses that have integrated into chromosomes of the host cell is a risk in the context of all mammalian cells, whether in culture or in the transgenic tissue of an animal. In particular, porcine retroviruses have been zoonotically transmitted to cultured human cells (20). However, there are no documented cases of zoonosis from therapeutics procured from slaughterhouse livestock such as insulin from bovine or porcine pancreas, Factor VIII from porcine plasma, or porcine cells and tissues (21). The threat of zoonosis of pathogens with unknown etiology—such as with spongiform encephalopathies—has gained worldwide attention, and some strains of Creutzfeldt-Jakob disease are believed to be closely related and transmitted from eating infected animal tissue (22). This appears to be a species-specific pathogen, and pigs have been shown to be resistant to infection through dietary challenge (23). Although the risk of zoonosis is a serious concern for xenograft transplantation of live cells and tissue, there are multiple pathogen inactivation barriers that can be installed into purification processing, which can yield acceptable pathogen reduction and inactivation for therapeutic protein products (24).

The effects of expression of recombinant proteins on the health of the host animal have also been detailed, and in some cases these effects can be adverse. For example, the expression of high levels of human growth hormone in the mammary gland and promiscuous expression in other tissues has been shown to cause sterility in females (25). However, in most cases, expression of recombinant protein has produced no significant physiologically adverse effects, even at g/L levels that noticeably change the composition of milk. The physiological effects of the expression of high levels of recombinant protein are usually studied in transgenic mice before moving on to larger animals such as livestock.

2.4. Bioreactor Efficiency

Analysis of the main throughput issues of classical mammalian cell-culture bioreactors reveals that cell densities are too low and there are rate limitations in post-translational processing and secretory pathways in most cell types. The productivity of transgenic livestock bioreactors lies in the secretion of protein by cells at a high density in the mammary gland tissue into the harvested milk. Annual milk volume that can be harvested from the pig, sheep, goat, and cow are approx 300, 500, 800, and 8000 L/yr, respectively *(26)*. The mammary epithelia are at about 10^9 cells/mL, whereas most large-scale mammalian cell reactors are about 10^6 to 10^7 cells/L. By design, the mammary gland is a tissue that has less restrictive limitations in intracellular secretory pathways in terms of proteins secreted into milk. For example, caseins constitute as much as 30–40 g/L and are the major milk proteins. Most mammalian cell lines secrete complex proteins such as γ-carboxylated vitamin K-dependent proteins at less than 1 pg/cell/d *(27)*. Complex proteins such as recombinant human protein C and fibrinogen have been made in the 1–5 g/L range, which translates to greater than 10 pg/cell/d *(28)*. Thus, the mammary gland is also frequently higher in its rate of synthesis on a per-cell basis than most secretory cells grown in culture.

3. Post-Translational Processing of Complex Proteins in Transgenic Animals

Transgenic animals can provide a unique source of recombinant proteins with the equivalent or improved bioactivities relative to native proteins in either replacement or other therapeutic applications. As is the case for synthesis recombinant proteins in cell culture, the bioactivities and also in vivo half-life can be highly dependent upon the PT signature left by the host livestock cell. The heterogeneity of recombinant proteins produced in transgenic animal bioreactors can be the result of species-specific differences between the animal's endogenous PT processing enzymes and the exogenous human protein substrate, and rate limitations in PTMs as protein secretion rates overwhelm the cellular PT processing capabilities in the endoplasmic reticulum (ER) and Golgi. The resultant isoforms may have bioactivities that are equally diverse. Optimization of any bioreactor requires a thorough knowledge of rate-limiting steps and, in the case of the transgenic bioreactor, species-specific differences in PT processing in the mammary gland. Improving the efficiency of PTMs made in the transgenic animal bioreactors must be accomplished before their potential can be fully realized. By improving the efficiency of upstream biosynthesis, downstream purification and processing will also be greatly facilitated because isoforms may have very different physical-chemical characteristics that complicate product recovery, and inactive or harmful (e.g., immunogenic) isoforms will not have to be removed.

The majority of efforts in the field of transgenic animal bioreactors have been dedicated to understanding the molecular biology of inducing high protein-expression levels. Relatively little effort has been dedicated to understanding how the biochemistry of the mammary epithelial cells impacts PTM processing of an exogenous protein, or what animal species has the best ability for performing particular PTMs. Here we review the production of several recombinant proteins in different transgenic animal species, and comment on two important PTMs—glycosylation and γ-carboxylation—which can dramatically affect biological activity and pharmacokinetics.

3.1. Glycosylation of Recombinant Proteins

Glycosylation affects three-dimensional protein structure, solubility, biological activity, and circulation half-life *(29)*. The complexity of oligosaccharide composition and structure is evident even in native plasma proteins made by the liver that circulate as a collection of subpopulations of several glycoforms *(30)*. Two glycan linkages occur: N-linked and O-linked. The consensus peptide sequence for the attachment of N-linked oligosaccharides is Asn-X-Ser/Thr. There is no consensus sequence for the attachment of O-linked oligosaccharides at Ser or Thr residues. The formation of glycosylated structures results from a complex sequence of enzymatic steps in the ER and Golgi. Briefly, N-linked glycans are generated by initial transfer of a lipid-linked oligosaccharide ($Glc_3Man_9GlcNAc_2$-P-P-Doli) in the ER *(31)*. The oligosaccharides are trimmed by exoglycosidases in the ER, and further modified in the Golgi by exoglycosidases and glycosyltransferases. N-linked oligosaccharides are classified as (i) high mannose; (ii) complex; and (iii) hybrid (Fig. 1). O-linked glycans are generated by the initial attachment of Gal-NAc to a Ser or Thr residue. Nucleotide sugars are the substrates for stepwise addition to O-linked glycans. However, unlike the human liver, recombinant cells such as human kidney 293 cells and mammary epithelial cells tend to substitute GalNAc for Gal *(10)*.

Proteins with high mannose glycans or with desialylated glycans with terminal Gal or GalNAc moieties are rapidly cleared from circulation by hepatic cells *(31)*, and so the presence of desialylated or high-mannose glycans will result in much higher doses needed to achieve a therapeutic affect. Thus the biochemistry of PTMs in the mammary epithelial cells can play an equally important role as the overall expression level in determining the productivity of the transgenic animal. Complicating factors in the glycosylation of recombinant proteins made in transgenic animals are (i) the species-specific glycosylation pattern inherent to the animal *(31)*; (ii) the tissue-specific glycosylation pattern in the mammary epithelial cells, which may differ from hepatic cells *(31)*; (iii) the potential rate limitations in either the transfer of the core oligosaccharide in the ER or modifications of the core oligosaccharide made in the ER and Golgi; and (iv) the potential for changing glycosylation patterns as lactation progresses *(32)*.

There is growing evidence that the species-specific nature of the glycosylation pattern made by mammary epithelial cells on endogenous milk proteins is also imparted on recombinant proteins. Unfortunately, published reports with detailed glycan structure and composition data of either recombinant or endogenous milk proteins from transgenic livestock animal species are still sparse. However, lactoferrin (LF) is an endogenous milk protein that ubiquitous to the goat, cow, and pig, and its glycosylation pattern has been characterized across these species (Table 1). The number of potential glycosylation sites and structures in LF has been shown to be species-specific. The human contains two sites, the pig contains one site, the cow contains five, and the goat contains four *(33,34)*. Both goat and bovine LF contained two sites with high-mannose oligosaccharides only. Human and porcine LF did not contain high-mannose glycans. Evidence thus far indicates that the mammary epithelial cells of ruminants tend to incorporate more high mannose glycans on recombinant proteins than the pig. For example, recombinant protein C made in the milk of transgenic pigs contained only low-mannose stuctures that were similar to the human plasma-derived protein C *(35a,b)*. More work is needed to establish whether or not species-specific differences in glycosylation are significant issues in determining which animal would be the best bioreactor for a given protein. Thus, the challenge in producing a recombinant glycoprotein in transgenic animals is to determine whether or not the glycan heterogeneity is

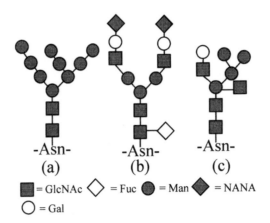

Fig. 1. Three classes of N-linked oligosaccharides: (**a**) high-mannose, (**b**) complex, (**c**) hybrid.

Table 1
Summary of Published Glycan Structure and Composition for Recombinant and Endogenous Milk Proteins from Transgenic Livestock Animals

Source	Protein	Glycan structure	Glycan composition information
Goat	Recombinant ATIII		
	Asn 96	Complex	Fucosylation, monosialylated, NGNA substitution for NANA
	Asn 192	Complex	Fucosylation, monosialylated, NGNA substitution for NANA
	Asn 155	High mannose	
	Recombinant LATPA		Low NANA, low Gal, GalNAc present
Human	Endogenous lactoferrin	2 Complex (no high mannose)	α-1,6 and α-1,3 fucosylation no GalNAc
Cow	Endogenous lactoferrin	2 High mannose 3 Complex	α-1,6 fucosylation GalNAc substitution for Gal
Pig	Endogenous lactoferrin	Complex (no high mannose)	α-1,6 fucosylation GalNAc substitution for Gal
Goat	Endogenous lactoferrin	2 High-mannose 2 Complex	α-1,6 fucosylation no GalNAc

within satisfactory limits that ensure overall product consistency, safety, and efficacy. We now present two case studies in which recombinant human plasma glycoproteins have undergone human clinical trials. These proteins were produced in the milk of transgenic ruminants (goat and sheep), and we will comment on the glycosylation signature reported.

3.2. Antithrombin III

Antithrombin III (ATIII) is a 58-kDa serine protease inhibitor that is synthesized in the liver and circulates in plasma at a concentration of 170–290 µg/mL *(36)*. ATIII acts as an anticoagulant by inhibiting the proteolytic activities of thrombin and Factor Xa. The activation of ATIII requires interaction with heparin, which causes a conformational change in ATIII and increases the binding affinity with thrombin. ATIII replacement therapy is indicated for patients with congenital deficiency or transient deficiencies such as during cardiopulmonary bypass (CPB) surgery *(37a,b)*. Recent reports have also studied rATIII as an anti-sepsis therapeutic *(38)*. The estimated clinical demand for ATIII is 75 kg/yr *(26)*. PTMs made to ATIII include three disulfide bridges and the attachment of four N-linked glycans.

GTC Biotherapeutics is producing rATIII in the milk of transgenic goats using a β-casein promoter system. The rATIII expression levels achieved in goats and used for producing material for clinical trials range between 1.5 and 3.5 g/L over a 300-d lactation. With an average milk yield of about 2 L/d, and a reported purification yield of ~50%, a single goat can produce about 1.2 kg rATIII/yr. GTC was granted orphan drug status for using rATIII, and it was the first transgenic animal-derived therapeutic that successfully entered and completed a Phase III clinical trial study (for treating heparin resistance during CPB surgery). However, before a second Phase III study was completely finished, GTC decided to stop further rATIII development for economic reasons, and is now pursuing the production of monoclonal antibodies (MAbs) and fusion proteins *(39)*. Although rATIII may never come to market, one of the most important results from these trials is that a transgenic animal-derived protein was determined to be safe and efficacious *(40)*. These results should pave the way for future approval of therapeutics with larger market demands, such as MAbs.

Edmunds et al. have characterized the glycosylation of rATIII produced in goat milk *(41)*. ATIII is glycosylated at four Asn residues: Asn-96, Asn-135, Asn 155, and Asn 192. In human plasma-derived ATIII (pd-ATIII), the glycans are all complex, and biantennary, contain no fucoseor GalNAc, and are disialylated *(42)*. Edmunds et al. found that rATIII produced in goats had a higher level of heterogeneity in glycan structure than plasma-derived ATIII (Table 1). The major glycans at Asn-96 and Asn-192 were fucosylated, biantennary, and monosialylated. Minor glycans at Asn-96 and Asn-192 were found to be disialylated, and have NGNA substitution for NANA and GalNAc substitution for Gal. Glycans at Asn-155 were high mannose, and glycans at Asn-135 could not be characterized. For comparison, monosaccharide composition was reported for recombinant long-acting tissue plasminogen activator made in goat milk *(43)*. GalNAc was found in rLA-TPA from goat milk, and NANA content was lower than expected. Several studies have indicated that ATIII Asn-155 glycan composition and structure determine heparin affinity *(44a,b)*. The rATIII had a fourfold higher affinity for heparin than pd-ATIII, which was attributed to lower sialic acid content, especially at Asn-155. The pharmacokinetics of rATIII was acceptable, with a half-life of about 43 h, which was comparable to some formulations of pd-ATIII *(45)*.

3.3. Alpha-1-Antitrypsin

Alpha-1-antitrypsin (AAT) is a serine protease inhibitor (52 kDa) synthesized in the liver, which circulates in plasma at a concentration of 1–2.5 mg/mL. The primary target of AAT is neutrophil elastase, a serine protease stored in neutrophils. Deficiencies in functional AAT result in liver disease in children and emphysema in adults *(46)*. Estimated clinical demand for AAT is 5000 kg/yr *(26)*. Plasma-derived AAT (pd-AAT) has three N-linked glycans at Asn-46, Asn-83, and Asn-247 *(47)*. Six major glycoforms of pd-AAT have been recently identified, all containing a heterogeneous mixture of complex sialylated bi- and tri-antennary sialylated oligosaccharides with varying degree of fucosylation. Native glycosylation of AAT is required for an adequate circulation half-life *(48)*, but does not appear to affect its anti-elastase activity. Unglycosylated rAAT produced in prokaryotes and hypermannosylated rAAT produced in yeast retain anti-elastase activity *(49a,b)*.

PPL Therapeutics has generated lines of transgenic sheep that secrete high levels (up to 35 g/L) of rAAT in milk *(50a,b)*. PPL completed phase II clinical trials for transgenic-derived rAAT as a treatment for cystic fibrosis *(51)*. The safety of rAAT was established, and PPL and Bayer have now joined to enter into Phase III clinical trials with an aerosol formulation of rAAT. The published data on biochemical characterization of rAAT produced in sheep milk are not as detailed as data for rATIII. Details on glycan structure are not reported—only results that demonstrate that all three N-linked sites are occupied, are removed by treatment with PNGase F (indicating complex glycan structure), and probably have lower sialic acid content than pd-AAT (as deduced from isoelectric focusing experiments) *(52)*. Carver et al. did report that approx 10% of rAAT had a post-translational proteolytic cleavage of the first five amino acids, similar to minor subpopulations of pd-AAT *(53)*.

3.4. Production of Recombinant VKD Proteins in Transgenic Animals

The Vitamin K-dependent (VKD) family of proteins made by the human liver are important members of the coagulation cascade. VKD proteins are distinguished by γ-carboxylation of glutamate to γ-carboxyglutamate (Gla), an enzymatic process that requires vitamin K as a cofactor *(54)*. Vitamin K-dependent proteins in plasma include Protein C (PC), Factor IX (FIX), Factor X (FX), Factor VII (FVII), prothrombin (PT), and Protein S (PS). Protein C is a central anticoagulant, and its annual clinical demand in the United States to treat sepsis and disseminated intravascular coagulation is estimated at 100 kg/yr. Factor IX is an important procoagulant and its deficiency causes Hemophilia B. The worldwide clinical demand for FIX is estimated at 1–2 kg/yr *(35b)*.

A functional Gla domain is required for biological activity-binding of calcium ions to the Gla domain results in a conformational change that facilitates the VKD protein's interaction with the phospholipid membrane of the endothelial cells and platelets *(55)*. The salient feature of the enzyme system that modifies Glu to Gla is its complex sequence of redox reactions, which make it difficult to simply add the enzymes to prokaryotic and lower-order eukaryotic cells. VKD carboxylase is an integral membrane protein in the endoplasmic reticulum (ER) that adds CO_2 to the γ-carbon of glutamate to convert it to γ-carboxyglutamate. Oxygen and reduced vitamin K are required cofactors for this process. Reduced vitamin K is oxidized to the alkoxide ion form by the epoxidase domain of the carboxylase enzyme, which is a strong base that extracts the hydrogen from the γ-carbon of glutamate *(56)*. After carboxylation of the glutamate, the vitamin K epoxide is subsequently reduced by the vitamin K epoxide reductase system (VKOR) *(57)*. The propeptide is required for γ-carboxylation of plasma proteins, and functions as a binding domain to "dock" the VKD protein with the carboxylase *(58a,b)*. Subsequent removal of the propeptide is also required prior to secretion for biological activity *(59)*. The protease PACE/furin, which recognizes pairs of basic amino acids, can accomplish this proteolytic cleavage in recombinant production systems *(60a,b)*. Carboxylase is a processive enzyme, capable of modifying a number of glutamates in a single enzyme-substrate-binding event *(61)*. The clinical demand for PC and FIX has inspired extensive research into the production of these proteins in transgenic animals. It appears that the pig is more efficient at γ-carboxylation than other species.

3.5. Protein C and Factor IX—Rate Limitations in γ-Carboxylation

Although endogenous milk proteins do not require γ-carboxylation of Glu, endogenous γ-glutamyl carboxylase is expressed in the mammary epithelial cells during lactation, and has avidity for the substrate. However, production of rFIX and rPC in transgenic animals indicates the existence of rate limitations and species-specific differences in γ-carboxylation capabilities. By using variations of the mouse WAP promoter and the Protein C transgene (cDNA and genomic), we generated transgenic mouse and pig lines with daily expression levels ranging from 1 µg/mL to 2 mg/mL *(62a,b,c)*. In both pigs and mice, the percentage of γ-carboxylated rPC was measured as the percentage of protein that bound to a conformational metal-dependent MAb directed against the Gla domain (7D7B10) *(63)*. We found that the amount of γ-carboxylated rPC was inversely proportional to the expression level (Table 2).

Published data also indicate species-specific and substrate-specific differences in the activity of the animal's endogenous γ-glutamyl carboxylase system for recombinant human VKD proteins. Approximately 44% of rFIX produced in transgenic mouse milk was recoverable by purification using a Gla-domain specific MAb (Table 1) *(64)*. We have found that transgenic pigs appear to have greater γ-carboxylation efficiency for rFIX at moderate expression levels (< 200 µg/mL) *(65)*. Transgenic sheep produced rFIX at very low levels (25 ng/mL), and only 2% of the rFIX was recovered by affinity chromatography *(66)*. Although the very small amounts of rFIX recovered were biologically active, the overall efficiency of γ-carboxylation of sheep cannot be determined from these data. For comparison, CHO cells were reported to produce 1.5% fully carboxylated rFIX *(67)*, along with partially and un-carboxylated forms *(68)*. Thus, porcine mammary epithelial cells appear to have a higher carboxylase activity vs murine and ovine epithelial cells, and rFIX may be a better substrate for porcine and murine carboxylase than rPC.

4. Conclusion

Transgenic animal bioreactors have been used to produce many therapeutic recombinant proteins in the mammary epithelial cells and to secrete the product into milk. Mammary epithelial cells are by nature designed for high protein synthesis rates during lactation, and milk is easily harvested. The combination of efficient mammary-specific transgene promoters and a high cell density in the mammary gland (~10^9 cells/mL) results in high protein production rates, with g/L/h recombinant protein concentrations now routine. Transgenic animal production also offers the advantage of safety when compared with donor tissue- or plasma-derived sources. In addition, scale-up of production is less expensive than for mammalian cell culture. These characteristics have made transgenic animals a viable alternative to production of proteins, particularly plasma proteins and humanized MAbs. Rate limitations and species-specific differ-

Table 2
Estimated γ-Carboxylation Efficiency of Recombinant VKD Proteins Produced in Transgenic Animals. γ-Carboxylation Efficiency is Measured by the Percentage of Protein that Either Bound to Conformation-Dependent Monoclonal Antibodies or that was Biologically Active in In Vitro Coagulation Assays

Protein	Recombinant source	Expression level	γ-Carboxylation efficiency
Protein C	Mouse	> 20 µg/mL	0–5%
	Pig	50–400 µg/mL	20–35%
		900–1000 µg/mL	10%
Factor IX	Mouse	60 µg/mL	~44%
	Sheep	25 ng/mL	Not determined
	Pig	100–200 µg/mL	~100%
	CHO cells	100 µg/mL	~2%

ences in PTMs made to complex proteins point to the next level of engineering the mammary gland for more efficient PT processing. Re-engineering the mammary gland will require detailed knowledge of endogenous PT processing enzymes and their interaction with human protein substrates.

References

1. Garber, K. (2001) Biotech Industry Faces New Bottleneck. *Nat. Biotechnol.* **19(3)**, 184–185.
2a. Houdebine, L. M. (2000) Transgenic animal bioreactors. *Transgenic Res.* **9(4–5)**, 305–320.
2b. Kerr, D. E., Liang, F., Bondioli, K. R., Zhao, H., Kreibich, G., Wall, R. J., et al. (1998) The bladder as a bioreactor: urothelium production and secretion of growth hormone into urine. *Nat. Biotechnol.* **16(1)**, 75–79.
3. Points to consider in the manufacture and testing of therapeutic products for human use derived from transgenic animals. Docket No. 95D-0131. U.S. Food and Drug Administration, Center for Biologics Evaluation and Research (1995).
4. "PPL Therapeutics plc ("PPL") Announces Status of Phase III AAT Trial" PPL press release, October 15, 2001.
5. "Genzyme Transgenics, Genzyme General start recombinant antithrombin III pivotal clinical trial, first transgenic therapeutic to reach phase III study." Genzyme Transgenics press release, May 13, 1998.
6. Van den Hout, J. M., Reuser, A. J., de Klerk, J. B., Arts, W. F., Smeitink, J. A., and Van der Ploeg, A. T. (2001) Enzyme therapy for pompe disease with recombinant human alpha-glucosidase from rabbit milk. *J. Inherit. Metab. Dis.* **24(2)**, 266–274.
7. Han, K. K. and Martinage, A. (1992) Post-translational chemical modification(s) of proteins. *Int. J. Biochem.* **24(1)**,19–28.
8. Yan, S. C., Grinnell, B. W., and Wold, F. (1989) Post-translational modifications of proteins: some problems left to solve. *Trends Biochem. Sci.* **14(7)**, 264–268.
9. Lerouge, P., Bardor, M., Pagny, S., Gomord, V., and Faye, L. N-glycosylation of recombinant pharmaceutical glycoproteins produced in transgenic plants: towards an humanisation of plant N-glycans. *Curr. Pharm. Biotechnol.* **1(4)**, 347–354.
10. Grinnell, G. W., Walls, J. D., Gerlitz, B., Berg, D. T., McClure, D. B., Ehrlich, H., et al. (1990) Native and modified human protein C: function, secretion, and post-translational modifications. Protein C and Related Anticoagulants (Bruley, D. F. and Drohan,W. N., eds.), Gulf Publishing Co., Houston, TX, pp. 29–63.

11. Wall, R. J. (2002) New gene transfer methods. *Theriogenology* **57(1)**, 189–201.

12. Bondioli, K. R., Biery, K. A., Hill, K. G., Jones, K. B., De Mayo, F. J. (1991) Production of transgenic cattle by pronuclear injection. *Biotechnology* **16**, 265–273.

13. Hammer, R. E., Pursel, V. G., Rexroad, C. E. Jr., Wall, R. J., Bolt, D. J., and Ebert, K. M. (1985) Production of transgenic rabbits, sheep and pigs by microinjection. *Nature* **315(6021)**, 680–683.

14. Wall, R. J. (1996) Transgenic livestock: progress and prospects for the future. *Theriogenology* **45**, 57–68.

15. Wall, R. J., Kerr, D. E., and Bondioli, K. R. (1997) Transgenic dairy cattle: genetic engineering on a large scale. *J. Dairy Sci.* **80(9)**, 2213–2224.

16. Niemann, H. and Kues, W. A. (2000) Transgenic livestock: premises and promises. *Anim. Reprod. Sci.* **60-61**, 277–293.

17. Renard, J. P., Zhou, Q., LeBourhis, D., Chavatte-Palmer, P., Hue, I., Heyman,Y., et al. (2002) Nuclear transfer technologies: between successes and doubts. *Theriogenology* **57(1)**, 203–222.

18a. Schnieke, A. E., Kind, A. J., Ritchie, W. A., Mycock, K., Scott, A. R., Ritchie, M., et al. (1997) Human factor IX transgenic sheep produced by transfer of nuclei from transfected fetal fibroblasts. *Science* **278(5346)**, 2130–2133.

18b. Campbell, K. H., McWhir, J., Ritchie, W. A., and Wilmut, I. (1996) Sheep cloned by nuclear transfer from a cultured cell line. *Nature* **380(6569)**, 64–66.

19. Points to consider in the manufacture and testing of monoclonal antibody products for human use. Docket No. 94D-0259. U.S. Department of Health and Human Services, Food and Drug Administration, Center for Biologics Evaluation and Research (1997).

20. Fiane, A. E., Mollnes, T. E., and Degre, M. (2000) Pig endogenous retrovirus—a threat to clinical xenotransplantation? *APMIS* **108(4)**, 241–250.

21. Cunningham, D. A., Herring, C., Fernandez–Suarez, X. M., Whittam, A. J., Paradis, K., and Langford, G. A. (2001) Analysis of patients treated with living pig tissue for evidence of infection by porcine endogenous retroviruses. *Trends Cardiovasc. Med.* **11(5)**, 190–196.

22. Bruce, M., Chree, A., McConnell, I., Foster, J., Pearson, G., and Fraser, H. (1995) Transmission of bovine spongiform encephalopathy and scrapie to mice: strain variation and the species barrier. *Phil. Trans. R. Soc. Lond. B.* **343**, 405–411.

23. Ryder, S. J., Hawkins, S. A., Dawson, M., and Wells, G. A. (2000) The neuropathology of experimental bovine spongiform encephalopathy in the pig. *J. Comp. Pathol.* **122(2-3)**, 131–143.

24. Busby, T. F. and Miekka, S. I. (2000) Viral inactivation, emerging technologies for human blood products. Encyclopedia of Cell Technology. (Spier, R. E., ed.), Wiley, pp. 1173–1182.

25. Devinoy, E., Thepot, D., Stinnakre, M. G., Fontaine, M. L., Grabowski, H., Puissant, C., et al. (1994) High level production of human growth hormone in the milk of transgenic mice: the upstream region of the rabbit whey acidic protein (WAP) gene targets transgene expression to the mammary gland. *Transgenic. Res.* **3(2)**, 79–89.

26. Rudolph, N. S. (1999) Biopharmaceutical production in transgenic livestock. *Trends Biotechnol.* **17(9)**, 367–374.

27. Kaufman, R. J., Wasley, L. C., Furie, B. C., Furie, B., and Shoemaker, C. B. (1986) Expression, purification, and characterization of recombinant gamma-carboxylated factor IX synthesized in Chinese hamster ovary cells. *J. Biol. Chem.* **261(21)**, 9622–9628.

28. Subramanian, A., Paleyanda, R. K., Lubon, H., Williams, B. L., Gwazdauskas, F. C., Knight, J.W., et al. (1996) Rate limitations in posttranslational processing by the mammary gland of transgenic animals. *Ann. NY Acad. Sci.* **782**, 87–96.

29. Reviewed in:

   a. Jenkins, N. and Curling, E. M. (1994) Glycosylation of recombinant proteins: problems and prospects. *Enzyme Microb. Technol.* **16(5)**, 354–364.

   b. Goochee, C. F., Gramer, M. J., Andersen, D. C., Bahr, J. B., and Rasmussen, J. R. (1991) The oligosaccharides of glycoproteins: bioprocess factors affecting oligosaccharide structure and their effect on glycoprotein properties. *Biotechnology* (NY) **9(12)**, 1347–1355.

c. Wright, A. and Morrison, S. L. (1997) Effect of glycosylation on antibody function: implications for genetic engineering. *Trends Biotechnol.* **15(1)**, 26–32.

d. Jenkins, N., Parekh, R. B., and James, D. C. (1996) Getting the glycosylation right: implications for the biotechnology industry. *Nat. Biotechnol.* **14(8)**, 975–981.

e. Bhatia, P. K. and Mukhopadhyay, A. (1999) Protein glycosylation: implications for in vivo functions and therapeutic applications. *Adv. Biochem. Eng. Biotechnol.* **64**, 155–201.

30. For examples see:

a. Gelfi, C., Righetti, P. G., and Mannucci, P. M. (1985) Charge heterogeneity of human protein C revealed by isoelectric focusing in immobilized pH gradients. *Electrophoresis* **6**, 373–376.

b. Heeb, M. J., Schwarz, H. P., White, T., Lammle, B., Berrettini, M., and Griffin, J. H. (1988) Immunoblotting studies of the molecular forms of protein C in plasma. *Thromb. Res. Suppl.* **52(1)**, 33–43.

c. Miletich, J. P. and Broze, G. J. Jr. (1990) Beta protein C is not glycosylated at asparagine 329. The rate of translation may influence the frequency of usage at asparagine–X–cysteine sites. *J. Biol. Chem.* **265(19)**, 11,397–11,404.

d. Frebelius, S., Isaksson, S., and Swedenborg, J. (1996) Thrombin inhibition by antithrombin III on the subendothelium is explained by the isoform AT beta. *Arterioscler. Thromb. Vasc. Biol.* **16(10)**, 1292–1297.

e. Peterson, C. B. and Blackburn, M. N. (1985) Isolation and characterization of an antithrombin III variant with reduced carbohydrate content and enhanced heparin binding. *J. Biol. Chem.* **260(1)**, 610–615.

31. Reviewed in several articles, the following reference is specific to production of recombinant proteins in different expression systems. Goochee, C. F., Gramer, M. J., Andersen, D. C., Bahr, J. B., and Rasmussen, J. R. (1991) The oligosaccharides of glycoproteins: bioprocess factors affecting oligosaccharide structure and their effect on glycoprotein properties. *Biotechnology* (NY) **9(12)**, 1347–1355.

32. Vijay, I. K. (1998) Developmental and hormonal regulation of protein N-glycosylation in the mammary gland. *J. Mammary Gland Biol. Neoplasia.* **3(3)**, 325–336.

33. Spik, G., Coddeville, B., Mazurier, J., Bourne, Y., Cambillaut, C., and Montreuil, J. (1994) Primary and three–dimensional structure of lactotransferrin (lactoferrin) glycans. *Adv. Exp. Med. Biol.* **357**, 21–32.

34. Wei, Z., Nishimura, T., and Yoshida, S. (2000) Presence of a glycan at a potential N-glycosylation site, Asn-281, of bovine lactoferrin. *J. Dairy Sci.* **83(4)**, 683–689.

35a. Morcol, T. M. (1995) Potential sources for the large scale production of human protein C. PhD Dissertation, Virginia Tech.

35b. Lubon, H., Paleyanda, R. K., Velander, W. H., and Drohan, W. N. (1996) Blood proteins from transgenic animal bioreactors. *Transfus. Med. Rev.* **10(2)**, 131–143.

36. Pratt, C. W. and Church, F. C. (1991) Antithrombin: structure and function. *Semin. Hematol.* **28(1)**, 3–9.

37a. Menache, D. (1991) Replacement therapy in patients with hereditary antithrombin III deficiency. *Semin. Hematol.* **28(1)**, 31–38.

37b. Levy, J. H., Weisinger, A., Ziomek, C. A., and Echelard, Y. (2001) Recombinant antithrombin: production and role in cardiovascular disorder. *Semin. Thromb. Hemost.* **27(4)**, 405–416.

38. Minnema, M. C., Chang, A. C., Jansen, P. M., Lubbers, Y. T., Pratt, B.M., Whittaker, B.G., et al. (2000) Recombinant human antithrombin III improves survival and attenuates inflammatory responses in baboons lethally challenged with *Escherichia coli. Blood* **95(4)**, 1117–1123.

39. Genzyme Transgenics Announces Expectations for rhATIII. PRNewswire. Feb 6, 2001.

40. Levy, J. H., Weisinger, A., Ziomek, C. A., and Echelard, Y. (2001) Recombinant antithrombin: production and role in cardiovascular disorder. *Semin. Thromb. Hemost.* **27(4)**, 405–416.

41. Edmunds, T., Van Patten, S. M., Pollock, J., Hanson, E., Bernasconi, R., Higgins, E., et al. (1998) Transgenically produced human antithrombin: structural and functional comparison to human plasma-derived antithrombin. *Blood* **91(12)**, 4561–4571.

42. Edmunds, T., Van Patten, S. M., Pollock, J., Hanson, E., Bernasconi, R., Higgins, E., et al. (1998) Transgenically produced human antithrombin: structural and functional comparison to human plasma-derived antithrombin. *Blood* **91(12)**, 4561–4571.

43. Denman, J., Hayes, M., O'Day, C., Edmunds, T., Bartlett, C., Hirani, S., et al. (1991) Transgenic expression of a variant of human tissue–type plasminogen activator in goat milk: purification and characterization of the recombinant enzyme. *Biotechnology* (NY) **9(9)**, 839–843.

44a. Olson, S. T., Frances-Chmura, A. M., Swanson, R., Bjork, I., and Zettlmeissl, G. (1997) Effect of individual carbohydrate chains of recombinant antithrombin on heparin affinity and on the generation of glycoforms differing in heparin affinity. *Arch. Biochem. Biophys.* **341(2)**, 212–221.

44b. Turk, B., Brieditis, I., Bock, S. C., Olson, S. T., and Bjork, I. (1997) The oligosaccharide side chain on Asn-135 of alpha-antithrombin, absent in beta–antithrombin, decreases the heparin affinity of the inhibitor by affecting the heparin–induced conformational change. *Biochemistry* **36(22)**, 6682–6691.

45. Lu, W., Mant, T., Levy, J. H., and Bailey, J. M. (2000) Pharmacokinetics of recombinant transgenic antithrombin in volunteers. *Anesth. Analg.* **90(3)**, 531–534.

46. Schwaiblmair, M. and Vogelmeier, C. (1998) Alpha 1-antitrypsin. Hope on the horizon for emphysema sufferers? *Drugs Aging* **12(6)**, 429–440.
Carrell, R. W., Lomas, D. A., Sidhar, S., and Foreman, R. (1996) Alpha 1-antitrypsin deficiency. A conformational disease. *Chest* **110(6 Suppl)**, 243S–247S.

47. Mills, P. B., Mills, K., Johnson, A. W., Clayton, P. T., and Winchester, B. G. (2001) Analysis by matrix assisted laser desorption/ionisation-time of flight mass spectrometry of the posttranslational modifications of alpha 1-antitrypsin isoforms separated by two-dimensional polyacrylamide gel electrophoresis. *Proteomics* **1(6)**, 778–786.

48. Casolaro, M. A., Fells, G., Wewers, M., Pierce, J. E., Ogushi, F., Hubbard, R., et al. (1987) Augmentation of lung antineutrophil elastase capacity with recombinant human alpha-1-antitrypsin. *J. Appl. Physiol.* **63(5)**, 2015–2023.

49a. Straus, S. D., Fells, G. A., Wewers, M. D., Courtney, M., Tessier, L. H., Tolstoshev, P., et al. (1985) Evaluation of recombinant DNA-directed *E. coli* produced alpha 1-antitrypsin as an anti-neutrophil elastase for potential use as replacement therapy of alpha 1-antitrypsin deficiency. *Biochem. Biophys. Res. Commun.* **130(3)**, 1177–1184.

49b. Kang, H. A., Sohn, J. H., Choi, E. S., Chung, B. H., Yu, M. H., and Rhee, S. K. (1998) Glycosylation of human alpha 1-antitrypsin in Saccharomyces cerevisiae and methylotrophic yeasts. *Yeast* **14(4)**, 371–381.

50a. Wright, G., Carver, A., Cottom, D., Reeves, D., Scott, A., Simons, P., et al. (1991) High level expression of active human alpha-1-antitrypsin in the milk of transgenic sheep. *Biotechnology* (NY) **9(9)**, 830–834.

50b. Carver, A. S., Dalrymple, M. A., Wright, G., Cottom, D. S., Reeves, D. B., Gibson, Y. H., et al. (1993) Transgenic livestock as bioreactors: stable expression of human alpha-1-antitrypsin by a flock of sheep. *Biotechnology* (NY) **11(11)**, 1263–1270.

51. Tebbutt, S. J. (2000) Technology evaluation: transgenic alpha-1-antitrypsin (AAT), PPL therapeutics. *Curr. Opin. Mol. Ther.* **2(2)**, 199–204.

52. Carver, A., Wright, G., Cottom, D., Cooper, J., Dalrymple, M., Temperley, S., et al. (1992) Expression of human alpha 1 antitrypsin in transgenic sheep. *Cytotechnology* **9(1-3)**, 77–84.

53. Jeppsson, J. O., Lilja, H., and Johansson, M. (1985) Isolation and characterization of two minor fractions of alpha 1-antitrypsin by high-performance liquid chromatographic chromatofocusing. *J. Chromatogr.* **327**, 173–177.

54. Reviewed in:

a. Wu, S. M., Stanley, T. B., Mutucumarana, V. P., and Stafford, D. W. (1997) Characterization of the gamma-glutamyl carboxylase. *Thromb. Haemost.* **78(1)**, 599–604.

b. Furie, B. C. and Furie, B. (1997) Structure and mechanism of action of the vitamin K-dependent gamma-glutamyl carboxylase: recent advances from mutagenesis studies. *Thromb. Haemost.* **78(1)**, 595–598.

c. Suttie, J. W. (1993) Synthesis of vitamin K-dependent proteins. *FASEB J.* **7(5)**, 445–452.

55. Sunnerhagen, M., Drakenberg, T., Forsen, S., and Stenflo, J. (1996) Effect of Ca2+ on the structure of vitamin K-dependent coagulation factors. *Haemostasis* **26(suppl 1)**, 45–53.

56. Reviewed in:

a. Furie, B. C. and Furie, B. (1997) Structure and mechanism of action of the vitamin K-dependent gamma-glutamyl carboxylase: recent advances from mutagenesis studies. *Thromb. Haemost.* **78(1)**, 595–598.

b. Dowd, P., Ham, S. W., Naganathan, S., and Hershline, R. (1995) The mechanism of action of vitamin K. *Annu. Rev. Nutr.* **15**, 419–440.

57. Cain, D., Hutson, S. M., and Wallin, R. (1997). Assembly of the warfarin-sensitive vitamin K 2,3-epoxide reductase enzyme complex in the endoplasmic reticulum membrane. *J. Biol. Chem.* **272**, 29,068–29,075.

58a. Stanley, T. B., Wu, S. M., Houben, R. J., Mutucumarana, V. P., and Stafford, D. W. (1998) Role of the propeptide and gamma-glutamic acid domain of factor IX for in vitro carboxylation by the vitamin K-dependent carboxylase. *Biochemistry* **37(38)**, 13,262–13,268.

58b. Li, S., Furie, B. C., Furie, B., and Walsh, C. T. (1997) The propeptide of the vitamin K-dependent carboxylase substrate accelerates formation of the gamma-glutamyl carbanion intermediate. *Biochemistry* **36(21)**, 6384–6390.

59. Bristol, J. A., Freedman, S. J., Furie, B. C., and Furie, B. (1994) Profactor IX: the propeptide inhibits binding to membrane surfaces and activation by factor XIa. *Biochemistry* **33(47)**, 14,136–14,143.

60a. Rehemtulla, A., Roth, D. A., Wasley, L. C., Kuliopulos, A., Walsh, C. T., Furie, B., et al. (1993) In vitro and in vivo functional characterization of bovine vitamin K-dependent gamma-carboxylase expressed in Chinese hamster ovary cells. *Proc. Natl. Acad. Sci. USA* **90(10)**, 4611–4615.

60b. Wasley, L. C., Rehemtulla, A., Bristol, J. A., and Kaufman, R. J. (1993) PACE/furin can process the vitamin K-dependent pro-factor IX precursor within the secretory pathway. *J. Biol. Chem.* **268(12)**, 8458–8465.

61. Stenina, O., Pudota, B. N., McNally, B. A., Hommema, E. L., and Berkner, K. L. (2001) Tethered processivity of the vitamin K-dependent carboxylase: factor IX is efficiently modified in a mechanism which distinguishes Gla's from Glu's and which accounts for comprehensive carboxylation in vivo. *Biochemistry* **40(34)**, 10,301–10,309.

62a. Subramanian, A., Paleyanda, R. K., Lubon, H., Williams, B. L., Gwazdauskas, F. C., Knight, J. W., et al. (1996) Rate limitations in posttranslational processing by the mammary gland of transgenic animals. *Ann. NY Acad. Sci.* **782**, 87–96.

62b. Velander, W. H., Johnson, J. L., Page, R. L., Russell, C. G., Subramanian, A., Wilkins, T. D., et al. (1992) High-level expression of a heterologous protein in the milk of transgenic swine using the cDNA encoding human protein C. *Proc. Natl. Acad. Sci. USA* **89(24)**, 12,003–12,007.

62c. Van Cott, K. E., Lubon, H., Russell, C. G., Butler, S. P., Gwazdauskas, F. C., Knight, J., et al. (1997) Phenotypic and genotypic stability of multiple lines of transgenic pigs expressing recombinant human protein C. *Transgenic Res.* **6(3)**, 203–212.

63a. Subramanian, A., Paleyanda, R. K., Lubon, H., Williams, B. L., Gwazdauskas, F. C., Knight, J. W., et al. (1996) Rate limitations in posttranslational processing by the mammary gland of transgenic animals. *Ann. NY Acad. Sci.* **782**, 87–96.

63b. Van Cott, K. E., Williams, B., Velander, W. H., Gwazdauskas, F., Lee, T., Lubon, H., and Drohan, W. N. (1996) Affinity purification of biologically active and inactive forms of recombinant human protein C produced in porcine mammary gland. *J. Mol. Recognit.* **9(5-6)**, 407–414.

64. Yull, F., Harold, G., Wallace, R., Cowper, A., Percy, J., Cottingham, I., et al. (1995) Fixing human factor IX (fIX), correction of a cryptic RNA splice enables the production of biologically active fIX in the mammary gland of transgenic mice. *Proc. Natl. Acad. Sci. USA* **92(24)**, 10,899–10,903.

65. Van Cott, K. E., Butler, S. P., Russell, C. G., Subramanian, A., Lubon, H., Gwazdauskas, F. C., et al. (1999) Transgenic pigs as bioreactors: a comparison of gamma-carboxylation of glutamic acid in recombinant human protein C and factor IX by the mammary gland. *Genet. Anal.* **15(3-5)**, 155–160.

66. Clark, A. J., Bessos, H., Bishop, J. O., Brown, P., Harris, S., Lathe, R., et al. (1989) Expression of human anti-hemophilic factor IX in the milk of transgenic sheep. *Bio-Technology* **7**, 487–492.

67. Kaufman, R. J., Wasley, L. C., Furie, B. C., Furie, B., and Shoemaker, C. B. (1986) Expression, purification, and characterization of recombinant gamma-carboxylated factor IX synthesized in Chinese hamster ovary cells. *J. Biol. Chem.* **261(21)**, 9622–968.

68. Bond, M., Jankowski, M., Patel, H., Karnik, S., Strang, A., Xu, B., et al. (1998) Biochemical characterization of recombinant factor IX. *Semin. Hematol.* **35(2 Suppl 2)**, 11–17.

4

Mammalian Cell Culture

Process Development Considerations

Steven Rose, Thomas Black, and Divakar Ramakrishnan

1. Introduction

As discussed in the previous chapters of this book, mammalian cell-culture technology facilitates the production of complex therapeutic proteins (biopharmaceutical drugs). The mammalian cell-culture process is performed with a host cell line that has been transfected with DNA plasmids encoding for the protein of interest. The host cell line is selected based on the post-translational modifications necessary for protein efficacy. The most common production host-cell lines reported are Chinese hamster ovary (CHO) cell lines or murine myeloma lines (SP2/0, NS0). The site and number of plasmids inserted into the host cell line during transfection is somewhat of a random event and generates multiple variations in the initial population pool. Single-cell clones are isolated from the initial population pool of transfected cells. A single-cell clone for production is then selected based on performance in screening studies, and is expanded, dispensed into 1–2-mL vials, and then stored in a quiescent state (generally frozen or freeze dried). These vials are the master cell bank (MCB) that supplies the cells necessary to produce the development, clinical, launch, and commercial supplies.

Figure 1 shows a schematic diagram of a typical mammalian cell culture process to produce a protein such as a monoclonal antibody (MAb). The manufacturing process begins with a vial from the MCB. To extend the life of the MCB, a vial from the MCB is expanded, dispensed into 1–2-mL vials, and then stored in a quiescent state (generally frozen or freeze-dried) as the manufacturer's working cell bank (MWCB). MWCB are created as needed from the MCB. For each production lot, a MWCB vial is broken out and expanded in shaker flasks and/ or spinners and then in controlled bioreactors to provide sufficient inoculum or seed culture for the production-scale bioreactor step. The production-scale bioreactor can range from 1000–20,000 L in volume and may be operated as batch, fed-batch, or continuous perfusion mode. The protein is harvested from the production reactor either continuously (perfusion mode), semi-continuously, or at the end of the production run. At the time of harvest, the cells are removed from the culture broth, and the clarified and conditioned broth is then purified through a series of two to four chromatography steps along with unit operations to inactivate/ remove potential viral contaminants, host-cell protein, and DNA. The purified product

From: *Handbook of Industrial Cell Culture: Mammalian, Microbial, and Plant Cells*
Edited by: V. A. Vinci and S. R. Parekh © Humana Press Inc., Totowa, NJ

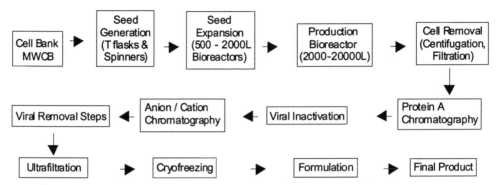

Fig. 1. Schematic of a typical cell culture based protein-manufacturing process.

is then typically formulated into vials containing the final product. Final formulation is impacted by many factors associated with the pharmaceutical delivery method (i.e., injectable vs inhaled) and anticipated storage and use conditions.

For each process step described here, there are validated operating ranges and criteria that need to be met for succession to the proceeding step. This is to ensure consistency in Safety, Identity, Strength, Purity and Quality (SISPQ) of the product. Failure to maintain the process within the specified operating ranges or failure to pass criteria prompts an investigation to evaluate the impact that the deviation has on the SISPQ of the product and to determine whether the relevant lot is suitable for release. The task of the process developer is to determine the appropriate operating conditions that produce a robust, cost-effective process without impacting the safety, identity, strength, purity, or quality of the protein product.

This chapter also provides an industrial perspective on some commonly discussed but important topics that impact mammalian cell-culture process development strategy and methodology (which have not been discussed in the previous three chapters on mammalian cell-culture technology). These topics include:

- Business Drivers and Process Development Strategy
- Science and Techniques of Cell-Culture Process Development
- Choice of Bioreactor Processes and Engineering/Scale-Up Considerations
- Cell Removal and Downstream Issues

2. Business Drivers and Process Development Strategy

The development of biopharmaceutical products is an expensive process, which can easily take about 6 yr *(1)* from preclinical process development work to product launch for a typical protein therapeutic, with a significant risk that the product will not be shown to be efficacious in clinical studies. For instance, it has been reported *(2)* that average success values for new biopharmaceutical drug products range as follows:

Clinic Phase I   ~ 15%
Clinic Phase II   ~ 40%
Clinic Phase III   ~ 80%

Considering these success rates and timelines, it should not be surprising that the cost of developing a new drug can be quite high. In fact, a recent press report (Nov 30th, 2001) by the Tufts Center for the Study of Drug Development (http://

www.tufts.edu/med/csdd) estimates the average development cost of a new drug to be as high as $802 million. This is an increase from $231 million in 1987 (equivalent to $318 million in 2000 dollars). The increased cost is attributed primarily to higher clinical trial and associated R&D costs. In addition, an outpacing of development costs compared to the growth in sales *(3)* has resulted in pressure to reduce cycle times and development costs for drugs in general. Of the total drug development costs, *(1)* up to 40–60 % of costs are associated with Process Development and clinical manufacturing costs for biopharmaceuticals (such as those produced using mammalian cell-culture technology). Contrary to popular belief, process development and clinical manufacturing costs can equal or exceed the costs of clinical trials. Moreover, process development and clinical manufacturing costs increase as a drug advances through clinical development, where it is estimated that up to 50–60% of costs are incurred during Phase III. More recently, the demands on mammalian cell-culture-based biopharmaceutical manufacturing have increased thanks in part to the "tsunami" wave of potential therapeutics arising from the likes of the Human Genome project and also in part because of improved safety and efficacy through protein engineering and expression technologies *(4)*.

As a result of these trends in R&D costs and the probability of success of drug candidates in the different stages of drug development, process development efforts in the early stages of drug development, such as Phase I trials, have primarily focused on the use of platform/"assembly line" type processes to minimize the timelines for development and reduce development costs. When a drug moves from Phase I to Phase II clinical trials and beyond, process development efforts then shift the focus to factors such as yield improvement and process robustness and ability to meet estimated market demand with existing/planned manufacturing capabilities and capacities *(5)*.

3. Science and Techniques of Process Development

The challenge with process optimization is to achieve a robust process that meets business needs in a timely fashion. The business needs are dependent on the stage that the developing process is in relative to pre-clinical and clinical trials. For processes in early phase development, it is more cost effective to produce material for testing as quickly as possible, rather than to define the optimal process. For the later phase processes, speed in producing material is less critical compared to defining a high-yield, robust process.

Although not necessary, it is ideal for the early-phase process to represent the later-phase process. This could be achieved by performing systematic screening studies of the factors that tend to significantly affect cell-culture productivity. These studies will identify critical factors and ranges that could improve process productivity, from which designed studies could identify acceptable operating set points.

Cell-culture productivity is affected by the cell-specific productivity and the viable cell concentration. In many cases, the cell-specific productivity is dependent on the cell-specific growth rate *(6)*. The optimal productivity may not exist at the operating conditions that maximize the specific productivity, the specific growth rate, or the maximum cell concentration (or the time-integral area of the viable cell-mass profile). However, an understanding of how process conditions affect the specific growth rate or specific productivity may lead to improved performance. The variability in characteristics between cell lines and between cell clones leads to unique optimal conditions that must be demonstrated by experimentation. The following subsections provide an overview of how some of the key process parameters affect productivity and culture growth. This is followed by an overview of experimental design methods that could minimize the time and resources needed to successfully develop a process.

3.1. Science of Cell-Culture Media Design

Cell-culture medium contains glucose, amino acids, vitamins, trace metals, buffer, inorganic salts, and proteins to meet the nutrient demands and process-control demands of the mammalian cell culture. The composition of these components can affect cell growth, protein production, protein quality, and downstream protein purification. Medium formulations may be prone to depletion of critical nutrients that could lead to cell death, reduced productivity, or reduced post-transcription protein processing. Alternatively, the medium could contain too high a concentration of nutrients that could shift metabolism, causing the toxic accumulation of byproducts such as lactate and ammonia. Therefore, the cell-culture medium offers opportunities for optimization in commercial cell-culture processes. The following sections provide a review of the major functions of cell-culture medium with a focus on process optimization.

In 1955, Eagle characterized the basic nutrient requirements for mammalian cells. Eagle showed that mammalian cells can survive on glucose, amino acids, vitamins, and inorganic salts *(7)*. To support cell proliferation, the defined medium required animal serum or animal extract supplements. Variations of the serum-containing medium (complex medium) have been developed. Typically for a new process, using one of these complex media would require less development time for achieving adequate cell-culture performance. However, removing the animal additives reduces the risk of viral contamination, increases process robustness (by reducing exposure to lot-to-lot variability of animal-sourced ingredients), reduces medium costs, and simplifies downstream purification *(8–10)*. Also, serum and/or animal source material (ASM) free processes are becoming an expectation for new processes filed to the regulatory agencies. Moreover, there are even examples of synthetic media (serum-free) demonstrating improved process performance compared to complex medium *(11)*.

3.1.1. Serum Properties

Serum provides sterols, fatty acids, growth factors, protein stability, protein transporters, trace metals, vitamins, and shear protection to mammalian cell cultures. Removing serum without addressing these functions could result in poor cell growth and performance. However, not all of these functions are critical for every cell line. Successful serum replacements reported in the literature should be explored in statistically design experiments to determine the main or interacting factors that are required for the process medium. Literature reports cite successful development of serum-free medium by supplementing the medium with sterols and/or fatty acids. Cholesterol with methyl β-cyclodextrin *(11)*, *cis*-unsaturated fatty acids *(12,13)*, ethanolamine and yeast extracts (13) have been shown to result in increased growth and productivity in serum-free medium. Another method to increase the fatty acids for cells in serum-free medium is to supply insulin, which stimulates the biosynthesis of fatty acids *(14)*. Insulin also stimulates the biosynthesis of nucleic acids, the GLUT-4 glucose transporter, and glucose metabolism *(14)*. Serum may stimulate transporters other than the

GLUT-4 glucose transporter; therefore, the removal of serum may reduce the cellular uptake rate of these other nutrients. If these other nutrients are diffusion-limited (which is known to happen mostly in aggregated and microcarrier-based systems), the uptake rate can be improved by increasing the concentrations of these nutrients in the medium. Diffusion-limited nutrients are those that cannot pass through the cell membrane as quickly as the cell demands, and the diffusion rate limits the consumption rate of that nutrient. The contributions of trace metals by serum *(15)* may also need to be supplemented to the serum-free medium. In addition to metabolic effects, serum is also known to provide shear protection to the cells, thereby significantly reducing cell death in sparged systems. However, the use of synthetic block copolymer surfactants such as Pluronic F-68 is known to provide equivalent protection in serum-free systems *(16,17)*.

3.1.2. TCA Cycle: Glucose, Glutamine, and Amino Acids

Glucose, glutamine, and amino acids are required for producing energy, cellular proteins, and the protein product. Inefficient use of these nutrients leads to toxic accumulation of byproducts such as lactate and ammonia. The consumption rates of these nutrients and the formation rates of byproducts are interrelated through the TCA cycle (Fig. 2). This section reviews the effect that the concentrations of glucose, glutamine, and the amino acids have on the metabolic network to provide some insight into methods of optimizing the nutrient concentrations.

Glucose is the primary source of energy in mammalian cell-culture medium. It is also used for the synthesis of nonessential amino acids (Table 1), nucleic acids, carbohydrates, and fatty acids. Glucose is transported into the cell through five glucose transporters: GLUT -1, 2, 3, 4, and 5 (18). The GLUT4 glucose transporter is stimulated by insulin *(19)* to increase the transport of glucose into the cell.

Once in the cytoplasm, glucose is metabolized to generate two NADPH and pyruvate, or is used for the synthesis of nucleic acids, carbohydrates, or serine and glycine. Pyruvate can enter the TCA cycle to generate more energy or generate nonessential amino acids, synthesize fatty acids and sterols, or form lactate or alanine byproducts. A significant amount of pyruvate is directed to fatty acids and sterols *(20)*, or lactate and alanine *(21)*. Increased glucose concentrations lead to higher accumulations of lactate *(22)* and alanine (meta-analysis from published data *[23–26]*). Lactate accumulation has been shown to adversely affect cell-culture productivity when it exceeds 22–55 mM *(27–32)*. Ammonia is concurrently produced with alanine. Ammonia inhibits cell growth and productivity when it exceeds 1.8–33 mM *(27–30,32–35)*. Reducing the flux of glucose into the cells can reduce the lactate and alanine formulation. The flux could be reduced by reducing the glucose concentration *(36,37)*, reducing the medium pH *(38)*, substituting galactose for glucose *(39)*, or reducing the insulin concentration. For batch cultures, reducing the glucose concentration may make glucose the limiting nutrient *(40)* to the point where glucose starvation may shorten the culture time before lactate toxicity would occur. In fed-batch cultures, robust adaptive control strategies must be implemented to maintain low glucose concentrations with minimal risk of glucose depletion. Reducing the glucose concentration results in an increased glutamine consumption *(41)*, which could increase byproduct formation from glutaminolysis.

Glutamine is used to synthesize proteins, and can enter the TCA cycle to produce energy *(42)* or nonessential amino acids *(7)*. Glutamine affects the consumption rate of

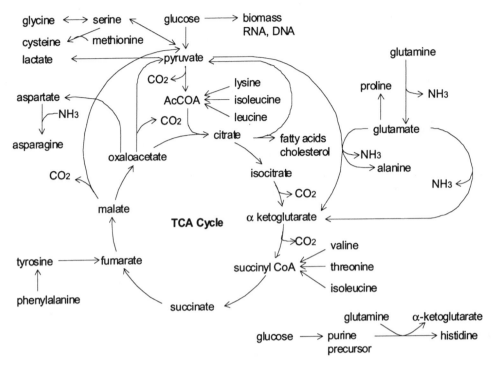

Fig. 2. The TCA cycle.

Table 1
The Essential and Nonessential Amino Acids
for Mammalian Cell Systems

Essential amino acids	Nonessential amino acids
glutamine	alanine
isoleucine	arginine
leuceine	asparagine
lysine	aspartic acid
methionine	cystine
phenylalanine	glutamic acid
threonine	glycine
tryptophan	histidine
valine	proline
	serine
	tyrosine

glucose *(41)*. Higher glutamine concentrations reduce the glucose consumption rate. However, the increased glutamine concentrations increase the production of proline, alanine, and ammonium byproducts ([*see* ref. *40*], and meta analysis from published data [refs. *23–26*]). The production of proline requires energy; therefore, the accumulation of proline suggests inefficient glutamine metabolism *(43)*. Alanine is produced from the reaction with pyruvate and glutamate. An accumulation of alanine minimizes the amount of pyruvate directed to fatty acids and sterols synthesis, which is also undesirable. Ammonia is produced concurrently with alanine, from the spontaneous degradation of glutamine in the medium and from glutaminolysis as well. Accumulation or build-up of ammonia inhibits cell growth and productivity at levels exceeding 1.8–33 m*M* *(27–30,32–35)*. The byproducts from glutaminolysis can be minimized by using a glutamate-based medium for cells containing glutamine synthetase *(39,44)*, using glutamine dipeptides *(45)*, or by maintaining a low concentration of glutamine. For batch cultures, reducing the glutamine concentration may make glutamine the limiting nutrient *(40)* to the point where glutamine starvation may shorten the culture time before ammonium toxicity would occur. In fed-batch cultures, robust and adaptive control strategies must be implemented to maintain low glutamine concentrations with minimal risk of glucose depletion. Reducing the glutamine concentration can also result in an increased glucose consumption *(41)*, which could increase byproduct formation from glycolysis.

Aside from glucose and glutamine, amino acids also comprise a critical part of cell culture media. There are 20 amino acids used for protein synthesis. Nine of these amino acids are essential amino acids (Table 1). Depletion of any of essential amino acids induces apoptosis of the culture. The other eleven amino acids (nonessential amino acids) can be synthesized from glycolysis, glutaminolysis, or the TCA cycle (Fig. 2). However, supplying the nonessential amino acids to the culture reduces the amount of glucose and glutamine that needs to be supplied in the medium *(46–51)*, and the lower glucose and glutamine concentrations can lead to reduced byproduct formation *(36,37,52)*. Depletion of any of the nonessential amino acids results in a drop in the cell-culture viability *(53)* possibly because of the increased burden on the glycolysis, glutaminolysis, or the TCA cycle to produce the nonessential amino acid, resulting in the depletion of any of the essential amino acids or glucose. Supplying the medium with higher concentrations of amino acids improves culture performance *(31,36,54,55)*. However, the consumption or production rates of the amino acids are dependent on the amino acid concentrations *(24,56,57)*. Increasing or decreasing the concentration of the amino acids independently can change the TCA cycle flux and may possibly cause a depletion of an essential amino acid or accumulation of a metabolic byproduct. An optimal amino acid concentration would maximize productivity and robustness, with low medium costs.

3.1.3. Medium Osmolality

Medium osmolality significantly affects cell-culture productivity. Increased medium osmolality has demonstrated decreased specific cell-growth rate and increased specific production rate *(31,38,58)*. The medium osmolality can be predicted from the medium formulation. The amount of interaction between medium components typically does not make the osmolality significantly different from the sum of each component's contribution. The osmotic contributions of selected medium components are tabulated here (Table 2). Note that the effect of medium component(s) concentration and medium osmolality can be decoupled in experimental studies by adjusting the potassium chloride and/or sodium chloride concentration(s) to achieve the desired medium osmolality. The shift in the metabolic rates in response to variations in osmolality may cause some nutrients in the medium formulation to become limiting or overabundant. Ide-

Table 2
Osmotic Contributions
of Medium Components[a]

Medium components	mOsm/g/L
$CaCl_2$	19.42
$CuSo_4 \cdot 5H_2O$	7.23
KCl	25.15
MgCl	15.00
$MgSO_4$	5.62
NaCl	34.74
NaH_2PO_4	18.23
$NaHCO_3$	23.27
$ZnS_4\text{-}7H_2$	8.63
Glucose	6.49
L-glutamine	6.84
Amino acid pools	8.59
NaOH	50.00
Pluronic F-68	0.00
FBS	2.64

[a]Note that the amino acid pool is an estimate of osmotic contribution from a typical amino acid composition and would differ depending on the composition of the pool.

ally, the interaction between medium formulation, osmolality, pH, and temperature should be incorporate in a statistically designed experiment.

3.2. Culture pH

The pH setpoint can significantly affect the cell-culture performance. Medium pH affects the intracellular enzymes' activities *(59)*. Lowering the pH reduces the specific glucose consumption rate and reduces the specific lactate production rate *(29,38,60)*, reducing the risk of glucose depletion or toxic levels of lactate. Medium pH also affects the specific growth rate *(38,60)* and the specific production rate *(60)*, which ultimately affects the overall culture productivity. The pH can be optimized to minimize lactate accumulation or maximum volumetric productivity. The shift in metabolic rates in response to different pH set points may cause some nutrients in the medium formulation to become limiting. Ideally, the interaction between medium formulation, osmolality, pH, and temperature should be captured in a designed experiment.

3.3. Culture Temperature

Lower temperatures reduce metabolic demand, reduce culture growth and death rates, and can result in an overall increase in culture productivity *(61,62)*. In growth-dependent producing systems, an optimal temperature modifies the specific growth rate and specific production rate in such a way as to maximize volumetric productivity.

The lower temperature probably affects the enzyme activities in the cell as modeled by the Arrhenius equation *(63)* and as demonstrated in the oxygen uptake rate *(64)*. In growth-independent producing systems, a temperature-shift methodology is often employed to maximize the culture productivity *(65)*. This suggests that culture temperature, pH, osmolality, and medium formulation will have high-order interacting effects on the culture performance. Ideally, the interaction between medium formulation, osmolality, pH, and temperature should be determined and described.

3.4. Process Operation Methods

There are four types of processes that can deliver the nutrient requirements to the cell culture: batch, fed-batch, chemostat, and perfusion. The process methods are listed in the order of operational complexity, with the batch process the simplest and the perfusion process the most complex. The increased operational complexity enables greater control of the nutrient and byproduct concentrations in the bioreactor, and potentially results in improved productivity. The increased operational complexity usually requires a greater number of experiments to develop, and increases the complexity of performing the experiments in scale-down models.

3.4.1. Batch

The batch process is the simplest to operate and develop in scale-down systems. In the batch process, after the cells are inoculated into fresh medium, nothing is added or removed from the culture except gasses and alkali (for pH control). The nutrient concentration in batch medium must be sufficiently high to extend culture life without leading to the toxic accumulation of byproducts (*see* preceding sections) *(40,54)*. At the end of the run, the reactor is harvested for the protein of interest. The concentration of nutrients, byproducts, cells, and product varies over time. Although the batch culture is simple to operate, there are limited opportunities (i.e., initial medium formulation, operating temperature, and operating pH) to improve the metabolic rate through rational medium design.

3.4.2. Fed Batch

The fed-batch process is more difficult to operate and scale-down than the batch process. In the fed-batch process, once the cells are inoculated into fresh medium, medium is added at defined intervals, or continuously to prevent nutrient depletion or maintain a desired nutrient concentration *(26)*. Nothing is removed from the batch-fed culture. At the end of the run, the reactor is harvested for the protein of interest. The concentration of byproducts, cells, and product, and possibly nutrients, varies over time. The byproduct concentration tends to be significantly less than the batch culture because the nutrients are delivered over time and do not need to be at high initial concentrations. This leads to increased cell mass and productivity.

The rate of medium feed can be determined empirically *(31,36)* or in proportion to the oxygen uptake rate *(66–68)*, cell density, or nutrient measurements *(26,52,69,70)*. The medium feed formulation can be determined empirically *(31)* or rationally, based on stoichiometric balances *(46–51,71,72)*, metabolic flux analysis (MFA) *(20,21,23,56,73–76)*, or cybernetic modeling *(77–81)*. The models used for determining feed requirements in the fed-batch process use assumptions to reduce the rank of the mathematical model (and thus the complexity of the calculations), an activity that may be inappropriate for some systems. Intermediate component concentrations are assumed to be constant in the stoichiometric and MFA models. This may be an inappropriate assumption in dynamic systems. Metabolic pathways are grouped together. The stoichiometric requirements for some of the components are estimated. One study has demonstrated that there was a depletion of phosphate because of incorrect assumptions in the phosphate requirements *(68)*. The models improve as more metabolic intermediates and pathways can be measured. However, empirical studies may need to be done to confirm or better define the current assumptions.

3.4.3. Chemostat

The chemostat process is more difficult to operate and scale-down than the fed-batch process. In the chemostat process, once the cells are inoculated into fresh medium, feed medium is continuously added, and spent medium containing cells is continuously removed at the same rate as the feed. The concentration of nutrients, byproducts, cells, and product remain constant over time and are dependent on the cell concentration and dilution rate. The chemostat is not a commercial process. However, the chemostat is a valuable research tool because the steady-state system allows mass balance analysis of the metabolic fluxes. The metabolic-flux analysis can then be used to define fed-batch or perfusion-feeding protocols.

3.4.4. Perfusion

The perfusion process is more sophisticated to operate and scale-down than the other processes. However, it is used for industrial recombinant protein manufacture because of the ability to achieve significantly high volumetric production rates compared to other methods of culture. In the perfusion process, once the cells are inoculated into fresh medium, feed medium is continuously added and spent medium is continuously removed at the same rate as the feed. The cells are selectively retained in the perfusion vessel by cell-retention devices. This system typically permits the accumulation of higher biomass concentrations than fed-batch systems because of the ability to feed nutrients and remove (potentially toxic) metabolites/by-products continuously *(35,82)*.

3.5. Techniques for Process Optimization

The challenge with process optimization is in identifying the critical factors that have an effect on the process outcome in the presence of complex factor interactions and process noise (variance). A number of techniques have been developed to enable a comparison of the measured effect to the system noise *(83)*. From the numerous factors that define an operating condition (i.e., medium formulation, temperature, pH, seeding density, dissolved CO_2 [pCO_2], and dissolved oxygen [DO]), only the factors that have measured effects that are significantly greater than the system noise are considered to significantly affect the process outcome. The level of these significant factors is adjusted to optimize the process outcome. The sensitivity of this test improves with the number of replicate runs or measurements performed, or as the system noise is decreased by reducing sources of variability within the studies.

The large number of factors that are involved in defining a cell-culture process and the high level of interaction between these factors lead to an iterative procedure for process development. There are several available process optimization strategies which are described in Table 3. The strategies that tend to become overused—borrowing and

Table 3
Process Optimization Strategies

Strategy	Concept	Advantages	Disadvantages
Borrowing	Obtain process from literature	– Simple and easy – Good place to start	– May not apply to your situation – Many to choose from – Lab media may not have industrial rebalance – Not necessarily good performance process
Component swapping	Swap one for another at same usage level	– Not limited to predefined components – Can try large numbers of components easily	– No study of concentration effects – No interactions
Biological mimicry	Mass balance used to determine composition of medium	– Medium not short of any ingredient – Good check of composition	– Accurate composition data needed – Complex ingredients can vary in batch-to-batch concentrations
One at a time	Adjust one factor at a time, maintaining the others	– Simple and easy to perform	– No interactions investigated – Optimum can be missed – Large number of experiments required
Statistical Experimental Design			
Full factorial	All interactions quantifiable	– Best research space coverage	– Large number of experiments
Partial factorial	Subset of factorial design; some interactions quantifiable	– Compromise in coverage to reduce number of experiments	– Some interactions not investigated
Plackett-Burman	Design in measuring N-1 main effects in N runs	– Good first screening tool to identify important factors	– No interactions
Central composite	Partial factorial of 3n used to estimate 2nd order response model	– Estimates curvature of effects – Prediction anywhere in research space	– Moderate number of experiments – Nonsequential experiment – Less coverage than factorial

(continued)

Table 3 *(continued)*

Strategy	Concept	Advantages	Disadvantages
Box-Behnken	Partial factorial of 3n used to estimate 2nd order response model	– Prediction anywhere in research space – Spherical construction	– Nonsequential experiment – Less coverage than central composite
Optimization Techniques			
Response surface methodology	Systematic coverage to map research space	– Through 3D map of research space	– Moderate number of experiments
Steepest ascent	Follow steepest gradient to find maxima / minima	– Quick way to find local max/min	– Requires iterative experimental approach
Multiple linear regression	Making polynomial fit to data	– Widely used and available	- Computational needs high - Incomplete coverage of space - Correlation between variables (little interaction information)
Computational Methods			
Neural networks	Mimics the learning ability of the brain	– Handles large amounts of data – Good at recognizing patterns – No mechanistic knowledge needed	– Requires computational expertise – Does not handle duplicate data well – Not tolerant of missing data
Fuzzy logic	Series of "rules"	– Tolerant of noisy or missing data	– Needs process experts to define rules – Requires computational expertise
Genetic algorithms	Uses evolutionary selection process	– Process steadily improves – Not biased by preconceived notions	– No insight from process experts used – Improvement may be slightly incremental

Adapted from Kennedy and Krouse (*83*).

"one factor at a time"—do not enable the process developer to characterize the interactions that can exist between the cell-culture operating parameters. Statistically designed experiments have a much greater utility for cell-culture process development. Its sequential and predictive nature allows a process developer to identify and optimize the factors that have the greatest impact on process performance with a minimal number of experiments (*see* Fig. 3).

3.5.1. Initial Process Development Platform

The first step in performing the process development is to identify an appropriate medium platform that sustains growth and productivity. Several commercial or in-house media may be available as an initial medium to use for process development. Any of these media will be an acceptable starting point if they support growth and productivity. An initial screening study of the available media (and possibly operating parameters such as temperature or pH) is often performed to select the better medium for growth and productivity. The medium and operating parameters that are identified as having acceptable performance are selected as an initial starting point

3.5.2. Experimental Factors and Responses

Experimental factors are the process inputs, and the responses are the measured outputs. There are two different strategies for selecting the factors that will be examined in the studies. The closed strategy examines factors in the initial process platform (i.e., initial medium ingredients). The open strategy examines factors that are not included in the initial process platform (i.e., additional medium ingredients). The advantage of the closed strategy is that the optimization path is fairly straightforward, because the possible factors are fixed. However, open strategies could identify cost-effective medium formulations, simpler process operations, materials with improved lot-to-lot consistencies, or more readily available materials.

The responses that can be optimized are any measurable process parameter, including growth rate, titer, cell mass, specific productivity, protein quality, or operational cost. Depending on the development stage, different responses may be optimized. For early development processes, process yield and availability of resources for quick implementation are important. For later development processes, the volumetric productivity, process robustness, and product quality are more important.

3.5.3. Process Development Scale

The large number of factors that are involved in defining a cell-culture process and the high level of interaction between these factors lead to an iterative procedure for identifying critical factors and optimizing these factors to improve process performance. Typically, the objective of the first few studies performed is to identify which of the numerous factors can significantly affect the process outcome. This is normally done by a series of main-effect screening studies performed at small-scale (i.e., flasks). There is more value in identifying factors that have a greater impact on process performance than having a complete data package on fewer factors. Therefore, only the most critical responses should be measured (for example, final volumetric productivity). More responses that characterize the process are measured when the number of factors (and experimental runs) decreases. Once the number of factors being examined has been reduced, studies are performed to characterize the response surface generated by

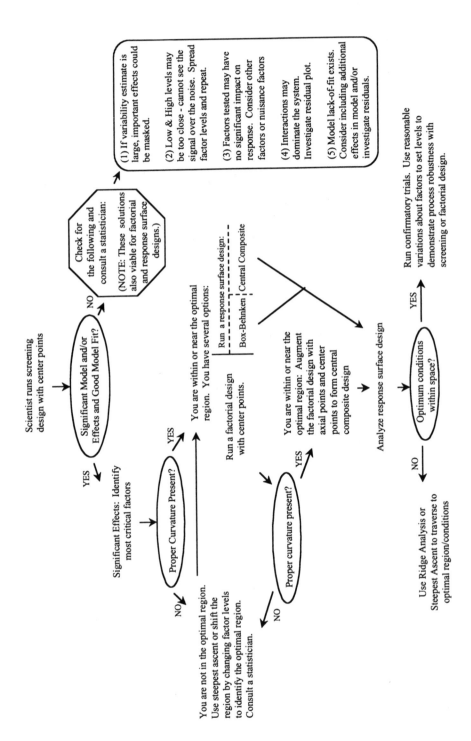

Fig. 3. Flowchart for sequential experimentation (Source: personal and non-confidential communication, Gary Sullivan [Eli Lilly]).

the significant factors and to locate operating ranges that improve process performance. These studies are typically performed in flasks or small-scale reactors, depending on the factors being examined (pH or DO is not controlled in flasks). The final step is to identify the critical-process parameters that affect process robustness or product quality, and to specify ranges that will keep the process in control. These studies are typically performed in larger-scale or production-scale bioreactors.

The impact of process scale on the measured responses should be understood to minimize scale effects. The scale effect is usually an offset in the measured responses. In this case, the experimental outcome may demonstrate a rank order of the conditions that impact the process outcome. In some cases, the relative differences in process outcome scale well. However, it is important to consider the risk of the scale effect interacting with factors in the experimental design that could lead to an inaccurate estimate of the process outcome at scale. Because it is impractical to perform the screening studies at scale, these risks must be accepted, or the scale-down model would need to be evaluated to minimize this risk.

3.5.4. Experimental Design and Analysis For Process Optimization

Process optimization strategies are numerous and varied (Table 3). They run the gamut from the classical strategy of optimizing the level of one ingredient at a time to the heavy use of computing power and mathematics to design and model complicated searches such as genetic algorithms. A review of the various strategies can be found in Kennedy and Krouse *(83)*. A discussion of some of the methods currently explored in the literature follows. For more depth and detail, *see* refs. *84,85,104.*

3.5.4.1. ONE FACTOR AT A TIME

"One factor at a time" optimization is frequently used *(86–91)* and may be the most common method used for process optimization *(92)*. Using this method, the process developer determines the optimal level of a single factor at a time by holding all others constant and manipulating the level of that one factor (Fig. 4). Significant differences are observed when changes in the factor level result in a difference in the response that is significantly greater than the noise demonstrated in replicate runs. This is a straightforward, easy-to-design, and easy-to-interpret method. However, it provides no understanding of the possible effects that interactions may have on the process outcome, and does not offer the most efficient use of resources.

3.5.4.2. STATISTICAL DESIGNS: ANALYSIS OF VARIANCE

Designed experiments for analysis of variance (ANOVA) provide a greater understanding of factor interactions and require fewer experiments to identify improved operating conditions than the one factor at a time approach. The designed experiment differs from the one factor at a time approach in that all the factors that are studied are examined simultaneously by testing different combinations of high and low levels of the factors. It is important to balance the design space so that for every factor there are an equal number of conditions run at the low level and at the high level. It is also important to randomize the runs, minimize random noise by closely controlling the conditions under which the runs are performed, and avoid the introduction of confounding factors. For example, if all the low levels of factor A are run by one scientist and all of the high levels of factor A are performed by another scientist, then the effect that factor A has on the process outcome is confounded with scientist techniques. All

One ingredient at a time: outcome

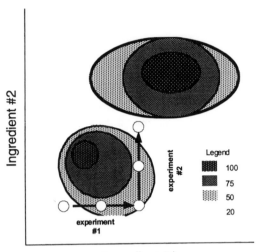

Fig. 4. Outcome of the "One Factor at a Time" approach. With two experiments, the investigator has identified an ingredient combination that yields an outcome of 50, but several more productive combinations still exist in the research space. The second contour area in this figure shows that more than one "optimum" space can exist in a given process.

Factorial design: outcome

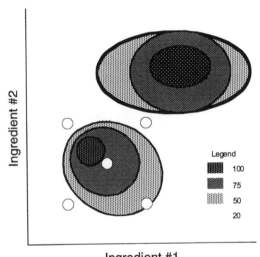

Fig. 5. Outcome of a factorial (statistical) design. The investigator has used the same number of flasks, but has identified a higher-producing region *(75)* and has shown that "curvature" exists in the research space (an example of the value and importance of the center point). This suggests further experimentation.

possible sources of variance (analyst, time of day, equipment) must be identified and minimized. To avoid a source of variance confounding with the effect a factor has on

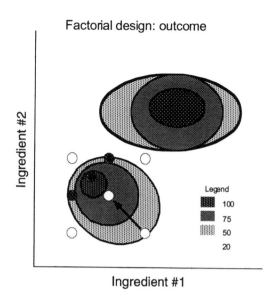

Factorial design: outcome

Fig. 6. Results of further statistical experimentation design. The investigator has added several flasks to his previous design (replicating two of the previous flasks and adding those shown in gray) and identified a "peak" in the research space. He has run two iterations of experiments, using 10 flasks, and has doubled his productivity compared to the "one at a time" approach.

the process outcome, the order in which the experiment is conducted should be randomized. Examples of designed experiments can be found in the literature *(13,93–102)*. The two-level factorial design is the most commonly used designed experiment (Fig. 5).

The minimum number of runs necessary to characterize all the main and interacting effects within the design space of a two-level factorial study is 2^n, where n is the number of factors to be tested. One advantage of the balanced arrangement of this experimental design technique is that it is easy to add replicates to the design to further clarify the measured effects (Fig. 6).

When evaluating a large number of factors, the minimum number of studies needed to perform a full factorial design becomes overwhelming. Two techniques can be used to manage the number of experiments in a study: fractional factorial designs and blocking.

3.5.4.3. FACTIONAL FACTORIAL SCREENING STUDIES

The first technique is to reduce the order of the factorial study to a fractional factorial study. The number of studies for a fractional factorial study is 2^{n-m}, where n is the number of factors to be tested and $1/2^m$ is the fraction applied. The benefit of a fractional factorial study is the reduced number of runs needed to perform the analysis. However, in a fractional factorial study, the higher-order interactions are confounded with other interactions and possibly the main effects. Typically, the higher-order interactions do not have as great an effect as the lower interactions, but this may not always be the case. When planning fractional factorial designs, it is important to avoid confounding interactions that are suspected to have a great effect on the process outcome. Most experimental design software will indicate which effects are directly confounded, allowing the investigator to modify the design as necessary. Figure 7 shows the rela-

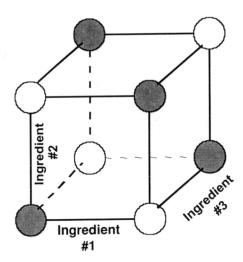

Fig. 7. Full factorial and Fractional Factorial design. The fractional factorial design (a half factorial) is illustrated by the gray nodes. It allows the estimation of the main effects, but not the two-factor interactions (the two-factor interactions are confounded with the main effects). In practice, several center points should also be included into the full or fractional factorial design to allow estimation of random error and curvature.

tionship between a small factorial design and its half fraction. Additional flasks should be run at the midpoint (center point) of those factors, which allows some estimate of error and curvature. Once the fractional factorial is complete, it may be possible to tease out significant interactions that were confounded in the design, depending on the how many independent factors are found to be insignificant. Alternatively, additional trials can be added to it to complete the factorial design for the significant factors. There are many examples of fractional factorials in the literature *(95,97)*.

3.5.4.4. PLACKETT-BURMAN SCREENING STUDIES

Another screening design used with regularity is the Plackett-Burman design *(103)* (Table 4). Compared to the fractional factorial design, the Plackett-Burman design further reduces the number of experimental runs that are needed to determine factors that have a significant main effect on the process outcome. However, it is impossible to determine the interacting effects from this design. Evaluation of the Plackett-Burman design is straightforward, and a clue can be taken from the arrangement of the positive and negative coding of the design. One can see that each factor is (+) in exactly half of the flasks and (–) in the other half. Because all of the factors are treated similarly, one can estimate the effect of going from the (–) level to the (+) level of any factor by adding the experimental outcomes of the flasks with (+) level and subtracting the outcomes of the flasks with (–) level. If the inclusion of the ingredient had no impact on the outcome, then this estimate would be not significantly different from zero. If the inclusion of the ingredient had a positive effect, this estimate would be positive and the larger the effect, the larger the estimate would be. In contrast, a negative number indicates a negative impact on the outcome by moving from the (–) level to the (+) level, and the magnitude of the estimate is still indicative of the impact.

Table 4
Plackett–Burman Table for 12 Factors[a]

Flask	A	B	C	D	E	F	G	H	I	J	K
						Factor					
1	+	−	+	−	−	−	+	+	+	−	+
2	+	+	−	+	−	−	−	+	+	+	−
3	−	+	+	−	+	−	−	−	+	+	+
4	+	−	+	+	−	+	−	−	−	+	+
5	+	+	−	+	+	−	+	−	−	−	+
6	+	+	+	−	+	+	−	+	−	−	−
7	−	+	+	+	−	+	+	−	+	−	−
8	−	−	+	+	+	−	+	+	−	+	−
9	−	−	−	+	+	+	−	+	+	−	+
10	+	−	−	−	+	+	+	−	+	+	−
11	−	+	−	−	−	+	+	+	−	+	+
12	−	−	−	−	−	−	−	−	−	−	−

[a]Each flask has the same number of "+" and "−" levels; thus, the experiment is "balanced." If a particular factor has an influence on the fermentation outcome, it can be quantified by looking at the difference of the average "+" vs "−" outcomes. If one or more of the factors are left "unassigned," they can be used to quantify the random variation ("noise") in the experiment. The design order should be randomized.

3.5.4.5. BLOCKING

If the number of runs in an experiment cannot be run concurrently, or the inclusion of a possible variation in the experimental design is inevitable (for example, potency analysis must occur on two separate days), one can design the experiment so that the experiment can be divided into manageable groups. This technique is known as blocking. Essentially, the block is included in the design as a factor. The estimation of the effect of the block can then be calculated and modeled.

3.5.4.6. DEFINING THE DESIGN SPACE

Factor levels to be tested must be reasonably bold, in order to map the research space in a realistic manner. An example of an overly conservative choice is seen in Fig. 8. On the other hand, too broad a research space can limit the amount of useful data (Fig. 9). The investigator must use process knowledge and expertise to determine reasonable levels. In our experience, scientists must be encouraged to be bold in selecting appropriate ranges. It is also recommended that several center points are included in this design to estimate curvature and systematic error.

Once the functional factorial design has been completed, curvature is evaluated. If the center point yields the most desirable outcome (i.e., maximum potency), then an optimal region exists between the factor levels previously chosen; if not, factor levels must be adjusted to find an optimal region. An example of curvature is provided in Fig. 10. One method of determining how to find this optimal region is called the direction of steepest ascent. One moves the factor levels perpendicular to the contours developed from the earlier model. A detailed example of use of the response surface and steepest-ascent methodology can be found in Box et al. *(104)*. Examples from the primary literature include Sadhukhan et al. *(96)*.

Too small a search area: outcome

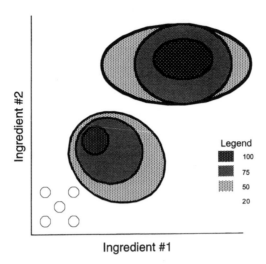

Fig. 8. The results of selecting too small a research space to examine. The investigator interprets this to mean that increasing levels of Ingredient #1 or Ingredient #2 do not affect to the productivity of the system.

To wide a search area: outcome

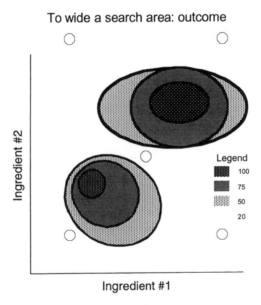

Fig. 9. Too broad a research space. The investigator has selected too broad a research space in this example. Again, his interpretation is that changing levels of Ingredient #1 and Ingredient #2 have no impact on the productivity of the system.

3.5.4.7. DATA ANALYSIS OF DESIGNED EXPERIMENTS AND FOLLOWUP

Once the design is created and the experiments are completed, the design must be analyzed to determine significance. Box et al. *(104)* explains in great detail how the

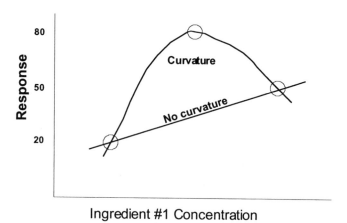

Ingredient #1 Concentration

Fig. 10. An illustration of curvature. The data above are taken from the Ingredient #1 data in Fig. 5. By including the center point, one can see that in between the low and high levels of Ingredient #1 a maximum exists. If curvature did not exist, one would expect that the maximum existed outside of the research space.

analysis is accomplished; however, several software packages (such as JMP® from SAS Institute, Inc.) are commercially available and make the analysis reasonably simple. These programs develop a model of the effects of the factors on the measured response. If the model describes a large portion of the variation in the experimental outcome, the factors are fairly important and can be further investigated. There are several reasons why a model may not fit well. These include a large amount of variability in the system—if the system is not very reproducible, the "noise" will mask important effects. Also, the low and high levels for the factors may not be spaced widely enough (not "bold" enough). In addition, the factors chosen for the design may have very little impact on the measured response, in which case the investigator needs to pick new factors and try again. Although unlikely, interactions could be more important than some main effects, which are confounded and not easily analyzed.

If center points were included in a factorial or fractional factorial design, the model would develop an idea of curvature. A response surface design can be completed by adding axial points and additional center points to the design already completed. An example can be seen in Fig. 11. With this technique, the performance boundaries—those spaces where very small changes in ingredients can cause large changes in outcome—can be identified. This can be used to develop an idea of the "robustness" of the medium designed.

This is an important consideration in a production facility. For example, a common industrial practice is to try to use ingredients in the amounts in which they typically are packaged and shipped. The Plant Manager saves money on repackaging or an increased packaging fee from the manufacturer when unusual quantities are not required to be delivered to the plant. Inventory control is also much easier. Therefore, if the media design calls for 24.5 kg of Ingredient #1 and this ingredient is packaged in 25-kg lots, the Manufacturing group is likely to add the additional 0.5 kg to each fermentation instead of removing 0.5 kg from each lot of Ingredient #1 and requiring storage for it. A response surface map of the research space can predict whether this small difference in concentration of Ingredient #1 is expected to affect productivity in a substantial way.

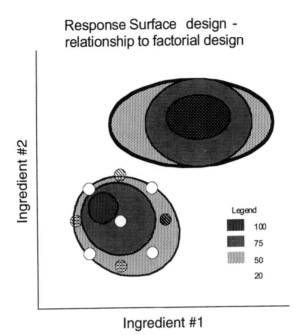

Fig. 11. Converting a Full Factorial design into a Response Surface design. The points in gray are the additional flasks run to convert the full factorial design into a response-surface design. The "axial" points for a two-dimensional response surface design are √2 distance from the center point (the vertices are the distance of 1 from the center point).

Once the process optima has been found, the investigator can vary the levels in a calculated manner to determine the exact process robustness. Through the process knowledge gained during the optimization, the investigator can usually identify several parameters of the fermentation that are critical to the success of the process. These parameters are known as Critical Process Parameters (CPPs). It is often very useful to know the impact of variation of the operating ranges of these CPPs on the performance (both productivity and product quality) of the fermentation. Consideration should be given to the most likely sources of variation or error that will occur in the manufacturing setting. Then, a fractional factorial experiment can be designed and executed for the capable or expected range. This experimentation will demonstrate the effect of the probable operating range of this CPP on performance of the culture. In this way data can be generated that identify Proven Acceptable Ranges (PARs) for the medium ingredient concentrations and process operating conditions.

As stated previously, process development using this approach is an iterative process. A flowchart illustrating this can be found in Fig. 3.

3.5.4.8. A Medium Development Example

Figure 4 represents the outcome or coverage of this space by an investigator using the one at a time approach. He ran two experiments, using a total of six flasks (with one replicate), improved medium performance from 20 units to 50 units, and found three different formulations that resulted in this improvement. However, this design gave him no idea of which interactions exist between the media components. With a balanced statistical design approach, the investigator does one experiment with six flasks, one each at the vertices, and two in the center (to determine error and curvature) and improved the fermentation titer from 20 to 75 units (Fig. 5).

Also, in his analysis of the outcome, he saw that as the levels of Ingredient #1 decreased and Ingredient #2 increased, he achieved an increase of productivity up to a point, followed by a marked drop of titer. This suggested that he more fully explore the space around the center of the design along the arrow (Fig. 6). Because he had done his previous experiments in a controlled fashion, he easily added to the original design the four additional flasks marked in gray (including one of the previous conditions to assess possible systematic error) and found the most productive region in his research space (Fig. 6). In this contrived experiment, our investigator, using a "classical" approach, improved his medium performance by 250%. Using a statistical design approach with no additional resources, he improves it by 375%. With three (four) additional flasks in a subsequent experimental run, he raises the improvement to 500%. He then more fully maps the research space using a response-surface design, which allows him to make reasonable estimates of the PARs for these two medium ingredients (Fig. 11).

3.5.5. Computational Optimization Methods

Statistical optimization processes utilize linear estimation techniques (least squares estimation) to produce models that describe the research space. The growth in speed and power of computers in the last 30 years has allowed the development of computer algorithms that use nonlinear optimization techniques to develop these models. Computational optimization methods such as genetic algorithms, neural networks, and particle swarm optimization (PSO) have shown some promise in developing optimization strategies. Their advantages include the ability to define hidden data patterns, work well with large amounts of data, and they require no knowledge of the mechanisms utilized by the end user. Their disadvantages include a reliance on high quality and complete data, a certain degree of mathematical aptitude required to make full use of these tools, and the fact that a process model may not be possible from the computation optimization methods because of lack of orthogonality. Also, incremental progress can be seen in statistical design methodology, a consideration that can convince investigators and management that progress is being made on the project. This is a luxury that some of the computational optimization methods do not afford.

3.5.5.1. GENETIC ALGORITHMS

Use of a genetic algorithm is one method of performing process optimization. Optimization is performed by coding the ingredients as "genes" at a specific place on the "chromosome" (the experiment) and measuring the response (fitness). The higher the fitness, the more likely the chromosome is to be retained during subsequent experimentation (selection). Variation is induced by allowing the successful chromosomes to exchange genes (crossover), and genes are randomly changed (mutation). In this way, a new round of experiments are obtained. Commercial software packages such as Genehunter (Ward Systems Group, Inc.) are available, and universities such as MIT also make available genetic algorithm software. Examples from the literature include Zuzek et al. *(105)*, Pinchuk et al. *(106)* with a review in Weuster-Botz *(92)* and Chatterjee et al. *(107)*.

3.5.5.2. NEURAL NETWORKS

Neural networks again use a biologically based concept to design a computation strategy. The software is "trained" on a sample data set, and builds a hidden nonlinear model of the connections or patterns hidden within the dataset. After the software is trained and a hidden model is built, new medium combinations can be tried and their response predicted. Examples in the literature include *(108)*.

3.5.5.3. PARTICLE SWARM OPTIMIZATION

PSO is defined more in social terms rather than purely biological ones, but the concept is the same. The population of member experiments are run, and the individuals remember the good solutions as they explore the research space. After every iteration, the best solutions are shared within the population, factors are updated, and another round of experimentation is run. One advantage of this search algorithm is that it can explore a large search area and not totally fixate on a local maximum, unlike response-surface methodology. For example, a PSO is likely to find the second process optima seen in Fig. 4. A direct comparison of PSO and statistical design can be found in the literature *(109,110)*. Both optimization techniques produced similar titer increases after similar numbers of shake-flask experiments, but the statistical design allowed the simplification of the medium (14 ingredients to 5). However, the PSO has defined additional regions of the research space in which increased titer could be found at marked differences of ingredient levels, an unexpected outcome.

4. Choice of Bioreactor Processes and Engineering Considerations

Cell-culture processes can be broadly divided into suspension systems and adherence-dependent systems such as microcarriers (*see* ref. *111*). Currently, a large majority of industrial cell-culture processes comprise suspension cultures, and this chapter limits its focus to suspension-culture systems *(112)*.

4.1. Engineering Considerations for Stirred Tank Systems

Suspension systems are mostly grown in curved or domed-bottom stainless steel vessels (up to about 20,000 L) with aspect ratios (tank height to diameter ratios) ranging from 1 to 3 *(113)*. While it can be demonstrated that the 20,000L scale does not represent the upper limit of scale for cell-culture stirred-tank reactors, we believe that business drivers and manufacturing strategy have played an important role in the determination of the current upper scale of operation. Adequate mixing in stirred-tank systems is provided mostly through the use of low-shear agitators such as marine impellers, although there are published reports of novel impeller designs *(114)*. Oxygen delivery and carbon dioxide removal is typically facilitated by a gas sparge device (porous micro-sparger or open pipe). During scale-up and scale-down of bioreactor processes, the important engineering aspects that must be considered are: mixing times, mass transfer (of oxygen and carbon dioxide), and shear effects.

Historically, shear was considered to be a significant challenge with mammalian cell-culture systems; however, it is presently understood that mammalian cells can usually withstand relatively high levels of agitation *(115)*. Shear-associated cell death is caused by hydrodynamic shear sensitivity largely caused by bubbles, which can be overcome with the use of surfactants (such as Pluronic F-68) *(116,117)*. This concern about hydrodynamic damage by large bubbles has resulted in the wide use of porous microspargers to maximize oxygen gas-liquid mass-transfer efficiencies. Although this solution was successful in maximizing oxygen mass transfer with minimal aeration, it also caused difficulty in carbon dioxide *(112)*. Of late, there have been reports to resolve this problem through either the use of supplemental gas sparging *(118)* or the design of "gas-liquid mass transfer in-efficient" spargers to balance oxygen supply with stripping of CO_2 *(82,119)*. Apart from carbon dioxide stripping, dissolved carbon dioxide levels are also influenced by factors such as the total base addition (used for pH control) in reactors *(82)*; thus, there is scope for additional and more comprehensive methods of addressing dissolved carbon dioxide control, if necessary. Aside from shear and mass transfer considerations, the impact of scaling on mixing time has been identified as an equally important design parameter for large-scale reactors *(115,120)*. This parameter is particularly important for cell-culture reactors because of the potential for pH gradients resulting from concentrated nutrient feed additions (especially in fed-batch systems) and base feed utilized for pH control. Mixing-time issues can be resolved by several methods *(115)*: (a) using larger tanks with lower aspect ratios (not recommended), (b) use of multipoint feed additions and optimal location of feed additions near impellers, and (c) scaling-up agitation rates based on mixing time harmonization (if multipoint feeding is not sufficient).

To summarize, the three major issues that affect scale-up are (a) shear, (b) mass transfer, and (c) mixing time. Shear can be addressed through the use of surfactants (such as Pluronics). Mass transfer can be addressed by using gas sparge systems that balance oxygen mass-transfer efficiency with carbon dioxide stripping. Mixing time can be addressed by optimal location and use of multipoint feed additions and sufficient/adequate agitation rates. Although agitation rates in stirred-tank reactors can be determined through scale-up based on different criteria such as impeller-tip speed, power per unit volume or the specific pumping rate of the impeller *(115,121)*, our collective experience suggests that scale-up based on power per unit volume with geometric similarity offers a reasonable compromise for cell-culture scale-up.

For perfusion stirred-tank systems, the choice and scale-up of the cell retention device poses an additional technical challenge beyond the issues described here for stirred-tank reactors *(82,122)*. Although the topic of scalability and engineering expertise for the design and implementation of cell-retention devices has been identified as important, perfusion reactors are known (unconfirmed industry sources) to be operationally feasible at scales as high as 5,000 L, thus, indicating that engineering solutions do exist for scale-up of perfusion technology.

4.2. Choice of Processes: Fed-Batch vs Perfusion

Currently, most commonly used large-scale cell-culture systems can be classified either as fed-batch processes or continuous perfusion processes, depending on their mode of operation. As described earlier, a batch culture is a closed system in which a growth of culture comprises a lag phase (in some cases) followed by exponential, stationary, and decline phases. The productivity of such processes can be further increased or optimized through the control feeding of key nutrients. According to a recent report by J.P. Morgan securities, fed-batch systems comprise about 90% of commercial production systems for biologics manufacturing *(5)*. Perfusion culture systems, although less reported, are equally viable commercial production systems. Asides from product quality consistency considerations, perfusion reactors are particularly attractive because they offer a productivity gain advantage (mg/L/d) of at least fourfold as compared to a fed-batch operation in the same reactor volume, because of their ability

Table 5
A Comparison of Batch/Fed Batch and Perfusion Cell Culture Systems

Fed-batch processes	
Pros	Cons
Low development costs because of short development cycle times and less labor-intensive	More prone to changing product quality because of proteolytic degradation, aggregation and desialaytion
Well understood scale-up up to about 15,000 L	Requires large capacity tanks for low-productivity cell lines; less efficient use of fixed capital
Reduced operational failure rate relative to perfusion, because these symptoms require fewer interventions/manipulations per reactor run and are less complicated and thus more robust with respect to process deviations	Inefficient for growth rate-associated protein production systems
Better tolerance of unstable expression systems because of fewer population doublings or shorter-run duration.	

Perfusion processes	
Pros	Cons
More efficient use of installed/fixed capital because of higher productivity (mg/L/d)	Higher development costs/time because of longer run durations
Better suited to products prone to post-secretion modifications (and will result in less product-quality variability)	Less scheduling flexibility in a multiproduct manufacturing facility, because of longer run durations
Suitable for both growth and non-growth rate-associated protein production systems	Requires higher level of operator training because of complexity and intensity of the process
Process economics better suited for production of 100+ kg/yr ofproduct	Scale-up and engineering experience is limited to a fewer companies as compared to fed-batch processes

to sustain biomass levels and run duration of at least 2x higher than fed-batch reactors. These systems comprise a means of continuously perfusing cells with fresh culture medium. In such systems, cell retention devices such as spin filters, ultrasonic separators, and gravity sedimentation devices are used *(82,122–124)*. A comparison of the pros and cons of each system is provided in Table 5.

Although both fed-batch and perfusion systems have pros and cons (Table 5), in reality the choice of the cell-culture process is usually made by practical considerations such as the installed capacity, in-house process expertise, and scale-up experience with these technologies. In industrial environments that have expertise and experience with both technologies, the decision between fed-batch and perfusion systems can be made based from a decision tree (Fig. 12) that incorporates knowledge about product quality

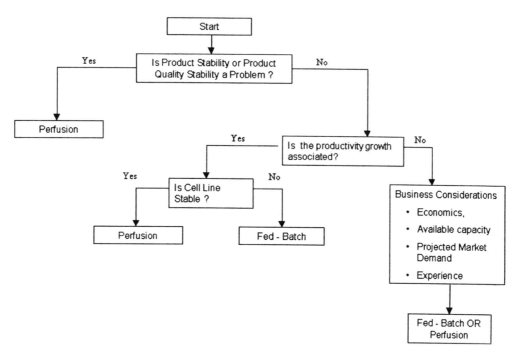

Fig 12. A Simplistic Decision Tree to Determine Choice of Cell-Culture Process (Source: non-confidential discussions among R. Taticek [Genentech], D. Ramakrishnan [Eli Lilly]).

and stability, growth-rate dependency on production, cell-line stability, and process economics considerations (i.e., available manufacturing capacity, projected market demand, and Cost Of Product Sold analyses).

5. Primary Clarification

The purification of the protein from the conditioned cell-culture broth is performed in a series of chromatography columns as diagrammed in Fig. 1. The life and efficiency of these columns are significantly reduced by cell debris present in the broth charged on the columns. Therefore, primary clarification is typically performed to remove whole cells and cellular debris from the harvested fermentation broth prior to loading on a chromatography column. This section focuses on issues that should be addressed during the selection and development of the primary clarification process. The objective of the primary clarification process is to prepare conditioned broth for downstream purification. The downstream purification efficiencies decrease with increased concentrations of DNA, host-cell proteins, lipids, or cell debris in the conditioned broth. The downstream purification efficiencies increase with increased titer in fermentation/cell-culture broth. Both fermentation conditions and the primary clarification conditions affect the concentrations of DNA, host-cell proteins, lipids, cell debris, or titer in the conditioned broth. Ideally, the effect that fermentation conditions have on the primary clarification and purification processes should be examined during the fermentation process development to identify the most productive process. In general, the cell-culture viability and titer should be used as criteria for harvest to achieve high purification yields. For example, culture viability decreases with fermentation age. As

Table 6
Parameters to Consider for Optimization
During Development of the Primary Clarification Step

Response	Objective = ratio of postclarification/preclarification
Relative particle size distribution	Minimize
Relative turbidity	Minimize
Relative residual DNA	Minimize
Relative residual LDH	Minimize
Residual host-cell protein	Minimize
Cell viability	Maximize
Titer	Maximize

the culture viability decreases, the amount of DNA, host-cell proteins, lipids, and cell debris in the broth increases. The resulting decreased purification yields may overshadow the higher titers achieved in the extended fermenter run, reducing the overall process productivity.

Four types of primary clarification processes are practiced in industry: dead-end depth filtration, tangential flow filtration, continuous centrifugation, and expanded-bed chromatography. The process types are listed in order of initial capital costs, with dead-end depth filters having the lower initial capital costs and expanded bed chromatography having the higher initial capital costs. Repeated operations tend to be more expensive with the depth filters and less expensive with the expanded bed chromatography. The type of process that is ideal for an application can be determined from a cost analysis. The cost analysis should consider the initial capital costs, disposable goods, utilities, manpower, process yield, and scalability of scale-down process development models (development costs). For each of the different types of primary clarification processes, the parameters that could be measured as responses in designed experiments include the following (Table 6):

While the objective of this section is to highlight the importance of primary clarification considerations, more design details about the different primary recovery processes can be found elsewhere (depth filters *[125]* filter theory *[126–129]* tangential flow filtration *[130–138]* centrifuge *[139,140]*, expanded-bed adsorption *[141–146]*).

6. Conclusion

In summary, this chapter presents an overview of a typical mammalian cell-culture process and provides an industrial perspective on frequently discussed and important topics that impact the cell-culture bioreactor process development and manufacturing strategy. These topics include discussions about (a) current business drivers and resultant process development strategy, (b) science and techniques of process development, (c) choice of bioreactor processes and engineering/scale-up considerations, and (d) cell removal and downstream considerations during cell-culture bioreactor process development.

References

1. Rosenberg, M. (2000) Development costs and process economics of recombinant proteins produced in *E. Coli*, mammalian cell culture, and natural products derived products: A compara-

tive analysis, in *Proceedings of the IBC Conference on the Economics of Biopharmaceuticals*. San Diego, CA.

2. Zabriskie, D. W. (2000) Regulatory trends related to process validation, in *Biopharmaceutical Process Validation* (Sofer, G. and Zabriskie, D. W., eds.), Marcel Dekker, New York, NY pp. 1–15.

3. Carey, J. and Barrett, A. Drug prices: What's Fair?, in *Business Week*. Dec, 10, 2001, pp. 61–70.

4. Ezzell, C. (2001) Magic bullets fly again. *Sci. Am.* **Oct 2001**, pp. 35–41.

5. Moldowa, D. T., Shenouda, M. S., and Meyers, A. P. Industry analysis: the state of biologics manufacturing, equity research, in *J.P. Morgan Securities Inc.* March 12, 2001.

6. Miller, W. M., Blanch, H. W., and Wilke, C. R. (1988) A kinetic analysis of hybridoma growth and metabolism in batch and continuous suspension culture: effect of nutrient concentration, dilution rate, and pH. *Biotechnol. Bioeng.* **32**, 947–965.

7. Eagle, H. (1955) Nutrient needs of mammalian cells in tissue culture. *Science* **122(3168)**, 501–540.

8. Barnes, D. (1987) Serum–free animal cell culture. *Biotechniques* **5**, 534–542.

9. Jayme, D. W., Epstein, D. A., and Conradt, D. R. (1988) Fetal bovine serum alternatives. *Nature* **334(6182)**, 547–548.

10. McKeehan, W. L., et al. (1990) Frontiers in mammalian cell culture. *In Vitro Cell. Dev. Biol.* **26**, 9–23.

11. Gorfien, S., et al. (2000) Growth of NS0 cells in protein-free, chemically defined medium. *Biotechnol. Prog.* **16**, 682–687.

12. Butler, M. and Huzel, N. (1995) The effect of fatty acids on hybridoma cell growth and antibody production in serum-free cultures. *J. Biotechnol.* **39**, 165–173.

13. Liu, C. H., Chu, I. M., and Hwang, S. M. (2001) Factorial designs combined with the steepest ascent method to optimize serum-free media for CHO cells. *Enzyme Microb. Tech.* **28**, 314–321.

14. Komolov, I. S. and Fedotov, V. P. (1978) Influence of insulin on mitotic rate in cultivated mammalian cells. *Endocrinol. Exp.* **12(1)**, 43–48.

15. Veillon. C., et al. (1985) Characterization of a bovine serum reference material for major, minor, and trace elements. *Anal. Chem.* **57**, 2106–2109.

16. Meier, S. J., Hatton, T. A., and Wang, D. I. C. (1999) Cell death from bursting bubbles: role of cell attachment to rising bubbles in sparged reactors. *Biotechnol. Bioeng.* **62(4)**, 468–478.

17. Dey, D. and Emery, N. (1999) Problems predicting cell damage from bubble bursting. *Biotechnol. Bioeng.* **65(2)**, 240–245.

18. Shepherd, P. R. and Kahn, B. B. (1999) Mechanisms of disease: glucose transporters and insulin action—implications for insulin resistance and diabetes mellitus. *New Engl. J. Med.* **341(4)**, 248–257.

19. Gould, G. W. and Holman, G. D. (1993) The glucose transporter family: structure, function, and tissue-specific expression. *Biochem. J.* **295**, 329–341.

20. Bonarius, H. P. J., et al. (1996) Metabolic flux analysis of hybridoma cells in different culture media using mass balances. *Biotechnol. Bioeng.* **50(3)**, 299–318.

21. Bonarius, H. P. J., et al. (2001) Metabolic-flux analysis of continuously cultured hybridoma cells using $^{13}CO_2$ mass spectrometry in combination with ^{13}C–lactate nuclear magnetic resonance spectroscopy and metabolic balancing. *Biotechnol. Bioeng.* **74(6)**, 528–538.

22. Zielke, H. R., et al. (1978) Reciprocal regulation of glucose and glutamine utilization by cultured human diploid fibroblasts. *J. Cell Physiol.* **95**, 41.

23. Follstad, B. D., et al. (1999) Metabolic flux analysis of hybridoma continuous culture steady state multiplicity. *Biotechnol. Bioeng.* **63(6)**, 675–683.

24. Linz, M., et al. (1997) Stoichiometry, kinetics, and regulation of glucose and amino acid metabolism of a recombinant BHK cell line in batch and continuous cultures. *Biotechnol. Prog.* **13(4)**, 453–463.

25. Cruz, H. J., et al. (2000) Metabolic shifts do not influence the glycosylation patterns of a recombinant fusion protein expressed in BHK cells. *Biotechnol. Bioeng.* **69(2),** 129–139.

26. Kurokawa, H., et al. (1994) Growth characteristics in fed–batch culture of hybridoma cells with control glucose and glutamine concentrations. *Biotechnol. Bioeng.* **44,** 95–103.

27. Glacken, M. W. (1987) Development of mathematical description of mammalian cell culture kinetics for the optimization of fed-batch bioreactors, in *Thesis*. Massachusetts Institute of Technology, Cambridge, MA.

28. Reuveny, S., et al. (1987) Factors affecting monoclonal antibody production in culture. *Dev. Biol. Stand.* **66,** 169–175.

29. Miller, W. M., Wilke, C. R., and Blanche, H. W. (1988) Transient responses of hybridoma cells to lactate and ammonia pulse and step changes in continuous culture. *Bioproc. Eng.* **3,** 113–122.

30. Ozturk, S. S. and Palsson, B. O. (1992) Effects of ammonia ion and extracellular pH on hybridoma growth, metabolism and antibody production. *Biotechnol. Bioeng.* **39,** 418–431.

31. Bibila, T. A., et al. (1994) Monoclonal antibody process development using medium concentrates. *Biotechnol. Prog.* **10,** 87–96.

32. Cruz, H. J., et al. (2000) Effects of ammonia and lactate on growth, metabolism, and productivity of BHK cells. *Enzyme Microb. Tech.* **27(1-2),** 43–52.

33. MacQueen, A. and Bailey, J. E. (1990) Effect of ammonium ion and extracellular pH on hybridoma cell metabolism and antibody production. *Biotechnol. Bioeng.* **35,** 1067–1077.

34. Truskey, G. A., et al. (1990) Kinetic studies and unstructured models of lymphocyte metabolism in fed-batch culture. *Biotechnol. Bioeng.* **36,** 797–807.

35. Yang, M. and Butler, M. (2000) Effects of ammonia on CHO cell growth, erythropoietin production, and glycosylation. *Biotechnol. Bioeng.* **68(4),** 370–380.

36. Zhou, W., Rehm, J., and H., W. S. (1995) High viable cell concentration fed-batch cultures of hybridoma cells through on-line nutrient feeding. *Biotechnol. Bioeng.* **46,** 579–587.

37. Gambhir, A., Europa, A. F., and and H. U., W. S., Alteration of cellular metabolism by consecutive fed-batch cultures of mammalian cells. *J. Biosci. Bioeng.* **87(6),** 805–810.

38. DeJesus, M. J., et al. (2001) The influence of pH on cell growth and specific productivity of two CHO cell lines producing human anti-RhD IgG, in *17th ESACT Meeting*. Tylosand, Sweden.

39. Altamirano, C., et al. (2000) Improvement of CHO cell culture medium formulation: simultaneous substitution of glucose and glutamine. *Biotechnol. Prog.* **16,** 69–75.

40. Sanfeliu, A., et al. (1996) Analysis of nutrient factors and physical conditions affecting growth and monoclonal antibody production of the hybridoma KB–26.5 cell line. *Biotechnol. Prog.* **12,** 209–216.

41. Hu, W. S., et al. (1987) Effect of glucose on the cultivation of mammalian cells. *Devel. Biol. Stand.,* **66,** 155–160.

42. Zielke, H. R., et al. (1980) Lactate: a major product of glutamine metabolism by human diploid fibroblasts. *J. Cell Physiol.* **104,** 433–441.

43. Eigenbrodt, E., Fister, P., and Reinacher, M. (1985) New perspectives on carbohydrate metabolism in tumor cells, in *Regulation of Carbohydrate Metabolism* (B. R., ed.), CRC Press, Boca Raton, FL, pp. 141–169.

44. Altamirano, C., Cairó, J. J., and Gòdia, F. (2001) Decoupling cell growth and product formation in Chinese hamster ovary cells through metabolic control. *Biotechnol. Bioeng.* **74(4),** 351–360.

45. Christie, A. and Butler, M. (1999) The adaptation of BHK cells to a non-ammoniagenic glutamate-based culture medium. *Biotechnol. Bioeng.* **64(3),** 298–309.

46. Xie, L. and Wang, D. I. C. (1993) Stoichiometric analysis of animal cell growth and its application in medium design. *Biotechnol. Bioeng.* **43,** 1164–1174.

47. Xie, L. and Wang, D. I. C. (1994) Fed-batch cultivation of animal cells using different medium design concepts and feeding strategies. *Biotechnol. Bioeng.* **43(11),** 1175–1189.

48. Xie, L. and Wang, D. I. C. (1996) High cell density and high monoclonal antibody production through medium design and rational control in a bioreactor. *Biotechnol. Bioeng.* **51(6),** 725–729.
49. Xie, L. and Wang, D. I. C. (1996) Material balance studies on animal cell metabolism using a stoichiometrically based reaction network. *Biotechnol. Bioeng.* **52(5),** 579–590.
50. Xie, L. and Wang, D. I. C. (1996) Energy metabolism and ATP balance in animal cell cultivation using a stoichiometrically based reaction network. *Biotechnol. Bioeng.* **52(5),** 591–601.
51. Xie, L. and Wang, D. I. C. (1997) Integrated approaches to the design of media and feeding strategies for fed-batch cultures of animal cells. *Trends Biotechnol.* **15(3),** 109–113.
52. Siegwart, P., et al. (1999) Adaptive control at low glucose concentration of HEK-293 cell serum–free cultures. *Biotechnol. Progr.* **15,** 608–616.
53. Simpson, N. H., et al. (1998) In hybridoma cultures, deprivation of any single amino acid leads to apoptotic death, which is suppressed by the expression of the bcl–2 gene. *Biotechnol. Bioeng.* **59,** 90–98.
54. Jo, E. C., Park, H. J., and Kim, K. H. (1990) Balance nutrient fortification enables high–density hybridoma cell culture in batch culture. *Biotechnol. Bioeng.* **36,** 717–722.
55. Franek, F. (1995) Starvation-induced programmed death of hybridoma cells: prevention by amino acid mixtures. *Biotechnol. Bioeng.* **45,** 86–90.
56. Cruz, H. J., Moreira, J. L., and C., M. J. T. (1999) Metabolic shifts by nutrient manipulation in continuous cultures of BHK cells. *Biotechnol. Bioeng.* **66(2),** 104–113.
57. Dowd, J. E., Kwok, K. E., and Piret, J. M. (2000) Increased t-PA yields using ultrafiltration of an inhibitory product from CHO fed-batch culture. *Biotechnol. Prog.* **16,** 786–794.
58. Kimura, R. and Miller, W. M. (1996) Effects of elevated pCO_2 and/or osmolality on the growth and recombinant tPA production of CHO cells. *Biotechnol. Bioeng.* **52(1),** 152–160.
59. Chen, G., Fournier, R. L., and Varanasi, S. (1997) Experimental demonstration of pH control for a sequential two-step enzymatic reaction. *Enzyme Microb. Tech.* **21,** 491–495.
60. Sauer, P. W., et al. (2000) A high-yielding generic fed–batch cell culture process for production of recombinant antibodies. *Biotechnol. Bioeng.* **67(5),** 585–597.
61. Chuppa, S., et al. (1997) Fermentor temperature as a tool for control of high–density perfusion cultures of mammalian cells. *Biotechnol. Bioeng.* **55(2),** 328–338.
62. Ducommun, P., et al. (2002) Monitoring of temperature effects on animal cell metabolism in a packed bed process. *Biotechnol. Bioeng.* **77(7),** 838–842.
63. Boon, M. A., Janssen, A. E. M., and van 't Riet, K. (2000) Effect of temperature and enzyme orgin on the enzymatic synthesis of oligosacchrides. *Enzyme Microb. Tech.* **26,** 271–281.
64. Jorjani, P. and Ozturk, S. S. (1999) Effects of cell density and temperature on oxygen consumption rate for different mammalian cell lines. *Biotechnol. Bioeng.* **64(3),** 349–356.
65. Horvath, B., Gu, X., and Rosenberg, M. (2001) Optimization of temperature shift to increase fed–batch productivity of a CHO culture, in *ACS Spring National Meeting.* San Diego, CA.
66. Oeggerli, A., Eyer, K., and Heinzle, E. (1995) On-line gas analysis in animal cell cultivation: I. control of dissolved oxygen and pH. *Biotechnol. Bioeng.* **45,** 42–53.
67. Eyer, K., Oeggerli, A., and Heinzle, E. (1995) On-line gas analysis in animal cell cultivation: II. methods for oxygen uptake rate estimation and its application to controlled feeding of glutamine. *Biotechnol. Bioeng.* **45,** 54–62.
68. deZengotita, V. M., et al. (2000) Phosphate feeding improves high-cell-concentration NSO myeloma culture performance for monoclonal antibody production. *Biotechnol. Bioeng.* **69(5),** 566–576.
69. Konstantinov, K., et al. (1996) Control long-term perfusion Chinese hamster ovary cell culture by glucose auxostat. *Biotechnol. Prog.* **12,** 100–109.
70. Sapre, A. G., et al. (2000) Analysis of micromolar concentrations of glucose by an interference free flow injection based biosensor. *Biotechnol. Lett.* **22,** 569–573.

71. Herwig, C., Marison, I. W., and vonStockar, U. (2001) On-line stoichiometry and identification of metabolic state under dynamic process conditions. *Biotechnol. Bioeng.* **75(3),** 345–354.
72. Paredes, C., et al. (1998) Estimation of the intracellular fluxes for a hybridoma cell line by material balances. *Enzyme Microb. Tech.* **23,** 187–198.
73. Christensen, B., et al. (2001) Simple and robust method for estimation of the split between the oxidative pentose phosphate pathway and the Ebden-Meyerhof-Parnas pathway in microorganisms. *Biotechnol. Bioeng.* **74(6),** 517–523.
74. Dauner, M., Bailey, J. E., and Sauer, U. (2001) Metabolic flux analysis with a comprehensive isotopomer model in *Bacillus subtilis. Biotechnol. Bioeng.* **76(2),** 144–156.
75. Europa, A. F., et al. (2000) Multiple steady states with distinct cellular metabolism in continuous culture of mammalian cells. *Biotechnol. Bioeng.* **67(1),** 25–34.
76. Forbes, N. S., Clark, D. S., and Blanch, H. W. (2001) Using isotopomer path tracing to quantify metabolic fluxes in pathway models containing reversible reactions. *Biotechnol. Bioeng.* **74(3),** 196–211.
77. Varner, J. and Ramkrishna, D. (1999) The non–linear analysis of cybernetic models. Guidelines for model formulation. *J. Biotechnol.* **71,** 67–104.
78. Varner, J. and R., D. (1998) Application of cybernetic models to metabolic engineering: investigation of storage pathways. *Biotechnol. Bioeng.* **58(2-3),** 282–291.
79. Varner, J. and Ramkrishna, D. (1999) Mathematical model of metabolic pathways. *Curr. Opin. BioTechnol.* **10,** 146–150.
80. Varner, J. and Ramkrishna, D. (1999) Metabolic engineering from a cybernetic perspective. 1. Theoretical preliminaries. *Biotechnol. Prog.* **15,** 407–425.
81. Varner, J. and Ramkrishna, D. (1999) Metabolic engineering from a cybernetic perspective. 2. Qualitative investigation of nodal architechtures and their response to genetic perturbation. *Biotechnol. Prog.* **15,** 426–438.
82. Ozturk, S. S. (1996) Engineering challenges in high density cell culture systems. *Cytotechnology* **22,** 3–16.
83. Kennedy, M. and Krouse, D. (1999) Strategies for improving fermentation medium performance: a review. *J. Ind. Microbiol. Biotechnol.* **23,** 456–475.
84. Myers, R. H. (1995). Response surface methodology : process and product optimization using designed experiments. J. Wiley and Sons, Inc.. New York, NY.
85. Haaland, P. D. (1989). Experimental design in biotechnology. Marcel Dekker, New York, NY.
86. Schrader, K. K. and Blevins, W. T. (2001) Effects of carbon source, phosphorus concentration, and several micronutrients on biomass and geosmin production by *Streptomyces halstedii. J. Ind. Microbiol. Biotechnol.* **26,** 241–247.
87. Calvente, V., de Orellano, M. E., Sansone, G., Benuzzi, D., and de Tosetti, M. I. Sanz. (2001) Effect of ntrogen source and pH on siderophore production by Rhodotorula strains and their application to biocontrol of phytopathogenic moulds. *J. Ind. Microbiol. Biotechnol.* **26,** 226–229.
88. Osek, et al. (1995) Improved medium for large-scale production of recombinant cholera toxin B subunit for vaccine purposes. *J. Microbiol Meth.* **24(2),** 117–123.
89. Peterson, L. A., et al. (2001) Effects of amino acid and trace element supplementation on pneumocandidn production by *Glarea lozoyensis*: impact on titer, analogue levels, and the identification of new analogues of pneumocandin B_0. *J. Ind. Microbiol. Biotechnol.* **26,** 216–221.
90. Souza, M. F. V. Q., Lopes, C. E., and Pereira, N. Jr. (1997) Medium optimization for the production of actinomycin-D by *Streptomyces parvulus. Arq. Biol. Tecnol.* (Curitiba). **40(2),** 405–411.
91. Stiens, L. R., et al. (2000) Development of serum–free bioreactor production of recombinant human thyroid stimulating hormone receptor. *Biotechnol. Prog.* **16(5),** 703–709.
92. Weuster-Botz, D. (2000) Experimental design for fermentation media development: statistical design or global random search? *J. Biosci. Bioeng* **90(5),** 473–483.

93. Jacques, P., et al. (1999) Optimization of biosurfactant lipopeptide production from *Bacillus subtilis* S499 by Plactett-Burman design. *Appl. Biochem. Biotechnol.* **77-79,** 223–233.

94. Park, et al. (1996) Medium optimization for recombinant protein production by *Bacillus subtilis*. *Biotechnol. Lett.* **18(6),** 737–740.

95. Ooijkaas, L. P., et al. (1999) Medium optimization for spore production of *Coniothyrium minitans* using statistically-based experimental designs. *Biotechnol. Bioeng.* **64(1),** 92–100.

96. Sadhukhan, A. K., et al. (1999) Optimization of mycophenolic acid production in solid state fermentation using response surface methodology. *J. Ind. Microbiol. Biotechnol.* **22,** 33–38.

97. Gaertner, J. G. and Dhurjati, P. (1993) Fractional factorial study of hybridoma behavior: 2. Kinetics of nutrient uptake and waste production. *Biotechnol. Prog.* **9(3),** 309–316.

98. Lee, S. H. and Rho, Y. T. (1999) Improvement of tylosin fermentation by mutation and medium optimization. *Lett. Appl. Microbiol.* **28(2),** 142–144.

99. Kim, E. J., Kim, N. S., and Lee, G. M. (1999) Development of a serum–free medium for dihydrofolate reductase-deficient Chinese hamster ovary cells (DG44) using a statistical design: beneficial effect of weaning cells. *In Vitro Cell. Dev. Biol.* **35,** 178–182.

100. Lee, G. M., et al. (1999) Development of a serum–free medium for the production of erythro-poietin by suspension culture of recombinant Chinese hamster ovary cells using a statistical design. *J. Biotechnol.* **69,** 85–93.

101. El-Helow, E. R., Sabry, S. A., and Khattab, A. A. (1997) Reduction of beta-mannanase by *B. subtilis* from agro-industrial by-products: Screening and optimization: Mannanase from wastes. *Antonie Leeuwenhoek* **71(3),** 189–193.

102. Castro, P. M. L., et al. (1992) Application of a statistical design to the optimization of culture medium for recombinant interferon-gamma production by Chinese hamster ovary cells. *Appl. Microbiol. Biotechnol.* **38(1),** 84–90.

103. Plackett, R. L. and Burman, J. P. (1946) The design of optimum multifactorial experiments. *Biometrika* **33,** 305–325.

104. Box, G. E. P., Hunter, W. G., and Hunter, J. S. (1978) Statistics for experimenters. J. Wiley and Sons, Inc., New York, NY, p. 653.

105. Zuzek,, M., Friedrich, J., Cestnik, B., Karalic, A., and Cimerman, A. (1996) Optimization of fermentation medium by a modified method of genetic algorithms. *Biotechnol. Tech.* **10(12),** 991–996.

106. Pinchuk, R. J., et al. (2000) Modeling of biological processes using self-cycling fermentation and genetic algorithms. *Biotechnol. Bioeng.* **67(1),** 19–24.

107. Chatterjee, S., Laudato, M., and Lynch, L. A. (1996) Genetic algorithms and their statistical applications: an introduction. *Computational Statistics and Data Analysis* **22,** 633–651.

108. Liu, C. H., Hwang, C. F., and Liao, C. C. (1999) Medium optimization for glutathione production by *Saccharomyces cerevisiae*. *Proc. Biochem.* **34(1),** 17–23.

109. Cockshott, A. R. and Hartman, B. E. (2001) Improving the fermentation medium for *Echinocandin* B production part II: Particle swarm optimization. *Proc. Biochem.* **36(7),** 661–669.

110. Cockshott, A. R. and Sullivan, G. R. (2001) Improving the fermentation medium for *Echinocandin*-B production. Part I: sequential statistical experimental design. *Proc. Biochem.* **36(7),** 647–660.

111. Reuveny, S. (1990) Microcarrier culture systems, in *Large-Scale Mammalian Cell Culture Technology* (Lubiniecki, A. S., ed.), Marcel Dekker, New York, NY pp. 271–341.

112. Chu, L. and Robinson, D. K. (2001) Industrial choices for protein production by large-scale cell culture. *Curr. Opin. Biotechnol.* **12,** 180–187.

113. Griffiths, B. (2001) Scale up of suspension and anchorage-dependent animal cells. *Mol. Biotechnol.* **17(3),** 225–238.

114. Tolbert, W. R. and Feder, J. (1983) Large scale cell culture technology. *Annu. Rep. Ferment. Proc.* **6,** 35–74.

115. Varley, J. and Birch, J. (1999) Reactor design for large scale suspension animal cell culture. *Cytotechnology* **29,** 177–205.

116. Goochee, C. F. and Murhammer, D. W. (1990) Sparged animal cell bioreactors: mechanism of cell damage and pluronic F-68 protection. *Biotechnol. Prog.* **6,** 391–397.

117. Zhang, Z., Al-Rubai, M., and Thomas, C. R. (1992) Effect of pluronic F-68 on the mechanical properties of mammalian cells; protection of hybridoma cell culture against shear. *Enzyme Microb. Technol.* **14(12),** 980–983.

118. Pattison, R. N., et al. (2000) Measurement and control of dissolved carbon dioxide in mammalian cell culture processes using an *in situ* fiber optic chemical sensor. *Biotechnol. Prog.* **16,** 769–774.

119. Zhou, W. C., et al. (1996) Large scale production of recombinant mouse and rat growth hormone by fed-batch GS-NS0 cell cultures. *Cytotechnology* **22(1-3),** 239–250.

120. Amanullah, A., et al. (2001) Scale-down model to simulate spatial pH variations in large-scale bioreactors. *Biotechnol. Bioeng.* **73(5),** 390–399.

121. Chisti, Y. (1993) Animal cell culture in stirred bioreactors: Observations on scale-up. *Bioprocess Eng.* **9,** 191–196.

122. Batt, B. C., Davis, R. H., and Kompala, D. (1990) Inclined sedimentation for selective retention of viable hybridomas in a continuous suspension bioreactor. *Biotechnol. Prod.* **6(6),** 458–64.

123. Ryll, T., et al. (2000) Performance of small-scale CHO perfusion cultures using an acoustic cell filtration device for cell retention: characterization of separation efficiency and impact on perfusion on product quality. *Biotechnol. Bioeng.* **69(4),** 400–449.

124. Clayton, G. R., et al. (1997) Using an external vortex flow filtration device for perfusion culture. *Pharm. Technol.* **21,** 10,116,188,122–123.

125. Cuno Co., E. P. F. (1992) Charge modified depth filter—technology and its evolution. *Filtration & Separation* **29(3),** 221–226.

126. Brose, D. J., Cates, S., and Hutchison, F. A. (1994) Studies on the scale-up of microfiltration membrane devices. *PDA J. Pharm. Sci. Tech.* **48(4),** 184–188.

127. Badmington, F., et al. (1995 September) Vmax testing for practical microfiltration train scale-up in biopharmaceutical processing. *BioPharm* 46–52.

128. Grace, H. P. (1956) Structure and performance of filter media. *AIChE J.* **2(3),** 307–315.

129. Hermia, J. (1982) Constant pressure blocking filtration laws—application to power-law non-newtonian fluids. *Trans. IChemE* **60,** 183–187.

130. Bouchard, C. R., et al. (1994) Modeling of ultrafiltration: predictions of concentration polarization effects. *J. Membrane Sci.* **97,** 215–229.

131. Burns, D. B. and Zydney, A. (1995) Effect of solution pH on protein transport through ultrafiltration membranes. *Biotechnol. Bioeng.* **64(1),** 27–37.

132. Foley, G., MacLoughlin, P. F., and Malone, D. M. (1995) Membrane fouling during constant flux crossflow microfiltration of dilute suspensions of active dry yeast. *Sep. Sci. Technol.* **30(3),** 383–398.

133. Levesley, J. A. and Hoare, M. (1999) The effect of high frequency backflushing on the microfiltration of yeast homogenate suspensions for the recovery of soluble proteins. *J. Membrane Sci.* **158,** 29–39.

134. Liew, M. K. H., Fane, A. G., and Rogers, P, L. (1997) Fouling of microfiltration membranes by broth-free antifoam agents. *Biotechnol. Bioeng.* **56,** 89–98.

135. McCarthy, A., Walsh, P. K., and Foley, G. (1996) On the relation between filtrate flux and particle concentration in batch crossflow microfiltration. *Sep. Sci. Technol.* **31(11),** 1615–1627.

136. Maruyama, T., et al. (2001) Mechanism of bovine serum albumin aggregation during ultrafiltration. *Biotechnol. Bioeng.* **75(2),** 233–238.

137. Meacle, F., et al. (1999) Optimization of the membrane purification of a polysaccharide-protein conjugate vaccine using backpulsing. *J. Membrane Sci.* **161,** 171–184.

138. Vyas, H. K., Bennett, R. J., and Marshall, A. D. (2000) Influence of operating conditions on membrane fouling in crossflow microfiltration of particulate suspensions. *Int. Dairy J.,* **10,** 477–487.

139. Kempken, R., Preissmann, A., and Berthold, W. (1995) Assessment of a disc stack centrifuge for use in mammalian cell separation. *Biotechnol. Bioeng.* **46(2),** 132–138.

140. Tebbe, H., et al. (1996) Gentle separation of hybridoma cells by continuous disc stack cetrifugation. *Cytotechnology* **22(1-3),** 119–127.

141. Dainiak, M. B., Galaev, I. Y., and Mattiasson, B. (2001) Direct capture of product from fermentation broth using a cell-repelling ion exchange column. *J. Chromatography A* **942,** 123–131.

142. Fahrner, R. L., Blank, G. S., and Zapata, G. A. (1999) Expanded bed protein A affinity chromatography of a recombinant humanized monoclonal antibody: process development, operation, and comparison with a packed bed method. *J. Biotechnol.* **75,** 273–280.

143. Fernandez-Lahore, H. M., et al. (2000) The influence of cell adsorbent interactions on protein adsorption in expanded beds. *J. Chromatography A* **873,** 195–208.

144. Feuser, J., et al. (1999) Interaction of mammalian cell broth with adsorbents in expanded bed adsorption of monocolonal antibodies. *Proc. Biochem.* **34,** 159–165.

145. Hjorth, R. (1997) Expanded-bed adsorption in industrial bioprocessing: recent developments. *Trends Biotechnol.* **15,** 230–235.

146. Lutkemeyer, D., et al. (1999) Estimation of cell damage in bench- and pilot-scale affinity expanded-bed chromatography for purification of monoclonal antibodies. *Biotechnol. Bioeng.* **65(1),** 114–119.

II

MICROBIAL CELL CULTURE

5

Natural Products

Discovery and Screening

Matthew D. Hilton

1. Introduction

The discovery and development of penicillin for human medical treatment ushered in an era of cultivating microbes for the discovery of natural products that continues today. Although many technical details have changed during this era, the fundamental potential remains: nature has developed valuable but yet-to-be-recognized drugs during its 3 billion years of evolution. As long as humans believe this potential exists and that it can be translated into commercial products, there will be a need for scientists and engineers to cultivate microbes for natural products discovery.

Discovery of natural products is the human process of associating a secondary metabolite from nature with a biological activity. Many methods have been used to make these associations, beginning with the trial and error that led to the traditional medical practices of many cultures; the methods evolved tremendously during the twentieth century, with the many advances in basic and medicinal science. Yet even today, natural products discovery is akin to discovery of a deposit of diamonds in the earth. The deposit already exists, but someone must find it and recognize its value to humans for that value to be manifested. Similarly, secondary metabolites and their biological activities already exist; the discovery occurs when humans recognize those pre-existing relationships.

Secondary metabolites are the precious gems of our evolutionary heritage. Traditional medicines have focused on plants, probably because they were obvious and their exploration as possible sources of food would have led to de facto safety and efficacy trials. Awareness of microbes during the last century, along with the commercial success of early antibiotics, has driven the expansion of discovery efforts to encompass microbes and marine invertebrates in addition to continued evaluation of plants, although in a more systematic and compound-oriented fashion. This chapter focuses on cultivation of microbes, especially the filamentous microbes. However, many of the concepts apply equally to plants, with the potential to cultivate plant cells in vitro for the expression of secondary metabolites.

The term "secondary metabolite" stands in contrast to the primary metabolites. Primary metabolites are essential to growth of living cells and therefore are critically important. Primary metabolites include the 20 conserved L-amino acids, the nucleo-

From: *Handbook of Industrial Cell Culture: Mammalian, Microbial, and Plant Cells*
Edited by: V. A. Vinci and S. R. Parekh © Humana Press Inc., Totowa, NJ

sides, vitamins, and carbohydrates that provide for the structure and energy management of life. Secondary metabolites have been described and defined in many ways. The simplest and perhaps most accurate definition is that they are metabolites that are not primary. As a group, secondary metabolites are abundant. They can be found in most—probably all—species of microbes and multicellular organisms. As individual chemical entities, there is tremendous variety among the secondary metabolites, with more than 100,000 known today. With this diversity of chemical structures comes diversity of biological activity. The "purpose" for secondary metabolites has been studied and subjected to extensive rationalization and speculation (for reviews, *see* refs. *1,2*). In many instances, secondary metabolites appear to be involved in differentiation (sporulation or fruiting) of certain bacteria and fungi. In other instances, a secondary metabolite has been shown to "communicate" with another species of organism. Berdy, a renowned natural products scientist, generalized secondary metabolites as the "the chemical interface between microbes and the rest of the world" *(3)*. Attempts to establish the function or benefit of a secondary metabolite to its producer are often rationalizations to justify the energetic burden of the genes and gene products responsible for synthesis of the secondary metabolites. These rationalizations are sometimes productive and sometimes futile. However, they have value in helping to reveal possible places and ways to search for new secondary metabolites.

In major pharmaceutical companies, a common question is "why do natural products discovery?", since synthetic chemistry methods have recently expanded dramatically. Curiously, the public, at least in the United States, has simultaneously embraced herbal or traditional natural remedies (presumably there are active secondary metabolites in these herbals that are efficacious) to an extent not previously seen. Considering the changeable nature of humanity, it is useful to review the basics of natural products to keep expectations realistic and apply technologies with reasonable expectations. The answer to the question of "why continue doing natural products discovery?" can be reduced to two basic arguments. First, because they have proven incredibly valuable in the past, so it is reasonable to continue searching for the undiscovered gems that remain. The first beta-lactam antibiotic was discovered about a century ago, yet the market today exceeds $11 billion per year. All commercial secondary metabolites sold for their antibiotic activity combined constitute a $28 billion market worldwide *(4)*. Other natural-product-based drugs add up to another $10 billion per year. In addition, many synthetic drugs had their structural roots in a natural product. The second argument for continued searching is the preponderance of circumstantial evidence that would compel most to believe there is a great deal more biodiversity yet to be discovered and that their yet-to-be-discovered metabolites could point to activity motifs that cannot yet be imagined by humans *(5)*.

2. Overview of the Practice of Natural Products Discovery

As mentioned in the introduction, natural products discovery is the process of associating secondary metabolites from natural sources with their intrinsic biological activity. Thus, the process can be divided into two convergent paths. One path is the biodiversity source and sample preparation, and the second is selection and development of an assay or screen to distinguish biologically active samples. These are followed by isolation and characterization of the active compound (Fig. 1).

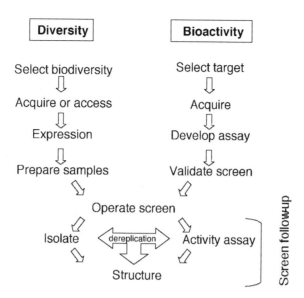

Fig. 1. Major steps in natural products discovery. Natural products discovery is the convergence of two paths. Biodiversity is tested for bioactivity, leading to isolated active compounds in association with their intrinsic activity and chemical structure.

The first steps in both lines are analogous decisions. Specifically, a decision must be made regarding which biodiversity sources are to be put into the screen and which activity target will be used as the basis for the biological screen. A few years ago, targets were relatively short in supply and a limited amount of the world's biodiversity was accessible, so both of these decisions were usually passive. Thus, essentially the same targets and the same general biodiversity—primarily the cultivable filamentous microbes—were accessed by most involved in drug discovery. However, genomics and the recent application of molecular genetic techniques to biodiversity surveys have changed the paradigm. Now we have evidence, in the form of DNA sequences, for far more potential screening targets and biodiversity sources than can be prepared for screens. Thus, the questions of which target, which screening samples and how many samples, have become critical. These represent strategic questions that are likely to determine the success or failure of individual screening programs, and possibly the next cycle of natural products discovery by humans.

The next stage is the acquisition of the physical material required for screening. The acquisition of targets is outside of the scope of this chapter, but today is often done by cloning the DNA sequence or "phone-cloning" the gene that codes for the selected target in an appropriate expression vector and its production in a surrogate host. Depending on the specific protein target, the expression of recombinant proteins can be relatively simple or extremely difficult. Those difficulties can be technical—such as the expression of active integral membrane receptors—or can be a legal hurdle, as when the gene of interest has been patented by someone else. The acquisition of biodiversity has similar technical and nontechnical barriers. Traditionally, acquisition involved collecting biological materials and bringing them back to the discovery laboratory to be catalogued and placed in a library. Those biological materials might have been collected bulk materials such as plant parts, mushrooms, or sponges, or cultivable microorganisms. Cultivable microbes were favored during the last century and continue to present the clear advantage that if a new metabolite of interest is found, the production process can be improved and scaled up to high levels, and production can be in a controlled factory fermentor. However, there are alternatives today. Molecular genetics has made it possible to transfer several genes for a secondary metabolite from its natural producer to a surrogate "domesticated" host *(6–8)*. In theory, the pathways for secondary metabolites of microbes that cannot be cultivated can be transferred to a surrogate host and expressed. A growing body of evidence indicates that this theory can be a reality, at least for some pathways and certain hosts.

Once the key materials for a screen have been gathered, the next step is to develop a process for the effective use of those materials. For the selected target, this means developing an assay that has the potential to be introduced into an automated—and typically, high-throughput—format for screening operations. For the biodiversity source, this means developing multiple methods to prepare the material to be compatible with the screening format. Achieving compatibility is technically complex, and ideally, is a core mission for those who prepare samples and those who develop the screens. Success is dependent on the ability to raise the signal of the active component relative to the noise of the system. Natural materials are chemically complex, and thus it would be very expensive to purify every chemical constituent from every diversity source, but purification would maximize the activity signal-to-noise ratio. Also, more of the metabolites of life are common (e.g., primary metabolites and a few widely distributed secondary metabolites) than special (i.e., rare secondary metabolites). Thus, sample preparation is always a balance between cost, yield, and selectivity. For bulk biological materials, this balance usually translates into solvent extraction and some fractionation. For culturable sources, the program typically seeks to raise the signal, or concentration of secondary metabolites, via cultivation experiments to increase expression of secondary metabolites, and to develop methods for selective enrichment of subsets of these compounds. The improved, cultivated broths are solvent-extracted and are often subjected to some fractionation, since whole broths commonly carry significant chemical noise from residual media components as well as many metabolites produced during the fermentation.

These method-development activities typically evolve directly into the final stages, the "industrialization" of each line before their convergence. New discoveries are rare; thus, large numbers of samples must be run through screens at high rates to be competitive. The assay developed must be translated for high-throughput automated operations, and validated for operational fidelity before it can be considered a "screen." The preparation of samples for screening is usually also done in a high-throughput mode, and is typically automated to whatever extent the sponsor institution can afford.

Prepared samples are typically run through screens in factory-like operations, and those that are active are followed up to determine the structure of the active compound. The factory-like component of high-throughput screening is in the standardization, automation, throughput, monitoring, and control of the process. Much of this is accomplished by implementation of an integrated and automated process. Raw materials go in, in the form of samples, targets, and substrates, and result in product data. The key data are often referred to as "hits." A hit is a sample position with either an inhibition or activation score above the background. The position of the hit then points, via a database, back to a specific biodiversity source and the procedure used to prepare that sample. Factory-like statistical process control is used to monitor screen operational fidelity. Positive hits from the screen results are repeated, to eliminate false-positives, and the source biodiversity is then introduced into a follow-up process. Follow-up is a catchall, since screening samples are each unique chemical mixtures. Even when screen samples have been highly fractionated to minimize complexity, it cannot be assumed a priori that the most abundant compound is responsible for the activity. It is very common for a quantitatively minor compound, perhaps considered a contaminant, to be responsible for an observed biological activity. Thus, it is necessary to follow the screen with an iterative sequence of isolation and assay steps to establish which chemical entity is responsible for the observed activity. The goal of this follow-up is to associate a single chemical entity with dose-dependent activity and chemical structure. This follow-up can be easy when the activity is the result of a relatively common metabolite held in pure form, or if the compound is present in the screening sample at a concentration that is easily detected by spectroscopic methods. This rapid detection is commonly called "dereplication," reflecting the generally valid assumption that most active samples are the result of a "replicate" of a previous discovery. However, the follow-up process can be extremely difficult if the active component is very potent or does not behave as a discrete entity during fractionation. Both can result in spectroscopic invisibility and the need to iterate supply and isolation until the concentration can be raised above the threshold of the spectral detectors. Although such samples may only represent a small percentage of all samples, they often demand the majority of a discovery group's resources, since they initially have exactly the qualities sought in the screen—i.e., apparent potency and potential novelty.

This chapter explores the basics of natural products discovery described in this overview and focuses on microbes as biodiversity sources and their relationship to other components of the discovery process. Natural products discovery involves the detection of unexpected biological activities (signals) from unusual, or secondary, metabolites in a background of biochemical and assay variation (noise). This process is conceptually simple, but complex in practice, because secondary metabolites are found in every living system. Natural products discovery has historically focused on bacteria, fungi, plants, and marine animals. This chapter focuses on two microbial groups—especially filamentous bacteria and fungi that grow in submerged (liquid) culture—because these are the cultivatable groups with significant history and literature. The discussion is divided into key processes for obtaining secondary metabolites from biodiversity and the current practices for getting those metabolites into screens and isolation for structure determination. The discussion is limited to the relatively small scales associated with the discovery stages, since scale-up is treated elsewhere. In recent times, the successes in plant-cell culture and insect-cell culture make it conceivable that in the future, cells of virtually any kingdom could be grown in culture for production of secondary metabolites. A look through the literature on fungi, actinomycetes, and plant-cell culture suggests many parallels in the principles of eliciting secondary metabolites between the kingdoms. The most frequent key trigger appears some form of sublethal stress (*see* refs. *9–12*).

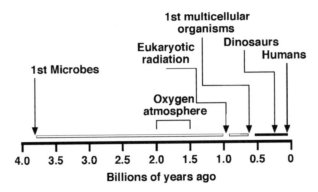

Fig. 2. Timeline of life on earth.

3. Sources of Biodiversity

Active natural products can probably be discovered from anything living. Because of limitations in technologies and history, objective surveys of all life have not been done, so questions about where it is best to look for new natural products are largely speculative. Until the detection of microbes by von Leuwenhoek in the seventeenth century and the first proof of their significance by Pasteur in the nineteenth century, only plants and animal life were recognized by humans, so traditional remedies mostly came from plants. Although the "out-of-sight, out-of-mind" perspective is still maintained by many today, the discovery of penicillin in 1928 from bread mold led many humans to imagine filamentous microbes as sources of new antibiotics. Up to that time, science had focused on microbes as agents of disease, and humankind was virtually unaware of this vast majority of biodiversity until the closing years of the twentieth century. With the advancement of molecular genetics tools of the last 20 years, we can begin to see life in more balance. First, it has become clear that most of the history of life belonged to microbes alone, and thus not surprisingly, most of the diversity of life is microbial (Fig. 2). Second, the majority of life, especially prokaryotic microbes, are not culturable by the conventional methods we know today (*13–15*). This invisible biodiversity, known as unculturable, uncultured, or, more hopefully, not-yet-cultured, are now detected by PCR amplification of a gene, usually the large ribosomal RNA gene, for determination of sequence relatedness to other known species. The conclusion is that most of what is amplified out of many common niches is unknown—i.e., not previously cultured and sequenced. Since the products of PCR amplification do not necessarily quantitatively represent the organisms initially present in a sample, these sequences have been used to generate specific hybridization probes and to quantitate the relative abundance of new and uncultured microorganisms using methods such as FISH (fluorescence *in situ* hybridization (*15*). The pattern observed is that the microorganisms cultivated from natural samples usually represent only a small percentage of the apparently viable microorganisms present. Further, there are more than a million named species, and tens of millions of unnamed species may exist (*16,17*). This is more species than any organization, public or private, can reasonably hope to capture and hold in a collection.

Although most can agree that the majority of life has not been surveyed for production of secondary metabolites, two current issues remain on which there is no consensus. First, how much of life must be surveyed to find the valuable metabolites? And second, how can we obtain representative metabolites (or data indicating their presence and nature) when we do not know how to cultivate most of the population? These questions cannot be answered satisfactorily here, but this chapter describes the current state-of-the-art and some general perspectives. The first question is how much of life (how many of the species) must be screened before we can assume that we have detected most secondary metabolites, especially those that are sufficiently different to have potential medical or other economic value. Some would answer this question from a biological diversity perspective, arguing that tremendous unscreened biodiversity exists, and assume that biological diversity will predict metabolite diversity. This assumption seems generally valid at high levels, at least among the traditional five kingdoms. Thus, plants are very different than fungi and from prokaryotes, and thus each can be expected to make metabolites that are different from one another. However, horizontal gene transfer, especially common among microbes, is now also well-documented between microbes and multicellular organisms *(18–24)*. Plus, microbes associated with a multicellular organism often produce secondary metabolites *(25)*. The extraction of the microbes and their products with the multicellular organism makes the identity of the true producer difficult to determine without tracking down the biosynthetic genes. These factors, which demonstrate the fluidity of genetics and ambiguity of biology, have led scientists to turn their attention to several areas. Some have focused on the diversity of niches and geographies rather than focusing on diversity in evolutionary history. Some have focused on DNA, without attempting to cultivate all the diversity present. Some, including the natural products discovery group at Eli Lilly and Co., have chosen to focus on the secondary metabolites directly. This strategy is descended conceptually from the "chemical screening" of Zahner *(26)*.

3.1. Plants and Animals as Sources of Chemical Diversity

Although not the primary focus of this chapter, macroorganisms are important sources of secondary metabolites. Plants have been better surveyed overall because of their accessibility and ease of classification relative to microbes. Studies have revealed that a large proportion of known natural products are derived from plants. This may appear to be a disproportionately large number when compared to known species, but plants have not been exhausted as a source of natural products. In fact, recent strategies have expanded their potential. Plant-tissue culture, root culture, and hydroponic cultivation are now available for culturing many plants in addition to the traditional greenhouse cultivation, field cultivation, and collection from the wild *(27–31)*. The first methods in this list all have the potential to be done in controlled settings, akin to the process control found in industrial fermentations. Also analogous to microbial fermentation is the opportunity to introduce environmental or chemical stresses in a controlled fashion as elicitors of secondary metabolism *(12,32,33)*. This procedure has the potential to stimulate expression of previously undetected secondary metabolites from well-characterized plant species. As with microbes, not all plant cells can be cultivated in vitro.

Marine invertebrates are perhaps the most heavily studied group in the animal kingdom *(34)*. Researchers often rationalize that since the earth is mostly covered by water, marine biodiversity must offer tremendous diversity if things are roughly proportional. What is certain is that it is generally costly to acquire samples for screening and even more costly to resupply material if a preliminary discovery is made. This is because the samples must be collected from research ships using diving gear, with costs rising rapidly as collection moves further from shore and deeper. And, as with all collected samples put into screening, no guarantee can be made that a metabolite of interest will be present when recollected, even when recollection is from the same species in the same vicinity as the original collection. Insects may be the most accessible major reservoir of new natural products. Some screening has been cited in literature and meetings, but not enough is published to determine its potential impact. It is widely believed that most insects have unique microbes associated with them, thus expanding their value as a readily harvested screening source. Considering the number and diversity of insect species *(16)*, it is easy to imagine a highly productive program being developed to survey the secondary metabolites of insects and isolate the microbes (or their DNA) associated with the insects for a parallel sample preparation and screening program. In theory, the tissue or cells of other animals could also be evaluated for secondary metabolites; however, few studies have reported of this for the purpose of drug discovery. Obviously, the biochemistry community has discovered many secondary metabolites (e.g., endorphins, transferrin, cecropins) motivated by basic biological and medical sciences.

3.2. Isolation and Cultivation of Unknown Microbes

Until recently, most large natural products discovery groups held isolation and screening of new microbes at the center of their program. Microbes were often their sole source of chemical diversity. For many years, isolation of filamentous microbes, including the actinomycetes (filamentous prokaryotes) and filamentous fungi, dominated discovery. The approach to cultivating new microbes drew upon two precedents from traditional microbiology: enrichment techniques pioneered by Winogradski and its most common applied extreme, the selective cultivation of pathogens in clinical diagnosis. The key to enrichment or selection was to identify conditions that strongly favored the group of microbes desired. A sample composed of many microbes could be cultured and ultimately plated, and under favorable conditions, the desired group would represent a significant fraction of outgrowing colonies although they had been numerically minor in the original sample. The type of microbes pursued changed over the years as major discoveries and literature led the community's perspective on where to look. Some popular methods included pretreatment of soils by desiccation to eliminate rapidly growing, ubiquitous "weed" microbes such as the pseudomonads, heat treatment to select for spores, or selection of resistance to a particular antibiotic *(35,36)*. A notable example of innovation resulting from the principles of enrichment was the discovery of the monobactam group of antibiotics *(37)*. This discovery resulted from a conscious effort to explore a new microbial niche.

It is wise to assume that a newly isolated culture will not be axenic following enrichment and plating under selective conditions, even when that culture is obtained from what appears to be a single colony of a single microorganism. Selective media often favor one group of microbes over others, but do not necessarily kill all other microbes. What appears to be a well-isolated single-organism colony may be contaminated with other microbes that have been suppressed by the selective conditions. Typically, to assure axenic culture prior to preservation, colonies from the selective plating are streaked on a nonselective agar medium, and then a well-isolated colony is picked for outgrowth and preservation from this second isolation. Several forms of risk are introduced during these laboratory cultivation steps. At the heart of this concern is that all culture conditions are selective, whether intentionally or not. Conditions called "non-selective" typically select for a rapid-growth phenotype under a specific set of laboratory conditions. Picking one isolated colony (cloning) clearly limits the diversity from a given type of organism found in nature. Cloning two times sequentially—first under intentionally selective conditions—then under incidentally selective conditions is likely to result in clones that no longer represent the original population in the natural sample. It can be estimated that thirty-five to forty generations result from the original sample to preservation (based on the assumption that picking a colony carried a half a million cells forward was followed by two cycles of colony growth and possibly some submerged vegetative growth). This number of generations provides ample opportunity for a spontaneous mutation to appear in every gene, and then for any mutants with favorable traits (i.e., rapid growth in lab conditions) to grow to dominance *(38)*. Although each individual of a particular organism type can be presumed to have the potential to produce the same secondary metabolites, considerable phenotypic variation in expression levels is common. Highly productive microbes are much more likely to yield positive results in screens purely because of the improved signal to noise, but may also be more likely to produce other new metabolites *(39)*.

3.3. Engineered Sources of Diversity

Thus far I have focused on obtaining natural secondary metabolites from naturally evolved organisms. Current technologies facilitate engineering of chimeras at both the genetic and cellular level. A reasonable solution, in theory, for capturing the secondary metabolites of the "unculturable" for the sake of secondary metabolite production would be to transfer the gene cluster for the pathway to some domesticated microbe, or a surrogate production host. The appeal comes from sidestepping the slow process of empirically exploring possible media and environmental conditions to cultivate new groups of microorganisms. These surrogate host technologies have been proven in principle, but not necessarily to the point where we can state what the limits might be *(6,7,40,41)*. A variation on this potential to create genetic chimeras is to fuse portions of different pathways to construct chimeric enzymes and pathways and thus metabolites. This has been done, and has been the basis for the formation of several biotech companies *(42–49)*.

3.4. Issues in Bringing New Cultures into a Collection

Replicate cultures are common when environmental samples are initially plated. Thus, some kind of attempt is invariably made, at this pick-and-streak stage, to take only one colony of each morphological type from a given plating source. After isolation and cloning, two options arise. The clones can be preserved immediately, assuming they are worth the effort, or they can be evaluated prior to the preservation decision. The assumption of worth is common if the cultures were expensive in money to an external provider or cost in interal time and effort. The alternative has been to give each culture an "entrance exam," typically in the form of some screen applied to a sample prepared from a fermentation extract. The screen may be based on biological activies or chemical composition. Only those that pass the test are preserved. In either instance, the major objectives of building a culture library are analogous to those of the traditional library of books. For example, there is a physical inventory, hopefully including contents sought by the library's clients, which must be held safely and indexed so the history and locations of the acquisitions can be maintained and queried.

The physical inventory is usually either lyophilized or cryopreserved to retain viability over long periods of time. Lyophilization of cells, especially as spores, is probably the best-tested method for very long-term storage, but it suffers from the need for much labor in preservation and in resurrection. Furthermore, an ampoule is good for only a single resurrection, so multiple ampoules must be preserved and tracked. The popular alternative is storage in the presence of a cryopreservative, such as glycerol or dimethyl sulfoxide (DMSO), in the vapor phase above liquid nitrogen. Preservation can be rapid, easy, and potentially automated. Most cultures can be preserved either as vegetative cells or spores, and can be thawed, sampled, and refrozen at least several times. The obvious need for vigilance in assuring that the liquid nitrogen dewars are not allowed to run out of liquid nitrogen requires that tanks be filled according to a routine schedule or monitored and filled automatically. Extensive use of cryopreserved cultures results in much greater use of liquid nitrogen, so fill frequency cannot rely on a simple schedule. In addition, safety precautions must be used when large volumes of nitrogen are handled because of the potential for insidious suffocation.

The greatest barrier to preservation of cultures is the lack of physical homogeneity. Typically, this is not a problem if the organisms are eubacteria such as *Bacillus* or pseudomonads, but since it is often vegetative cultures of filamentous microbes that are to be preserved, the lack of physical homogeneity is a frequent issue. Certain fungi that grow as ball-like colonies and reach diameters up to a centimeter under some conditions are especially problematic; these cultures are impossible to pipet and inoculum taken from them are variable. One solution for difficult cultures is to macerate them to a fine homogeneous mix, but this is labor-intensive and introduces the additional risk of contamination during the extensive handling. A better solution, which is more amenable to mass-production operations, is to cultivate problematic microbes in the presence of one of several materials that appear to nucleate more—and thus smaller—growth centers *(50,51)*.

Isolation and preservation of cultures is the start of a collection. Also required is a database to capture the history, and thus potential value of each organism, and to point to the location or locations of the preserved seed stock. The database may be minimal or extensive, depending on the strategy of the organization. A minimal database is likely to include where the culture came from (isolation technique, who did the isolation, culture conditions, or preservation method), when it was deposited, and an inventory of use. At best, the database captures or points to everything known about a culture. This might include taxonomic data, media used in previous fermentations, results of any screens run on the culture's extracts or fractions of those extracts, and any known compounds produced by the culture. If the number of cultures in the collection is relatively small, the database can be managed using a standard desktop PC and software (e.g., Claris Filemaker Pro™ or Microsoft Access™). However, the data should be migrated to a professionally managed database (e.g., Oracle™ software on a backed-up server) when the numbers of entries grows large or where the cost of data loss or corruption would be unacceptable.

The quality of a microbiological collection is crucial to its value. Quality can be divided into microbiological (diverse, axenic, viable, and stable), informatic (accurate, useful information), and practical (the collection is useful). Some measures of biological diversity and usefulness are contextual and perhaps subjective, so these issues are not dealt with extensively here. However, other qualities can be measured objectively *(52)* (Table 1). First, axenicity (pure culture) is a requirement that is not as easily determined as is often believed. The first challenge is that typical sampling methods, developed for clinical labs with eubacterial targets, are not very sensitive. For example, a common microbiological loop used to streak a plate will hold approx 10 µL. If the total culture sampled is 10 mL, that 10 µL constitutes only 0.1% of the sample. It is common for a contaminant to be present at a frequency below 0.1% and thus to go undetected. Further, a cryopreserved stock can be thawed, used, and refrozen repeatedly, but each handling event introduces a risk of new, low-level contamination, as well as propagation of preexisting but low-level contaminants. It is extremely difficult to assure that a preserved sample is not contaminated without extensive or repeated analyses. Practically, it is necessary to monitor contamination until harvest of fermentations derived from the seed stock. During this time, many generations of growth may bring a contaminant's frequency up to detectable levels. Then again, if detected at harvest, the source may or may not have been the seed stock. Although it is often impossible to unequivocally determine the original source of a contamination event, routine monitoring at each use of a seed and after any expansion steps is justified to help diagnose contamination events.

4. Fermentation of Microbes for Screening

Up to this point, the discussion has focused on making diverse microbes available to grow in culture for a natural products discovery program. The next challenge is to select which organisms and conditions to use for cultivation to prepare samples for screens. Cultivation at this stage has a different objective than cultivation for preservation. Although viable cells or spores are the goal of cultivation for preservation, the objective here is the expression of secondary metabolites. Culturing microbes for production of metabolites is commonly referred to as fermentation. This common modern usage of the word "fermentation" is different than Pasteur's original use to denote cultivation in which energy for growth is derived anaerobically. Industrial fermentations today are mostly aerobic.

High cell mass is a prerequisite for significant metabolite production in fermentations, so some type of vegetative culture or growth stage to expand cell mass always precedes fermentation. Since secondary metabolites are not essential for growth, they are usually not produced during the more rapid, vegetative growth stage. Rather, a stage of slow growth following the exponential stage is often necessary to accumulate secondary metabolites. This slow-growth stage, sometimes called the idiophase, may not result in any detectable increase in cell numbers in some cases. Processes in which both vegetative growth and slow growth occur in a single vessel can be devised. Such processes often constitute the final process stage of an optimized fermentation. These concepts of a growth phase and a production phase are central to fermentation, but the transition between them may or may not be discrete and obvious.

Table 1
Issues to Consider in Introduction of New Cultures
in a Biological Diversity Library

Physical inventory
    Representing targeted diversity
    Best producers of each biotype
    Genetically pure (clonal)
    Stable preservation and method for reproducible outgrowth and performance
    Unique identifier (bar code)
Database information
    Source location
    Isolation method
    Who isolated the culture?
    Any taxonomic information
        Kingdom (minimally)
        Genus species (ideally)
        16s sequence
        Taxa-specific hybridization pattern results
    Any other information
        Third-party supplier and any linking identifiers
        Any restrictions on use of the organism or its products
            Contractual
            USDA
            International biodiversity treaty
            Known pathogen (animal or plant)
    Preservation method
        Vegetative culture-medium composition
        Culture duration and conditions
        Harvest stage and methods
        Cryopreservatives/cryoprotectants included
    Link of bar code to a location for retreival
        Tank
        Rack
        Box
        Position
Insurance against failures
    High temperature alarm
    Back-up copy of collection (ideally at a second physical site)
    Limited access to facility
    SOPs and training for use of libraried samples
    Check-in, check-out use tracking

Fermentation conditions are best developed by initially matching theory with the process objectives to limit the number of possible combinations, followed by empirical optimization within the process design limitations *(53,54)*. In an ideal world, the empirical experiments and the fermentation process would be tailored to each organism individually. In practice, this can be justified once a microbe is known to make a

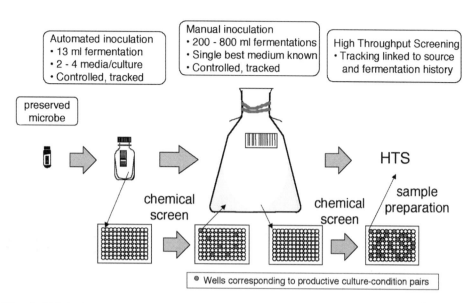

Fig. 3. Lilly's integrated microbial culture and fermentation condition evaluation scheme. Each microbe is stored in a bar-coded cryotube. Each transfer is tracked by time-stamp linkage of the bar codes of the source and recipient vessels. A subset of the 13-mL fermentation culture-condition pairs are selected based on a rapid chemical screen *(58)* for scale–up to flask fermentations. The flask fermentations are reanalyzed by a second chemical screen to confirm productivity upon scale-up and for data archives.

valuable product, but at the discovery stage some type of one-size-fits-all or grouping strategy (analogous to small, medium, and large options in clothing) is pursued to manage costs. One approach is to use clusters of microbes based on taxonomy and to assume that they have common nutritional requirements and triggers of secondary metabolism. An extreme that is not fully practical is to assume nothing, but to evaluate each organism under many cultivation conditions for their expressiveness. A significant challenge here, beyond doing all of the fermentations, is choosing indicators of secondary metabolism and obtaining the required assay capacity. Pigment formation or bioassays may serve as indicators if the scope of the discovery program is limited. Bioassays, especially of relatively crude extracts, suffer from many uninteresting positive results (false-positives), so the results can potentially lead to optimization of conditions for uninteresting outcomes. So called "chemical screening" *(26,55,56)* is another option that is growing in favor with advances in technology *(57,58)*. These measures of goodness can be applied to the selection of which microbes to favor in screens and also to guide the development of conditions for eliciting secondary metabolites *(59)*. Lilly scientists assembled an integrated strategy for simultaneously evaluating cultures and conditions to identify the rare, highly productive culture-condition pairs (Fig. 3).

Fermentation media evaluation and development for efficient production of compounds typically starts in one of several places: precedents from research experience with similar microorganisms, or theory. Studies of specific organisms and products are extensive, and these methods are summarized by Connor (*see* Chapter 7). Development of media and conditions to produce yet undiscovered secondary metabolites is more difficult, because the producer organisms and their product compounds are unknown. Thus, experience, theory, intuition, and empiricism are commonly employed. When working with many unknowns, it is reasonable to start with a few assumptions. First, cultures rarely produce secondary metabolites under conditions that are optimal for growth. Second, cultures must have continuous supplies of energy and substrates for biosynthesis. These two concepts dictate the fermentation media. Or, from another perspective, the cells typically must be growing, but under some type of stress that results in suboptimal growth. From the medium composition side, that stress can be a limitation of a key nutrient. This type of stress is simple to manage in standard industrial fermentations that allow a limiting nutrient to be fed at a controlled rate, but a reproducible nutrient limitation is much more difficult to manage in the small shaken-batch fermentations typical of a natural products discovery laboratory. Although phosphate and nitrogen availability are most easily limited, it can be tricky in practice if the secondary metabolite has nitrogen or phosphorus in its structure. Carbon limitation can work, but first principles dictate that a continuous source of carbon must be be available as substrate and energy supply (unless the organism is an autotroph such as an alga). In practice, the use of an oil such as soybean oil has proven very effective in supporting secondary metabolism. The oil apparently limits growth, since it is present in excess, through a kinetic limitation. In theory, trace elements can be limiting, but the reproducibility of such a limitation is difficult to manage in practice because glassware and dust can contribute significantly to trace-element availability.

The concept of elicitors is more popular in obtaining secondary metabolites from plants *(10)*, but applies to microbes as well. The conceptual distinction is that an elicitor is a trigger compound, possibly specific to the producer organism. Exogenous addition of a known signal of secondary metabolism, such as A-factor *(60)* or the somewhat analogous quorum-sensing factors associated with other secondary traits *(61)*, may be potential elicitors. There are reports of cell-wall material from microbes acting as elicitors of plant secondary metabolism *(10)* and enzyme expression in bacteria *(62)*. It could be rationalized that the target of an antibiotic might elicit the production of that antibiotic.

Many physical parameters, such as temperature, shear, and oxygen transfer rate are possible triggers for the production of secondary metabolites. In the gray zone between medium substrate changes and physical parameters are media ingredients that affect the physical environment of the cells, such as osmolarity, water activity, and ionic strength. In practice, such parameters are often tested empirically without worrying about the mechanism of action (MOA). Generally, several media are used per microbe screened simply because once the inoculation step begins, inoculating two or more media is not much more costly than inoculating a single medium. The key of the screen sample-generation stage of a discovery program is to find a few specific media and incubation condition pairs that work for as many of the available microbes as possible. Ideally, when several conditions are run in parallel, they will elicit secondary metabolism from different organisms and complement one another. However, extracts from a single organism grown under several conditions often result in the same activity profile, so it is always a balance between the number of organisms and the number of conditions per organism *(63)*.

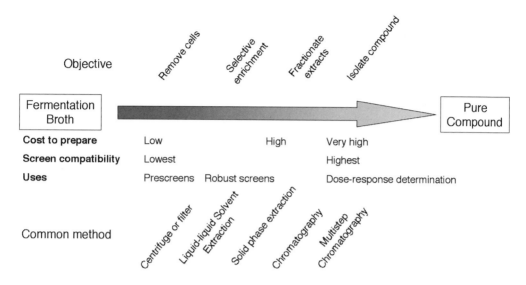

Fig. 4. Sample preparation and compound isolation are a continuum that requires decisions to balance the requirements for the intended purpose against costs. Often, it comes down to quantity vs quality.

5. Harvest and Sample Preparation for Screening

As with the previous steps, harvest and sample preparation can be minimal or complex, depending on the specific objectives (Fig. 4). The minimum would be either cell removal by centrifugation or filtration, direct extraction of the broth with a solvent, or some combination. The maximum would be isolation and recovery of individual compounds. Today, most natural products discovery groups choose one of the many intermediate routes to allow for more preparation to achieve compatibility with an increasingly diverse array of screening targets while simultaneously meeting the increasing pressure to follow-up quickly on screen hits. All of this is balanced against costs. In practice, researchers agree on the need to fractionate to purity when a compound is proven to be of interest, but they usually cannot agree on how to decide what is of interest and how much preparation is required to identify biological samples with compounds of interest present. Therefore, it is not uncommon for the cost of the original sample to drive the extent of preparation investment. The underlying assumption seems to be that if it is rare or costs more, it is worth the investment. The best strategy today appears to be a compromise with minimal preparation for a prescreen that is designed to point to the valuable samples that receive extensive preparation.

The basic goal of sample preparation is to get secondary metabolites into a state in which they can be detected (discovered) in a screen (Table 2). When it is assumed that the available biodiversity has discovery-worthy compounds, the exercise of asking what would prevent discovery of those compounds can be reduced to the signal-to-noise ratio—either the concentration of compounds is too low (weak signal) or there are too many compounds per sample (high background or noise). The basis for the former is self-evident, since all assay methods have limits of detection. Complex samples sometimes prevent discovery because the sum of all compounds limits further concentra-

Table 2
Typical Objectives of Sample Preparation

Stop further biological processes
Concentrate compounds of potential interest
Selectively enrich compounds of potential interest
Eliminate nuisance compounds and other interferences,
    especially cell mass and biopolymers
Minimize losses of compounds of interest
Minimize sample preparation costs
Stabilize compounds to chemical changes

tion, thus keeping the signal low. Alternatively, with many compounds present there may be a few nuisance compounds, such as the ubiquitous fatty acids, in the same sample as a compound with real potential. The nuisance compound is detected first in a process called dereplication and further follow-up would be stopped, although there may be multiple root sources of activity in the sample. The fatty acids are a particularly troublesome nuisance class because low concentrations of individual fatty acids are not likely to be active in screens, but if enough different fatty acids are present, they can act together to create the appearance of activity (false-positive). These two root causes are difficult to dissociate in practice. Purification generally dilutes samples, creating a need to concentrate. Conversely, without some selective isolation, further concentration is often very limited because of the high nonvolatiles (polymers such as starch, cell-wall materials, proteins, or nucleic acids) content commonly found in fermentations. Although both of these could be optimized for samples for which the compound of interest is known, the very nature of early-stage discovery is that you cannot know what that compound is. Thus you cannot optimize except on a probabilistic basis. Even the management of probabilities is dependent on rationalizations and assumptions.

Selective enrichment of some compounds over others, followed by a concentration step, is at the heart of essentially all sample preparation methods. Of course, the word "selective" means that although some compounds are enriched, others are reduced in quantity and may be effectively eliminated. Although this is the objective, it is important to remain aware of the selectivity of the specific methods as they are practiced, since it is often not the textbook ideal. Basic sample preparation often starts with centrifugation to remove cells and any other particulates; these are invariably not the objective in a drug-discovery program. Centrifugation is noninvasive and relatively economical in materials, but is labor-intensive and not easily automated. Filtration can sometimes substitute for centrifugation, but typically only works for some media or some cultures because others rapidly clog the filter and render the step untenable. If clogging is not a common problem for a set of cultures, filtration can generally be performed in parallel or be automated, thus reducing the cost of labor in the step. Both of these size-based separations can be done directly on the entire fermentation broth, and will result in enrichment of excreted, and generally water-soluble compounds in the supernatant and discrimination against cell-associated or otherwise particulate-associated compounds. This discrimination can include hydrophobic compounds at high concentrations, since they may form micelles or aggregates that behave like particles.

An alternative to immediate particle removal is to first add a solvent to the whole broth in an attempt to move more compounds into the soluble fraction. A water-miscible solvent such as methanol or ethanol can modify the solvency modestly and aid in a liquid-solid extraction of some compounds from the cells. Yet a water-miscible solvent adds volume, thus diluting soluble compounds. Water-miscible solvents are generally less selective than immiscible solvents; a possible virtue in minimizing prejudice against any class of compounds, but also a liability because some proteins and other macromolecules are not excluded. Solvents such as ethyl acetate or butanol do not fully mix with the aqueous layer so they can be considered to perform both liquid-liquid extraction and liquid-solid extraction to pull compounds from cells. These solvents are intermediate in selectivity, but this selectivity depends greatly on the pH of the broth because many compounds can be hydrophobic when they are free acids or free bases, but hydrophilic when ionized. Many more-hydrophobic and more-selective solvents are used alone or as mixtures. All organic solvents forming two layers will suffer from the pH-dependent-selectivity mentioned here. Although this can be used to great advantage when a specific compound is being isolated, it creates a strategic challenge in generic discovery preparation methods. Extraction with organic solvents often results in formation of emulsions, especially if the medium make-up includes any oils. Emulsions make recovery of either layer inefficient or even impossible. Centrifugation can help minimize the volume consumed by the emulsion but it adds labor.

Solvent extraction has been successfully automated, but typically depends on samples that generally form minimal emulsions. We have had some success with use of a nylon sieve to retain emulsions, thus allowing an ethyl acetate extract to be withdrawn using a robot programmed to a constant depth. Concentration of organic extracts is easy by evaporation if they have a high vapor pressure at ambient temperatures. Reducing the pressure around the samples can accelerate evaporation, but this reduction must be done with caution to avoid boiling. Elevation of the sample temperature is another option, but it carries the risk of increased degradation of chemically unstable compounds. The ease of evaporation is undermined if the sample volume is very large because the surface area is small compared to the liquid volume; this dictates long evaporation times or more effort and use of special equipment such as a rotary evaporator.

The next alternative with more selectivity is solid-phase extraction *(64–66)*. In essence, this is crude chromatography. The sample is loaded, those compounds that interact with the resin are retained, and those that do not significantly interact are washed through. Conditions are then changed in the wash buffer or solvent to "bump" the bound compounds off. This can either be a simple load, step-elute sequence, or multistep with a series of intermediate steps before the highest elution-strength strip is reached. As more and smaller steps are introduced, and the resin is packed in a column with a higher length/diameter ratio, the method eventually becomes chromatography. The distinction in when a method is considered chromatography is a combination of intent and performance. The choice of resin and solvents is diverse and common with chromatography. The two are always chosen as a system, since the nature of compound interactions with the resin will determine which solvent conditions favor binding and which favor release of a given type of compound. Thus, the mobile and stationary phase pair determines selectivity. Hydrophobic interactions, as found with C-18 activated resins and many proprietary polymeric resins, are popular stationary phases for natural products sample preparation. These hydrophobic resins are commonly used with methanol or acetonitrile as solvents. These systems are believed to be effective in binding compounds with drug-like polarity characteristics. As with liquid-liquid extraction, the pH of the system also influences the selectivity.

The full array of cation- and anion-exchange resins, special resins, and associated solvent systems that are available for chromatography can potentially be used for solid-phase extraction. Each has advantages for certain classes of compounds. Many resins can be obtained prepacked in small cartridges that are designed explicitly for sample preparation, although they are often developed and marketed for purposes other than natural products discovery. Automated solid-phase extraction systems are also marketed. The commercial devices are often limited in their target yield, so if more extract is needed, custom systems or parallel operations and combining effluents may be required. Functionalized membrane discs, rather than packed bead resins, may provide an alternative to packed resins and may—based on some theoretical advantages—come into frequent use in the next few years *(67)*.

Today, chromatography is the most extensive method of sample preparation and is used to separate a natural extract into many fractions. Although a chemist's wish is a collection of samples with only one compound per sample (and, of course, known structure, 100% purity, high chemical diversity…), that wish is typically unfulfilled because of cost. Chromatography, the method of choice for fractionation, was historically reserved for isolation of compounds of interest (i.e., shown to be active) but was not considered for sample preparation. The high cost of using chromatography can only be justified if the samples are believed to be of high value. Speculation about how to assess the value of unknown natural products extracts is common, sometimes exciting, and often controversial. What is generally agreed is that chromatographic fractionation can significantly improve the handling characteristics of samples in screens. First, it allows concentration of compounds many-fold above their initial concentration; this is often enough to allow compounds present to show their activity reproducibly. Second, nuisance compounds are generally localized during chromatography to one or a few fractions; this fractionation frees many other compounds from the presence of the interfering compound.

6. Screen Follow-Up

When samples are prepared and a screen is validated, screen operations begin (Fig. 1). Active samples, or hits, from the screen operations are routinely reassayed, and those samples found to be reproducibly active are put into a queue for screen follow-up. Screen follow-up is a process of determining the factors responsible for the observed activity. The hope is that a single novel chemical entity with high potency is responsible; often the activity is the result of a compound that is already known. If a known compound is seen often or is very nonspecific, it is considered a nuisance compound. Thus, the first step in follow-up is typically fractionation of the active sample—e.g. by high performance liquid chromatography (HPLC)—and bioassay of the resulting fractions. The goal is to associate a spectral handle of the active compound with the bioactivity. Spectrophotometers (UV and visible ranges) and electrospray mass spectrometers are the methods most commonly used to monitor compounds in the HPLC effluent. This step is often called dereplication, because the first action after associating an active fraction with a spectrum is to search commercially available databases for a spectral match. If found, it may indicate that the active compound is already known and this sample is a replicate discovery. These samples would typically be set aside with no further follow-up.

Some subset of the hits typically pass dereplication, either because the spectrum observed is unknown or because the signal is too weak to assign a spectrum to the active compound. In either case, the solution is most often a repeat fermentation for production of more sample material, so more active compound can be isolated. At this stage, the primary objective is simply to reproduce the prior fermentation and preparation that produced the active screening sample. However, since the sample has passed repeated tests and did not appear to be a replicate on first analysis, this repeat fermentation should be performed with an eye toward improved productivity.

6.1. Productivity Improvements for Screen Follow-Up

Fresh cultures isolated from nature are generally sought for generation of screening samples, but cultures from nature often produce secondary metabolites at low levels that become even lower during further passages of the culture in the laboratory. In many instances, that level of synthesis is inadequate for the immediate or anticipated needs. Sometimes it is so low that initial isolation cannot deliver sufficient material for unambiguous structural and activity determination after the initial screening hit is identified. If interest in a new compound continues after its structure is known, then a never-ending need for higher productivity tends to develop. Two complementary approaches are available, akin to nature and nurture. Specifically, the genetics can be improved with regard to expression using traditional empirical means or genetic engineering (Fig. 5). Or, expression can be improved by experimentally determining, then controlling, the critical environmental parameters that influence expression of the secondary metabolite (nurture). Both approaches are well-precedented and successful if done well. The two can be done iteratively, in parallel or in series. The method that is most rapid or requires least effort depends on many specific factors, and thus cannot be stated as a generalization.

Empirical screening, which requires little background knowledge of the specific organism, is often the favorite starting place. However, the prerequisite is a method to obtain clones after mutagenesis and a means of quantifying the product compound from many small-volume cultures. Cultivation to evaluate clones can be solid-state (e.g., Petri plates) or liquid, submerged culture at early stages of strain improvement. Gains usually occur in relatively large steps early, e.g., several-fold per mutagenesis and selection, when absolute productivity is low. This stage, in which large increases are possible, tolerates assay systems with high variability. For a product with antibiotic activity (and many compounds discovered for activity against a human enzyme or receptor have antibiotic activity against some microbe), crude bioassays can be very effective. For example, the sensitive microbe can be sprayed or overlayed in a soft agar over the top of the colonies of the mutagenized producer organism. Mutants are detected by the size of the zone of inhibition, although it is important to remember that the zone volume (not diameter) will be proportional to concentration. This detection scheme may seem relatively crude in the high-tech world of today, but it is often extremely effective and fast. As productivity increases, the sensitivity of the assay can be reduced

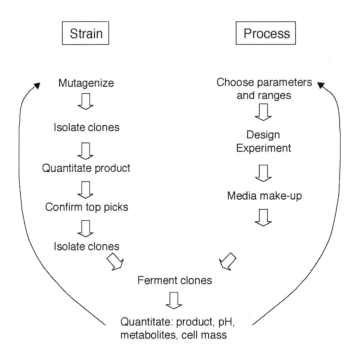

Fig. 5. Improvement of production of secondary metabolites can be nature (strain development) or nurture (media development and process control).

through the use of higher quantities of the sensitive indicator strain or use of a less sensitive indicator strain to extend the number of cycles before sophisticated analytical methods are needed. A liquid fermentation analog to this simple zone-of-inhibition assay system is akin to traditional antibody-titer determination. Whole broth from liquid cultures of mutant clones of the producer organism are delivered to microwell plates and diluted in a series. The sensitive-indicator strain and growth medium are added and the plates are incubated. Higher producers will inhibit growth through more steps of the dilution series. Although more effort is required than for the agar zone-of-inhibition method, these steps are very compatible with the readily available liquid handling equipment available for 3" × 5" microwell plates, so the method is easily automated.

7. Recent Trends in Natural Products Discovery

The fundamentals of natural products discovery never change, but the technology and politics-of-the-day change continuously. The last few years have brought tremendous technological change. Advances in automation and mass spectroscopy have had the greatest impact on natural products discovery. Use of isolated macromolecular targets in screens, sometimes called mechanism-based screening, has broadened natural products screens beyond antibiosis over the last two decades. Genomics has become a major, although indirect, force on natural products discovery by making many more targets available for mechanism-based screening. Microbial genomics has begun to be a direct force, through expanded human understanding of secondary metabolism in an ever more complete genetic and physiological context. It has only begun to deliver on the natural extension of this understanding, a field known as metabolic engineering. On the political side, pressure from intense competition plus worldwide pressure to limit drug costs have combined to drive an industry-wide push for much faster drug discovery and development.

The drive for higher and higher throughput discovery has affected natural products discovery in many ways. First, the supply of samples of synthetic compounds greatly increased during the 1990s because of combinatorial synthesis *(68–71)*, so demand and associated perceived value for all samples was diminished. Second, natural products discovery has had a history of being very slow, often taking a year or more from initial hit to confirmed active compound with structure. This slow discovery rate may be attributed to the fact that natural products come as mixtures of compounds at various and unknown concentrations as well as our hope to find the unexpected. These conditions dictate that researchers must be very rigorous in associating unknowns with activity. Some organizations have concluded that the investment in money and time for this rigor is too great, and have ended their natural products discovery business. Others evaluated process requirements and modified workflow to achieve greater speed and effectiveness. The most common modification in process was prefractionation of natural products samples to create less complex (fewer compounds per sample) than traditional natural products screening samples; effectively, purification was begun before the screening step. Since prefractionation is expensive and throughput is often limited, another response was to select a subset of samples for prefractionation to maximize their value. The following sections provide brief overviews of these emerging drug-discovery technologies and their relationship to natural products discovery.

7.1. Combinatorial Chemistry

Compounds synthesized in the laboratory have always been an alternative to natural products samples as a source of screening samples. The synthetic compounds were often analogs of known drugs, whether synthetic or natural at their hearts (the chemical heart of a compound is called a pharmacophore), and they were synthesized as by-products of activities to establish a structure-activity relationship, commonly called an SAR. Since they were intentionally related compounds, the general view was that it was adequate to screen a few representatives from each series against a new target. Their synthesis was typically iterative, beginning with a core pharmacophore and an SAR hypothesis leading to synthesis of an analog or series to test the hypothesis. Based on assay of that series, a new hypothesis was developed and a new series was synthesized. This process could be repeated through many cycles. That same basic process is still done today as part of a discovery stage often called "lead optimization." However, today's SAR usually involves more parallel, or combinatorial, syntheses for speed and more computational methods to guide hypothesis evaluation and generation between each round of synthesis.

In the early 1990s, the concept of parallel or combinatorial synthesis, also known as combinatorial chemistry, was introduced (for review, *see* refs. *69–71*). This crucial development resulted in much greater efficiencies in synthesizing new compounds. With combinatorial methods, one chemist could synthesize hundreds or thousands of compounds per day rather than just a few. The view emerged that these methods could enable an unlimited number of compounds to be synthesized for screening. There have been many discussions—often heated—of the benefits and limitations of these methods in creating drug discovery potential. Although these are not reviewed or debated here, it is important to note that combinatorial chemistry and the promise of unlimited numbers of cheap screening samples has had a tremendous impact on drug discovery in general. That promise of unlimited combinatorial samples has affected natural products discovery by virtue of the intrinsic competition between the two sample sources for screen developers' attention and screening capacity. Foremost, combinatorial synthesis has fueled a desire for much higher throughput in screening operations.

7.2. High-Throughput Screening

Screening today is almost entirely considered high-throughput screening (HTS) or ultrahigh throughput screening (UHTS) because data points are generated at rates that are 10- to 100-fold higher than just a decade ago. The advances in HTS were dependent on advances in automation and miniaturization and were driven by two orthogonal technological developments—combinatorial chemistry and genomics—combined with an increasingly competitive pharmaceutical industry. Combinatorial chemistry was producing much larger numbers of samples for screening. Genomics was expected to deliver human genes to science; each gene product was imagined as a possible new site of drug action and thus was a potential screening target. The view was that whoever could put a target into a screen fastest with the most compounds would win. This combination of an expectation of unlimited targets and unlimited samples motivated a desire for very high-throughput screening. Advances in screening technologies made this a reality. The key factors allowing higher screening rates were reduced volumes, simplified assays requiring fewer operations, and automation to include parallel operations. Most of the required inventions existed for reducing volumes and automating. Those efforts settled on the 3-inch by 5-inch plastic multiwell plates that historically contained 96 wells in an 8×12 matrix and which were first used for antibody titer determinations. The miniaturization trend has continued toward more and smaller wells but has retained the $3" \times 5"$ footprint allowing standardization of many robotic operations. Moving up by factors of four, 384-well (16×24 wells) and 1536-well (32×48) plates are now common, with the volume per assay dropping in inverse proportion to the number of wells per plate. These sample volumes are now in the 10s of microliters, and progress continues toward microliter-scale assays.

The second key to high-throughput, simplified assays has depended on significant innovation. Most striking has been the move to "homogeneous assays." This term reflects the fact that the signal can be measured without separation of the reactants. For example, although traditional assays often required mixing, binding, filtering, washing, or eluting and reading, the homogeneous assay typically only requires the first and last steps (72). The inventive portion was the development of methods to detect when two components were in close proximity. This was done by taking advantage of several physical realities. For example, Amersham™ commercialized the scintillation-proximity assay, which depends on scintillant in beads coated with the target protein (73). A known ligand is labeled with tritium because of the short range of its emitted beta particles; this short range is what makes it a proximity assay. The target-coated beads and tritium-labeled ligand are added to all wells. Screening compounds or samples are added to individual wells, and activity is detected by the reduction in scintillation emis-

Table 3
Homogeneous Assay Methods

Method	Comments	References
Scintillation proximity	SPA, scintillant in beads or plate bottom	*(73)*
Fluorescence polarization	FP, depends on the difference in kinetics of molecular rotation	Reviewed by *(80)*
Fluorescence resonance energy transfer	FRET, depends on resonance coupling of 2 flours in close proximity	*(81–83)*

sions, as the tritiated ligand is prevented from close proximity to the beads because of competition from active compound in the screening sample. Analogous proximity detectors that depend on fluorescence quenching, fluorescence depolarization, or fluorescence resonance energy transfer have shown similar value in allowing very simple operations and reproducible screening results (Table 3).

7.3. Prefractionation

Although natural products discovery has not driven the major drug-discovery trends described here, it has responded to the wave. Historically, the discovery screens were developed for natural products extracts and the natural products follow-up process was integrally linked to the screening operation in what was descriptively termed assay- or activity-guided isolation. This era effectively ended with the advent of high-throughput screening. These screens were developed for high-throughput foremost and often with synthetic compounds in mind, sometimes assuming one compound per sample. High-throughput limits operational flexibility, so iteration and variation became more difficult, often driving HTS and follow-up assays into separate organizational groups. Plus, the genomics-driven goal of a continuous stream of new targets has made going back to previous screens after one cycle of high-throughput operations more difficult to schedule. These realities, and the general focus on speed in discovery, have prompted a change in the operation of most natural products discovery groups. Overall, the major trends have moved toward more fractionated natural products samples presented in automation-compatible formats that allow them to behave more like synthetic compounds in HTS. A concomitant trend has emerged—to be more selective about which samples are put forward for fractionation, since the additional fractionation drives up the cost of preparation substantially.

The clearest public example of the automation of fractionation has been the development and marketing by Analyticon™ of a series of automated chromatography systems. The most complex is sold as the SepBox and is primarily intended to deliver a generic method to isolate most compounds *(74)*. The concept is to deliver a turnkey or generic solution for fractionation of complex mixtures. Since a single method is used, compounds can often be isolated to high purity in about 1 d, saving the time of methods development and associated delays for bioassay or other analytical data required to develop a customized isolation procedure. Although systems of various complexities are sold, the most complex system is based around dual-stage reversed-phase chromatography. The eluate of the primary separation-column gradient is diluted in-line with water and switched according to a time-based program between a series of reversed-phase capture columns. The compounds thus captured are then batch-eluted and diluted to begin a second resolution column that is selected to correspond to the polarity range of the primary column elution time (and thus solvency) window. The first column provides initial fractions divided into polarity windows, and the second column can often resolve individual compounds. Once loaded, the process is automatic, which minimizes labor and time. In contrast to traditional activity-guided isolation, no assays are done until the system has fully fractionated a bulk sample. A "light" version based essentially on a single-resolution column has been marketed and can be expected to provide samples that are immediately suitable for high-throughput screening. Although Analyticon has done the development work to integrate the two-stage system, all parts of the system are generally available, and other chromatography systems can be assembled to deliver similar outcomes or customized to a specific scale or strategy.

7.4. Selection of Cultures and Samples for Screens

As natural products discovery programs grow, at some point the availability of cultures far exceeds available time or resources for fermenting and preparing screening samples from them. Prioritization or selection of the best cultures therefore becomes essential. This need has become more common because many discovery programs now include extensive and expensive fractionation of the fermentation broths as a method for sample preparation. The significant investment associated with fractionation justifies careful selection among available fermentation broths and the evaluation of strategic need for fermentation broths relative to other available sources of biodiversity (e.g., plants or marine invertebrates). The big—perhaps the billion-dollar—question is: what criteria can the selection be based on? Discoveries are rare and appear to be random. Some actions may improve the probability of, but not determine, new discoveries. So—what makes sense?

Maximize source diversity to maximize potential chemical diversity. Although there is clearly not a one-to-one relationship between taxonomic, geographic, or ecological diversity and the chemical diversity from the organisms, it seems probable that the range of chemical diversity will be broadened with greater biological source diversity, however measured. However, the price is high, as each new source will require an investment in learning to cultivate and elicit secondary metabolism. Also, new groups—regardless of how they are defined—are typically not monolithic in their conditions for optimal growth and expression of secondary metabolites. Thus, it is not a single optimum that is sought, but the association of subpopulations with their respective optima.

The trend toward greater fractionation in sample preparation has two opposite impacts on the number of samples available for screening. First, it tends to raise the number of screening samples per biodiversity source, since one is split into many. On the other hand, fractionation is often relatively slow and costly, so that the number of biodiversity sources (species and strains) processed may be lower than previous rates when the strategic focus was on more and diverse organisms as natural product sources, and when only limited effort was put into preparation. Obviously, having diverse sources is critical, since preparation cannot introduce new compounds, but it can only help unmask them. However, the need for extensive preparation can become rate-limiting and thus discovery-limiting. The responses to this have been mixed, but a favorite solution has been to introduce a prescreen or "entrance exam" for potential biodiversity sources. Prescreens can be biological assays. However, targets change with time and are not predictable a priori, so several groups have developed prescreens based purely on evidence, specifically spectral evidence, for the presence and diversity of compounds in biodiversity sources *(57,58,75)*.

Zahner *(26)* pioneered the concept of screening for unique compounds first, and testing for biological activity only when purity was relatively high and the concentration of the compound was known. He called the strategy chemical screening. The basic tool used when this research was initiated was thin-layer chromatography, which was limited in information content and throughput. However, advances in mass spectroscopy—particularly electrospray MS because it can be used as a detector for liquid chromatography—in synergy with advances in information technologies in the closing years of the twentieth century—set the stage for remaking chemical screening at another level. We put together a particular combination of these tools to evaluate many biodiversity sources rapidly, then selected a small portion of the most productive and chemically diverse for fractionation. The ultimate effectiveness of this strategy is still unkown.

7.5. Access to the Genetic Potential of Uncultured Microorganisms

One alternative to this traditional microbiological strategy is the idea of gaining access to the secondary metabolic capacity of many organisms without cultivating them. Thus, the genetic cluster responsible for secondary metabolite production can be moved to a domesticated host and expressed. This idea clearly works at the individual gene level, since many genes have been cloned and expressed in surrogate hosts in the biotechnology field. It thus seems probable that the same can be done for a pathway responsible for synthesis of secondary metabolites. The open questions are: how difficult will it be and how frequent will success be? Expression of single recombinant genes often occurs through use of a chimeric gene, where expression signals natural to the surrogate host are combined with the amino-acid-coding region of a well-characterized foreign gene. In the case of natural products discovery, advocates of this strategy are proposing the cloning of entire pathways, where expression signals native to the pathway would need to function at some minimal level in the surrogate host. These pathways are commonly clustered on 20 to 100 kbp of DNA, with dozens of genes and complex regulatory systems that involve both global and specific regulatory mechanisms. Many secondary metabolites are toxic; thus, resistance genes are often clustered with the synthetic genes and must be expressed in balance with the expression of the toxic metabolite's synthesis. Although all of this seems entirely possible, the question remains: how probable is it on a general basis? I don't have an answer. I do hope the groups pursuing these strategies are successful, since the appeal is tremendous. The approach could potentially make all of nature's secondary metabolites available for testing in the same era that the human genome is opened to our view as potential targets of drug actions.

Several tactics have been pursued to capture the genes of diversity in surrogate hosts. The most advanced seems to be the construction of libraries from DNA isolated from samples gathered from unusual niches without culturing any microbes from that niche.

That sample could be collected material, such as soil, sediment, or scrapings from local surfaces. One clever variation is use of glass slides as microbe traps that are placed *in situ* and later collected for extraction of the DNA of the microbes that attached. This method takes advantage of the natural tendency of many microbes to adhere to any available surface, and has the advantage of delivering a relatively clean and reproducible sample for the cell-breakage and DNA-isolation methods required. This is especially important because the objective is to clone fragments as big as—and ideally much bigger than—the desired gene clusters. Since clusters of 100 kb are precedented, this forces vectors into the artificial chromosome size range, and demands special methods to protect the DNA extracted from shear or degradation in any of the process leading up to ligation and transformation of the surrogate host. A very different tactic within the same strategy has been to use protoplast fusion of unknown microbes to domesticated partner protoplasts. Both approaches suffer from difficulties in estimating, and thus improving on, the efficiency of the recovery of unknown genes.

8. Concluding Remarks

It took Nature 3.8 billion years of evolution to create today's diverse life and chemistry of life. Humans have spent tens- or perhaps hundreds-of-thousands of years exploiting natural materials empirically and surrounding those exploitations with superstition. In the last few centuries, progress in science ushered in the modern concept of a drug as a discrete chemical entity. In 1928, penicillin was discovered from a mold, and the modern notion of natural products discovery was born from the realization that microbes could produce discrete chemical entities and some could be drugs. The field of natural products discovery became the child of serendipity and science. Following penicillin came the glory days of the "Antibiotic Era," when many diseases were conquered (at least for a while) as science was continuously improving, and serendipity was common. As is inevitable in new areas, progress slowed and some contended that natural products discovery had ended. Then Umezawa (76) and others led science, and serendipity, to enzymes as targets of the action of natural products, and the drug discovery world realized that natural products could be useful for something other than antibiosis. Secondary metabolites with anti-cancer, cholesterol-lowering, immunosuppression, and other properties have been discovered and developed, and now improve the lives of many (4,77). Natural products discovery has seen multiple cycles of popular favor and disfavor, and this cycle is likely to continue in the future. The question for practitioners during each period of low favor should be: which dogma or lore prevents us from seeing the next wave of opportunity. Some possibilities now exist. Many active natural products are viewed as too large (some contend that anything larger than 600 m.w. is too large). Given the progress in delivering therapeutic proteins via novel routes, such as oral, pulmonary or trans-dermal, it seems very likely that those technologies will be applied to many of the larger natural products with the best potency, although they are not good candidates for oral delivery. Another traditional problem for crude natural product samples has been their chemical complexity—i.e., the number of compounds per sample. Innovative new screening technologies take advantage of the tremendous selectivity of mass spectroscopy and allow complex samples to be a potential discovery advantage (78,79). The key is the mass of the ligand that binds to a target is the screen result. Once known, isolation of the unknown ligand can be guided by mass spectroscopy, a highly effective process. These are two examples of the kinds of technology that may fuel another wave of natural products innovation. Given less than a century of intentional natural products discovery for compounds that evolved over almost 4 billion years, it is doubtful that we have unearthed all of the precious gems.

References

1. Demain, A. L. and Fang, A. (2000) The natural functions of secondary metabolites. *Adv. Biochem. Eng.-Biotechnol.* **69**, 1–39.
2. Firn, R. D. and Jones, C. G. (2000) The evolution of secondary metabolism—a unifying model. *Mol. Microbiol.* **37**, 989–994.
3. Berdy, J. (1989) The discovery of new bioactive microbial metabolites: screening and identification, in *Bioactive Metabolites from Microorganisms*, vol. 27 (Bushell, M. E. and Grafe, U., eds.) Elsevier, Amsterdam, pp. 3-35.
4. Demain, A. L. (2000) Small bugs, big business: the economic power of the microbe. *Biotechnol. Adv.* **18**, 499–514.
5. Henkel, T., Brunne, R. M., Muller, H., and Reichel, F. (1999) Statistical investigation into the structural complementarity of natural products and synthetic compounds. *Angew. Chem. Int. Ed.* **38**, 643–647.
6. Seow, K.-T., Meurer, G., Gerlitz, M., Wendt-Pienkowski, E., Hutchinson, C. R., and Davies, J. (1997) A study of iterative type II polyketide synthases, using bacterial genes cloned from soil DNA: a means to access and use genes from uncultured microorganisms. *J. Bacteriol.* **179**, 7360–7368.
7. Handelsman, J., Rondon, M. R., Brady, S. F., Clardy, J., and Goodman, R. M. (1998) Molecular biological access to the chemistry of unknown soil microbes: a new frontier for natural products. *Chem. Biol.* **5**, R245–R249.
8. Kealey, J. T., Liu, L., Santi, D. V., Betlach, M. C., and Barr, P. J. (1998) Production of a polyketide natural product in nonpolyketide-producing prokaryotic and eukaryotic hosts. *Proc. Natl. Acad. Sci. USA* **95**, 505–509.
9. Clark, G. J., Langley, D., and Bushell, M. E. (1995) Oxygen limitation can induce microbial secondary metabolite formation: investigations with miniature electrodes in shaker and bioreactor culture. *Microbiology* **141**, 663–669.
10. Dornenburg, H. and Knorr, D. (1995) Strategies for the improvement of secondary metabolite production in plant cell cultures. *Enzyme Microb. Technol.* 17, 674–684.
11. Burgess, J. G., Jordan, E. M., Bregu, M., Mearns-Spragg, A., and Boyd, K. G. (1999) Microbial antagonism: a neglected avenue of natural products research. *J. Biotechnol.* **70**, 27–32.
12. Pereira, A. M. S., Bertoni, B. W., Camara, F. L. A., Duarte, I. B., Queiroz, M. E. C., Leite, V. G. M., et al. (2000) Co-cultivation of plant cells as a technique for the elicitation of secondary metabolite production. *Plant Cell Tissue Organ Cult.* **60**, 165–169.
13. Ward, D. M., Weller, R., and Bateson, M. M. (1990) 16S rRNA sequences reveal numerous uncultured microorganisms in a natural community. *Nature* **345**, 63–65.
14. Woese, C. R. (1996) Phylogenetic trees—whither microbiology. *Curr. Biol.* **6**, 1060–1063.
15. Hill, G. T., Mitkowski, N. A., Aldrich-Wolfe, L., Emele, L. R., Jurkonie, D. D., Ficke, A., et al. (2000) Methods for assessing the composition and diversity of soil microbial communities. *Applied Soil Ecology* **15**, 25–36.
16. Bull, A. T., Goodfellow, M., and Slater, J. H. (1992) Biodiversity as a source of innovation in biotechnology. *Annu. Rev. Microbiol.* **46**, 219–252.
17. Lewis, B. (1997) Environmental Associates, Academy of Natural Sciences, Philadelphia, PA.

18. Nelson, K. E., Clayton, R. A., Gill, S. R., Gwinn, M. L., Dodson, R. J., Haft, D. H., et al. (1999) Evidence for lateral gene transfer between *Archaea* and *Bacteria* from the genome sequence of Thermotoga maritima. *Nature* **399,** 323–329.

19. Lawrence, J. G. (1999) Gene transfer, speciation, and the evolution of bacterial genomes. *Current Opinion In Microbiology* **2,** 519–523.

20. Walton, J. D. (2000) Horizontal gene transfer and the evolution of secondary metabolite gene clusters in fungi: An hypothesis. *Fungal Genet. Biol.* **30,** 167–171.

21. Eisen, J. A. (2000) Horizontal gene transfer among microbial genomes: new insights from complete genome analysis. *Curr. Opin. Genet. Dev.* **10,** 606–611.

22. Brown, J. R. and Doolittle, W. F. (1997) Archaea and the prokaryote-to-eukaryote transition. *Microbiol. Mol. Biol. Rev.* **61,** 456–502.

23. Doolittle, W. F. (1999) Phylogenetic classification and the universal tree. *Science* **284,** 2124–2129.

24. Salzberg, S. L., White, O., Peterson, J., and Eisen, J. A. (2001) Microbial genes in the human genome: lateral transfer or gene loss? *Science* **292,** 1903–1906.

25. Kobayashi, J. and Ishibashi, M. (1993) Bioactive metabolites of symbiotic marine microorganisms. *Chem. Rev.* **93,** 1753–1769.

26. Zahner, H., Drautz, H., and Weber, W. (1982) Novel approaches to metabolite screening, in *Bioactive Microbial Products: Search and Discovery*, (Bu'lock, J. D., Nisbet, L. J., and Winstanley, D. J., eds.), Academic Press, London, UK, pp. 51–70.

27. Bisaria, V. and Panda, A. (1991) Large-scale plant cell culture: methods, applications and products. *Curr. Opin. Biotechnol.* **2,** 370–374.

28. Ramakrishnan, D., Salim, J., and Curtis, W. R. (1994) Inoculation and tissue distribution in pilot-scale plant root culture bioreactors. *Biotechnol. Tech.* **8,** 639–644.

29. Zhong, J. J., Yu, J. T., and Yoshida, T. (1995) Recent advances in plant cell cultures in bioreactors. *World J. Microbiol. Biotechnol.* **11,** 461–467.

30. Canto-Canche, B. and Loyola-Vargas, V. M. (1999) Chemicals from roots, hairy roots, and their application. *Adv. Exp. Med. Biol.* **464,** 235–275.

31. Furusaki, S. (1999) Plant cell culture system. *Kagaku Kogaku Ronbunshu* **25,** 131–135.

32. Hui, C. and Feng, C. (2000) Effects of yeast elicitor on the growth and secondary metabolism of a high-tanshinone-producing line of the Ti transformed *Salvia miltiorrhiza* cells in suspension culture. *Process Biochem.* **35,** 837–840.

33. Wu, J. Y., Wang, C. G., and Mei, X. G. (2001) Stimulation of taxol production and excretion in *Taxus* spp cell cultures by rare earth chemical lanthanum. *Journal of Biotechnology* **85,** 67–73.

34. Faulkner, D. J. (2000) Marine natural products. *Nat. Prod. Rep.* **17,** 7–55.

35. Bull, A. T. (1992) Isolation and screening of industrially important organisms, in *Recent Advances in Biotechnology*, Vol. 210 (Vardar-Sukan, F. and Sukan, S. S., eds.), Kluwer Academic Publishers, Dordrecht, pp. 1–17.

36. Hunter-Cevera, J. C. and Belt, A. (1994) Bacteria as a source of novel therapeutics, in *The Discovery of Natural Products with Therapeutic Potential* (Gullo, V. P., ed.), Butterworth-Heinemann, Stoneham, MA, pp. 31–47.

37. Wells, J. S., Hunter, J. C., Astle, G. L., Sherwood, J. C., Ricca, C. M., Trejo, W. H., et al. (1982) Distribution of beta-lactam and beta-lactone producing bacteria in nature. *J. Antibiot.* **35,** 814–821.

38. Shoham, Y. and Demain, A. L. (1990) Effect of medium composition on the maintenance of a recombinant plasmid in *Bacillus subtilis*. *Enzyme Microb. Technol.* **12,** 330–336.

39. Monaghan, R. L., Polishook, J. D., Pecore, V. J., Bills, G. F., Nallin-Omstead, M., and Streicher, S. L. (1995) Discovery of novel secondary metabolites from fungi—is it really a random walk through a random forest? *Can. J. Bot. Suppl.* **1B,** S925–931.

40. Bedford, D. J., Schweizer, E., Hopwood, D. A., and Khosla, C. (1995) Expression of a functional fungal polyketide synthase in the bacterium *Streptomyces coelicolor* A3(2). *J. Bacteriol.* **177,** 4544–4548.

41. Wang, G. Y. S., Graziani, E., Waters, B., Pan, W. B., Li, X., McDermott, J., et al. (2000) Novel natural products from soil DNA libraries in a streptomycete host. *Organic Letters* **2**, 2401–2404.

42. Kao, C. M., Katz, L., and Khosla, C. (1994) Engineered biosynthesis of a complete macrolactone in a heterologous host. *Science* **265**, 509–512.

43. McDaniel, R., Hutchinson, C. R., and Khosla, C. (1995) Engineered biosynthesis of novel polyketides—analysis of tcmn function in tetracenomycin biosynthesis. *J. Am. Chem. Soc.* **117**, 6805–6810.

44. Oliynyk, M., Brown, M. J., Cortes, J., Staunton, J., and Leadlay, P. F. (1996) A hybrid modular polyketide synthase obtained by domain swapping. *Chem. Biol.* **3**, 833–839.

45. Kuhstoss, S., Huber, M., Turner, J. R., Paschal, J. W., and Rao, R. N. (1996) Production of a novel polyketide through the construction of a hybrid polyketide synthase. *Gene* **183**, 231–236.

46. Staunton, J. (1998) Combinatorial biosynthesis of erythromycin and complex polyketides. *Current Opinion in Chemical Biology* **2**, 339–345.

47. Bentley, R. and Bennett, J. W. (1999) Constructing polyketides: from collie to combinatorial biosynthesis. *Annu. Rev. Microbiol.* **53**, 411–446.

48. McDaniel, R., Thamchaipenet, A., Gustafsson, C., Fu, H., Betlach, M., and Ashley, G. (1999) Multiple genetic modifications of the erythromycin polyketide synthase to produce a library of novel ''unnatural'' natural products. *Proc. Natl. Acad. Sci. USA* **96**, 1846–1851.

49. Carreras, C. W. and Ashley, G. W. (2000) Manipulation of polyketide biosynthesis for new drug discovery. *Exs* **89**, 89–108.

50. Hobbs, G., Frazer, C. M., Gardner, D. C. J., Cullum, J. A., and Oliver, S. G. (1989) Dispersed growth of *Streptomyces* in liquid culture. *Appl. Microbiol. Biotechnol.* **31**, 272–277.

51. Gibbs, P. A., Seviour, R. J., and Schmid, F. (2000) Growth of filamentous fungi in submerged culture: problems and possible solutions. *Crit. Rev. Biotechnol.* **20**, 17–48.

52. Dietz, A. (1982) Culture preservation and instability, in *Bioactive Microbial Products: Search and Discovery* (Bu'lock, J. D., Nisbet, L. J., and Winstanley, D. J., eds.), Academic Press, London, UK, pp. 27–35.

53. Hutter, R. (1982) Design of culture media capable of provoking wide gene expression, in *Bioactive Microbial Products: Search and Discovery* (Bu'Lock, J. D., Nisbet, L. J., and Winstanley, D. J., eds.), Academic Press, London, UK, pp. 37–50.

54. Parekh, S., Vinci, V. A., and Strobel, R. J. (2000) Improvement of microbial strains and fermentation processes. *Appl. Microbiol. Biotechnol.* **54**, 287–301.

55. Nakagawa, A. (1992) Chemical screening, in *The Search for Bioactive Compounds from Microorganisms* (Omura, S., ed.), Springer Verlag, New York, NY, pp. 263–280.

56. Grabley, S., Wink, J., and Zeeck, A. (1992) Chemical screening as applied to the discovery and isolation of micobial secondary metabolites, in *Biotechology Focus 3* (Finn, R. K., Prave, P., Schlingmann, M., Crueger, W., Esser, K., Thauer, R., et al, eds.), Hanser Publishers, Munich, pp. 359–370.

57. Julian, R. K., Higgs, R. E., Gygi, J. D., and Hilton, M. D. (1998) A method for quantitatively differentiating crude natural extracts using high-performance liquid chromatography electrospray mass spectrometry. *Anal. Chem.* **70**, 3249–3254.

58. Higgs, R. E., Zahn, J. A., Gygi, J. D., and Hilton, M. D. (2001) Rapid method to estimate the presence of secondary metabolites in microbial extracts. *Appl. Environ. Microbiol.* **67**, 371–376.

59. Zahn, J. A., Higgs, R. E., and Hilton, M. D. (2001) Use of direct-infusion electrospray mass spectrometry to guide empirical development of improved conditions for expression of secondary metabolites from actinomycetes. *Appl. Environ. Microbiol.* **67**, 377–386.

60. Beppu, T. (1996) Genes, enzymes and secondary metabolites in industrial microorganisms. The 1995 Thom Award Lecture. *J. Ind. Microbiol.* **16**, 360–363.

61. Whiteley, M., Lee, K. M., and Greenberg, E. P. (1999) Identification of genes controlled by quorum sensing in Pseudomonas aeruginosa. *Proc. Natl. Acad. Sci. USA* **96**, 13,904–13,909.

62. Broenneke, V. and Fiedler, F. (1994) Production of bacteriolytic enzymes by *Streptomyces globisporus* regulated by exogenous bacterial cell walls. *Appl. Environ. Microbiol.* **60,** 785–791.

63. Yarbrough, G. G., Taylor, D. P., Rowlands, R. T., Crawford, M. S., and Lasure, L. L. (1993) Screening microbial metabolites for new drugs—theoretical and practical issues. *J. Antibiot.* **46,** 535–544.

64. Allanson, J. P., Biddlecombe, R. A., Jones, A. E., and Pleasance, S. (1996) The use of automated solid phase extraction in "96 well" format for high throughput bioanalysis using liquid chromatography coupled to tandem mass spectrometry. *Rapid Commun. Mass Spectrom.* **10,** 811–816.

65. Schmid, I., Sattler, I., Grabley, S., and Thiericke, R. (1999) Natural products in high throughput screening: automated high-quality sample preparation. *J. Biomolecular Screening* **4,** 15–25.

66. Thiericke, R. (2000) Drug discovery from Nature: automated high-quality sample preparation. *Journal of Automated Methods and Management in Chemistry* **22,** 149–157.

67. Josic, D., Reusch, J., Loster, K., Baum, O., and Reutter, W. (1992) High-performance membrane chromatography of serum and plasma membrane proteins. *J. Chromatogr.* **590,** 59–76.

68. Eichler, J. and Houghten, R. A. (1995) Generation and utilization of synthetic combinatorial libraries. *Mol. Med. Today.* **1,** 174–180.

69. Houghten, R. A. (2000) Parallel array and mixture-based synthetic combinatorial chemistry: tools for the next millennium. *Annu. Rev. Pharmacol. Toxicol.* **40,** 273–282.

70. Ohlstein, E. H., Ruffolo, R. R., and Elliott, J. D. (2000) Drug discovery in the next millennium. *Annu. Rev. Pharmacol. Toxicol.* **40,** 177–191.

71. Thompson, L. A., and Ellman, J. A. (1996) Synthesis and applications of small molecule libraries. *Chem. Rev.* **96,** 555–600.

72. Fernandes, P. B. (1998) Technical advances in high-throughput screening. *Current Opinion in Chemical Biology* **2,** 597–603.

73. Bosworth, N. and Towers, P. (1989) Scintillation proximity assay. *Nature* **341,** 167–168.

74. God, R., Gumm, H., Juschka, M., and Heuer, C. (1999) G.I.T. *Laboratory Journal* 3/99, pp. 188–191.

75. Devlin, J. P. (1997) Microcollection of plants for biochemical profiling, in *Natural Products as a Discovery Resource* (Devlin, J. P., ed.), Marcel Dekker, Inc., New York, NY, pp. 49–76.

76. Umezawa, H. (1982) Low molecular weight enzyme inhibitors of microbial origin. *Annu. Rev. Microbiol.* **36,** 75–99.

77. Demain, A. L. (1999) Pharmaceutically active secondary metabolites of microorganisms. *Applied Microbiology And Biotechnology* **52,** 455–463.

78. van Breemen, R. B., Huang, C.-R., Nikolic, D., Woodbury, C. P., Zhao, Y.-Z., and Venton, D. L. (1997) Pulsed ultrafiltration mass spectrometry: a new method for screening combinatorial libraries. *Anal. Chem.* **69,** 2159–2164.

79. Xavier, K., Eder, P., and Giordano, T. (2000) RNA as a drug target: methods for biophysical characterization and screening. *Trends Biotechnol.* **18,** 349–356.

80. Owicki, J. C. (2000) Fluorescence polarization and anisotropy in high throughput screening: perspectives and primer. *Journal of Biomolecular Screening* **5,** 297–306.

81. Boisclair, M. D., McClure, C., Josiah, S., Glass, S., Bottomley, S., Kamerkar, S., et al. (2000) Development of a ubiquitin transfer assay for high throughput screening by fluorescence resonance energy transfer. *Journal of Biomolecular Screening* **5,** 319–328.

82. Jones, J., Heim, R., Hare, E., Stack, J., and Pollok, B. A. (2000) Development and application of a GFP-FRET intracellular caspase assay for drug screening. *J. Biomolecular Screening* **5,** 307–318.

83. Matsumoto, C., Hamasaki, K., Mihara, H., and Ueno, A. (2000) A high-throughput screening utilizing intramolecular fluorescence resonance energy transfer for the discovery of the molecules that bind HIV-1 TAR RNA specifically. *Bioorg. Med. Chem. Lett.* **10,** 1857–1861.

6

Genetic Engineering Solutions for Natural Products in Actinomycetes

Richard H. Baltz

1. Introduction

1.1. Applications for Yield Enhancement

There are a number of genetic and molecular methods that are useful for the improvement of production of natural products in actinomycetes. These include traditional chemical mutagenesis and selection, transposition mutagenesis, targeted deletions and duplications by genetic engineering, and genetic recombination by protoplast fusion. With the recent developments in microbial genomics, transcriptome analysis, proteomics, metabolic reconstruction, and metabolite flux analysis, these new technologies are becoming valuable tools to aid in strain development. A robust strain development program can benefit from the synergy provided by coupling several of these approaches. This chapter limits its discussion to genetic engineering approaches to yield enhancement. More classical mutation and recombination approaches have been reviewed elsewhere *(1–6)*.

1.2. Applications for Novel Secondary Metabolites

Genetic engineering or combinatorial biosynthesis approaches in actinomycetes are based in part upon successful examples of the production of novel secondary metabolites by mutagenesis, mutasynthesis, mutational biosynthesis, hybrid biosynthesis, bioconversions, and in vitro enzymatic conversion of natural or unnatural substrates *(1)*. These approaches demonstrated that some enzymes in secondary metabolism have relaxed substrate specificities, thus facilitating the incorporation of related or modified substrates into novel products. Current genetic engineering approaches to produce novel secondary metabolites are also based upon more recent observations that the genes encoding multidomain subunits of the giant type I polyketide synthases (PKSs) *(7,8)* and nonribosomal peptide synthetases (NRPSs) *(9,10)* can be modified by molecular genetic manipulations to generate functional hybrid enzymes. Also, individual subunits of type II PKSs can be exchanged, and the hybrid multi-component enzymes are often functional *(11,12)*. In addition, the "tailoring enzymes", including glycosyltransferases, methyltransferases and hydroxylases, for example, can also be manipulated to produce

From: *Handbook of Industrial Cell Culture: Mammalian, Microbial, and Plant Cells*
Edited by: V. A. Vinci and S. R. Parekh © Humana Press Inc., Totowa, NJ

novel secondary metabolites. Examples of the use of genetic engineering to modify polyketide, peptide and tailoring functions are covered in this chapter.

2. Genetic Engineering Tools

2.1. Gene Transfer Methods

A number of molecular genetic technologies are needed to carry out manipulations to genetically engineer secondary metabolite biosynthetic pathways in actinomycetes for yield improvement or for production of novel products. These include methods to identify and clone antibiotic biosynthetic genes, cloning vectors, methods to introduce cloned genes into actinomycetes, and methods to exchange gene sequences in precise locations in the chromosome. Some of the key elements needed to successfully genetically engineer actinomycetes are detailed in the following paragraphs. Many of the procedures to carry out genetic engineering tasks can be found in the manual "Practical Streptomyces genetics" *(2)*.

2.1.1. Introduction of Plasmid DNA into Actinomycetes

The most common method to introduce DNA into actinomycetes is by protoplast transformation *(2,13)*. Protoplast transformation works well in some actinomycetes, especially those that do not express significant levels of restriction *(13–15)*. In some cases, the use of single-strand plasmid DNA has improved the outcome of protoplast transformation, particularly for gene replacement *(16,17)*. A number of methods to implement protoplast transformation have been described *(2,13,* and references therein). These methods employ different media and growth conditions for protoplast formation and regeneration—key steps in the process. It is advisable to try several different procedures to identify one that works well for the specific actinomycete strain being studied.

Another method to introduce plasmid DNA into actinomycetes is by conjugation from *Escherichia coli (18)*, a procedure that appears to circumvent restriction barriers *(14,19–21)*. Conjugation is mediated by the transfer functions of the broad host range Gram-negative plasmid RP4 (RK2). Conjugation has been used to introduce plasmids into a number of actinomycetes, including *Streptomyces nanchangensis*, a strain that does not yield protoplasts suitable for transformation *(22)*.

Bacteriophage FP43-mediated transduction can also be used to introduce plasmid DNA into a wide array of streptomycetes *(23–25)*. Transduction can help circumvent restriction barriers by: (i) transducing cells grown under conditions that minimize the expression of restriction; (ii) preparing transducing lysates on hosts that modify the plasmid for specific restriction systems; and (iii) transducing at very high multiplicities of infection *(23–25)*. Transduction can be used to rapidly transfer cloned DNA into a large number of unrelated streptomycetes, or into many mutants of a given species. Although high-frequency transduction of plasmid DNA is currently limited to streptomycetes and to pIJ702-derived plasmids containing the cloned *pac* site from bacteriophage FP43, it should in principle be applicable to other headfull packaging bacteriophages and plasmids from other actinomycetes.

Plasmid DNA has also been introduced into *Streptomyces* species and into *Saccharopolyspora erythraea* by electroporation *(26–29)*. Electroporation may not present any general advantages over protoplast transformation and conjugation for the introduction of plasmid DNA into actinomycetes. However, it may be a powerful method to facilitate Tn5 transposition mutagenesis (*see* **Subheading 3.1.**).

2.1.2. Cloning Vectors

There are many cloning vectors available for *Streptomyces* species, including: (i) self-replicating vectors of high and low copy number; (ii) vectors for site-specific integration into phage or plasmid attachment sites; (iii) vectors for conjugal transfer from *E. coli* or from other streptomycetes; (iv) cosmid and bacterial artificial chromosome (BAC) vectors for cloning large blocks of genes; (v) vectors for insertion of large blocks of genes into the chromosome; (vi) vectors for high-level expression of cloned genes; (vii) temperature-sensitive vectors for transposon delivery and for homologous insertion of genes into the chromosome; (viii) vectors that facilitate gene replacement, including single-strand vectors; and (ix) vectors for insertion of genes into neutral chromosomal sites by homologous recombination *(2,14,16,19,20,30–35)*. The conjugal vectors that contain the origin for plasmid transfer (*oriT*) from RK2 are useful to bypass restriction barriers in actinomycetes *(14,15,36)*. Plasmids that integrate site-specifically into plasmid or phage attachment sites, or into neutral genomic sites, are useful to generate stable recombinants that exert little or no negative effects on secondary metabolite production *(15,19)*. Cosmids are useful to clone large blocks of secondary metabolite biosynthetic genes, up to ~40kb *(31)*. BAC vectors are useful to clone much larger clusters of genes, up to ~140 kb *(35)*. Because it is not uncommon for antibiotic biosynthetic pathways to span 50–100 kb *(1)*, BACs are useful to clone complete antibiotic biosynthetic pathways *(37)*. Plasmids containing combinations of key features are particularly useful *(19,20)*.

2.1.3. Cloning Procedures for Antibiotic Biosynthetic Genes

The procedures to clone antibiotic biosynthetic genes *(34,38–40)* are based on the observation that antibiotic biosynthetic and resistance genes are generally clustered *(1,34,41)*. If one gene in the pathway can be cloned, it can be used as a probe to identify cosmids or BACs spanning the whole pathway. Cloning the initial gene in a pathway has been accomplished by: (i) purifying an enzyme in the biosynthetic pathway, obtaining partial amino acid sequence, and preparing a degenerate probe to identify the gene in a library; (ii) cloning an antibiotic resistance gene; (iii) complementing a mutation in an antibiotic biosynthetic gene; (iv) identifying an antibiotic biosynthetic gene by transposon mutagenesis and cloning the sequences flanking the transposon insertion; (v) using a gene or sequences designed from related genes as hybridization probes; and (vi) purifying a protein from a two-dimensional gel expressed during secondary metabolism, but not expressed in a secondary metabolite defective mutant. Characterization of the secondary metabolic genes can be carried out by DNA sequence analysis, insertional mutagenesis, complementation of mutants blocked in secondary metabolite biosynthesis, and heterologous expression of individual genes or the whole pathway. There are many examples of secondary metabolite biosynthetic gene clusters that have been cloned and analyzed by these methods *(1,2,34,41)*.

2.1.4. Gene Replacement

To modify secondary metabolite biosynthetic genes, DNA segments are joined by recombinant DNA techniques, and segments of DNA are inserted into the chromosome by recombination. Replacing a segment of DNA by double-crossover recombination can be tedious. Recombinants containing a single crossover inserting a plasmid into the chromosome by homologous recombination are easily identified by selecting for an antibiotic resistance marker on the plasmid. However, identification of progeny that have undergone double crossovers often requires screening a large number of colonies derived from the initial recombinant for the loss of antibiotic resistance associated with the plasmid, and confirmation of the recombinant genotype by Southern hybridization or PCR analysis. Selecting for double crossovers that lead to gene disruption can be simplified by inserting an antibiotic resistance gene within the gene to be disrupted. A good example of this method is the targeted disruption of tylosin genes *(42,43)*. In this system, a hygromycin resistance cassette is inserted within structural genes and delivered to *S. fradiae* by conjugation from *E. coli* S17-1 on pOJ260, a nonreplicating plasmid. The use of single-stranded DNA in protoplast transformations also yields high frequencies of the desired double crossovers during the initial recombinational event *(16,17,44)*.

Direct selection for the desired double crossover during the initial recombinational event is feasible only if the recombinant can be selected. Replacement of a functional gene for another functional gene at a precise location of the chromosome requires a system to select for the first crossover inserting a plasmid, followed by selection for a second crossover and loss of the plasmid vector. One system that addresses the two-stage selection uses the *rpsL* gene (45). The key feature of the *rpsL* gene is that it is the locus for mutation to streptomycin resistance (SmR). Since streptomycin sensitivity (SmS) is dominant over SmR, double crossovers can be selected by placing a SmR mutation in the chromosome and placing the wild-type (SmS) *rpsL* gene on the plasmid. Gene replacement can then be accomplished by selecting for the first crossover that inserts a temperature-sensitive plasmid expressing an antibiotic resistance marker into the chromosome by homologous recombination. Selection for SmR progeny after growth at the nonpermissive temperature for plasmid replication yields recombinants that have eliminated the plasmid (SmS) by a second crossover. Although some of the progeny will have lost the plasmid by a single crossover at the original site of insertion, a high percentage of the progeny will have undergone gene replacement. This procedure is simplified when an antibiotic resistance gene is present in the gene targeted for replacement, since recombinants containing double crossovers can be selected during the initial recombinational event. The desired recombinant will be SmR and sensitive to the antibiotic initially present in the gene to be replaced. Another advantage of the *rpsL* system is that certain SmR mutants are associated with enhanced antibiotic biosynthesis *(46)*, so recombinants containing gene replacements should require no further genetic manipulations to maintain secondary metabolite production. This clearly requires screening several SmR mutants of the desired production strain for secondary metabolite production yields to identify a suitable strain for gene-replacement experiments.

The glucose analog, 6-deoxy-D-glucose (Dog), is converted to a toxic metabolite by glucose kinase. Strains with no glucose kinase activity are resistant to Dog (DogR). In such strains, Dog can be used to select DogR recombinants that have eliminated a functional glucose kinase gene (*glkA*) located on a phage or plasmid *(47–49)*. This system can be used to select for double crossovers in a manner similar to the rpsL system. However, it is not known if *glkA*-defective strains are proficient in producing high levels of secondary metabolites.

2.1.5. Heterologous Expression from Plasmid Vectors

Another method to carry out genetic engineering or combinatorial biosynthesis to generate novel secondary metabolites is to express secondary metabolite biosynthetic genes from plasmids introduced into a heterologous host. This procedure has been validated in *S. lividans* using engineered PKS genes expressed from one, two, or three compatible plasmids *(50–52)*, and will be discussed in **Subheading 4.1.2.**

2.2. Site-Directed Mutagenesis and DNA Shuffling

There are a number of methods available for molecular evolution of enzymes and other proteins that may be relevant to the genetic engineering of novel secondary metabolite biosynthetic pathways *(53–61)*. One very powerful method is DNA shuffling *(55,59–61)*. This procedure employs error-prone polymerase chain reaction (PCR) and random recombination of DNA fragments to rapidly evolve proteins with improved characteristics. DNA shuffling of single genes has been used to: (i) improve the catalytic activity of an enzyme *(59)*; (ii) change the substrate specificity of an enzyme *(62)*; (iii) improve the folding of a protein in a heterologous host *(63)*; and (iv) improve the function of a multigene pathway *(64)*. An even more powerful procedure is DNA family shuffling *(65–67)*, in which a family of related genes is shuffled under conditions of error-prone PCR. This technology may be applied to secondary metabolite tailoring enzymes (e.g., acyltransferases, glycosyltransferases, haloperoxidases, hydroxylases, and methyltransferases) to alter substrate specificity or, in some cases, to alter cofactor specificity (e.g., NDP-sugar specificity for glycosyltransferases and acyl-CoA specificity for acyltransferases). DNA family shuffling should also have applications in generating hybrid PKS and NRPS multienzymes, and in optimizing the catalytic properties of the hybrid enzymes.

2.3. Sources of Genes for Genetic Engineering and Combinatorial Biosynthesis

Many genes that encode antibiotic and other secondary metabolite biosynthetic pathways have been cloned and sequenced *(1,2,34,68)*. It appears that many actinomycetes encode multiple clusters of secondary metabolite genes *(22,69,70)*. At the current rapid pace of cloning and sequencing of secondary metabolite genes, many more secondary metabolite biosynthetic genes will be available for genetic engineering and combinatorial biosynthesis studies. These include type I and type II PKS genes, NRPS genes, sugar biosynthetic and glyosyltransferase genes, and genes that encode tailoring functions such as acyltransferases, methyltransferases, hydroxylases, and haloperoxidases. In addition, genes with novel functions, such as the spinosad genes that encode polyketide cross-bridging functions *(71)*, will continue to be discovered and exploited.

3. Secondary Metabolite Yield Enhancement by Genetic Engineering

3.1. Transposon Mutagenesis

3.1.1. Insertional Inactivation of Regulatory Genes

Both positive and negative regulatory elements can influence the production of natural products *(13,73)*. An excellent example is the tylosin system, in which several positive regulatory genes and one negative regulatory gene may be involved in controlling product yield *(74–76)*. Transposon mutagenesis is a potentially powerful method to disrupt regulatory genes. Disruption of negative regulatory genes should enhance product yields, whereas disruption of positive regulatory genes should reduce yields. In either case, the regulatory gene can be isolated by cloning the transposon marker and the chromosomal DNA flanking the transposon. Negative regulatory genes can be deleted and positive regulatory genes can be duplicated in neutral sites to enhance production.

This transposition approach is facilitated in streptomycetes by IS493-derived transposons that employ hygromycin or apramycin resistance markers *(77–81)*. Both markers are selectable in streptomycetes and in *E. coli*. A number of other transposons with different features have been developed for streptomycetes. Their properties and uses are described by Kieser et al. *(2)*. Of particular interest is Tn4560, which has been used recently to generate a "megalibrary" of about 100,000 mutants of *S. coelicolor* A3(2) *(82)*. Tn*4560* inserts randomly with no apparent target sequence specificity.

A very promising new transposition method is the use of Tn5 derivatives to carry out transposition in vitro *(83)*. Plasmids containing transposon inserts are first identified in *E. coli*, then transferred to *S. coelicolor* by protoplast transformation or by conjugation from *E. coli*. Even more promising is the direct introduction of the transposome mix into actinomycete cells by electroporation. In this case, transposition takes place in the target organism, rather than in *E. coli*. The successful application of this approach has been demonstrated in *Rhodococcus rhodochrous* and in a wild-type *Rhodococcus* soil isolate *(84)*. There are at least two advantages to this approach. First, many primary transposants can be generated in a single electroporation experiment. Secondly, the small size of the transposon helps avoid restriction barriers.

3.1.2. Insertional Inactivation of Competing Pathways

Transposons can also be used to inactivate pathways that compete for precursors, cofactors, reducing power, and energy supply needed for the production of natural products. The spectrum of mutations induced by transposon insertions is likely to be different than that induced by chemical mutagens. Chemical mutagens generally produce subtle changes by single-nucleotide substitutions, whereas transposons can make radical alterations by disrupting genes and operons.

3.1.3. Random Insertion of Promoters

Transposons derived from IS493 have outward reading promoter activity *(85)*. Therefore, IS493-derived transposons can be used to randomly insert a promoter at different sites in the chromosome. This concept can be expanded by inserting strong or regulated outward-reading promoters into different transposons. Random transposition mutagenesis in the *Streptomyces roseosporus* with Tn*5099* resulted in the identification of mutants enhanced in daptomycin production *(86)*. The mutant genotypes were not characterized, but could have been the result of the expression of the portable promoter, the inactivation of negative regulatory elements, or inactivation of competing metabolic processes.

3.2. Targeted Gene Duplications and Promoter Fusions

Gene duplications can arise as spontaneous or induced mutations, but have not been developed as a general method to improve product yields. Duplication or amplification of secondary metabolite gene clusters is associated with yield improvements in oxytet-racycline and lincomycin fermentations *(87,88)*. Gene cloning and insertion of genes in neutral chromosomal sites offers an effective method to carry out directed gene duplications. These methods have an additional advantage: the recombinants are generally stable and lack self-replicating plasmids, which often reduce product yields *(14,15,19)*.

The general approach for targeted gene duplication is to: (i) identify and clone a neutral site in the chromosome where genes can be inserted without altering the fermentation properties of the strain; (ii) insert the gene to be duplicated into the neutral site in a plasmid vector; (iii) introduce the vector into the actinomycete host; and (iv) insert the gene to be duplicated into the chromosomal neutral site by homologous double crossover *(14,15,19)*. The first crossover, which integrates the plasmid into the chromosome, can be selected using an antibiotic resistance marker on a temperature-sensitive plasmid that can be cured at elevated temperature. Examples of such temperature-sensitive vectors are derivatives of pGM160 *(20,89)* and pMT660 *(90,91)*. The second crossover can be screened or selected using a counter selectable marker, such as *rpsL* (*45, see* **Subheading 2.1.4.**). Targeted gene duplication by a double crossover into a neutral site generates recombinants that contain no heterologous DNA, which can enhance production *(14)*.

An example of targeted gene duplication in a neutral site by double crossover is that of the *tylF* gene in *Streptomyces fradiae*. The *tylF* gene encodes macrocin O-methyltransferase, the rate-limiting enzyme in tylosin biosynthesis *(92,93)*. In this example, the neutral site was identified by Tn*5099* transposition mutagenesis, and cloned by selecting for the transposon marker in *E. coli* *(81,94)*. Recombinants containing two copies of the *tylF* gene converted macrocin to tylosin very efficiently, thus improving the tylosin yields substantially *(81,94)*.

Another method for targeted duplication is to introduce the cloned gene(s) into a plasmid that can insert site-specifically into a neutral site *(19,95)*. A good example of this method is the insertion of the pristinamycin biosynthetic genes *Sna*A and *Sna*B, under transcriptional control of the *ermE** promoter, into the plasmid pSAM2 *attB* site in *Streptomyces pristinaespiralis* *(96)*. The recombinant converted pristinamycin IIB to IIA very efficiently, resulting in increased product yields.

Targeted tandem duplications of larger blocks of genes can be accomplished by introducing the genes on a conjugal cosmid vector lacking actinomycete replication functions *(21,97)*. Recombinants are formed by single crossovers of circular plasmids or double crossovers of linear plasmid concatemers *(95)*, duplicating ~40 kb of DNA. This method has been used to duplicate a block of spinosyn biosynthetic genes in *Saccharopolyspora spinosa*, thus facilitating the rapid conversion of the precursor pseudoaglycone to spinosad *(98)*.

3.3. Genomics Approach to Yield Improvement

Genomic sequence analysis, metabolic reconstruction, transcriptome analysis, and proteomics offer the most important new techniques to augment molecular genetic approaches to strain improvement. Genomic sequence analysis and metabolic reconstruction *(99–104)* is advantageous because it provides a primary blueprint of all genes and their structural organization in the producing actinomycete. This sets the stage for comparative analysis with all other genomes in databases. There are many complete genomic sequences available in databases, including several key genomes from actinomycetes (*Streptomyces coelicolor, Thermomonospora fusca, Corynebacterium glutamicum, Mycobacterium tuberculosis*, and other *Mycobacterium* species at the time of this writing). Having the genomic sequence of a secondary metabolite-producing actinomycete allows you to do the following: (i) identify all secondary metabolite structural genes and linked regulatory genes; (ii) identify all potential sources for precursors and cofactors required for secondary metabolite production; (iii) identify potential competing pathways; (iv) identify regulatory genes and regulatory elements unlinked to the secondary metabolite gene cluster (e.g., streptomycete antibiotic regulatory proteins (SARPS), two component regulatory systems, sigma factors, other global regulatory genes, and promoters for subsequent manipulations); (v) identify missing genes/pathways that may be introduced from other actinomycetes; and (vi) identify genes and genetic elements to develop as tools for genetic manipulation (e.g., plasmids, bacteriophages, insertion sequence (IS) elements, transposons, neutral-site gene candidates, and restriction/modification systems). This information facilitates the selection of candidate genes for gene cloning by PCR amplification, gene disruption (to identify those relevant to secondary metabolite production), and gene duplication or gene modification (e.g., by changing or modifying a promoter region) to enhance secondary metabolite yields.

Transcriptome analysis is becoming a routine method to quantify differences of mRNA levels of bacteria grown under different nutritional or other environmental conditions. *(105–110)*. The advantage is that essentially all transcripts can be analyzed simultaneously. Transcriptome analysis provides a means to determine which genes are expressed during critical stages in the fermentation cycle, which genes are expressed at higher levels in production media relative to other media, and which genes are expressed at higher levels in highly productive stains. This approach is a logical extension of genomic sequence analysis and metabolic reconstruction, and provides relevant information to help choose candidate genes for disruption, duplication, or promoter modification. Limitations of transcriptome analysis are that it does not necessarily predict protein levels, does not predict post-translational modification of proteins, and does not distinguish between wild-type and mutant genes (i.e., active or inactive protein products).

Proteomics *(111–114)*, coupled with genomic sequence analysis and transcriptome analysis, can help in choosing genes for disruption and duplication analysis. The levels of expression of a large number of proteins can be studied in different strains (e.g., wild-type and highly productive mutants), in different media, and at different times in the fermentation. One current limitation of proteomics is that low-abundance proteins may not be detected. Thus, proteomics may not be useful in quantifying differences in the levels of some key regulatory proteins. Transcriptome analysis may be better suited for this application. A key advantage of proteomics over transcriptome analysis is that protein levels may be more indicative of what is going on in the fermentation, and the subtleties of post-translational modification of proteins can be explored and factored into the molecular genetic approaches to improve yields. Ideally, genomic sequence analysis, proteomics, transcriptome analysis, metabolic reconstructions, and metabolic flux analysis *(115)* will be fully integrated in the future to give a robust picture of the overall metabolism during secondary metabolite production, and will help guide molecular genetic approaches to improve secondary metabolite yields.

4. Production of Novel Secondary Metabolites

4.1. Hybrid Polyketides

4.1.1. Type I Polyketides

The genetics and biochemical mechanisms of biosynthesis of complex polyketides are relatively well understood *(7,8,11,116–133)*. Type I PKSs are involved in the biosynthesis of many relatively reduced polyketide structures, including the macrolide antibiotics erythromycin, tylosin, spiramycin, josamycin, and rifamycin, and other secondary metabolites such as avermectin, spinosyns, rapamycin, and FK506. The modular type I PKSs are composed of a small number of large multifunctional enzymes, each composed of modules that contain all of the enzymatic functions required to process the individual fatty-acid building units. These enzyme functions specify the side-chain length, the stereochemistry at chiral centers, the reduction level at each condensation step, and the macrolide ring size *(7,8,11,116,134)*. An important feature of the type I PKS mechanism of polyketide assembly is that each enzymatic reaction is carried out by separate enzymatic domain within a module. Modular PKS multienzymes are encoded by modular genes. Implicit in the genetic and biochemical mechanisms is that the biosynthetic code can be changed by focused deletions, additions, exchanges, rearrangements and translocations of modules; by engineering individual modules; and by manipulating the macrolide ring formation. The most extensive genetic engineering studies have been carried out on the erythromycin PKS, so this system serves as a model for how PKS genes can be manipulated.

4.1.2. Erythromycin and Ketolides

Erythromycin is a good model for genetic engineering or combinatorial biosynthesis applications, because derivatives of erythromycin continue to be important agents to treat respiratory pathogens *(135–143)*. The biosynthetic steps for erythromycin biosynthesis beyond polyketide assembly have been known for some time (144). The genetic control and biosynthesis of erythronolide, the 14-member polyketide ring of erythromycin, is now well-established *(7,8,11,116,118,121,123,125,126,128–130,145)*. This knowledge has facilitated a number of genetic engineering experiments to modify the PKS to produce novel derivatives of erythronolide or erythromycin in *S. erythraea*, *S. coelicolor*, or *S. lividans (7,8,11,120,124,146)*.

The Abbott group *(7)* demonstrated in *S. erythraea* that deletions in the enoylreductase of module 4 or in the ketoreductase (KR) of module 5 of the 6-deoxyerythronolide B synthase (DEBS)-coding region result in the biosynthesis of novel 14-member macrolides, including a derivative with increased acid stability, an important feature of second-generation erythromycins *(135)*. They also made acyltransferase-domain substitutions to generate desmethyl derivatives *(147)* and an ethyl-substituted derivative of erythromycin *(148)*. The production of 6-ethylerythromycin A required further engineering of *S. erythraea* by adding a crotonyl-CoA reductase gene (*ccr*) from *Streptomyces collinus* to provide the butyryl-CoA needed to initiate biosynthesis of the ethyl-substituted polyketide.

Genetic engineering of erythronolide biosynthesis was also carried out in *S. coelicolor* by expressing the PKS genes on a self-replicating plasmid *(149)*. Translocation of the thioesterase (TE) domain from the end of module 6 to the end of module 2 caused the production of 6-member triketides *(134,150)* and translocation of the TE domain to the end of module 5 caused the production of a 12-member macrolide *(150)*. The results of these and other studies *(132,151,152)* demonstrated that the side-chain length of building units, the reduction level at specific positions of the polyketide, and the macrolide ring size can be genetically engineered by deletion mutation by moving domains or modules within a given pathway, or by substituting domains or modules from heterologous pathways. In addition, Rowe et al. *(132)* have demonstrated ring expansion of the 14-member ring erythronolide to a 16-member ring compound in *S. erythraea*.

Expanding on these principles, McDaniel et al. *(50)* generated combinatorial libraries of erythronolide derivatives by modifying one to three carbon centers. They characterized 61 novel erythronolide derivatives. This combinatorial approach to genetic engineering has been streamlined by using two *(51)* or three plasmid systems *(52)*. The latter has provided a means to mix all combinations of molecular genetic changes in the three individual giant PKS genes encoding the three subunits of the erythronolide PKS. A drawback to this heterologous expression system is that the recombinant product yields are often very low *(1,51,52)*.

Another method to generate novel polyketide structures is to genetically engineer the starter or loading module. The avermectin loading module has the capability to accept a large number of analogs related to normal branched-chain fatty acids *(153,154)*. Marsden et al. *(155)* exchanged the avermectin-loading module for the erythronolide-loading module in *S. erythraea*, and the recombinant produced six derivatives of natural erythromycin factors. This finding demonstrated that changing the starter unit did not interfere with the subsequent steps in polyketide assembly or in the two glycosylations required for antibiotic activity. When the recombinant *S. erythraea* strain was fed short-chain fatty acids, an additional 12 erythromycin derivatives were generated *(156)*. This approach could be coupled with the other modifications the PKS genes to generate large libraries of erythromycin derivatives or of other polyketides.

An important application of genetic engineering of erythronolide is the production of ketolides. Ketolides are semisynthetic 14-member-ring derivatives of erythromycin that have a keto group at the 3 position *(137)*. Ketolides have potent antibacterial activity against many Gram-positive and upper-respiratory Gram-negative pathogens *(136–143)*. Ketolides are prepared chemically from erythromycin A by acid cleavage of cladinose, oxidation of the 3-hydroxyl group to a keto group, and further chemical modifications *(137)*. An alternative route to produce 3-keto intermediates lacking desosamine was accomplished by engineering module 6 of the erythronolide gene cluster in *S. coelicolor (50)*. However, it is not known from these studies whether these molecules can be converted by to the glycosylated intermediate needed for chemical modification to produce ketolides. Since cladinose is added to erythronolide before desosamine *(144)*, the sugar required for ketolide activity, it is unlikely that *S. erythraea* is capable of glycosylating 3-keto-erythronolide. An alternative approach to producing appropriately glycosylated ketolides is discussed *(see* **Subheading 4.4.***)*.

4.1.3. Tylosin and Other 16-Member Macrolides

Tylosin is a macrolide antibiotic produced by *Streptomyces fradiae (93)*. Tylosin is composed of a 16-member polyketide and three sugars—mycaminose, mycarose and mycinose. Tylosin and the tylosin-derivatives 3-*O*-acetyl-4"-*O*-isovaleryltylosin and tilmicosin are used in veterinary medicine *(157,158)*. The related 16-member macrolides spiramycin and josamycin are used in human medicine *(157,159–161)*. Most clinical isolates of *S. pyogenes* and *S. pneumoniae* that are resistant to erythromycin derivatives are susceptible to josamycin and spiramycin *(162)*.

Tylosin is a good model for genetic engineering approaches to novel macrolides because: (i) tylosin can be bioconverted to highly active antibiotics by enzymatic acylations *(163)* (*see* **Subheading 4.5.**); (ii) mutants blocked in tylosin biosynthesis produced novel antibiotics *(93,164)* that can be enzymatically acylated *(165)*; (iii) the genes involved in tylosin biosynthesis have been cloned, sequenced, and analyzed in some detail *(42,43,74–76,166–174)*; (iv) the genes for self-resistance to tylosin in *S. fradiae* have been cloned, sequenced, and analyzed *(175–182)*; (v) the biosynthetic pathway to tylosin is relatively well understood *(93,183,184)*; (vi) the regulation of tylosin biosynthesis is fairly well understood *(74–76,171,172)*; and (vii) high-producing strains of *S. fradiae* are amenable to genetic engineering *(20,81,93,94)*.

Spiramycin is also a good model for molecular genetic manipulation of 16-member macrolides for many of the same reasons as tylosin. The producing organism, *S. ambofaciens*, is a particularly good recipient for gene transfer by protoplast transformation and conjugation from *E. coli (20,185)*, and genetically stable high-producing strains have been isolated *(186)*. Macrolide biosynthetic and regulatory genes have been cloned and analyzed in *S. ambofaciens (31,187–190)*. For instance, Kuhstoss et al. *(190)* exchanged the tylosin starter module, which specifies the incorporation of propionate, for the acetate starter module of platenolide in the spiramycin PKS in *S. ambofaciens*, and the recombinant produced 16-methyl platenolide I, a novel 16-member polyketide. Thus *S. ambofaciens* is a suitable host for genetic engineering of PKS genes.

4.1.4. Type II Polyketides

Actinorhodin and other polycyclic aromatic secondary metabolites such as daunorubicin, elloramycin, granaticin, jadomycin, urdamycin, landomycin, mithromycin, nogalamycin, tetracenomycin, and tetracycline are synthesized by type II PKSs *(11,191)*. The multienzyme type II PKSs are composed of several monofunctional proteins. Type II polyketides differ from the complex type I macrolides because they do not require extensive reduction or reduction and dehydration cycles *(116)*. Several type II PKS gene clusters have been cloned *(11,191–199)*, and many of the related enzymatic functions are interchangeable, allowing for the production of hybrid polyketides *(11,116, 200,201)*. This combinatorial biosynthesis approach has generated a library of compounds from five PKS clusters *(200)*. Many type II polyketides are glycosylated, and additional diversity has been generated by novel hybrid glycosylations (*see* **Subheading 4.4.**).

4.2. Hybrid Peptides

4.2.1. Peptides and Modular Organization of Nonribosomal Peptide Synthetases (NRPSs)

Many secondary metabolites, including β-lactam antibiotics (penicillins and cephalosporins), glycopeptides (vancomycin, chloroeremomycin, and teicoplanin), and cyclic peptides (daptomycin and pristinamycin I) contain a peptide backbone synthesized by NRPSs *(202–206)*. Like the modular type I PKS enzymes, the NRPS enzymes are composed of a small number of very large multifunctional enzymes composed of modules that express all enzymatic functions required to process individual amino acids *(205–210)*. NRPS genes are generally organized as linear modules corresponding to the modular functions on the NRPS enzymes. NRPS enzymes can synthesize peptides containing L-amino acids, D-amino acids, β-amino acids, hydroxy acids, and hydroxy-, N- and C-methylated amino acids, together comprising approximately 300 different residues *(204)*. The potential for genetic engineering and combinatorial biosynthesis of NRPSs is enormous.

4.2.2. Domain or "Minimal Module" Swaps

Bacillus species serve as models for how genetic engineering of NRPS genes may be applied in actinomycetes. In *Bacillus subtilis*, the adenylation domain of the Leu module of a monomodular subunit of the surfactin peptide synthetase-coding region was substituted with NRPS modules of bacterial and fungal origin, and the hybrid genes encoded functional NRPSs with altered amino acid specificities *(211)*. The minimal module, a fragment of a module comprising the adenylation and thiolation domains, was also exchanged in a trimodular subunit of the surfactin NRPS coding region, yielding functional surfactin NRPS with altered substrate specificity *(212)*.

4.2.3. Module Fusions

Mootz et al. *(213)* have demonstrated the successful fusion of modules by coupling modules within the linker region between thiolation and condensation domains. This approach minimizes some the effects of the intrinsic editing function that resides in the condensation domain *(214–217)*.

4.2.4. Site-Directed Modifications of the Adenylation Domain

The crystal structure of the phenylalanine activating subunit of the gramicidin synthetase 1 (PheA) of *Bacillus brevis* complexed with AMP and L-phe has been determined *(9,218)*. Using this information, Stachelhaus et al. *(219)* and Challis et al. *(220)* have determined the specificity-conferring code used by adenylation domains of NRPSs, based upon comparative studies of the amino acids lining the Phe binding site in the gramicidin PheA module and many other NRPS sequences. The 9 *(220)* or 10 *(219)* specificity-conferring amino acids are located in a 100 amino-acid stretch in the adenylation domain. Using the deduced code for amino acid specificity, Stachelhaus et al. *(219)* introduced two mutations by site-directed mutagenesis in the PheA-encoding module and converted it to specify Leu. They also converted an Asp-specifying module to one that specified Asn by a single mutation. This methodology causes only minor changes in the protein structure, minimizing potential deleterious effects on protein-protein interactions, and may be relevant for certain changes of NRPS module specificities in actinomycetes.

4.2.5. Rules for Amino Acid Coupling and Modification

Much progress has been made on understanding the rules associated with specificity of amino acid activation, amino acid modification, peptide coupling, and peptide cyclization/release from the NRPS *(214–230)*. For instance, it now appears that the condensation domains exhibit specificity for the downstream amino acid, but also for the D-

or L-configuration of the upstream amino acid. As the rules become more refined, it will become easier to predict a priori which combinations of site-directed amino acid code changes; domain and module exchanges; deletions and additions; amino acid modifications; and cyclization/release reactions are likely to yield functional NRPSs.

4.2.6. Peptide Antibiotics Produced by Actinomycetes

There are a number of important secondary metabolites synthesized by the NRPS mechanism in actinomycetes. This chapter examines three peptide antibiotic classes as candidates for genetic engineering to exemplify how this methodology might be used.

(i) Daptomycin is a cyclic lipopeptide produced by *S. roseosporus (231)*. It is composed of a thirteen amino-acid peptide cyclized to a ten amino-acid ring and a three amino-acid tail. It has a ten-carbon-unit fatty acid (decanoic acid) attached to the N-terminal Trp. Daptomycin is a potent antibiotic with bacteriocidal activity against important Gram-positive pathogens, including methycillin-resistant *S. aureus*, vancomycin-resistant enterococci, and penicillin-resistant *S. pneumoniae (231–233)*. The genes for daptomycin biosynthesis have been cloned and sequenced *(37,40)*. Since *S. roseosporus* can be manipulated genetically by a variety of techniques, including transposition mutagenesis, conjugation from *E. coli*, homeologous recombination, and gene replacement using the *rpsL* system *(40,45,86,234)*, it represents a suitable system to apply genetic engineering of NRPS genes to generate novel derivatives of daptomycin.

(ii) Chloroeremomycin and balhimycin are glycopeptide antibiotics that are related to vancomycin and are produced by *Amycolatopsis* species. The genes for the biosynthesis of chloroeremomycin and balhimycin have been cloned and sequenced *(235–239)*, and the NRPS genes have been identified. A transformation and gene replacement system has been developed for *Amycolatopsis mediterranei*, the producer of balhimycin *(238,239)*. *A. mediterranei* may prove to be a suitable model system to generate hybrid NRPS genes to produce peptide analogs of vancomycin, chloroeremomycin, balhimycin, and teicoplanin. Alternatively, the cloning and sequencing of the NRPS genes encoding the glycopeptide A47934 from *Streptomyces toyocaensis* could provide the basis to carry out genetic engineering of glycopeptide genes, since *S. toyocaensis* can be manipulated by molecular genetic techniques *(236,240)*.

(iii) Pristinamycins I (PI) and II (PII) are antibiotics produced by *Streptomyces pristinaespiralis (96,241)*. In combination, PI and PII act synergistically against staphylococci, streptococci, and *H. influenzae (96)*. PI is composed of cyclohexadepsipeptides belonging to the B group of streptogramins, and PII of macrolactones of the A group of streptogramins. The genes for PI biosynthesis have been cloned and sequenced *(241,242,243)*, and could provide the basis for genetic engineering of pristinamycins.

4.3. Hybrid Polyketide/Peptide Structures

There are several of examples of secondary metabolites produced by actinomycetes that contain core structures composed of polyketide and peptide components. Some examples include rifamycin *(122)*, pristinamycin II *(244)*, rapamycin *(245–247)* and bleomycin *(248,249)*. Mixed NRPS/PKS multidomain proteins encoding siderophores and other biologically active substances have been reported in Gram-negative bacteria (*see* refs. *248–250*). In the case of rapamycin, the addition of pipecolate to the completed polyketide chain is catalyzed by a separate NRPS enzyme encoded by the *rapP* gene *(247)*. Shen and colleagues have begun exploring the rules for the coupling of amino acids and fatty acids *(248–250)*. This promises to be a very productive avenue for research and future applications, because the potential chemical combinations generated by coupling the biosynthetic mechanisms of type I PKS and NRPS multienzymes are staggering.

Rifamycin is a good example of a hybrid molecule synthesized by a mixed NRPS/ PKS pathway that is potentially set up for engineering. Rifamycin initiates polyketide assembly with an unusual starter unit, 3-amino-5-hydroxybenzoic acid (AHBA). The loading module for AHBA is an adenylation-thiolation didomain of NRPS origin. The 90 kb cluster of genes involved in rifamycin biosynthesis in *Amyolatopsis mediterranei* has been cloned, sequenced, and partially analyzed *(122,124,251)*. The seven genes encoding the biosynthesis of AHBA have been expressed in *S. coelicolor* and the recombinant produced AHBA *(251)*. The loading module can accept a variety of substituted benzoates *(252,253)*, so it seems possible that engineering this novel loading module coupled with the expression of the AHBA genes could be used to generate additional chemical diversity in other polyketide and peptide pathways.

4.4. Hybrid Glycosylations

Many macrolide, glycopeptide, and aminoglycoside antibiotics have been identified as secondary metabolites that are produced by actinomycetes. Antibiotic activity is often strongly influenced by sugar residues. For instance, tylactone, the macrolide precursor of tylosin devoid of the normal three sugars, lacks any detectable antibiotic activity *(254)*. Piepersberg and Distler *(255)* have described over 100 different deoxyhexoses found in secondary metabolites. Many glycosyltransferase and sugar biosynthetic genes involved in secondary metabolism have been cloned and sequenced, and many sugar biosynthetic genes are clustered or partially clustered in actinomycetes *(42,131,144,196,237,239,256–261)*. It is now technically feasible to construct novel combinations of macrolides or peptides with various sugar biosynthetic and glycosyltransfer functions.

4.4.1. Glycosylation of Macrolides (Type I Polyketides)

A number of different 6-deoxy sugars are present on 14- and 16-member macrolides *(255,256,260)*. Tylactone (protylonolide), the non-glycosylated precursor to tylosin, can be taken up and glycosylated by *S. ambofaciens*, the producer of spiramycin, when the spiramycin polyketide biosynthesis is inhibited by cerulenin *(262)*. The product chimeramycin contained the lactone of tylosin, the forosamine of spiramycin, and the two sugars (mycaminose and mycarose) present on both tylosin and spiramycin. In *S. fradiae*, one of the glycosyltransferases can transfer mycarose instead of mycaminose to tylactone in tylB mutants *(263,264)*. This glycosyltransferase has either relaxed NDP-sugar specificity or relaxed macrolide substrate specificity, and may be a candidate to for combinatorial glycosylations. Since the the sugar biosynthetic and glycosyltransferase genes are well-characterized in *S. fradiae (42,169–171,173,174)*, and since high-producing strains of *S. fradiae* can be genetically engineered *(20,81,93,94)*, *S. fradiae* may be a useful host for genetic engineering of glycosylations to produce novel 16-member macrolide antibiotics.

Streptomyces venezuelae produces four macrolide compounds, the 14-member narbomycin and pikromycin, and the 12-member methymycin and neomethymycin.

Narbomycin and pikromycin, and methymycin and neomethymycin, differ from each other in their hydroxylation patterns *(257,265)*. Narbomycin and pikromycin are natural ketolides containing a 3-keto group and desosamine attached at the 5 position. A single glycosyltransferase enzyme is capable of adding the amino sugar desosamine to the precursors of all four compounds *(257)*. Since the desosaminyltransferase has a relaxed substrate specificity and does not require a glycosylated 14-member lactones as substrate, and since narbomycin and pikromycin are natural ketolides, the *S. venezuelae* desosaminyltransferase gene *desVII* was a logical choice for genetic engineering ketolide production. Tang and McDaniel *(266)* cloned the NDP-desosamine biosynthetic genes and the desosaminyltransferase gene in *S. lividans*. They also introduced plasmids that encoded modified PKSs into the strain, and generated 20 glycosylated macrolides with antibiotic activity. Some of compounds had the 3-keto functionality needed to generate ketolides. The yields were generally low, but this approach provides a potential route to synthesize ketolides biochemically that may be optimized by mutagenesis or by gene shuffling.

Leadlay and colleagues *(267)* have developed a system to generate novel glycosylated macrolides in *S. erythraea*, using various mutants blocked in glycosyltransferase activities or in polyketide synthetase and glycosyltransferase activities. Using this system, they have generated 3-O-rhamnosyl-erythronolide B by cloning in the oleG2 rhamnosyltransferase gene from *Streptomyces antibioticus*. They have also generated desosaminyl-tylactone by cloning the tylM2 gene from *S. fradiae* and feeding tylactone to the *S. erythraea* recombinant. This demonstrated that both oleG2 and tylM2 glycosyltransferases may be useful for combinatorial biosynthesis.

4.4.2. Glycosylation of Type II Polyketides

Hybrid glycosylated anthracyclines have been generated by cloning PKS genes from *Streptomyces purpurascens* and *Streptomyces nogalater* into strains of *Streptomyces galilaeus* and *Streptomyces steffisburgensis* *(268–270)*. Genetic engineering has also been used to generate novel glycosylated derivatives of tetracenomycin *(271)* and urdamycin *(272)*. Hoffmeister et al. *(273)* generated functional chimeric glycosyltransferase enzymes derived from two genes involved in the urdamycin pathway. There data identified a 31 amino-acid region in near the N-terminus that determines both sugar donor and acceptor substrate specificity. This sets the stage for more extensive genetic engineering, perhaps by DNA shuffling, to generate hybrid glycosyltransferases with novel combinations of nucleotide sugar and polyketide substrate specificities.

4.4.3. Glycosylation of Glycopeptides

Glycopeptide antibiotics are potent bacteriocidal compounds that are active against Gram-positive bacteria. The glycopeptide antibiotics differ from each other in the core crosslinked heptapeptide and in their glycosylation patterns *(274,275)*. The simplest "glycopeptides" are comprised of chlorinated or chlorinated and sulfated heptapeptides devoid of sugars *(274)*. These compounds are highly active as antibacterials in vitro, but lack sufficient in vivo activity to be pursued clinically. The clinically relevant glycopeptides, vancomycin and teicoplanin, contain two and three sugars, respectively *(274)*.

Chloroeremomycin, which is closely related to vancomycin *(274)*, contains the vancomycin heptapeptide and chlorine residues, but has an additional amino sugar. Both amino sugars present in chloroeremomycin are epivancosamine residues rather than vancosamine, the amino sugar present in vancomycin. Chloroeremomycin has modest antibacterial activity against vancomycin-resistant enterococci *(275)*, whereas the p-chloro-biphenyl derivative oritavancin is highly active against vancomycin-resistant enterococci and other Gram-positive pathogens *(275–280)*. The glycosylations of the heptapeptide are critical for in vivo activity, and also provide a foundation for further chemical modifications to shift the target from primarily inhibition of transpeptidase activity (D-ala-D-ala binding) to inhibition of transglycosylase activity to treat vanco-mycin-resistant enterococci *(275,281,282)*.

There are three heptapeptide patterns and a variety of glycosylation patterns for natu-ral glycopeptides *(274,275)*. Genetic engineering may be used to shuffle these patterns to produce novel compounds. Solenberg et al. *(235,236)* initiated studies to generate recombinant strains to glycosylate A47934. A47934 has the same heptapeptide as the clinically important teicoplanin, but contains three chlorine residues rather than two, and contains an additional sulfate residue. If the three glycosylations of the chloro-eremomycin heptapeptide could be made on the A47934 heptapeptide, genetic engi-neering could provide a novel starting material for chemical modification to generate additional novel glycopeptides.

Solenberg et al. *(235,236)* cloned the three glycosyltransferase genes from the chloroeremomycin-producer and two from the vancomycin producer. They showed by expression studies in *E. coli* that the glucosyltransferase enzymes encoded by the *gtfE* gene from the vancomycin-producer, and the *gtfB* gene from the chloroeremomycin-producer, could attach glucose to the vancomycin aglycone. The GtfE enzyme (but not the GtfB enzyme) could also glycosylate A47934. Insertion of the *gtfE* gene into the chromosome of *S. toyocaensis* under the control of the *ermEp** promoter (but not the natural promoter) caused the production of glucosyl-A47934. This demonstrated that the heterologous glycosylation of a peptide substrate, related to the natural substrate, is feasible in vivo. However, only one of the two glucosyltransferases had relaxed sub-strate specificity, and it required expression from a strong constitutive promoter.

Building on the observations of Solenberg et al. *(236)*, Losey et al. *(283)* cloned, expressed, and purified the products of the *gtfB* and *gtfC* genes from the chloroeremomycin producer, and the *gtfD* and *gtfE* genes from the vancomycin producer. They confirmed that the GtfE glucosyltransferase had relaxed substrate specificity by showing that it could glucosylate the vancomycin aglycone and the tycoplanin aglycone, whereas the GtfB glucosyltransferase had good activity only on the vancomycin aglycone. They further demonstrated that GtfC and GtfD enzymes were proficient in attaching epivancosamine to the vancomycin psuedoaglycone and to the glucosylated teicoplanin when UDP-β-L-4-*epi*-vancosamine was used as the sugar donor. Importantly, the GtfD enzyme, which usually adds vancosamine to the vancomycin pseudoaglycone, was able to catalyze a reaction using an unnatural accepter substrate and an unnatural nucleotide sugar donor.

These studies demonstrate the potential to develop hybrid glycopeptide antibiotics through genetic engineering. Further advancements will require the cloning and expres-sion of genes involved in the biosynthesis of 6-deoxy sugars such as vancosamine, epivancosamine, and dehydrovancosamine *(238)*. The genes required for the conver-sion of 4-keto-6-deoxy-D-glucose to epivancosamine have been cloned and sequenced *(237)*. Furthermore, the genes for teicoplanin biosynthesis *(284)* and for balhimycin biosynthesis *(238,239)* have been cloned and partially characterized. This sets the stage for further genetic engineering of glycopeptide biosynthesis.

4.4.4. Modification of NDP-Sugar Biosynthesis

Another promising genetic engineering method to is to modify the NDP-sugar biosynthetic pathway. Madduri et al. *(285)* demonstrated the feasibility of this approach by blocking the step in TDP-L-daunosamine biosynthesis encoded by the *dnmV* gene in *Streptomyces peucetius*, and complementing the mutation with the *avrE* gene from the NDP-oleandrose biosynthetic pathway from *S. avermitilis*, or with the eryBIV gene from the NDP-desosamine pathway in *S. erythraea*. Recombinants produced 4'-epidoxorubicin, containing the L-4-epidaunosamine. In this case, the glycosyltransferase had broad cofactor specificity, facilitating the use of either of the 4' epimers.

In another example, Borosova et al. *(286)* have made mutations in genes involved in NDP-desosamine biosynthesis, and some of the intermediates were used by the *S. venezualae* glycosyltransferase to generate novel derivatives of pikromycin and narbomycin. This study and others *(257,266)* indicate that the glycosyltransferase has broad macrolide substrate and broad NDP-sugar cofactor specificities.

4.5. Acylations

4.5.1. Acylation of 16-Member Macrolides

The activity of macrolide antibiotics can be modified by the acylation of hydroxyl groups on the polyketide or on a sugar attached to the polyketide. An example is 3-*O*-acetyl 4'-*O*-isovaleryltylosin (AIV), a derivative of tylosin used in veterinary medicine *(157)*. AIV is produced by a two-step fermentation/bioconversion process, using *S. fradiae* to produce tylosin and *Streptomyces thermotolerans* blocked in polyketide biosynthesis to acylate tylosin *(287)*. Arisawa et al. *(288,289)* cloned and sequenced genes from *S. thermotolerans* required for the acylations and introduced them into *S. fradiae*. They first introduced the macrolide 3-*O*-acyltransferase gene (acyA) on the self-replicating plasmid, the recombinant produced 3-*O*-acetyltylosin *(288)*. The acylation of the 4'-hydroxyl position of the deoxysugar mycarose was more problematic. The cloned *acyB1* (*carE*) gene that encodes the 4'-*O*-acyltransferase did not express enough enzyme to convert tylosin to the 4'-acylated product. Expression of *acyB1* in *S. fradiae* required the presence of *acyB2*, a regulatory gene adjacent to *acyB1* in *S. thermotolerans*. When *acyB1*, *acyB2*, and *acyA* were introduced into *S. fradiae* on a replicating plasmid, the recombinant produced 3-*O*-acetyltylosin and 3-*O*-acetyl 4'-*O*-acetyltylosin, each at ~10% of the control yields *(289)*. When leucine was added to the medium to enhance isovaleryl-CoA production, the predominant product was AIV, but the overall yield was <20% of control. Since self-replicating plasmids often cause reduced secondary metabolite production *(1,14,15)* the reduced overall yield is not surprising. Improved yields may be obtained by inserting the three genes in a chromosomal neutral site.

Epp et al. *(290)* demonstrated that the *carE* (*acyB1*) gene expressed from its own promoter was sufficient to catalyze the conversion of spiramycin to 4'-isovaleryl-spiramyicin in *S. ambofaciens*. Perhaps *S. ambofaciens* expresses an *acyB2*-like function that is not expressed in *S. fradiae*. This example demonstrates that various *Streptomyces* species can regulate the expression of heterologous genes differently. These differences may be minimized by fusing the genes of interest to a strong constituitive pro-moter such as *ermEP** *(19)*.

4.6. Other Tailoring Enzymes

There are a number of other enzymes that modify peptides, polyketides, or sugar components of secondary metabolites. These include methyltransferases, hydroxy-lases, reductases, and haloperoxidases. These modifications are often critical for bio-logical activity. Many genes that encode tailoring functions have been cloned and sequenced, and these could be used directly for genetic engineering in some cases, or become starting materials for DNA family shuffling to alter substrate specificity. Two recent examples demonstrate the potential of genetic engineering applications of methyltransferases. In the first example, Gaisser et al. *(291)* have used the *S. erythraea* system for hybrid glycosylation to further embellish the hybrid com-pounds by sugar methylation. They demonstrated through a combination of genetic engineering and bioconversions that they could generate *O*-methylated rhamnosyl-derivatives of erythronolide, using cloned *O*-methyltransferase genes from the spinosyn biosynthetic pathway *(71)*. In the second example, Palatto et al. *(292)* cloned three *O*-methyltransferase genes from *Streptomyces olivaceus*, the producer of elloramycin. They demonstrated that two of the enzymes could *O*-methylate glycosylated polyketides containing sugars related to their natural substrates. These studies demonstrated that these sugar *O*-methyltransferases can recognize their homologous sugar substrates attached to heterologous polyketides in one case, or heterologous sugars attached to homologous substrates in the other, providing opportunities for further tailoring of hybrid glycosylated polyketides.

References

1. Baltz, R. H. (2001) Molecular genetic and combinatorial biology approaches to produce novel antibiotics, in *Antibiotic Development and Resistance*, (Hughes, D. and Anderson, D. eds.), Harwood Academic Publishers, Amsterdam. pp. 233–257.
2. Kieser, T., Bibb, M. J., Buttner, M. J., Chater, K. F., and Hopwood, D. A. (2000) Practical *Streptomyces* Genetics, The John Innes Foundation, Norwich, UK.
3. Baltz, R. H. (1986) Mutagenesis in *Streptomyces*, in *Manual of Industrial Microbiology and Biotechnology*, (Demain, A. L. and Soloman, N. A. eds.), American Society for Microbiology, Washington, DC. pp. 184–190.
4. Baltz, R. H. (1986) Mutation in Streptomyces, in *The Bacteria*, Vol. IX. Antibiotic-Producing *Streptomyces*, (Queener, S. W. and Day, L. E. eds.), Academic Press, New York, NY. pp. 61–93.
5. Matsushima, P. and Baltz, R. H. (1986) Protoplast fusion, in *Manual of Industrial Microbiol-ogy and Biotechnology*, (Demain, A. L. and Solomon, N. A. eds.) American Society for Microbiology, Washington, DC. pp. 170–183.
6. Vinci, V. A. and Byng, G. (1999) Strain improvement by nonrecombinant methods, in *Manual of Industrial Microbiology and Biotechnology*, Second Edition, (Demain, A. L. and Davies, J. E. eds.), ASM Press, Washington, DC. pp. 103–113.
7. Katz, L. (1997) Manipulation of modular polyketide synthases. *Chem. Rev.* **97**, 2557–2575.
8. Khosla, C. (1997) Harnesing the biosynthetic potential of modular polyketide synthases. *Chem. Rev.* **97**, 2577–2590.
9. Marahiel, M. A., Stachelhaus, T., and Mootz, H. D. (1997) Modular peptide synthetases involved in nonribosomal peptide synthesis. *Chem. Rev.* **97**, 2651–2673.

10. von Dohren, H., Keller, U., Vater, J., and Zocher, R. (1997) Multifunctional peptide synthetases. *Chem. Rev.* **97**, 2675–2705.

11. Hopwood, D. A. (1997) Genetic contributions to understanding polyketide synthases. *Chem. Rev.* **97**, 2465–2479.

12. Khosla, C. and Zawada, R. J. X. (1996) Generation of polyketide libraries via combinatorial biosynthesis. *Trends Biotechnol.* **14**, 335–341.

13. Baltz, R. H. (1995) Gene expression in recombinant *Streptomyces. Bioprocess Technol.* **22**, 309–381.

14. Baltz, R. H. and Hosted, T. J. (1996) Molecular genetic methods for improving secondary-metabolite production in actinomycetes. *Trends Biotechnol.* **14**, 245–250.

15. Baltz, R. H. (1997) Molecular genetic approaches to yield improvement in actinomycetes. *Drugs Pharm. Sci.* **82**, 49–62.

16. Hillemann, D., Puhler, A., and Wohlleben, W. (1991) Gene disruption and gene replacement in *Streptomyces* via single stranded DNA transformation of integrative vectors. *Nucl. Acid Res.* **19**, 727–731.

17. Oh, S.-H. and Chater, K. F. (1997) Denaturation of circular or linear DNA facilitates targeted integration of *Streptomyces coelicolor* A3(2): possible relevance to other organisms. *J. Bacteriol.* **179**, 122–127.

18. Mazodier, P., Petter, R., and Thompson, C. (1989) Intergeneric conjugation between *Escherichia coli* and *Streptomyces* species. *J. Bacteriol.* **171**, 3583–3585.

19. Baltz, R. H. (1998) Genetic manipulation of antibiotic producing Streptomyces. *Trends Microbiol.* **6**, 76–83.

20. Bierman, M., Logan, R., O'Brien, K., Seno, E. T., Rao, R. N., and Schoner, B. E. (1992) Plasmid cloning vectors for the conjugal transfer of DNA from *Escherichia coli* to *Streptomyces* spp. *Gene* **116**, 43–49.

21. Matsushima, P., Broughton, C. M., Turner, J. R., and Baltz R. H. (1994) Conjugal transfer of cosmid DNA from *Escherichia coli* to *Saccharopolyspora spinosa*: effects of chromosomal insertions on macrolide A83543 production. *Gene* **146**, 39–45.

22. Sun, Y., Zhou, X., Liu, J., Zhang, G., Tu, G., Kieser, T., et al. (2001) *Streptomyces nanchangensis*, a producer of the insecticidal polyether antibiotic nanchangmycin and the antiparasitic macrolide meilingmycin, contains multiple polyketide gene clusters. *Microbiology* **148**, 361–371.

23. McHenney, M. A. and Baltz, R. H. (1988) Transduction of plasmid DNA in *Streptomyces* and related genera by bacteriophage FP43. *J. Bacteriol.* **170**, 2276–2282.

24. McHenney, M. A. and Baltz, R. H. (1989) Transduction of plasmid DNA in macrolide producing streptomycetes. *J. Antibiotics* **42**, 1725–1727.

25. Matsushima, P., McHenney, M. A., and Baltz, R. H. (1989) Transduction and transformation of plasmid DNA in *Streptomyces fradiae* strains that express different levels of restriction. *J. Bacteriol.* **171**, 3080–3084.

26. Mazy-Servais, C., Baczkowski, D., and Dusart, J. (1997) Electroporation of intact cells of *Streptomyces parvulus* and *Streptomyces vinaceus. FEMS Microbiol. Lett.* **15**, 135–138.

27. Fitzgerald, N. B., English, R. S., Lampel, J. S., and Vanden Boom, T. J. (1998) Sonication-dependent electroporation of the erythromycin-producing bacterium *Saccharopolyspora erythraea. Appl. Environ. Microbiol.* **64**, 1580–1583.

28. English, R. S., Lampel, J. S., and Vanden Boom, T. J. (1998) Transformation of *Saccharopolyspora erythraea* by electroporation of germling spores: construction of propionyl CoA carboxylase mutants. *J. Ind. Microbiol. Biotechnol.* **21**, 219–224.

29. Pigac, J. and Schrempf, H. (1995) A simple and rapid method of transformation of *Streptomyces rimosus* R6 and other streptomycetes by electroporation. *Appl. Environ. Microbiol.* **61**, 352–356.

30. Hopwood, D. A., Bibb, M. J., Chater, K. F., and Kieser, T. (1987) Plasmid and phage vectors for gene cloning and analysis in *Streptomyces*. *Methods Enzymol.* **153**, 116–166.

31. Rao, R. N., Richardson, M. A., and Kuhstoss, S. (1987) Cosmid shuttle vectors for cloning and analysis of *Streptomyces* DNA. *Methods Enzymol.* **153**, 166–198.

32. Kieser, T. and Hopwood, D. A. (1991) Genetic manipulation of *Streptomyces*: Integrating vectors and gene replacement. *Methods Enzymol.* **204**, 430–458.

33. Rowe, C. J., Cortes, J., Gaisser, S., Staunton, J., and Leadlay, P. F. (1998) Construction of new vectors for high-level expression in actinomycetes. *Gene* **216**, 215–223.

34. Meurer, G. and Hutchinson, C. R. (1999). Genes for the biosynthesis of microbial secondary metabolites, in *Manual of Industrial Microbiology and Biotechnology*, Second Edition, (Demain, A. L., and Davies, J. E. eds.), ASM Press, Washington, DC, pp. 740–758.

35. Sosio, M., Guisino, F., Cappellano, C., Bossi, E., Puglia, A. M., and Donadio, S. (2000) Artificial chromosomes for antibiotic-producing actinomycetes. *Nat. Biotechnol.* **18**, 343–345.

36. Flett, F., Mersinias, V., and Smith, C. P. (1997) High efficiency intergeneric conjugal transfer of plasmid DNA from *Escherichia coli* to methyl DNA-restricting streptomycetes. *FEMS Microbiol. Lett.* **155**, 223–229.

37. Miao, V., Coeffet-LeGal, M.- F., Silva, C., Penn, J., Whiting, A., Brost, R., et al. (2001) Utility of BAC libraries in the study of large biosynthetic pathways. Abstracts of the 12th International Symposium of the Biology of Actinomycetes, Vancouver, Canada, August 5–9.

38. Jones, G. (1989) Cloning of *Streptomyces* genes involved in antibiotic synthesis and its regulation, in *Regulation of Secondary Metabolism in Actinomycetes*, (Shipiro, S., ed.), CRC Press, Boca Raton, FL, pp. 49–73.

39. Mohrle, V., Roos, U., and Bormann, C. (1995) Identification of cellular proteins involved in nikkomycin production in *Streptomyces tendae* Tu901. *Mol. Microbiol.* **15**, 561–571.

40. McHenney, M. A., Hosted, T. J., DeHoff, B. S., Rosteck, P. R., Jr., and Baltz, R. H. (1998) Molecular cloning and physical mapping of the daptomycin gene cluster from *Streptomyces roseosporus*. *J. Bacteriol.* **180**,143–151.

41. Seno, E. T. and Baltz, R. H. (1989). Structural organization and regulation of antibiotic biosynthesis and resistance genes in actinomycetes, in *Regulation of Secondary Metabolism in Actinomycetes*, (Shipiro, S,. ed.), CRC Press, Boca Raton, FL, pp. 1–48.

42. Bate, N., Butler, A. R., Smith, I. P., and Cundliffe, E. (2000) The mycarose-biosynthetic genes of *Streptomyces fradiae*, producer of tylosin. *Microbiology* **146**, 139–146.

43. Butler, A. R., Flint, S. A., and Cundliffe, E. (2001) Feedback control of polyketide metabolism during tylosin production. *Microbiology* **147**, 795–801.

44. Onaka, H., Nagagawa, T., and Horinouchi, S. (1998) Involvement of two A-factor receptor homologues in *Streptomyces coelicolor* A3(2) in the regulation of secondary metabolism and morphogenesis. *Mol. Microbiol.* **28**, 743–753.

45. Hosted, T. J. and Baltz, R. H. (1997) Use of rpsL for dominance selection and gene replacement in *Streptomyces roseosporus*. *J. Bacteriol.* **179**, 180–186.

46. Hosoya, Y., Muramatsu, H., and Ochi, K. (1998) Acquisition of certain streptomycin-resistant (str) mutations enhances antibiotic production in bacteria. *Antimicrob. Agents Chemother.* **42**, 2041–2047.

47. Fisher, S. H., Bruton, C. J., and Chater, K. F. (1987) The glucose kinase gene of *Streptomyces coelicolor* and its use in selecting spontaneous deletions for desired regions of the genome. *Mol. Gen. Genet.* **206**, 35–44.

48. Buttner, M. J., Chater, K. F., and Bibb, M. J. (1990) Cloning, disruption, and transcriptional analysis of three RNA polymerase sigma factor genes of *Streptomyces coelicolor* A3(2). *J. Bacteriol.* **172**, 3367–3378.

49. van Wezel, G. P. and Bibb, M. J. (1996) A novel plasmid vector that uses the glucose kinase gene (glkA) for the positive selection of stable gene disruptions in *Streptomyces*. *Gene* **182**, 229–230.

50. McDaniel, R., Thamchaipenet, A., Gufstafsson, C., Fu, H., Betlach, M., Ashley, G., et al. (1999) Multiple genetic modifications of the erythromycin polyketide synthase to produce a library of novel "unnatural" natural products. *Proc. Natl. Acad. Sci. USA* **96,** 1846–1851.

51. Ziermann, R., and Betlach, M. C. (2000) A two vector system for the production of recombinant polyketides in *Streptomyces. J. Ind. Microbiol. Biotechnol.* **24,** 46–50.

52. Xue, Q., Hutchinson, C. R., and Santi, D. V. (1999) A multi-plasmid approach to preparing large libraries of polyketides. *Proc. Natl. Acad. Sci. USA* **96,** 11,740-11,745.

53. Encell, L. P., Landis, D. M., and Loeb, L. A. (1999) Improved enzymes for cancer gene therapy. *Nat. Biotechnol.* **17,** 143–147.

54. Matsumura, T., Mijai, K., Trakulnaleamsai, S., Yomo, T., Shima, Y., Miki, S., et al. (1999) Evolutionary molecular engineering by random elongation mutagenesis. *Nat. Biotechnol.* **17,** 58–61.

55. Patten, P. A., Howard, R. J., and Stemmer, W. P. C. (1997) Applications of DNA shuffling to pharmaceuticals and vaccines. *Curr. Opin. Biotechnol.* **8,** 724–733.

56. Skandalis, A., Encell, L. P., and Loeb, L. A. (1997) Creating novel enzymes by applied molecular evolution. *Chem. Biol.* **4,** 889–898.

57. Zhao, H. and Arnold, F. H. (1997) Optimization of DNA shuffling for high fidelity recombination. *Nucl. Acid Res.* **25,** 1307–1308.

58. Zhao, H., Giver, L., Shao, Z., Affholter, J. A., and Arnold, F. H. (1998) Molecular evolution by staggered extension process (StEP) in vitro recombination. *Nat. Biotechnol.* **16,** 258–261.

59. Stemmer, W. P. C. (1994) Rapid evolution of a protein in vitro by DNA shuffling. *Nature* **370,** 389–391.

60. Stemmer, W. P. C. (1994) DNA shuffling by random fragmentation and reassembly: In vitro recombination for molecular evolution. *Proc. Natl. Acad. Sci. USA* **91,** 10,747–10,751.

61. Coco, W. M., Levinson, W. E., Crist, M. J., Hektor, H. J., Darzins, A., Pienkos, P. T., et al. (2001) DNA shuffling method for generating highly recombined genes and evolved enzymes. *Nat. Biotechnol.* **19,** 354–359.

62. Zhang, J.-H., Dawes, G., and Stemmer, W. P. C. (1997) Directed evolution of fucosidase from a galactosidase by DNA shuffling and screening. *Proc. Natl. Acad. Sci. USA* **94,** 4504–4509.

63. Crameri, A., Whitehorn, E. A., Tate, E., and Stemmer, W. P. C. (1996) Improved green fluorescent protein by molecular evolution using DNA shuffling. *Nat. Biotechnol.* **14,** 315–319.

64. Crameri, A., Dawes, G., Rodriquez, E., Jr., Silver, S., and Stemmer, W. P. C. (1997) Molecular evolution of an arsenate detoxification pathway by DNA shuffling. *Nat. Biotechnol.* **15,** 436–438.

65. Crameri, A., Raillard, S.-A., Bermudez, E., and Stemmer, W.P.C. (1998) DNA shuffling of a family of genes from diverse species accelerates directed evolution. *Nature* **391,** 288–291.

66. Christians, F. C., Scapozza, A., Crameri, A., Folkers, G., and Stemmer, W. P. C. (1999) Directed evolution of thymidine kinase for AZT phosphorylation using DNA shuffling. *Nat. Biotechnol.* **17,** 259–264.

67. Chang, C.-C. J., Chen, T. T., Cox, B. W., Dawes, G. N., Stemmer, W. P. C., Punnonen, J., et al. (1999) Evolution of a cytokine using DNA family shuffling. *Nat. Biotechnol.* **17,** 793–797.

68. von Dohren, H. and Kleinkauf, H. (1997) Enzymology of peptide synthetases, in *Biotechnology of Antibiotics*, (Strohl, W. R., ed.), Marcel Dekker, New York, NY, pp. 217–240.

69. Socio, M., Bossi, E., Bianchi, A., and Donadio, S. (2000) Multiple peptide synthetase gene clusters in Actinomycetes. *Mol. Gen. Genet.* **264,** 213–221.

70. Ōmura, S., Ikeda, H., Ishikawa, J., Hanamoto, A., Takahashi, C., Shinose, M., et al. (2001) Genome sequence of an industrial microorganism *Streptomyces avermitilis*: deducing the ability of producing secondary metabolites. *Proc. Natl. Acad. Sci. USA* **98,** 12,215–12,220.

71. Waldron, C., Matsushima, P., Rosteck, P. R., Jr., Broughton, M. C., Turner, J., Madduri, K., et al. (2001) Cloning and analysis of the spinosad biosynthetic gene cluster of *Saccharopolyspora spinosa. Chem. Biol.* **8,** 487–499.

72. Baltz, R. H. (1995) Gene expression in recombinant *Streptomyces*. *Bioprocess Technol.* **22,** 309–381.

73. Bibb, M. (1996) 1995 Colworth Prize Lecture. The regulation of antibiotic production in *Streptomyces coelicolor* A3(2). *Microbiology* **142,** 1335–1344.

74. Bate, N., Butler, A. R., Gandecha, A. R. and Cundliffe, E. (1999) Multiple regulatory genes in the tylosin-biosynthetic cluster of *Streptomyces fradiae*. *Chem. Biol.* **6,** 617–624.

75. Stratigopoulos, G. and Cundliffe, E. (2001) Expression analysis of the tylosin-biosynthetic gene cluster: pivotal regulatory role of the tylQ product. *Chem. Biol.* **9,** 71–78.

76. Stratigopoulos, G. and Cundliffe, E. (2001) Inactivation of a transcriptional repressor during empirical improvement of the tylosin producer, *Streptomyces fradiae*. *J. Ind. Microbiol. Biotechnol.* **28,** 219–224.

77. Solenberg P. J. and Baltz R. H. (1991) Transposition of Tn5096 and other IS493 derivatives in *Streptomyces griseofuscus*. *J. Bacteriol.* **173,** 1096–1104

78. Hahn, D. R., Solenberg, P. J., and Baltz, R. H. (1991) Tn5099, a xylE promoter probe transposon for *Streptomyces* spp. *J. Bacteriol.* **173,** 5573–5577

79. Baltz, R. H., Hahn, D. R., McHenney, M. A., and Solenberg, P. J. (1992) Transposition of Tn5096 and related transposons in *Streptomyces* species. *Gene* **115,** 61–65.

80. Solenberg, P. J. and Baltz, R. H. (1994) Hyper-transposing derivatives of the streptomycete insertion sequence IS493. *Gene* **147,** 47–54.

81. Solenberg, P. J., Cantwell, C. A., Tietz, A. J., Mc Gilvray, D., Queener, S. W., and Baltz, R. H. (1996) Transposition mutagenesis in *Streptomyces fradiae*: identification of a neutral site for the stable insertion of DNA by transposon exchange. *Gene* **168,** 67–72.

82. Fowler, K. and Kieser, T. (2001) The *Streptomyces coelicolor* A3(2) transposon mutant "Megalibrary": a valuable tool for investigating gene function. Abstracts of the 12th International Symposium on the Biology of Actinomycetes, Vancouver, BC, p. 81.

83. Gehring, A. M., Nodwell, J. R., Beverly, S. M., and Losick, R. (2000) Genomewide insertional mutagenesis in *Streptomyces coelicolor* reveals additional genes involved in morphological differentiation. *Proc. Natl. Acad. Sci. USA* **97,** 9642–9647.

84. Fernandes, P. J., Powell, J. A. C., and Archer, A. C. (2001) Construction of Rhodococcus random mutagenesis libraries using Tn5 transposition complexes. *Microbiology* **147,** 2529–2536.

85. Baltz, R. H., McHenney, M. A., and Solenberg, P. J. (1993). Properties of transposons derived from IS493 and applications in streptomycetes, in *Industrial Microorganisms: Basic and Applied Molecular Genetics*, (Baltz, R. H., Hegeman, G., and Skatrud, P. L. eds.), American Society for Microbiology, Washington, DC, pp. 51–56.

86. McHenney, M. A. and Baltz, R. H. (1996) Gene transfer and transposition mutagenesis in *Streptomyces roseosporus*: mapping of insertions that influence daptomycin or pigment production. *Microbiology* **142,** 2363–2373.

87. Gravius, B., Glocker, D., Pigac, J., Pandza, K., Hranueli, D., and Cullum, J. (1994) The 387 kb linear plasmid of *Streptomyces rimosus* and its interactions with the chromosome. *Microbiology* **140,** 2271–2277.

88. Peschke, U., Schmidt, H., Zhang, H.-Z. and Piepersberg, W. (1995) Molecular characterization of the lincomycin-production gene cluster of *Streptomyces lincolnensis* 78-11. *Mol. Microbiol.* **16,** 1137–1158.

89. Muth, G., Nussbaumer, B., Wohlleben, W., and Pühler, A. (1989) A vector system with temperature-sensitive replicon for gene disruption and mutational cloning in streptomycetes. *Mol. Gen. Genet.* **219,** 341–348.

90. Birch, A. and Cullum, J. (1985) Temperature-sensitive mutants of the *Streptomyces plasmid* pIJ702. *J. Gen. Microbiol.* **131,** 1299–1303.

91. McHenney, M. A. and Baltz, R. H. (1991) Transposition of Tn5096 from a temperature-sensitive transducible plasmid in *Streptomyces* spp. *J. Bacteriol.* **173,** 5578–5581.

92. Seno, E. T. and Baltz, R. H. (1982) S-adenosyl-L-methionine: macrocin O-methyltransferase activities in a series of *Streptomyces fradiae* mutants which produce different levels of the macrolide antibiotic tylosin. *Antimicrob. Agents Chemother.* **21,** 758–763.

93. Baltz, R. H. and Seno E. T. (1988) Genetics of *Streptomyces fradiae* and tylosin biosynthesis. *Ann. Rev. Microbiol.* **42,** 547–574.

94. Baltz, R. H., McHenney, M. A., Cantwell, C. A., Queener, S. W., and Solenberg, P. J. (1997) Applications of transposition mutagenesis in antibiotic producing streptomycetes. *Antonie Leeuwenhoek* **71,** 179–187.

95. Baltz, R. H. (1998) New genetic methods to improve secondary metabolite production in *Streptomyces. J. Ind. Microbiol. Biotechnol.* **20,** 360–363.

96. Sezonov, G., Blanc, V., Bamas-Jacques, N., Friedman, A., Pernodet, J. –L. and Guerineau, M. (1997) Complete conversion of antibiotic precursor to pristinamycin IIA by overexpression of *Streptomyces pristinaespiralis* biosynthetic genes. *Nat. Biotechnol.* **15,** 349–353.

97. Baltz, R. H. (1999) Mutagenesis, in *Encyclopedia of Bioprocess Technology: Fermentation, Biocatalysis, and Separation*, (Flickinger, M. C. and Drew, S. W. eds.), Wiley, New York, NY, pp. 1819–1822.

98. Madduri, K., Waldron, C., Matsushima, P., Broughton, M. C., Crawford, K., Merlo, D. J., et al. (2001) Genes for the biosynthesis of spinosyns: applications for yield improvement in *Saccharopolyspora spinosa. J. Ind. Microbiol. Biotechnol.* **27,** 399–402.

99. Overbeek, R., Fonstein, M., D'Sousza, M., Pusch, G. D. and Malstev, N. (1999) The use of gene clusters to infer functional coupling. *Proc. Natl. Acad. Sci. USA* **96,** 2896–2901.

100. Overbeek, R., Larsen, N., Smith, W., Maltzev, N., and Selkov, E. (1997) Representation of function: the next step. *Gene* **191,** GC1–GC9.

101. Selkov, E., Jr., Grechkin, Y., Mikhailova, N., and Selkov, E. (1998) MPWW: the metabolic pathways database. *Nucleic Acids Res.* **26,** 43–45

102. Selkov, E., Maltzev, N., Olsen, G. J., Overbeek, R., and Whitman, W. B. (1997) A reconstruction of the metabolism of *Methanococcus jannaschii. Gene* **197,** GC11–GC26.

103. Covert, M. W., Schilling, C. H., Famili, I., Edwards, J. S., Goryanin, I. I., Selkov, E., et al. (2001) Metabolic modeling of microbial strains in silico. *Trends Biochem. Sci.* **3,** 179-186.

104. Overbeek, R., Larsen, N., Pusch, G. D., D'Souza, M., Selkov, E., Jr., Kyrpides, N., Fonstein, M., Maltsev, N., and Selkov, E. (2000) WIT: integrated system for high-throughput genome sequence analysis and metabolic reconstruction. *Nucleic Acids Res.* **28,** 123–125.

105. Tao, H., Bausch, C., Richmond, C., Blattner, F. R., and Conway, T. (1999) Functional genomics: expression analysis of *Escherichia coli* growing on minimal and rich media. *J. Bacteriol.* **181,** 6425–6440.

106. Pomposiello, P. J., Bennik, M. H. J., and Demple, B. (2001) Genome-wide transcriptional profiling of the *Escherichia coli* responses to superoxide stress and sodium salicylate. *J. Bacteriol.* **183,** 3890–3902.

107. Zheng, M., Wang, X., Templeton, L. J., Smulskoi, D. R., LaRossa, R. A., and Storz, G. (2001) DNA microarray-mediated transcriptional profiling of the *Escherichia coli* response to hydrogen peroxide. *J. Bacteriol.* **183,** 4562–4570.

108. DeLisa, M. P., Wu, C.-F., Wang, L., Valdes, J. J., and Bentley, W. E. (2001) DNA microarray-based identification of genes controlled by autoinducer 2-stimulated quorum sensing in *Escherichia coli. J. Bacteriol.* **183,** 5239–5247.

109. Luccini, S., Thompson, A., and Hinton, J. C. D. (2001) Microarrays for microbiologists. *Microbiology* **147,** 1403–1414.

110. Price, C. W., Fawcett, P., Ceremonie, H., Su, N., Murphy, C. K., and Youngman, P. (2001) Genome-wide analysis of the general stress response in *Bacillus subtilis. Mol. Microbiol.* **41,** 757–774.

111. VanBogelen, R. A., Schiller, E. E., Thomas, J. D., and Neidhardt, F. C. (1999) Diagnosis of cellular states of microbial organisms using proteomics. *Electrophoresis* **20,** 2149–2159.

112. Pandey, A. and Mann, M. (2000) Proteomics to study genes and genomes. *Nature* **405,** 837–846.

113. Voradsky, J., Li, X.-M., and Thompson, C. J. (1997) Identification of prokaryotic development stages by statistical analyses of two-dimentional gel patterns. *Electrophoresis* **18,** 1418–1428.

114. Voradsky, J., Li, X.-M., Dale, G., Folcher, M., Nguyen, L., Viollier, P. H., et al. (2000) Developmental control of stress stimulons in *Streptomyces coelicolor* revealed by statistical analyses of global gene expression patterns. *J. Bacteriol.* **182,** 4979–4986.

115. Schilling, C. H., Edwards, J. S., and Palsson, B. O. (1999) Toward metabolic phenomics: analysis of genomic data using flux balances. *Biotechnol. Prog.* **15,** 288–295.

116. Hutchinson, C. R. and Fujii, I. (1995) Polyketide synthase gene manipulation: a structure-function approach in engineering novel antibiotics. *Annu. Rev. Microbiol.* **49,** 201–238.

117. Haydock, S. F., Aparicio, J. F., Molnar, I., Schwecke, T., Khaw, L. E., Konig, A., et al. (1995) Divergent sequence motifs correlated with the substrate specificity of (methyl)malonyl-CoA:acyl carrier protein transacylase domains in modular polyketide synthases. *FEBS Lett.* **374,** 246–248.

118. Staunton, J., Caffrey, P., Aparicio, J. F., Roberts, G. A., Bethell, S. S., and Leadlay, P. F. (1996) Evidence for a double–helical structure for modular polyketide synthases. *Nat. Struct. Biol.* **3,** 188–192.

119. Aparicio, J. F., Molnar, I., Schwecke, T., Konig, A., Hayfcock, S. F., Khaw, L. E., et al. (1996) Organization of the biosynthetic gene cluster of rapamycin in *Streptomyces hygroscopicus*: analysis of the enzymatic domains in the modular polyketide synthase. *Gene* **196,** 9–16.

120. Leadlay, P. F. (1997) Combinatorial approaches to polyketide biosynthesis. *Curr. Opin. Chem. Biol.* **1,** 162–168.

121. Weissman, K. L., Timoney, M., Brycroft, M. C., Grice, P., Hanefeld, U., Staunton, J., et al. (1997) The molecular basis of Celmer's rules: the stereochemistry of the condensation step in chain elongation on the erythromycin polyketide synthase. *Biochemistry* **36,** 13,849–13,855.

122. August, P., Tang, L., Yoon, Y. J., Ning, S., Muller, R., Yu, T.-W., et al. (1998) Biosynthesis of the ansamycin antibiotic rifamycin: deductions from the molecular analysis of the rif biosynthetic gene cluster of Amycolatopsis mediterranei S699. *Chem. Biol.* **5,** 69–79.

123. Weissman, K. J., Brycroft, M., Cutter, A. L., Hanefeld, U., Frost, E. J., Timoney, M. C., et al. (1998) Evaluating precursor-directed biosynthesis towards novel erythromycins through in vitro studies on a bimodular polyketide synthase. *Chem. Biol.* **5,** 743–754.

124. Staunton, J. (1998) Combinatorial biosynthesis of erythromycin and complex polyketides. *Curr. Opin. Chem. Biol.* **2,** 339–345.

125. Cane, D. E., Walsh, C. T., and Khosla, C. (1998) Harnessing the biosynthetic code: combinations, permutations, and mutations. *Science* **282,** 63–68.

126. Bohm, I., Holzbaur, U., Cortes, J., Staunton, J., and Leadlay, P. F. (1998) Engineering of a minimal polyketide synthase, and targeted alteration of the stereospecificity of polyketide chain extension. *Chem. Biol.* **5,** 407–412.

127. Tang, L., Yoon, Y. J., Choi, C.-Y., and Hutchinson, C. R. (1998) Characterization of the enzymatic domains in the modular polyketide synthase involved in rifamycin B biosynthesis in Amycolatopsis mediterranei. *Gene* **216,** 255–265.

128. Gokhale, R. S., Tsuji, S. Y., Cane, D. E., and Khosla, C. (1999) Dissecting and exploiting intermodular communication in polyketide synthases. *Science* **284,** 482–485.

129. Gokhale, R. S., Hunziker, D., Cane, D. E., and Khosla, C. (1999) Mechanism and specificity of the terminal thioesterase domain from the erythromycin polyketide synthase. *Chem. Biol.* **6,** 117–125.

130. Holzbaur, I. E., Harris, R. C., Bycroft, M., Cortes, J., Bisang, C., Staunton, J., et al. (1999) Molecular basis of Celmer's rules: the role of the two ketoreductase domains in the control of chirality by the erythromycin modular polyketide synthase. *Chem. Biol.* **6**, 189–195.
131. Ikeda, H., Nonomiya, T., Usami, M., Ohta, T., and Omura, S. (1999) Organization of the biosynthetic gene cluster of the polyketide anthelmintic macrolide avermectin in *Streptomyces avermitilis*. *Proc. Natl. Acad. Sci. USA* **96**, 9509–9514.
132. Rowe, C. J., Bohm, I. U., Thomas, I. P., Wilkinson, B., Rudd, B. A. M., Foster, G., et al. (2001) Engineering a polyketide with a longer chain by insertion of an extra module into the erythromycin-producing polyketide synthase. *Chem. Biol.* **8**, 475–485.
133. Holzbaur, I. E., Ranganathan, A., Thomas, I. P., Kearney, D. J. A., Reather, J. A., Rudd, B. A. M., et al. (2001) Molecular basis of Celmer's rules: role of the ketosynthase domain in epimerization and demonstration that ketoreductase domains can have altered product specificity with unnatural substrates. *Chem. Biol.* **8**, 329–340.
134. Kao, C. M., Luo, G. L., Katz, L., Cane, D. E., and Khosla, C. (1995) Manipulation of macrolide ring size by directed mutagenesis of a modular polyketide synthase. *J. Am. Chem. Soc.* **117**, 9105–9106.
135. Kirst, H. A. (1991) New macrolides: expanded horizons for an old class of antibiotics. *J. Antimicrob. Chemother.* **28**, 787–790.
136. Agouridas, C., Bonnefoy, A., and Chantot, J. F. (1997) Antibacterial activity of RU 64004 (HMR 3004), a novel ketolide derivative active against respiratory pathogens. *Antimicrob. Agents Chemother.* **42**, 2149–2158.
137. Agouridas, C., Denis, A., Auger, J-M., Beneditti, Y., Bonnefoy, A., Bretin, F., et al. (1998) Synthesis and antimicrobial activity of ketolides (6-O-methyl-3-oxoerythromycin derivatives): a new class of antibacterials highly potent against macrolide-resistant and -susceptible respiratory pathogens. *J. Med. Chem.* **41**, 4080–4100.
138. Ednie, L. M., Spangler, S. K., Jacobs, M. R., and Applebaum, P. C. (1997) Susceptibilities of 228 penecillin- and erythromycin-susceptible and -resistant pneumococci to RU 64004, a new ketolide, compared with susceptibilities to 16 other agents. *Antimicrob. Agents Chemother.* **41**, 1033–1036.
139. Schulin, T., Wennersten, R. C., Moellering, R. C., Jr., and Eliopoulos, G. M. (1997) In vitro activity of RU 64004, a new ketolide antibiotic, against gram-positive bacteria. *Antimicrob. Agents Chemother.* **41**, 1196–1202.
140. Barry, A. L., Fuchs, P. C., and Brown, S. D. (1998) In vitro activities of the ketolide HMR 3647 against gram-positive clinical isolates and *Haemophilus influenzae*. *Antimicrob. Agents Chemother.* **42**, 2138–2140.
141. Barry, A. L., Fuchs, P. C., and Brown, S. D. (1998) Antipneumococcal activities of a ketolide (HMR 3647), a streptogramin (quinupristin-dalphopristin), a macrolide (erythromycin), and a lincosamide (clindamycin). *Antimicrob. Agents Chemother.* **42**, 945–946.
142. Barry, A. L., Fuchs, P. C., and Brown, S. D. (2001) In vitro activity of the ketolide ABT-773. *Antimicrob. Agents Chemother.* **45**, 2922–2924.
143. Goldstein, E. J. C., Conrads, G., Citron, D. M., Merriam, C. V., Warren, Y., and Tyrrell, K. (2001) In vitro activities of ABT-773, a new ketolide, against aerobic and anaerobic pathogens isolated from antral sinus puncture specimens from patients with sinusitis. *Antimicrob. Agents Chemother.* **45**, 2363–2367.
144. Seno, E. T., and Hutchinson, C. R. (1986). The biosynthesis of tylosin and erythromycin: model systems for studies of the genetics and biochemistry of antibiotic formation, in *The Bacteria*, vol. IX, Antibiotic-Producing *Streptomyces* (Queener, S. W. and Day, L. E., eds.), Academic Press, New York, NY, pp. 231–279.
145. Staunton, J. and Wilkinson, B. (1997) Biosynthesis of erythromycin and rapamycin. *Chem. Rev.* **97**, 2611–2629.

146. Rodriques, E. and McDaniel, R. (2001) Combinatorial biosynthesis of antimicrobials and other natural products. *Curr. Opin. Microbiol.* **4,** 526–534.

147. Ruan, X., Pereda, A., Stassi, D. L., Zeidner, D., Summers, R. G., Jackson, M., et al. (1997) Acyltransferase domain substitutions in erythromycin polyketide synthase yield novel erythromycin derivatives. *J. Bacteriol.* **179,** 6416–6425.

148. Stassi, D. L., Kakavas, S. J., Reynolds, K. A., Gunawardana, G., Swanson, S., Zeidner, D., et al. (1998) Ethyl-substituted erythromycin derivatives produced by directed metabolic engineering. *Proc. Natl. Acad. Sci. USA* **95,** 7305–7309.

149. Kao, C. M., Katz, L., and Khosla, C. (1994) Engineered biosynthesis of a complete macrolactone in a heterologous host. *Science* **265,** 509–512

150. Cortes, J., Wiesmann, K. E., Roberts, G. A., Brown, M. J., Staunton, J., and Leadlay, P. F. (1995) Repositioning of a domain in a modular polyketide synthase to promote specific chain cleavage. *Science* **268,** 1487–1489.

151. Bedford, D., Jacobson, J. R., Luo, G., Cane, D. E., and Khosla, C. (1996) A functional chimeric polyketide synthase generated via domain replacement. *Chem. Biol.* **3,** 827–831.

152. Oliynyk, M., Brown, M. J. B., Cortes, J., Staunton, J., and Leadlay, P. F. (1996) A hybrid modular polyketide synthase obtained by domain swapping. *Chem. Biol.* **3,** 833–839.

153. Denoya, C. D., Fedechko, R. W., Hafner, E., W., McArthur, H. A., Morgenstern, M. R., Skinner, D. D., et al. (1995) A second branched-chain alpha-keto acid dehydrogenase gene cluster (bkdFGH) from *Streptomyces avermitilis*: its relationship to avermectin biosynthesis and the construction of a bkdF mutant suitable for the production of novel antiparasitic avermectins. *J. Bacteriol.* **177,** 3504–3511.

154. Dutton, C. J., Gibson, S. P., Goudie, A. C., Holden, K. S., Pacey, M. S., Ruddock, J. C., et al. (1991) Novel avermectins produced by mutational biosynthesis. *J. Antibiot.* **44,** 357–365.

155. Marsden, F. A., Wilkenson, B., Cortes, J., Dunster, N. J., Staunton, J., and Leadlay, P. F. (1998) Engineering broader specificity into an antibiotic-producing polyketide synthase. *Science* **279,** 199–202.

156. Pacey, M. S., Dirlam, J. P., Geldart, R. W., Leadlay, P. F., McArthur, H. A., McCormick, E. L., et al. (1998) Novel erythromycins from a recombinant *Saccharopolyspora erythraea* strain NRRL 2338 pIG1. I. Fermentation, isolation and biological activity. *J. Antibiot.* **51,** 1029–1034.

157. Kirst, H. A. (1994) Semi-synthetic derivatives of 16-membered macrolide antibiotics. *Prog. Med. Chem.* **31,** 265–295.

158. Wilson, R. C. (1984) Macrolides in veterinary practice, in *Macrolide Antibiotics: Chemistry, Biology, and Practice*, (Ōmura, S., ed.), Academic Press, Tokyo, pp. 301–347.

159. Nakayama, I. (1984) Macrolides in clinical practice, in *Macrolide Antibiotics: Chemistry, Biology, and Practice*, (Ōmura, S., ed.), Academic Press, Tokyo, pp. 261–300.

160. Olafsson, S., Berstat, A., Bang, C. J., Nysaeter, G., Coll, P., Tefera, S., et al. (1999) Spiramycin is comparable to oxytetracycline in eradicating H. pylori when given with ranitidine bismuth citrate and metronidizole. *Aliment. Pharmacol. Ther.* **13,** 651–659.

161. Rubinstein, E. and Keller, N. (1998) Spiramycin renaissance. *J. Antimicrob. Chemother.* **42,** 572–576.

162. Klugman, K. P., Capper, T., Widdowson, C. A., Koornhof, H. J., and Moser, W. (1998) Increased activity of 16-membered lactone ring macrolides against erythromycin-resistant *Streptoccoccus pyogenes* and *Streptococcus pneumoniae*: characterization of South African isolates. *J. Antimicrob. Chemother.* **42,** 729–734.

163. Okamoto, R., Fukumoto, T., Nomura, H., Kiyoshima, K., Nakamura, K., and Takamatsu, A. (1980) Physicochemical properties of new acyl derivatives of tylosin produced by microbial transformation. *J. Antibiot.* **33,** 1300–1308.

164. Baltz, R. H. and Seno, E. T. (1981) Properties of *Streptomyces fradiae* mutants blocked in bisynthesis of the macrolide antibiotic tylosin. *Antimicrob. Agents Chemother.* **20,** 214–225.

165. Kirst, H. A., Debono, M., Willard, K. E., Trudell, B. A., Toth, T. E., Turner, J. R., et al. (1986) Preparation and evaluation of 3,4''-ester derivatives of 16-membered macrolide antibiotics related to tylosin. *J. Antibiot.* **39,** 1724–1735.

166. Cox, K. L., Fishman, S. E., Larson, J. L., Stanzak, R., Reynolds, P. A., Yeh, W. K., et al. 1986 The use of recombinant DNA techniques to study tylosin biosynthesis and resistance in *Streptomyces fradiae. J. Nat. Prod.* **49,** 971–980.

167. Fishman, S. E., Cox, K., Larson, J. L., Reynolds, P. A., Seno, E. T., Yeh, W.-K., et al. (1987) Cloning genes for the biosynthesis of a macrolide antibiotic. *Proc. Natl. Acad. Sci. USA* **84,** 8248–8252.

168. Beckmann, R. J., Cox, K., and Seno, E. T. (1989). A cluster of tylosin biosynthetic genes is interrupted by a structurally unstable segment containing four repeated sequences, in *Genetics and Molecular Biology of Industrial Microorganisms* (Hershberger, C. L., Queener, S.W., and Hegeman, G., eds.), American Society for Microbiology, Washington, DC, pp. 176–186.

169. Merson-Davies, L. and Cundliffe, E. (1994) Analysis of five tylosin biosynthetic genes from the tylIBA region of the *Streptomyces fradiae* genome. *Mol. Microbiol.* **13,** 349–355.

170. Gandecha, A. R., Large, S. L., and Cundliffe, E. (1997) Analysis of four tylosin biosynthetic genes from the tylLM region of the *Streptomyces fradiae* genome. *Gene* **184,** 197–203.

171. Wilson, V. and Cundliffe, E. (1998) Characterization and targeted disruption of a glycosyl-transferase gene in the tylosin producer, *Streptomyces fradiae. Gene* **214,** 95–100.

172. Butler, A. R., Bate, N., and Cundliffe, E. (1999) Impact of thioesterase activity on tylosin biosynthesis in *Streptomyces fradiae. Chem. Biol.* **6,** 287–292.

173. Bate, N. and Cundliffe, E. (1999) The mycinose-biosynthetic genes of *Streptomyces fradiae,* producer of tylosin. *J. Ind. Microbiol. Biotechnol.* **23,** 118–122.

174. Fouces, R., Mellado, E., Diez, B., and Barredo, J. L. (1999) The tylosin biosynthetic cluster from *Streptomyces fradiae*: genetic organization of the left region. *Microbiology* **145,** 855–868.

175. Birmingham, V. A., Cox, K. L., Larson, J. L., Fishman, S. E., Hershberger, C. L., and Seno, E. T. (1986) Cloning and expression of a tylosin resistance gene from a tylosin-producing strain of *Streptomyces fradiae. Mol. Gen. Genet.* **204,** 532–539.

176. Zalacain, M. and Cundliffe, E. (1989) Methylation of 23s rRNA caused by tlrA (ermSF), a tylosin resistance determinant from *Streptomyces fradiae. J. Bacteriol.* **171,** 4254–4260.

177. Keleman, G. H., Zalacain, M., Culebras, E., Seno, E. T., and Cundliffe, E. (1994) Transcriptional attenuation control of the tylosin-resistance gene tlrA in *Streptomyces fradiae. Mol. Microbiol.* **14,** 833–842.

178. Kovalic, D., Giannattasio, R. B., Jin, H.-J., and Weisblum, B. (1994) 23s rRNA domain V, a fragment that can be specifically methylated in vitro by the ermSF (tlrA) methyltransferase. *J. Bacteriol.* **176,** 6992–6998.

179. Rosteck, P. R., Jr., Reynolds, P. A., and Hershberger, C. L. (1991) Homology between proteins controlling *Streptomyces fradiae* tylosin resistance and ATP-binding transport. *Gene* **102,** 27–32.

180. Zalacain, M. and Cundliffe, E. (1991) Cloning of tlrD, a fourth resistance gene, from the tylosin producer, *Streptomyces* fradiae. *Gene* **97,** 137-142.

181. Gandecha, A. R. and Cundliffe, E. (1996) Molecular analysis of tlrD, an MLS resistance determinant from the tylosin producer, *Streptomyces fradiae. Gene* **180,** 173–176.

182. Fish, S. A. and Cundliffe, E. (1996) Structure-activity studies of tylosin-related macrolides. *J. Antibiot.* **49,** 1044–1048.

183. Baltz, R. H., Seno, E. T., Stonesifer, J., and Wild, G. M. (1983) Biosynthesis of the macrolide antibiotic tylosin: A preferred pathway from tylactone to tylosin. *J. Antiobiot.* **36,** 131–141.

184. Fish, S. A. and Cundliffe, E. (1997) Stimulation of polyketide metabolism in *Streptomyces fradiae* by tylosin and its glycosylated precursors. *Microbiology* **143,** 3871–3876.

185. Matsushima, P. and Baltz, R. H. (1985) Efficient plasmid transformation of *Streptomyces ambofaciens* and *Streptomyces fradiae* protoplasts. *J. Bacteriol.* **163,** 180–185.

186. Ford, L. M., Eaton, T. E., and Godfrey, O. W. (1990) Selection of *Streptomyces ambofaciens* mutants that produce large quantities of spiramycin and determination of optimal conditions for spiramycin production. *Appl. Environ. Microbiol.* **56,** 3511–3514.

187. Epp, J. K., Huber, M. L. B., Turner, J. R., Goodson, T., and Schoner, B. E. (1989) Production of hybrid macrolide antibiotics in *Streptomyces ambofaciens* and *Streptomyces lividans* by introduction of a cloned carbomycin biosynthetic gene from *Streptomyces thermotolerans*. *Gene* **85,** 293–301.

188. Richardson, M. A., Kuhstoss, S., Huber, M. L. B., Ford, L., Godfrey, O., Turner, J. R., et al. (1990) Cloning of spiramycin biosynthetic genes and their use in constructing *Streptomyces ambofaciens* mutants defective in spiramycin biosynthesis. *J. Bacteriol.* **172,** 3790–3798.

189. Geistlich, M., Losick, R., Turner, J. R., and Rao, R. N. (1992) Characterization of a novel regulatory gene governing the expression of a polyketide synthase gene in *Streptomyces ambofaciens*. *Mol. Microbiol.* **6,** 2019–2029

190. Kuhstoss, S., Huber, M., Turner, J. R., Paschal, J. W., and Rao, R. N. (1996) Production of a novel polyketide through the construction of a hybrid polyketide synthase. *Gene* **183,** 231–236.

191. Hutchinson, C. R. (1997) Biosynthesis of daunorubicin and tetracenomycin C. *Chem. Rev.* **97,** 2525–2535.

192. Binnie, C., Warren, M., and Butler, M. J. (1989) Cloning and heterologous expression in *Streptomyces lividans* of *Streptomyces rimosus* genes involved in oxytetracycline biosynthesis. *J. Bacteriol.* **171,** 887–895.

193. Han, L., Yang, K., Ramalingam, E., Mosher, R. H., and Vining, L. C. (1994) Cloning and characterization of polyketide synthase genes for jadomycin B biosynthesis in *Streptomyces venezuelae* ISP5230. *Microbiology* **140,** 3379–3389.

194. Decker, H., Rohr, J., Motamedi, H., Zahner, C. R., and Hutchinson, C. R. (1995) Identification of *Streptomyces olivaceus* Tu*2353* genes involved in the production of the polyketide elloramycin. *Gene* **166,** 121–126.

195. Decker, H. and Haag, S. (1995) Cloning and characterization of a polyketide synthase gene from *Streptomyces fradiae* Tu*2717*, which carries the genes for the biosynthesis of the angucycline antibiotic urdamycin A and a gene probably involved in its oxygenation. *J. Bacteriol.* **177,** 6126–6136.

196. Ichinose, K., Bedford, D. J., Tornus, D., Bechthold, A., Bibb, M., Revill, W. P., et al. (1998) The granaticin biosynthetic gene cluster of *Streptomyces violaceoruber* Tu22: sequence analysi sand expression in a heterologous host. *Chem. Biol.* **5,** 647–659.

197. Lombo, F., Brana, A. F., Mendez, C., and Salas, J. A. (1999) The mithramycin gene cluster of *Streptomyces argillaceus* contains a positive regulatory gene and two repeated DNA sequences that are located at both ends of the cluster. *J. Bacteriol.* **181,** 642–647.

198. Westrich, L., Domann, S., Faust, B., Bedford, D., Hopwood, D. A., and Bechthold, A. (1999) Cloning and characterization of a gene cluster from *Streptomyces cyanogenus* S136 probably involved in landomycin biosynthesis. *FEMS Microbiol. Lett.* **170,** 381–387.

199. Kantola, J., Kunnari, T., Hautala, A., Hakala, J., Ylinko, K., and Mantsala, P. (2000) Elucidation of anthraocyclinone biosynthesis by stepwise cloning of genes for anthracyclines from three different *Streptomyces* spp. *Microbiology* **146,** 155–163.

200. Tsoi, C. J. and Khosla, C. (1995) Combinatorial biosynthesis of unnatural natural products— the polyketide example. *Chem. Biol.* **2,** 355–362.

201. Khosla, C. and Zawada, R. J. X. (1996) Generation of polyketide libraries via combinatorial biosynthesis. *Trends Biotechnol.* **14,** 335–341.

202. Kleinkauf, H. and von Dohren, H. (1990) Bioactive peptides—recent advances and trends, in *Biochemistry of Peptide Antibiotics*, (Kleinkauf, H. and von Dohren, H., eds.), Walter de Gruyter, Berlin, pp. 1–31.

203. von Dohren, H. (1990) Compilation of peptide structures—a biogenetic approach, in *Biochemistry of Peptide Antibiotics*, (Kleinkauf, H., and von Dohren, H., eds.), Walter de Gruyter, Berlin, pp. 411–507.

204. Kleinkauf, H. and von Dohren, H. (1996) A nonribosomal system of peptide biosynthesis. *Eur. J. Biochem.* **236**, 135–151.

205. Marahiel, M. A., Stachelhaus, T., and Mootz, H. D. (1997) Modular peptide synthetases involved in nonribosomal peptide synthesis. *Chem Rev.* **97**, 2651–2673.

206. von Dohren, H., Keller, U., Vater, J., and Zocher, R. (1997) Multifunctional peptide synthetases. *Chem. Rev.* **97**, 2675–2705.

207. Stachelhaus, T. and Marahiel, M. A. (1995) Modular structure of genes encoding multifunctional peptide synthetases required for non-ribosomal peptide synthesis. *FEMS Microbiol. Lett.* **125**, 3–14.

208. Stein, T. and Vater, J. (1996) Amino acid activation and polymerization at modular multienzymes in nonribosomal peptide biosynthesis. *Amino Acids* **10**, 201–227.

209. Zuber, P. and Marahiel, M. A. (1997) Structure, function, and regulation of genes encoding multidomain peptide synthetases, in *Biotechnology of Antibiotics* (Strohl, W. R., ed.), Marcel Dekker, New York, NY, pp. 187–216.

210. Konz, D. and Marahiel, M. A. (1999) How do peptide synthetases generate structural diversity? *Chem. Biol.* **6**, 39–48.

211. Stachelhaus, T., Schneider, A., and Marahiel, M. A. (1995) Rational design of peptide antibiotics by targeted replacement of bacterial and fungal domains. *Science* **269**, 69–72.

212. Schneider, A., Stachelhaus, T., and Marahiel, M. A. (1998) Targeted alteration of the substrate specificity of peptide synthetases by rational module swapping. *Mol. Gen. Genet.* **257**, 308–318.

213. Mootz, H. D., Schwarzer, D., and Marahiel, M. A. (2000) Construction of hybrid peptide synthetases by module and domain fusions. *Proc. Natl. Acad. Sci. USA* **97**, 5848–5853.

214. Belshaw, P. J., Walsh, C. T., and Stachelhaus, T. (1999) Aminoacyl-CoAs as probes of condensation domain selectivity in nonribosomal peptide synthetases. *Science* **284**, 486–489.

215. Ehmann, D. E., Trauger, J. W., Stachelhaus, T., and Walsh, C. T. (2000) Aminoacyl-SNACs as small-molecule substrates for the condensation domains of nonribosomal peptide synthetases. *Chem. Biol.* **7**, 765–772.

216. Linne, U. and Marahiel, M. A. (2000) Control of directionality in nonribosomal peptide synthesis: role of the condensation domain in preventing misinitiation and timing of epimerization. *Biochemistry* **39**, 10,439–10,447.

217. Luo, L. and Marahiel, M. A. (2001) Kinetic analysis of three activated intermediates generated by the initiation module pheATE of gramacidin S synthetase. *Biochemistry* **40**, 5329–5337.

218. Conti, E., Stachelhaus, T., Marahiel, M. A., and Brick, P. (1997) Structural basis for the activation of phenylalanine in the non-ribosomal biosynthesis of gramacidin S. *EMBO J.* **16**, 4174–4183.

219. Stachelhaus, T., Mootz, H. D., and Marahiel, M. A. (1999) The specificity-conferring code of adenylation domains in nonribosomal peptide synthetases. *Chem. Biol.* **6**, 493–505.

220. Challis, G. L., Ravel, J., and Townsend, C. A. (2000) Predictive, structure-based model of amino acid recognition by nonribosomal peptide synthetase adenylation domains. *Chem. Biol.* **7**, 211–224.

221. Mootz, H. D. and Marahiel, M. A. (1999) Design and application of multimodular peptide synthetases. *Curr. Opin. Biotechnol.* **10**, 341–348.

222. von Dohren, H., Dieckmann, R., and Pavela-Vrancic, M. (1999) The nonribosomal code. *Chem. Biol.* **6**, R273–R279.

223. Doekel, S. and Marahiel, M. A. (2000) Dipeptide formation on engineered hybrid peptide synthetases. *Chem. Biol.* **7**, 373–384.

224. Stachelhaus, T. and Walsh, C. T. (2000) Mutational analysis of the epimerase domain in the initiation module pheATE of gramacidin S synthetase. *Biochemistry* **39**, 5775–5787.

225. Trauger, J. W., Kohli, R. M., Mootz, H. D., Marahiel, M. A., and Walsh, C. T. (2000) Peptide cyclization catalysed by the thioesterase domain of tyrocidin synthetase. *Nature* **407**, 215–218.

226. Kohli, R. M., Trauger, J. W., Schwarzer, D., Marahiel, M. A., and Walsh, C. T. (2001) Generality of peptide cyclization catalyzed by isolated thioesterase domains of nonribosomal peptide synthetases. *Biochemistry* **40**, 7099–7108.

227. Walsh, C. T., Chen, H., Keating, T. A., Hubbard, B. K., Losey, H. C., Luo, L., et al. (2001) Tailoring enzymes that modify nonribosomal peptides during and after chain elongation on NRPS assembly lines. *Curr. Opin. Chem. Biol.* **5**, 525–534.

228. Trauger, J. W., Kohli, R. M., and Walsh, C. T. (2001) Cyclization of backbone-substituted peptides catalyzed by the thioesterase domain from the tyrocidin nonribosomal peptide stnthetase. *Biochemistry* **40**, 7092–7098.

229. Marshall, C. G., Burkart, M. D., Keating, T. A., and Walsh, C. T. (2001) Heterocycle formation in vibriobactin biosynthesis: alternative substrate utilization and identification of a condensed intermediate. *Biochemistry* **40**, 10,655–10,663.

230. Linne, U., Doekel, S., and Marahiel, M. A. (2001) Portability of epimerization domain and role of peptidyl carrier protein on epimerization activity in nonribosomal peptide synthetases. *Biochemistry* **40**, 15,824–15,834.

231. Baltz, R. H. (1997) Lipopeptide antibiotics produced by *Streptomyces roseosporus* and *Streptomyces fradiae*, in *Biotechnology of Antibiotics*, (Strohl, W. R., ed.), Marcel Dekker, New York, NY, pp. 415–435.

232. Tally, F. P., Zeckel, M., Wasilewski, M. M., Carini, C., Berman, C. L., Drusano, G. L., et al. (1999) Daptomycin: a novel agent for Gram-positive infections. *Exp. Opin. Invest. Drugs* **8**, 1223–1238.

233. Tally, F. P. and DeBruin, M. F. (2000) Development of daptomycin for Gram-positive infections. *J. Antimicrob. Chemother.* **46**, 523–526.

234. Hosted, T. J. and Baltz, R. H. (1996) Mutants of *Streptomyces roseosporus* that express enhanced recombination within partially homologous genes. *Microbiology* **142**, 2803–2813

235. Solenberg, P. J., Matsushima, P., Stack, D. R., Wilkie, S. C., Thompson, R. C., and Baltz, R. H. (1997) Glycosyltransferase genes from *Amycolatopsis orientalis* and their use to produce novel glycopeptide antibiotics. *Dev. Ind. Microbial. Biotechnol.* **34**, 115–121.

236. Solenberg, P. J., Matsushima, P., Stack, D. R., Wilkie, S. C., Thompson, R. C., and Baltz, R. H. (1997) Production of hybrid glycopeptide antibiotics in vitro and in *Streptomyces toyocaensis*. *Chem. Biol.* **4**, 195–202.

237. van Wageningen, A. M. A., Kirkpatrick, P. N., Williams, D. H., Harris, B. R., Kershaw, J. K., Lennard, N. L., et al. (1997) Sequencing and analysis of genes involved in the biosynthesis of a vancomycin group antibiotic. *Chem. Biol.* **5**, 155–162.

238. Pelzer, S., Sussmuth, R., Heckmann, D., Recktenwald, J., Huber, P., Jung, G., et al. (1999) Identification and analysis of the balhimycin biosynthetic gene cluster and its use for manipulating glycopeptide biosynthesis in *Amycolatopsis mediterranei* DSM5908. *Antimicrob. Agents Chemother.* **43**, 1565–1573.

239. Rectenwald, J., Shawky, R., Puk, O., Pfennig, F., Keller, U., Wohlleben, W., et al. (2002) Nonribosomal biosynthesis of vancomycin-type antibiotics: a heptapeptide backbone and eight peptide synthetase modules. *Microbiology* **148**, 1105–1108.

240. Matsushima, P. and Baltz, R.H. (1996) A gene cloning system for '*Streptomyces toyocaensis*'. *Microbiology* **142**, 261–267.

241. Blanc, V., Gil, P., Bamas-Jacques, N., Lorenzon, S., Zagorec, M., Sheuniger, J., et al. (1997) Identification and analysis of genes form *Streptomyces pristinaespiralis* encoding enzymes

involved in the biosynthesis of the 4-diamino-L-phenylalanine precursor of pristinamycin I. *Mol. Microbiol.* **23**, 191–202.

242. Crecy-Lagard, V., Blanc, V., Gil, P., Naudin, L., Lorenzon, S., Famechon, A., et al. (1997) Pristinamycin I biosynthesis in *Streptomyces pristinaespiralis*: molecular characterization of the first two structural peptide synthetase genes. *J. Bacteriol.* **179**, 705–713.

243. Crecy-Lagard, V., Saurin, W., Thibaut, D., Gil, P., Naudin, L., Crouzet, J., et al. (1997) Streptogramin B biosynthesis in *Streptomyces pristinaespiralis* and *Streptomyces virginiae*: molecular characterization of the last structural peptide synthetase gene. *Antimicrob. Agents Chemother.* **41**, 1904–1909.

244. Thibaut, D., Ratet, N., Bisch, D., Faucher, D., Debussche, L., and Blanche, F. (1995) Purification of the two-enzyme system catalyzing the oxidation of the D-proline residue of pristinamycin IIA biosynthesis. *J. Bacteriol.* **177**, 5199–5205.

245. Schwecke, T., Aparicio, J. F., Molnar, I., Konig, A., Khaw, L. E., Haydock, S. F., et al. (1995) The biosynthetic gene cluster for the polyketide immunosuppressant rapamycin. *Proc. Natl. Acad. Sci. USA* **92**, 7839–7843.

246. Molnar, I., Aparicio, J. F., Haydock, S. F., Khaw, L. E., Schwecke, T., Konig, A., et al. (1996) Organization of the biosynthetic gene cluster for rapamycin in *Streptomyces hygroscopicus*: analysis of genes flanking the polyketide synthase. *Gene* **169**, 1–7.

247. Konig, A., Schwecke, T., Molnar, I., Bohm, G. A., Lowden, P. A. S., et al. (1997) The pipecolate-incorporating enzyme for the biosynthesis of the immunosuppressant rapamycin: nucleotide sequence analysis, disruption and heterologous expression of rapP from *Streptomyces hygroscopicus*. *Eur. J. Biochem.* **247**, 526–534.

248. Du, L., Sanchez, C., Chen, M., Edwards, D. J., and Shen, B. (2000) The biosynthetic gene cluster for the antitumor drug bleomycin from *Streptomyces verticillus* ATCC 15003 supporting functional interactions between nonribosomal peptide synthetases and a polyketide synthase. *Chem. Biol.* **7**, 623–642.

249. Du, L., Sanchez, C., and Shen, B. (2001) Hybrid peptide-polyketide natural products: biosynthesis and prospects toward engineering novel molecules. *Metabol. Eng.* **3**, 78–95.

250. Du, L. and Shen, B. (2001) Biosynthesis of hybrid peptide-polyketide natural products. *Curr. Opin. Drug Discov. Devel.* **4**, 215–218.

251. Yu, T. W., Muller, R., Muller, M., Zhang, X., Draeger, G., Kim, C. G., et al. (2001) Mutational analysis and reconstituted expression of the biosynthetic genes involved in the formation of 3-amino-5-hydroxybenzoic acid, the starter unit of rifamycin biosynthesis in *Amycolatopsis mediterranei* S699. *J. Biol. Chem.* **276**, 12,546–12,555.

252. Floss, H. G. (2001) Antibiotic biosynthesis: from natural to unnatural compounds. *J. Ind. Microbiol. Biotechnol.* **27**, 183–194.

253. Admiral, S. J., Walsh, C. T., and Khosla, C. (2001) The loading module of rifamycin synthase is an adenylation-thiolation didomain with substrate tolerance for substituted benzoates. *Biochemistry* **40**, 6116–6123.

254. Kirst, H. A., Wild, G. M., Baltz, R. H., Hamill, R. L., Ott, J. L., Counter, F. T., et al. (1982) Structure-activity studies among 16-membered macrolide antibiotics related to tylosin. *J. Antibiot.* **35**, 1675–1682.

255. Piepersberg, W. and Distler, J. (1997). Aminoglycosides and sugar components in other secondary metabolites, in *Biotechnology, 2nd ed., Vol. 7, Products of Secondary Metabolism*, (Rehm, H.-J., Reed, G., Pühler, A., and Stadler, P., eds.), VCH, Weinheim, pp. 399–488.

256. Kirshing, A., Bechtold, A. F.-W., and Rohr, J. (1997) Chemical and biochemical aspects of deoxysugars and deoxysugar oligosaccharides. *Top. Curr. Chem.* **188**, 1–84.

257. Xue, Y., Zhao, L., Liu, H. W., and Sherman, D. H. (1998) A gene cluster for macrolide antibiotic biosynthesis in *Streptomyces venezuelae*: architecture of metabolic diversity. *Proc. Natl. Acad. Sci. USA* **95**, 12,111–12,116.

258. Trefzer, A., Salas, J. A., and Bechtold, A. (1999) Genes and enzymes involved in deoxysugar biosynthesis in bacteria. *Nat. Prod. Rep.* **16,** 283–299.

259. Weitnauer, G., Muhlenweg, A., Trefzer, A., Hoffmeister, D., Sussmuth, R. D., Jung, G., et al. (2001) Biosynthesis of the orthosomycin antibiotic avilamycin A: deductions from the molecular analysis of the avi biosynthetic gene cluster of *Streptomyces* viridochromogenes Tu57 and production of new antibiotics. *Chem. Biol.* **8,** 569–581.

260. Olano, C., Lomovskaya, N., Fonstein, L., Roll, J. T., and Hutchinson, C. R. (1999) A two plasmid system for the glycosylation of polyketide antibiotics: Bioconversion of epsilon-rhodomycinone to rhodomycin. *Chem. Biol.* **6,** 845–855.

261. Wohlert, S., Lomovskaya, N., Kulowski, K, Fonstein, L., Occi, J. L., Gewain, K. M., et al. (2001) Insights about the biosynthesis of the avermectin deoxysugar L-oleandrose through heterologous expression of *Streptomyces* avermitilis deoxysugar genes in *Streptomyces lividans. Chem. Biol.* **7,** 681–700.

262. Ōmura, S., Sadakane, N., Tanaka, Y., and Matsubara, H. (1983) Chimeramycins: new macrolide antibiotics produced by hybrid biosynthesis. *J. Antibiot.* **36,** 927–930.

263. Ōmura, S., Sadakane, N., Kitao, C., Matsubara, H., and Nakagawa, A. (1980) Production of mycarosyl protolonolide by a mycaminose idiotroph from the tylosin-producing strain *Strep-tomyces fradiae* KA-427. *J. Antibiot.* **33,** 913-914.

264. Jones, N. D., Chaney, M. O., Kirst, H. A., Wild, G. M., Baltz, R. H., Hamill, R. L., et al. (1982) Novel fermentation products from *Streptomyces fradiae*: X-ray crystal structure of 5-O-mycarosyltylactone and proof of the absolute configuration of tylosin. *J. Antibiot.* **35,** 420–425.

265. Xue, Y., Wilson, D., Zhao, L., Liu, H. W., and Sherman, D. H. (1998) Hydroxylation of the macrolactones YC-17 and narbomycin is mediated by the pikC-encoded cytochrome P450 in *Streptomyces venezuelae. Chem. Biol.* **5,** 661 667.

266. Tang, L. and McDaniel, R. (2001) Construction of desosamine containing polyketide libraries using a glycosyltransferase with broad substrate specificity. *Chem. Biol.* **8,** 547–555.

267. Gaisser, S., Reather, J., Wirtz, G., Kellenberger, L., Staunton, J., and Leadlay, P. F. (2000) A defined system for hybrid macrolide biosynhtesis in *Saccharopolyspora erythraea. Mol. Microbiol.* **36,** 391–401.

268. Niemi, J., Ylihonko, K., Hakala, J., Parssinen, R., Kopio, R., and Mantsala, P. (1994) Hybrid anthracycline antibiotics: production of the new anthracyclines by cloned genes of *Streptomy-ces purpurascens* in *Streptomyces galilaensis. Microbiology* **140,** 1351–1358.

269. Ylihonko, K., Hakala, J., Kunari, T., and Mantsala, P. (1996) Production of hybrid anthra-cycline antibiotics by heterologous expression of *Streptomyces nogalater* nogalamycin bio-synthesis genes. *Microbiology* **142,** 1965–1972.

270. Kunnari, T., Tuikkanen, J., Hautala, A., Hakala, J., Ylihonko, K., and Mantsala, P. (1997) Isolation and characterization of the 8-methoxy steffimycins and generation of 2,8-demethoxy steffimycins in *Streptomyces steffisburgensis. J. Antibiot.* **50,** 496–501.

271. Blanco, G., Patallo, E. P., Brana, A. F., Trefzer, A., Bechthold, A., Rohr, J., et al. (2001) Identification of a sugar flexible glycosyltransferase from *Streptomyces olivaceus*, the pro-ducer of the antitumor polyketide elloramycin. *Chem. Biol.* **8,** 253–263.

272. Hofmeister, D., Ichinose, K., Domann, S., Faust, B., Trefzer, A., Drager, G., et al. (2000) The NDP-sugar co-substrate concentration and the enzyme expression level influence the substrate specificity of glycosyltransferases: cloning and characterization of deoxysugar biosynthetic genes of the urdamycin biosynthetic gene cluster. *Chem. Biol.* **4,** 821–831.

273. Hofmeister, D., Ichinose, K., and Bechthold, A. (2001) Two sequence elements of glycosyltransferases involved in urdamycin biosynthesis are responsible for substrate specificity and enzymatic activity. *Chem. Biol.* **8,** 557–567.

274. Lancini, C. and Cavalleri, B. (1990) Glycopeptide antibiotics of the vancomycin group, in *Biochemistry of Peptide Antibiotics*, (Kleinkauf, H. and von Dohren, H., eds.), Walter de Gruyter, Berlin, pp.159–178.

275. Nicas, T. I. and Cooper, R. D. G. (1997) Vancomycin and other glycopeptides, in *Biotechnology of Antibiotics*, (Strohl, W. R., ed.), Marcel Dekker, New York, NY, pp. 363–392.

276. Cooper, R. D. G., Snyder, N. J., Zweifel, M. J., Staszak, M. A., Wilkie, S. C., Nicas, T. I., et al. (1996) Reductive alkylation of glycopeptide antibiotics: synthesis and antibacterial activity. *J. Antibiot.* **49,** 575–581.

277. Baltch, A. L., Smith, R. P., Ritz, W. J., and Bopp, L. H. (1998) Comparison of inhibitory and bacteriocidal activities and postantibiotic effects of LY333328 and ampicillin used singly and in combination against vancomycin-resistant *Enterococcus faecium*. *Antimicrob. Agents Chemother.* **42,** 2564–2568.

278. Garcia-Garrot, F., Cercenado, E., Alcala, L. and Bouza, E. (1998) In vitro activity of the new glycopeptide LY333328 against multiply resistant gram-positive clinical isolates. *Antimicrob. Agents Chemother.* **42,** 2452–2455.

279. Kaatz, G. W., Seo, S. M., Aeschlimann, J. R., Houlihan, H. H., Mercier, R.-C., and Ribak, M. J. (1998) Efficacy of LY333328 against experimental methicillin-resistant *Staphylococcus aureus* endocarditis. *Antimicrob. Agents Chemother.* **42,** 981–983.

280. Saleh-Mghir A., Lefort, A., Petegnief, Y., Dautrey, S., Vallois, J.-M., Le Guludec, D., et al. (1999) Activity and diffusion of LY333328 in experimental endocarditis due to vancomycin-resistant *Enterococcus faecalis*. *Antimicrob. Agent Chemother.* **43,** 115–120.

281. Ge, M., Chen, Z., Onishi, H. R., Kohler, J., Silver, L. L., Kerns, R., et al. (1999) Vancomycin derivatives that inhibit peptidoglycan biosynthesis without binding D-ala-D-ala. *Science* **284,** 504–510.

282. Eggert, U., Ruiz, N., Falcone, B. V., Branstrom, A. A., Goldman, R. C., Silhavy, T. J., et al. (2001) Genetic basis for activity differences and glycolipid derivatives of vancomycin. *Science* **294,** 361–364.

283. Losey, H. C., Peczuh, M. W., Chen, Z., Eggert, U. S., Dong, S. D., Pelczer, I., et al. (2001) Tandem action of glycosyltransferases in the maturation of vancomycin and teicoplanin aglycones: novel glycopeptides. *Biochemistry* **40,** 4745–4755.

284. Sosio, M., Bianchi, A., Bossi, E., and Donadio, S. (2000) Teicoplanin biosynthesis genes in *Actinoplanes teichomyceticus*. *Antonie Leeuwenhoek* **78,** 379–384.

285. Madduri, K., Kennedy, J., Rivoli, G., Inventi-Solari, A., Zanuso, G., Colombo, A. L., et al. (1998) Production of the antitumor drug (4'-epidoxorubicin) and its precursor by a genetically engineered strain of *Streptomyces peucetius*. *Nat. Biotechnol.* **16,** 69–74.

286. Borosova, S. A., Zhao, L., Sherman, D. H., and Liu, H.-W. (1999) Biosynthesis of desosamine: construction of a new macrolide carrying a genetically designed sugar moity. *Org. Lett.* **1,** 133–136.

287. Okamoto, R., Fukumoto, T., Nomura, H., Kiyoshima, K., Nakamura, K., and Takamatsu, A. (1980) Physiochemical properties of new acyl derivatives of tylosin produced by microbial transformation. *J. Antibiot.* **33,** 1300–1308.

288. Arisawa A., Kawamura, N., Takeda, K., Tsunekawa, H., Okamura, K., and Okamoto, R. (1994) Cloning of the macrolide antibiotic biosynthetic gene acyA, which encodes 3-O-acyltransferase, from *Streptomyces thermotolerans* and its use for direct fermentative production of a hybrid macrolide antibiotic. *Appl. Environ. Microbiol.* **60,** 2657–2660.

289. Arisawa, A., Kawamura, N., Narita, T., Kojima, I., Okamura, K., Tsunekawa, H., et al. (1996) Direct fermentative production of acyltylosins by genetically-engineered strains of *Streptomyces fradiae*. *J. Antibiot.* **49,** 349–354.

290. Epp, J. K., Huber, M. L. B., Turner, J. R., Goodson, T., and Schoner, B. E. (1989) Production of hybrid macrolide antibiotics in *Streptomyces ambofaciens* and *Streptomyces lividans* by

introduction of a cloned carbomycin biosynthetic gene from *Streptomyces thermotolerans*. *Gene* **85,** 293–301.

291. Gaisser, S., Lill, R., Wirtz, G., Grolle, F., Staunton, J., and Leadlay, P. F. (2001) New erythromycin derivatives from *Saccharopolyspora erythraea* using sugar O-methyltransferases from the spinosyn biosynthetic gene cluster. *Mol. Microbiol.* **41,** 1223–1231.

292. Patallo, E. P., Blanco, G., Fischer, C., Brana, A. F., Rohr, J., Mendez, C., et al. (2001) Deoxysugar methylation during biosynthesis of the antitumor polyketide ellaromycin by *Streptomyces olivaceus*. *J. Biol. Chem.* **276,** 18,765–18,774.

7

Culture Medium Optimization and Scale-Up
for Microbial Fermentations

Neal C. Connors

1. Introduction to Microbial Fermentations

The art of fermentation has been around for centuries. Brewing and food-based fermentations were the beginnings of the fermentation industry as we know it today. At the turn of the twentieth century, fermentations were used to make chemicals such as glycerol for the manufacture of explosives. With all that history, the discovery of penicillin may have been the real "coming out" party for the fermentation industry. For the first time in its history, fermentation was used to produce a product with a tangible benefit—saving lives.

Since the dawn of the "penicillin age," the science of fermentation development has been influenced by advances in recombinant DNA technology, analytical techniques, computer sciences, and most recently, genomics and proteomics. This says nothing of the advances that have been made in the core disciplines of microbial physiology and biochemical engineering. Although the tools that are at the disposal of scientists and engineers have increased over the years, the fundamental problem that must be solved has remained the same; how to make large amounts of product at a sufficient scale in order to have a commercially viable process? Moreover, can this happen in a time frame that is suitable for a commercial success? This chapter first establishes a basic framework for the field of microbial fermentation development— some approaches, no matter how old, still succeed in developing a productive, commercially viable fermentation process. With this foundation in place, advancements of contemporary significance are brought into perspective.

The commercial fermentation process starts with an aliquot of a preserved culture (e.g., lyophilized or cryogenically preserved), which is used to inoculate a small-scale (e.g., shake-flask) inoculum. The inoculum is scaled-up through one or more shake-flask stages (of multiple flasks) before establishing an inoculum that is suitable for inoculation into a small pilot- or factory-scale fermentor. Once again, the inoculum is scaled-up through one or more fermentor stages before being suitable for inoculation into a final production fermentor (Fig. 1). Depending on the organism being cultivated, the fermentation is typically carried out at volumes ranging from a few hundred liters up to a few hundred thousand liters, and lasting for a period of several hours or up to a

From: *Handbook of Industrial Cell Culture: Mammalian, Microbial, and Plant Cells*
Edited by: V. A. Vinci and S. R. Parekh © Humana Press Inc., Totowa, NJ

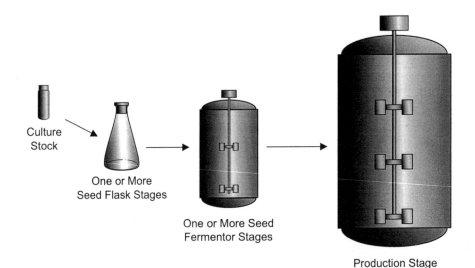

Fig. 1. Illustration of the stages involved in a typical pilot- or manufacturing-scale fermentation. Starting from a culture stock (e.g., cryo-preserved), biomass is generated through one or two shake-flask culture stages consisting of one or more flasks. The amount of biomass accumulated must be sufficient to support suitable growth in the first fermentor stage used for inoculum development. An additional stage of inoculum development at the fermentor scale (i.e., larger scale) may be necessary to accumulate the amount of biomass required to inoculate the final production-scale fermentation.

few weeks. At the end of the fermentation, the product of interest may be painstakingly isolated to ultra-high purity or may be used in its crudest form simply by using the fermentation broth directly or after spray drying.

An investigator (scientist or engineer) who takes on a fermentation development project must ask a handful of common sense, yet sometimes overlooked, questions. Addressing the following issues will quickly focus the development efforts so that time and resources are used most efficiently. Ultimately, the final measure of success of any process is how well it performs in the factory setting.

First, how is the product/metabolite of interest characterized in terms of market volume and value? Is the product a high-volume–low-value commodity chemical or enzyme? Or, at the opposite end of the spectrum, is it a low-volume–high-value pharmaceutical or therapeutic protein? Understanding the relationship between market size and value will determine the degree of latitude that the investigator has in developing a fermentation process.

Is the industry that the investigator is working in highly regulated? All commercial fermentations must be economically viable in order to succeed, but some—such as those used to produce pharmaceuticals or biologics—must operate within the guidelines of government agencies that are responsible for approving the use of the product in a particular country—e.g., the U.S. Food and Drug Administration (USFDA). There are also guidelines that govern the fermentation industry overall regarding safe, large-scale practices for nonrecombinant and recombinant organisms *(1)*.

What type of fermentation facility will be used to manufacture the product? The sizes and number of fermentors available will further determine what the seed train should be and what the productivity of the process must be. Is there computer control? Will the process require fed-batch capability, batch, or continuous sterilization of media or feeds? At the laboratory scale, there is great flexibility to develop a process of any description. However, whatever process rises from the laboratory, must "fit" the factory setting, or there must be a willingness to make the appropriate capital improvements to the facility.

Finally, is the process that is being developed limited to particular substrates (nutrients) or other feed streams? This is typical of integrated manufacturing operations in which a by-product or waste stream is fermented by bacteria or fungi to produce a value-added product. For example, the corn wet milling industry produces amino acids, organic acids, and ethanol from the by-products and waste streams of the corn refining process (e.g., corn steep liquor).

A number of different substrates (e.g., waste streams) have used to produce citric acid, a commodity chemical typically produced by fermentation of the fungus *Aspergillus niger*. Citric acid production has been carried out in submerged fermentation on simple substrates such as glucose and polymers such as starch *(2,3)*. In certain cases, solid substrates from crop residues (e.g., cassava bagasse and sweet potato) or waste streams from food processing (e.g., pineapple) have also been used for citric acid production by *Aspergillus niger (4–6)*.

2. Fermentation Medium Composition

Regardless of the nature of the process or whether the investigator is dealing with an inoculum stage or the final production stage, developing a suitable medium for culture growth and/or production of a product is paramount. For a production fermentation process, the goal of developing a culture medium is to maximize the amount of product being produced on a volumetric basis. This does not necessarily imply that the amount of biomass produced should be maximized, although the medium must support the growth of the organism. In fact, having the highest volumetric productivity possible with the smallest amount of biomass (i.e., high specific productivity) is highly desirable. However, in cases in which the product is the cell (e.g., single-cell protein) or in which product formation correlates directly with biomass levels, the goal of medium development would be adjusted to obtain high biomass levels.

2.1. Nutritional Requirements

All heterotrophic organisms have a basic requirement for sources of carbon, nitrogen, and inorganic salts, including phosphate, sulfate, and trace elements (e.g., zinc, iron, magnesium, potassium, calcium, copper, and manganese). Zabriskie et al. provide examples of the elemental composition of typical bacteria, yeast, and fungi, which illustrate the relative amounts of each of the previously listed components *(7)*. Although these "compositions" indicate how much carbon, nitrogen, or other elements must be supplied to the organism in order to achieve a specified level of biomass, they do little to elucidate how these components will impact the yield of the metabolite of interest. Nevertheless, elemental analysis provides the first approximation of the nutritional requirements of an organism.

Depending on the organism of interest, there may be additional requirements for amino acids, vitamins, purines, or pyrimidine. There may be a requirement for the addition of precursors for the metabolite of interest. Oxygen may or may not be required as the terminal electron acceptor, and the fermentation will need to be carried out at an appropriate pH.

The carbon source serves as the basic building block(s) for biomass and the major biomolecules that compose it (e.g., proteins, carbohydrates, lipids, and nucleic acids). It also serves as the building block(s) for the metabolite of interest, further emphasizing that biomass accumulation and product formation must be balanced. In addition to being a source of biosynthetic precursors, the carbon source (for many organisms) also serves as a source of energy through the production of ATP either by substrate-level phosphorylation or oxidative phosphorylation.

The type of carbon source and the level at which it is supplied plays a significant role in the metabolism of the organism. For organisms such as *Escherichia coli* and *Saccharomyces cerevisiae*, copious amounts of acetic acid and ethanol, respectively, are produced under conditions of excess glucose—also known as the Crabtree effect *(8,9)*. Certain species of *Streptomyces* have a similar "overflow metabolism," resulting in the excretion of acids such as pyruvate and alpha-ketoglutarate *(10,11)*. The carbon source can also play a significant role in the production of secondary metabolites. For example, a rapidly utilized carbon source such as glucose will repress enzymes involved in the synthesis of beta-lactam antibiotics produced by actinomycetes (e.g., cephamycin by *Streptomyces lactamdurans*) *(12)* and fungi (e.g., cephalosporin by *Penicillium chrysogenum*) *(13)*.

A source of nitrogen is required for the synthesis of all nitrogen-containing compounds in the cell, including amino acids, purines, and pyrimidines. The source of nitrogen chosen and the level at which it is supplied impacts the physiology of the organism. Nitrogen-source regulation or repression, brought about by the presence of a rapidly utilized nitrogen source, is a well-documented phenomenon *(14)* that can be the result of a complex network of regulatory genes *(15)*. Moreover, a large excess of carbon source in the presence of a limiting amount of nitrogen (high C/N ratio) enhances the synthesis of many microbial polysaccharides, particularly the production of xanthan gum by *Xanthomonas campestris (16)*.

2.2. Medium Components

The medium components used to satisfy an organism's nutritional requirements are partly influenced by the nature of the fermentation process being developed (as outlined in the Introduction). Carbon sources can be simple sugars (e.g., glucose or fructose), disaccharides (e.g., maltose, lactose, or sucrose), sugar polymers (e.g., starch or dextrins), polyols (e.g., glycerol or mannitol), or oils (e.g., soybean or rapeseed oils). Alcohols such as methanol are also utilized as carbon sources, especially for the production of recombinant proteins by methylotrophic yeast which utilizes methanol as an inducer and also as a carbon source *(17)*.

Carbon sources can be supplied as "pure" components or in feed-stocks that are rich in a particular carbon source. For example, cheese whey, a by-product of the dairy industry, serves as an excellent source of lactose, and molasses (beet or cane) serves as an enriched source of sucrose. Carbon sources which are polymeric require the organism to produce adequate amounts of hydrolytic enzymes—such as amylases and lipases for the utilization of starches and lipids—in order to make the carbon available for intermediary metabolism.

Nitrogen can be supplied in organic or inorganic and simple or complex forms. Inorganic nitrogen can be supplied as ammonium salts, nitrates, or nitrites. Nitrates and nitrites are oxidized forms of nitrogen that must be reduced by the organism to ammonia nitrogen by a specific reductase. Organic sources can be simple, such as amino acids or urea. Complex sources of organic nitrogen can be supplied by protein sources such as meals and flours. These protein sources must be hydrolyzed to amino acids or short peptides by proteases before being utilized by the organism.

The choice of media components comes down to a question of defined vs complex media. A defined medium is composed of discrete chemical entities present in specific amounts. Each component added satisfies a specific requirement for the growth of the cell and/or the synthesis of the product. Semi-defined is used to describe a medium that contains a single complex component—for instance, yeast extract—at a level of 1% (w/v) or less. With a defined medium, the impact of all of the components on the process can be accounted for. Moreover, the utilization of each component can, in theory, be monitored and controlled through on-line monitoring techniques and feeding strategies. Defined media that are soluble are necessary for metabolic modelling studies that require the investigator to be able to close the mass balance between what is supplied to the organism in the form of nutrients (e.g., carbon source) and what is produced (e.g., organic acids and carbon dioxide). Although not an absolute rule, defined media components typically show more lot-to-lot and vendor-to-vendor consistency, which is critical for the production of specific pharmaceuticals and biologics for which regulators (e.g., USFDA) require that the manufacturer demonstrate a consistent manufacturing process.

Complex media components are typically by-products of the food and agricultural industries, and provide one or more of an organism's nutritional requirements. For low-cost commodity fermentations such as organic acids, amino acids, or other bulk chemicals, the use of these components is essential because it reduces fermentation cost which is a primary determinant of economic feasibility. As mentioned in the Introduction, the goal may be to develop a process in which a waste stream from a particular industry is used as a growth substrate to produce a value-added product, such as the examples provided for citric acid production by *Aspergillus niger*. However, since complex components vary between vendors and are not produced specifically for the fermentation industry, the fermentation process performance that they support can be variable. This can be partially offset by establishing raw material specifications and carrying out small scale use-tests in order to determine the suitability of a particular medium component for large-scale production.

3. Fermentation Medium Development Methods

The initial approach to medium development is greatly influenced by how much is known about the organism under study or how much is known about a closely related species. A search of the literature can provide some thoughts on an initial medium formulation that supports the production of at least a small amount of the metabolite of interest. Also, culture-collection catalogs such as the American Type Culture Collection (ATCC) or reference texts such as the Handbook of Microbiological Media *(18)* can also provide leads for media formulations or specific nutritional requirements.

Starting from an initial medium formulation, the investigator is faced with the challenge of altering the medium composition along with other process parameters, such as pH and temperature, in a structured manner so that improvements in fermentation performance can be gained. The simplest approach is the "one-at-a-time" method, in which the concentration of each medium component supporting the best outcome is identified while the concentrations of each of the remaining components is held constant. Although this approach is widely described in the literature, it does not take into account the synergistic effects that medium components can have on process performance, and can result in a a "local" rather than "global" maximum.

Currently, most medium optimization strategies could be considered "structured empiricism." They are "structured" from the standpoint of having a defined path forward in order to improve the fermentation, and "empirical" because in most cases little is known about the extent of the physiological interactions that impact the desired process output. In the future, medium development will be greatly influenced by advancements in metabolic pathway modeling and the ability to tailor the medium to satisfy the intracellular mass balance. An example of this approach is the establishment of a stoichiometric model for the production of transglutaminase by *Strepverticillium mobaraense*, and the use of this model to increase enzyme activity production fourfold compared to the initial medium *(19)*.

It is important to note that medium development goes hand-in-hand with strain improvement (classical or recombinant DNA-based) to create a cycle of process improvement (Fig. 2). For each new culture that is generated from a strain improvement program, a new round of medium development must be instituted in order to realize the full potential of the newly generated strain. In addition to a new production medium formulation, a new screening medium is generated to identify future improved strains.

3.1. Statistical Experimental Design

Starting from an initial medium formulation, a number of powerful, time-tested, statistically based experimental design techniques can be employed in order to improve fermentation performance *(20)*. Software packages such as JMP (SAS Institute, Inc. Cary, NC), which is available for desktop computers, allow the investigator to design appropriate, statistically-based experiments and to analyze and graphically represent the data. These statistically based approaches can be divided into two categories: screening and optimization.

Independent variables (e.g., media components, initial pH, fermentation temperature, and fermentation time) are first screened to determine which have the most significant impact on the dependent variable (e.g., metabolite production). Plackett-Burman designs, for example, are classical methods used for evaluating a large number of variables (up to N-1 variables in N trials) at two levels to identify which have the most significant impact (using Student t tests) on the process (Table 1) *(21)*.

Fractional factorial designs can also be used to screen a large number of variables, and are particularly useful when one or more two factor interactions are suspected a priori. Typical fractional factorial designs consist of 2^{k-p} trials in which 2 is the number of levels (high and low), k is the number of factors being examined, and $(1/2)^p$ is the

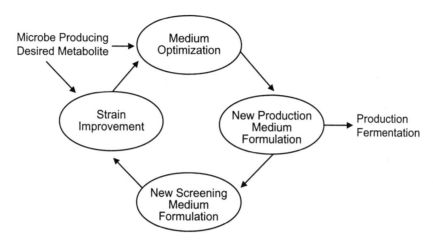

Fig. 2. Iterative and interactive nature of microbial strain and medium development. Once an organism producing something of interest (e.g., metabolite or enzyme) has been identified, strain improvement and medium optimization begin. In an ongoing process-improvement program, the best medium formulation serves as the basis for the production fermentation. After confirming superior production at the pilot- or manufacturing-scale, the new medium formulation serves as the basis for a new screening medium to identify improved mutants. In order to maximize productivity of the improved strain, a subsequent round of medium optimization is carried out.

Table 1
Plackett-Burman Design Used for Screening Fermentation Process Variables (e.g., medium components, pH, temperature) for Further Optimization Studies[a]

Trial	X_1	X_2	X_3	X_4*	X_5	X_6	X_7	X_8*	X_9	X_{10}	X_{11}*
1	+	+	−	+	+	+	−	−	−	+	−
2	−	+	+	−	+	+	+	−	−	−	+
3	+	−	+	+	−	+	+	+	−	−	−
4	−	+	−	+	+	−	+	+	+	−	−
5	−	−	+	−	+	+	−	+	+	+	−
6	−	−	−	+	−	+	+	−	+	+	+
7	+	−	−	−	+	−	+	+	−	+	+
8	+	+	−	−	−	+	−	+	+	−	+
9	+	+	+	−	−	−	+	−	+	+	−
10	−	+	+	+	−	−	−	+	−	+	+
11	+	−	+	+	+	−	−	−	+	−	+
12	−	−	−	−	−	−	−	−	−	−	−

[a]This design represents the testing of eight variables with three "dummy" variables used for the calculation of variance of the experimental system. Other tables are used to screen a larger number of variables *(21)*.

Significance of each variable is determined by comparing the response of the "+" and "−" levels using a Student t test. *Dummy variables not assigned any process parameter are used to calculate the variance of the system.

fraction of the full factorial design (i.e., all possible combinations of all factors at two levels) in which p is an integer less than k *(20,22)*. For example, a half fraction of a full factorial design for seven variables could be carried out in sixty-four trials (i.e., $2^{7-1} = 2^6 = 64$) with all the main effects and two factor interactions being estimated. Large two-factor interactions are an indication of "gross curvature" of the system. Because the 2-level Plackett-Burman and fractional factorial designs have points only at the corners of the design regions, it is advisable to run replicates at the center point, where the levels for all variables are midway and all values between the high and low values are covered. Of course, this can only be done when all the variables are continuous (i.e., numerical); however, when some are categorical (e.g., type of carbon or nitrogen source), more advanced procedures may be required.

If the mean response for the center point is significantly different (e.g., higher yield) from the mean of all the corner-point responses, then additional curvature exists in the system that may not be adequately accounted for by fitting the initial design. Under these circumstances, it may be necessary to investigate the region close to the center in greater detail by choosing smaller ranges for the variables and re-running the design in the focal region. Such structured sequential use of statistical designs is inherent in practical experimentation and is one strength of the methodology. If curvature is not large—so that first-order terms adequately approximate the data—then a steepest ascent method can be employed to move to a new design region of improved responses that possibly contain the optimum, with only a few additional runs.

Once a small number (two to five, typically) of the most important independent variables are identified, response surface methodology (RSM) can be used to optimize the levels of the independent variables. The Box-Behnken experimental design *(23)* or the Box-Wilson central composite experimental design *(24)* allow for the construction of a model, in the form of a polynomial equation, which describes the main effects of the individual variables, the interactions of any two of the variables, and any pure quadratic effects. It should be noted that two-factor interactions are quadratic effects, and represent curvature for continuous variables. For example, a Box-Behnken design for three variables evaluated at three levels would require thirteen trials (Fig. 3, Table 2), with the center point of the design being replicated for statistical purposes, to determine all main, interactive, and pure quadratic effects in the following polynomial equation:

$$Y = b_0 + b_1(x_1) + b_2(x_2) + b_3(x_3) + b_{12}(x_1x_2) + b_{13}(x_1x_3) + b_{23}(x_2x_3) + b_{11}(x_1)^2 + b_{22}(x_2)^2 + b_{33}(x_3)^2$$

where:

Y = independent variable (e.g., volumetric productivity, biomass level, etc.)

b_0 = constant coefficient for the mean overall effect

x_1, x_2, x_3 = independent variables

b_1, b_2, b_3 = linear (first-order) coefficients (i.e., main effects)

b_{12}, b_{13}, b_{23} = second-order interaction coefficients

b_{11}, b_{22}, b_{33} = pure quadratic coefficients

Thus, twenty-five trials, with additional replicates of the midpoint of the design, would be required for four variables evaluated at three levels with the appropriate terms added to the polynomial to account for each additional main, interactive, and quadratic effects.

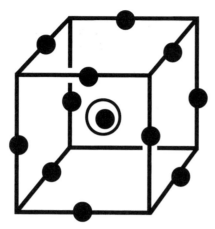

Fig. 3. Graphical representation of the experimental points in a Box-Behnken three-factor optimization study. The experimental points are located at the midpoint of each edge of the cube along with a design midpoint located in the middle of the cube.

Table 2
Box-Behnken Design for a Three Variable-
Three Level Fractional Factorial Study
(Coded Variables)

Trial	Independent variables		
	X_1	X_2	X_3
1	1	1	0
2	1	−1	0
3	−1	1	0
4	−1	−1	0
5	1	0	−1
6	1	0	−1
7	−1	0	1
8	−1	0	−1
9	0	1	1
10	0	1	−1
11	0	−1	1
12	0	−1	−1
13	0	0	0

−1, 0, 1 represent the coded values for the low, midpoint, and high levels for each variable.

It should be emphasized that for a Box-Wilson central composite design (CCD), fifteen trials are required in some cases to evaluate three variables, each at five levels (Table 3). This includes six axial trials which extend beyond each face of the cubic space (Fig. 4). However, it is important to note that the cube part of the design may have already been run as part of screening (i.e., corners of the screening experiment design region), so that only these additional axial trials may need to be run. This so-

Table 3
Box-Wilson Central Composite Design
for a Three Variable-Five Level Fractional
Factorial Study (Coded Variables)

Trial	Independent variables		
	X_1	X_2	X_3
1	−1	−1	−1
2	1	−1	−1
3	−1	1	−1
4	1	1	−1
5	−1	−1	1
6	1	−1	1
7	−1	1	1
8	1	1	1
9	−1.68	0	0
10	1.68	0	0
11	0	−1.68	0
12	0	1.68	0
13	0	0	−1.68
14	0	0	1.68
15	0	0	0

−1, 0, 1 represent the coded values for the low, midpoint, and high levels for each variable. The coded values for the axial points for each level is 1.68 or −1.68.

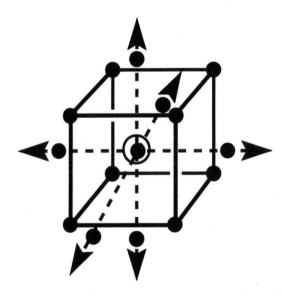

Fig. 4. Graphical representation of the experimental points in a Box-Wilson central composite design three-factor optimization study. The experimental points are located at each vertex of the cube with the midpoint located in the middle of the cube. Axial trials are experimental points that exist outside of the cubic space, and are represented by arrows.

called "sequential assembly" is one strength of the CCDs not shared by the Box-Behnken designs. For a full, classical CCD, twenty-five trials would be required for four variables evaluated at five levels. However, it is possible to reduce this to as few as sixteen trials using so-called Resolution III* designs *(24)*.

The number of trials required for the various response-surface designs is far fewer than the corresponding full factorial designs that are still often used. For example, full factorial designs for three or four variables evaluated at three levels would require twenty-seven (i.e., 3^3) and eighty-one (i.e., 3^4) trials, respectively.

Once an optimal medium formulation has been determined, an additional round of screening may be deemed necessary to determine whether independent variables considered unimportant in the first round of screening now influence the production process. Moreover, a new medium formulation can be implemented in an ongoing strain improvement program for the screening of higher producers (Fig. 2). Medium optimization for confirmed higher producers would then be carried out using the same statistical experimental designs in order to gain optimal performance from a strain.

3.2. Simplex Search

As mentioned, response-surface methods are effective at reducing the number of trials required to reach an optimum medium formulation and providing a mathematical description of the results as a function of the variables tested. Ordinarily, this work is carried out in shake flasks or other small-scale cultivation vessels (e.g., test tubes or well plates). These small-scale cultivation methods may not duplicate the conditions that exist in a fermentor, and thus the potential exists that optimization work done at the shake-flask scale, for example, may not translate into equal performance in laboratory-, pilot-, or factory-scale fermentors. In many instances, it may be advisable to carry out additional validation/optimization studies at the laboratory- or pilot-scale in fermentors with identical geometry and mixing characteristics; where variables such as nutrient concentrations, feeding strategies, pH, and temperature can be monitored and controlled better and can be optimized in an environment of shear stress, reduced mixing times/nutrient gradients, and sparged oxygen supply.

Screening and response-surface methods—especially those that use reduced numbers of runs—can also be exploited at the large-scale. However, it is often difficult to justify the large up-front investment required, especially when extensive work can be accomplished at the small scale. For this reason, other highly sequential approaches to fermentation improvement have been attempted. These proceed one run at a time to explore and optimize and thus do not require a commitment to a large prespecified number of runs. However, because of their inherent inability to account for overall response structure and to characterize variability, it may be necessary to actually run more experiments in the long run.

Among these sequential methods are a number of numerical optimization techniques—techniques developed for the optimization of mathematical functions without experimental error—that have been used in medium optimization. One such technique is simplex search. The variables can be either continuous (i.e., concentrations of a particular nutrient) or categorical (e.g., different nitrogen sources). The strategy is to move the geometric simplex (with N+1 vertices) through experimental space in the direction of an improved response by running an initial set of N+1 trials trial and then reflecting away from the vertices that generate the poorest results *(25)*. A simple spreadsheet is all that is required to calculate new levels of each independent variable for the subsequent trial. The procedure is repeated sequentially until a desired performance level is reached.

This simplex search approach does not require an exhaustive statistical treatment of the data, and if a problem occurs (e.g., equipment failure) causing the simplex to move away from an improved response, subsequent experiments will correct the direction. However, because simplex search does not explicitly account for experimental variability—as the statistical treatment does—it is sensitive to experimental noise. For this reason, it may be more desirable to use simplex designs in conjunction with RSM as a means of fine-tuning conditions as the transition is made from shake-flask to fermentor.

3.3. Evolutionary Algorithms

Another set of numerical optimization techniques that can be applied to medium development are the so-called "evolutionary" algorithms. They work by "evolving" a population of potential solutions (i.e., media formulations or process parameters) through experimental space. As with the simplex search procedure, no assumptions about the nature of the process being optimized or the shape of the solution landscape are required. Although statistical experimental designs require that the number of variables to be optimized be reduced through an initial screening phase, evolutionary algorithms accommodate large numbers of variables in the search for an overall optimum response. As with the simplex search, the problem with evolutionary algorithms is that they are subject to experimental variability. They also require a fair amount of mathematical and computer expertise and resources.

Genetic algorithms (GAs) are one type of evolutionary algorithm that were inspired by biological or "Darwinian" evolution *(26)*. Zuzek et al. offers an analogy between potential solutions and chromosomes containing genes *(27)*. Each potential solution is a "chromosome" and the medium component concentrations or process parameter settings (such as temperature and pH) are the individual "genes" that make up the chromosome. After each round of fermentation, the fitness (i.e., improvement in process performance) of each chromosome is evaluated, and replication of the chromosome occurs in proportion to the improvement using a linear scaling technique *(28)*. Thus, a larger proportion of the most fit solutions are carried over to the next round of experimentation. The chromosomes in the population are then subject to "crossover," which involves the exchange of individual genes in each chromosome. After the attributes of the solutions are exchanged (i.e., crossed over), "point mutations" are randomly introduced into the genes at a prescribed frequency *(29)*. After each round of experimentation, the process of replication, crossover, and mutation is carried out to generate a new population of solutions to be tested. Ultimately, after several rounds, the population converges on a single overall optimum.

The GA approach has been applied successfully in fermentation development. For instance, mevinolin production by *Aspergillus terreus* was increased threefold after four generations involving 10 medium components *(27)*. A 23% increase in the production of L-lysine by *Corynebacterium glutamicum* was achieved by manipulating 13 variables over the course of 468 experiments (i.e., solutions) *(30)*. Formate dehydrogenase activity produced by *Candida boidinii* was increased 40% by manipulating 14 variables over the course of 250 experiments *(31)*. Note that the large number of experiments ultimately run illustrates one potential shortcoming of sequential optimization.

Particle swarm optimization (PSO) is another example of an evolutionary algorithm that shows promise as a technique for medium optimization. Similar to GAs, PSO does not require assumptions about the topology of the solution space being explored. Cockshott and Hartman explain, that unlike GAs which function along "biological" lines (e.g., Darwinian evolution), PSO has more of a "social" context *(32)*. Each individual solution in the PSO population remembers its own best solution and the best overall solution of the population. Conversely, for GAs, past information is lost with each successive generation/iteration.

In PSO, each individual particle initially occupies a position in the solution space with coordinates that are the medium component concentrations, and process parameters such as pH and temperature. An initial range is set for each independent variable so that a reasonable amount of experimental space can be explored. After each iteration, an update equation is used to determine the levels for the independent variables for the next round of experimentation. This update equation includes terms that describe the "personal best" (i.e., $pbest_i$) for each individual and the "global best" (i.e., gbest) for the entire population. The "personal best" and "global best" terms provide the memory component for the individual that distinguishes it from GAs. After several iterations (approx 8–12), the individuals in the population converge on a single solution representing the "global best."

In an interesting comparison, statistical experimental design (SED) methodology was compared to PSO to optimize the fermentation medium for echinocandin B production by *Aspergillus nidulans (32,33)*. Fifteen medium components were considered at the outset for both optimization strategies. For the statistical experimental design approach, a 46% increase in echinocandin B titer was realized after 108 shake-flask fermentations carried out over 5 iterations. This included an initial round of variable screening (i.e., Plackett-Burman) to identify the five components that became the focus of the optimization studies. In comparison, for the PSO approach, 135 shake-flask fermentations carried out over 9 iterations resulted in a 41% increase in titer. It is interesting to note that although both optimization approaches yielded comparable increases in titer in a reasonably similar number of shake-flask fermentations, different optimum (or near optimum) medium formulations were obtained.

4. Fermentation Medium as a Research Tool

Although the goal of any media development program is to generate a formulation that will support a high level of product formation, it is important to remember that the organism's responses to different media or conditions can provide an investigator with a great deal of information on the underlying physiology and biochemistry relevant to the fermentation. In short, the components of any culture medium serve as the independent variables of a very powerful experimental system for studying microbial physiology.

Both batch and nutrient-limited chemostat cultivations are useful approaches. Simple manipulations of culture-medium conditions, combined with novel techniques for metabolic analysis, can reveal a great deal of physiological detail. While these experimental approaches may not lead directly to a factory process per se, the information provided by these types of studies can provide new leads for future media or strain-development programs.

The development of a fermentation process for the production of pneumocandins by the fungus *Glarea lozoyensis* provides two examples of how an investigator can "learn from the medium." In the first example, the pneumocadin B_0 titer and the level of the major, undesirable structural analog pneumocandin C_0, were shown to be influenced by the osmotic pressure of the medium *(34)*. High osmotic pressure supported acceptable pneumocandin B_0 titers, with low levels of pneumocandin C_0. Lower osmotic pressures resulted in higher pneumocandin B_0 titers but with unacceptably high levels of pneumocandin C_0. An examination of the underlying mechanism revealed a link between the relative amounts of *trans*-4 hydroxyproline and *trans*-3 hydroxyproline produced under the two osmotic pressure conditions. This line of experimentation provided critical information on the boundary conditions for the large-scale fermentation and provided insights into the physiology of the formation of an undesirable analog.

In a second example, supplementation with proline and arginine were shown to impact the pneumocandin titer and analog spectrum. Changes in the intracellular levels of *trans*-4 and *trans*-3 hydroxyprolines provided the key physiological link to the observations. These results suggest that the synthesis of the hydroxypolines may be part of a nitrogen regulatory circuit in *Glarea lozoyensis*. Moreover, supplementation of the production medium with zinc, cobalt, and nickel-trace elements known to inhibit α-ketoglutarate-dependent dioxygenases in vitro resulted in increases in the levels of pneumocandins that are differentiated by the absence of one or more hydroxyl groups around the cyclic hexapeptide. These results suggest the involvement of one or more α-ketoglutarate-dependent dioxygenases in the synthesis of pneumocandins *(35)*.

The alteration of media composition has also been useful in the area of metabolic flux analysis. *Penicillium chrysogenum* producing penicillin-G was grown in glucose-, ethanol-, and acetate-limited chemostat cultures to examine flux partitioning around the glucose-6-phosphate, 3-phosphoglycerate, mitochondrial pyruvate, and mitochondrial isocitrate nodes *(36)*. Although these three growth conditions resulted in different patterns of flux partitioning around these four nodes, no significant impact on the penicillin-G production was observed. In this same study, the cytosolic NADPH demand was altered by growing *Penicillium chrysogenum* on various combinations of glucose/xylose and ammonia/nitrate. Penicillin-G production was the highest when glucose and ammonia were used as the carbon and nitrogen sources, respectively (low NADPH demand), and xylose and nitrate supported the lowest production of penicillin-G. Together, these results suggest that the supply of NADPH may be the bottleneck in the production of penicillin-G rather than the supply of carbon precursors *(36)*.

In a similar approach, glucose- and ammonia-limited chemostat cultures of *E. coli* were examined using metabolic flux ratio (METAFoR) analysis, a technique based on two-dimensional ^{13}C-^1H NMR spectroscopy of fractionally labeled biomass (37). The results suggested that under ammonia limitations, a greater fraction of oxaloacetate was obtained from the carboxylation of phosphoenolpyruvate (PEP) than from the TCA cycle, and that there was a reduction in the amount of PEP generated by one transketolase reaction. This type of analysis allows an investigator to link fermentation condition (e.g., medium composition) to metabolic flux.

5. Fermentation Scale-Up—Issues and Opportunities

A large number of published reports have examined the fundamentals of fermentation scale-up, a topic that will not be described here *(38,39)*. Typically, a fermentation is scaled up in order to maintain a constant (relative to the laboratory fermentor scale) power per unit volume, volumetric mass transfer coefficient (i.e., K_La), impeller-tip speed, or dissolved oxygen tension. This serves to re-create, in a large-scale fermentor, those conditions identified at the laboratory scale as being conducive to good process performance. Although chemical factors such as medium composition can be held constant across scales, other physical factors change as a function of increasing scale.

Laboratory or pilot-scale fermentors, generally, provide a homogeneous environment for the organism because of very fast mixing of the broth in the vessel and inconsequential differences in hydrostatic pressure. Circulation times—the time required for a unit of biomass (e.g., cell or mycelial fragment) to depart and return to fixed point in the vessel—follow a distribution referred to as the circulation-time distribution, and represent an important parameter for scale-up (40). Mean circulation time increases as a function of scale as well as a function of fermentation broth viscosity, a situation common in mycelial fermentations run at the manufacturing scale (41). An increased circulation time results in heterogeneous conditions in which nutrient concentrations, dissolved oxygen, dissolved carbon dioxide and shear, show spatial variations within the reactor. When the oxygen uptake rate is faster than the circulation time, a unit of biomass can experience periods of anaerobiosis before returning to an oxygen-sufficient region in the fermentor. Bajpai and Reuss (42) proposed a two-zone model consisting of a well-mixed region of the fermentor around the impellers and suboptimal mixing for the remainder of the bulk liquid in the tank. In their model, a unit of biomass is exposed to various dissolved oxygen and carbon dioxide concentrations as it circulates through different regions of the fermentor. In the well-mixed region around the impellor, the biomass is exposed to high dissolved oxygen/low dissolved carbon dioxide concentrations. This is followed by a continual decline in dissolved oxygen concentration accompanied by an increase in dissolved carbon dioxide as the biomass passes through the macromixed bulk liquid. When a fermentation process will be run in existing factory fermentors with known (i.e., reduced) mixing capability, these variables must be considered during strain improvement, process or medium development, and scale-up to increase chances of success in the factory setting.

Despite the tools and techniques available for the scale-up of microbial fermentations, many large-scale fermentation processes yield lower than expected performances compared to experiments carried out in laboratory- or pilot-scale fermentors. Investigators are realizing the important interactions that exist between the fluid dynamics of fermentation broth and the physiological responses of microbes in large-scale fermentors (43). Over the past two decades, laboratory-scale fermentation models have been developed to simulate the conditions that an organism experiences in a large-scale fermentor. Carrying experimentation of this type out at the laboratory scale provides an efficient and economical alternative to evaluations at the manufacturing or pilot scales and allows for scale-up issues to be addressed early in a development program. Moreover, work in scale-down systems allows the investigator the opportunity to more fully establish process sensitivities that will allow scale-up problems to be more easily understood.

Sweere and colleagues describe the process for establishing a useful scale-down model that can be used to establish a productive large-scale process (44). Regime analysis is needed to identify the rate-limiting mechanisms of the process in terms of characteristic times (e.g., mean circulation time or oxygen transfer rate) for the full-scale fermentor and how these compare to the corresponding physiological properties of the organism (e.g., the uptake rate of substrate or oxygen-uptake rate). If one or more of these characteristic times is within an order of magnitude of a corresponding physiological parameter of the cell, then a rate-limiting mechanism worthy of investigation at the laboratory scale has been identified. The rate-limiting mechanism can then be simulated at the laboratory scale followed by re-optimization of the process, perhaps using one of the optimization techniques described here (e.g., simplex search). The re-optimized process can then be evaluated on a large scale.

Studies by Geraats on the scale-up of lipase production by *Pseudomonas alcaligenes* provides an excellent example of how a scale-down model can be used to identify and solve a scale-up problem *(45)*. In this work, increased levels of dissolved carbon dioxide were identified as being responsible for the reduction in lipase production at the 100 m^3 scale. This was first verified at the 10-L fermentor scale by blending various concentrations of carbon dioxide into the sparge air to generate carbon dioxide concentrations comparable to those present in a large-scale fermentor. With this information from the scale-down model, dissolved carbon dioxide at the 100 m^3 scale was reduced three-fold by reducing head pressure and increasing the airflow during the fermentation process.

Since biomass that circulates though a large-scale fermentor experiences different dissolved oxygen concentrations, Vardar and Lilly simulated in a 7-L fermentor the effects of dissolved oxygen cycling on penicillin production by *Penicillium chrysogenum* *(46)*. The dissolved oxygen concentration was continuously cycled around the critical concentration of 30% by changing the back pressure in the fermentor between 1 and 2 atmospheres, according to a sine wave function. The range of dissolved oxygen concentrations imposed on the culture was 23–37%. Although the mean dissolved oxygen concentration in this regime was 30%, the rate of penicillin production was reduced from 7400 U/gm DCW/h (constant 30% dissolved oxygen) to 5,300 U/gm DCW/h; similar to that of a culture maintained at a constant dissolved oxygen concentration of 26%.

Yegneswaran and colleagues conducted similar work with *Streptomyces clavuligerus* using 2-L fermentors *(40)*. Dissolved oxygen concentrations were cycled by supplying air to the fermentor for 5 s followed by a period of no aeration ranging from 8–44 s with a mean of 20 s. The period of no aeration was generated according to a Monte Carlo method which takes into account the circulation-time distribution typical of a large fermentor. The Monte Carlo-based cycling method was compared to a periodic cycling method in which the period of no aeration was a constant 20 s for each cycle. Both the periodic and Monte Carlo-based methods of oxygen deprivation influenced the growth rate but not the maximum amount of biomass. However, the Monte Carlo method resulted in lower cephamycin C titers than the periodic method as compared to the control.

An alternative approach to the study of large-scale fermentor heterogeneity is to use a two compartment system consisting of a fermentor or stirred-tank reactor (STR) and an external-loop or plug-flow reactor (PFR) in which the circulation time and volume can be varied (Fig. 5). Using this type of system, Larsson and Enfors demonstrated the kinetics of irreversible inhibition of *Penicillium chrysogenum* respiration with a circulation time of 5–10 min through an anaerobic PFR with a volume of 6% of the STR volume *(47)*. Moreover, this group was able to demonstrate the difference in inactivation kinetics between two different *P. chrysogenum* strains.

For certain fermentation processes, growth substrates (e.g., carbon source) are fed during the course of the fermentation, typically from the top of the reactor, which can result in nutrient concentration gradients in large-scale fermentors. The effects of

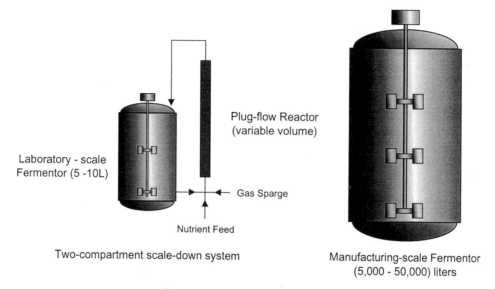

Fig. 5. Two-compartment scale-down system used to study, at laboratory scale, the physiological effects of nutrient gradients prevalent in large-scale fermentors. Whole broth is pumped from the fermentor (i.e., stirred-tank reactor) through a plug-flow reactor with a residence time that can be varied according to the mean circulation time of the large-scale fermentor being modeled. At the inlet of the plug-flow reactor, nutrients (e.g., glucose) or gases (e.g., O_2 or CO_2) can be introduced that subject the cells to transient increases or decreases in nutrient levels. Sampling can take place along the length of the plug-flow reactor or from the fermentor to identify physiological changes.

nutrient gradients on an organism's physiology can also be modeled and studied using scale-down methodologies.

A two-compartment system (STR and PFR) was used to study the impact of glucose-concentration oscillations on the metabolism of *Saccharomyces cerevisiae (48)*. As broth was circulated from a 15-L STR to an aerated 850-mL PFR containing static-mixer elements, molasses (sucrose source) was introduced at the inlet of the PFR, exposing the cells to repeated high concentrations of sugar. This was compared to a control reactor, in which molasses was introduced to the STR and mixing of the sugar was instantaneous (i.e., no concentration gradients). The two-compartment system (STR + PFR) produced less biomass and more ethanol (on a volumetric basis) than the STR reactor by itself. This overflow metabolism takes place because the yeast cells are exposed transiently to sugar concentrations, which are above the threshold concentration for ethanol formation. During the large-scale production of Baker's yeast, it is desirable to eliminate aerobic ethanol formation. This is typically accomplished by feeding sugar so that the concentration remains below the critical value that triggers ethanol formation. It is clear from this example that the nutrient gradients that exist in large-scale reactors will influence the amount of ethanol formed and reduce biomass formation.

6. Impact of Functional Genomics on Microbial Process Development

One of the most significant advances in the biological sciences over the last decade has been the ability to sequence the entire genome of any organism, including the human genome. A number of microbial genomes have been sequenced, including *E. coli*, *Saccharomyces cerevisiae*, *Bacillus subtilis*, *Haemophilus influenzae*, *Mycobacterium tuberculosis*, *Neisseria meningitidis*, and *Methanococcus jannischii* (for a listing of completed and in-progress sequencing projects, visit The Institute for Genomics Research web site, www.tigr.org/tdb/mdb/mdbcomplete.html). Currently, advances in sequencing technology and computer methods for the assembly of the sequencing data make it possible for the genome of a prokaryotic microbe to be sequenced in a very short period of time. This sequence information and subsequent open reading frame (ORF) annotation facilitates global gene expression analysis using microarray technology *(49)*.

Regardless of the approach utilized, the fermentation medium and process optimization strategies described here merely serve as a vehicle in which to maneuver through experimental design space and arrive at an optimum medium formulation and process parameter settings with process performance (e.g., metabolite production) as the only measured output. Although optimum conditions that maximize process performance may have been identified, a fermentation scientist may know very little about the organism's physiology and how it influences the fermentation process. Improved process performance as a function of a particular set of fermentation conditions is really the result of the organism's global physiological response to those conditions.

The importance of an organism's global physiological response as a function of fermentation condition or growth medium is illustrated by work carried out by Tao and colleagues *(50)*. In this work, gene-expression analysis was carried out on *E. coli* cultures grown in Luria broth plus glucose (i.e., rich medium) or minimal medium plus glucose. In rich medium, increased expression of translation-specific genes was observed, consistent with an increase in the rate of protein synthesis in rapidly growing cells. Conversely, cultures grown in minimal medium showed increased gene expression in several biosynthetic pathways and genes involved in acetate metabolism. Overall, the expression of genes in the two media types was consistent with the wealth of knowledge that has been accrued for this bacteria *(51)*.

Based on this study, gene-expression analysis carried out in concert with the types of experimental designs described in this section could enhance medium development studies by making them more informative and perhaps slightly less empirical. The different types of optimization strategies (as described above) provides the framework for the exploration of the experimental space under investigation, while the analysis of global gene regulation would extend the amount of information gleaned from each experiment. Because fermentation development is an iterative process (*see* Fig. 2), the "evolution" of gene expression as a function of fermentation condition could be identified. Along with metabolic modeling studies, this type of approach could be used to identify targets for genetic manipulation.

Global gene-expression analysis would also be a useful tool during scale-up of microbial fermentations in which the ultimate goal is to maintain, at a large scale, the physiological state of the organism identified during laboratory-scale experiments to be suitable for process performance. Although gene expression analysis could be carried out at the manufacturing scale, the ability to carry out expression analysis on a scale-down model of the process allows one to investigate a greater number of process parameters.

For example, the physiological impact of nutrient gradients prevalent in large-scale *E. coli* fed-batch fermentations was augmented by examining the expression of four stress genes (*clpB*, *dnaK*, *uspA*, and *proU*) and three genes known to respond to oxygen limitation or glucose excess (*pfl*, *frd*, and *ackA*) *(52)*. The fermentation systems employed were a 20,000-L fed-batch fermentation and a two-compartment scaled-down model. The outcome of this study indicated that *E. coli* responds to environmental changes (i.e., nutrient gradients or oxygen depletion) in a matter of seconds, shorter than the typical mean circulation times in large fermentors.

For the most part, the technical details of microarray experiments have been reduced to practice (e.g., array preparation, RNA purification, labeling, and hybridization). Yet variability is introduced at each step of the process and must be dealt with in order to generate a data set of sufficient quality for analysis especially where detecting small differences in expression is desired. As an example, Tseng and colleagues have carried out a systematic investigation of the intrinsic errors that are part of microarray analysis *(53)*. Quality filtering, normalization, and models of variation are applied to each individual slide used in a microarray experiment in order to maximize the statistical power of the analysis.

Once the raw microarray data has been processed to obtain a high-quality set of results, the challenge lies in turning the results into useful information. The techniques employed vary depending on what one wishes to accomplish. The goal may be to identify patterns of gene expression that classify a good vs bad fermentation performance, high vs low producing strain, or some other discriminating dependent variable. Clustering or partitioning techniques such as hierarchical clustering *(54)*, principal component analysis *(55)*, and self-organizing maps, *(56)* to name a few, have been used successfully to identify genes which are coordinately expressed. One of the drawbacks of clustering or partitioning methods is that they capture only a subset of the potential relationships that may exist among the genes studied and allow a gene to exist in a single cluster or group *(57)*. Moreover for time-course data, cluster analysis forms clusters of genes with expression patterns that agree across the entire range of time-points that are being analyzed. Pattern-discovery techniques may provide an alternative to the analysis of gene-expression data. When the TEIRESIAS pattern discovery algorithm was applied to a yeast gene expression data set previously analyzed by cluster analysis, several million patterns were identified, including all the gene groups identified in the original analysis. This included expression patterns that were similar only during the middle portion of the time-course *(57)*. Clearly advances in computational biology will continue to facilitate the analysis of the enormous amounts of data that are generated in gene-expression experiments. Yet, the burden is still on the investigator to identify the biological significance of the information generated.

7. Future Prospects

Some argue that fermentation is a "black box field" or "more art than science." Although the processes being developed usually work efficiently, very little is understood about why and how they work. The field of fermentation development continues to evolve as a result of technological advances. Genome sequencing and functional genomics clearly have utility beyond the boundaries of medicine and will serve to advance the science of fermentation development. Yet, fermentation development has a strong sense of history because many of the approaches that have been developed over the past two to three decades continue to remain en vogue. Continued success in this area will come as the disciplines of biology, genetics, engineering, and computer science continue to work in concert with one another.

Acknowledgments

The author wishes to thank Bert Gunter and Peter Salmon for critically reviewing the manuscript and to David Pollard and Professor James Liao for their helpful comments. This chapter is dedicated to the memory of Daniel J. Connors Jr.—a great father, teacher, and friend.

References

1. Frommer, W., Ager, B., Archer, L., Brunius, G., Collins, C., and Donikian, R. (1989) Safe biotechnology. III. Safety precautions for handling microorganisms of different risk classes. *Appl. Microbiol. Biotechnol.* **30,** 541–552.
2. Pazouki, M., Felse, P., Sinha, J., and Panda, T. (2000) Comparative studies on citric acid production by *Aspergillus niger* and *Candida lipolytica* using molasses and glucose. *Bioprocess. Eng.* **22,** 353–361.
3. Suzuki, A., Sarangbin, S., Kirimura, K., and Usami, V. (1996) Direct production of citric acid from starch by a 2-deoxyglucose-resistant mutant strain of *Aspergillus niger. J. Ferment. Bioeng.* **81,** 320–323.
4. Leangon, S., Maddox, I., and Brooks, J. (1999) Influence of the glycolytic rate on production of citric acid and oxalic acid by *Aspergillus niger* in solid substrate fermentation. *World J. Microbiol. Biotechnol.* **15,** 493–495.
5. Tran, C., Sly, L., and Mitchell, D. (1998) Selection of a strain of Aspergillus for the production of citric acid from pineapple waste in solid-state fermentation. *World J. Microbiol. Biotechnol.* **14,** 399–404.
6. Vandenberghe, L., Soccol, C., Pandey, A., and Lebeault, J. (2000) Solid-state fermentation for the synthesis of citric acid by *Aspergillus niger. Bioresour. Technol.* **74,** 175–178.
7. Zabriskie, D., Arminger, W., Phillips, D., and Albano, P. (1988) Fermentation medium formulation, in *Traders Guide to Fermentation Medium Formulation.* Traders Protein, Memphis, TN, pp. 1–44.
8. Han, K., Lim, H., and Hong, J. (1992) Acetic acid formation in *Escherichia coli* fermentation. *Biotechnol. Bioeng.* **39,** 663–671.
9. Van Dijken, J., Weusthuis, R., and Pronk, J. (1993) Kinetics of growth and sugar consumption in yeasts. *Antonie Leeuwenhoek* **63,** 343–352.
10. Dekeleva, M. and Strohl, W. (1987) Glucose stimulated acidogenesis by *Streptomyces peucetius. Can. J. Microbiol.* **33,** 1129–1132.
11. Madden, T., Ward, J., and Ison, A. (1996) Organic acid excretion by *Streptomyces lividans* TK24 during growth on defined carbon and nitrogen sources. *Microbiology* **142,** 3181–3185.
12. Cortes, J., Liras, P., Castro, J., and Martin, J. (1986) Glucose regulation of cephamycin biosynthesis in *Streptomyces lactamdurans* is exerted on the formation of alpha-aminoadipyl-cysteinyl-valine and deacetoxycephalosporin-C-synthase. *J. Gen. Microbiol.* **132,** 1805–1814.
13. Martin, J., Revilla, G., Zanca, D., and Lopez, N. (1982) Carbon catabolite regulation of penicillin and cephalosporin biosynthesis. *Trends Antibiot. Res. Genet. Biosynth., Actions, and New Subst. Proc. Int. Conf.* pp. 258–268.
14. Spizek, J. and Tichy, P. (1995) Some aspects of overproduction of secondary metabolites. *Folia Microbiol.* **40,** 43–50.
15. Caddick, M., Peters, D., and Platt, A. (1994) Nitrogen regulation in fungi. *Antonie Leeuwenhoek* **65,** 169–177.

16. Lo, Y., Yang, S., and Min, D. (1997) Effects on yeast extract and glucose on xanthan production and cell growth in batch culture of *Xanthomonas campestris*. *Appl. Microbiol. Biotechnol.* **47,** 689–694.

17. Gellissen, G. (2000) Heterologous protein production in methylotrophic yeasts. Appl Microbiol Biotechnol. 54, 741–750.

18. Atlas, R. and Parks, L. (1993) Handbook of Microbiological Media, CRC Press, Boca Raton, FL.

19. Zhu, Y., Rinzema, A., Tramper, J., and Bol, J. (1996) Medium design based on stoichiometric analysis of microbial transglutaminase production by *Streptoverticillium mobaraense*. *Biotechnol. Bioeng.* **50,** 291–298.

20. Strobel, R. and Sullivan, G. (1999) Experimental design for improvement of fermentations, in *Manual of Industrial Microbiology and Biotechnology* (Demain, A. and Davies, J., eds.), American Society for Microbiology, Washington, DC, pp. 80–93.

21. Plackett, R. and Burman, J. (1946) The design of optimum multifactorial experiments. *Biometrika* **33,** 305–325.

22. Box, G., Hunter, W., and Hunter, J. (1978) Fractional factorial designs at two levels, in: *Statistics for Experimenters*. John Wiley, New York, NY.

23. Box, G. and Behnken, D. (1960) Some three level variable designs for the study of quantitative variables. *Technometrics* **2,** 455–475.

24. Myers, R. and Montgomery, D. (1995) *Response Surface Methodology: Process and Product Optimization Using Designed Experiments*. John Wiley, New York, NY.

25. Demming, S. and Morgan, S. (1973) Simplex optimization of variables in analytical chemistry. *Anal. Chem.* **45,** 278–283.

26. Weuster-Botz, D. (2000) Experimental design for fermentation media development: statistical design or global random search. *J. Biosci Bioeng.* **90,** 473–483.

27. Zuzek, M., Friedrich, J., Cestnik, B., Karalic, A., and Cimerman, A. (2000) Optimization of fermen-tation medium by a modified method of genetic algorithms. *Biotechnol. Tech.* **10,** 991–996.

28. Goldberg, D. (1989) *Genetic Algorithms in Search, Optimisation, and Machine Learning*. Addison-Wesley, Reading.

29. Back, T. (1993) Optimal mutation rates in genetic search, in *Proceedings of the Fifth International Conference on Genetic Algorithms* (Forrest, S., ed.), Morgan Kauffman Pub., San Mateo, CA.

30. Weuster-Botz, D., Kelle, R., Frantzen, M., and Wandrey, C. (1997) Substrate controlled fed-batch production of l-lysine with *Corynebacterium glutamicum*. *Biotechnol. Prog.* **13,** 387–393.

31. Weuster-Botz, D. and Wandrey, C. (1995) Medium optimisation by genetic algorithm for continuous production of formate dehydrogenase. *Process Biochem.* **30,** 563–571.

32. Cockshott, A. and Hartmann, G. (2001) Improving the fermentation medium for echinocandin B production part II: particle swarm optimization. *Proc. Biochem.* **36,** 661–669.

33. Cockshott, A. and Sullivan, G. (2001) Improving the fermentation medium for echinocandin B production part I: sequential statistical experimental design. *Proc. Biochem.* **36,** 647–660.

34. Connors, N., Petersen, L., Hughes, R., Saini, K., Olewinski, R., and Salmon, P. (2000) Residual fructose and osmolality affect the levels of pneumocandins B_0 and C_0 produced by *Glarea lozoyensis*. *Appl. Microbiol. Biotechnol.* **54,** 814–818.

35. Petersen, L., Hughes, D., Hughes, R., DiMichele, L., Salmon, P., and Connors, N. (2001) Effects of amino acid and trace element supplementation on pneumocandin production by *Glarea lozoyensis*: impact on titer, analogue levels, and the identification of new analogues of pneumocandin B_0. *J. Industr. Microbiol. Biotechnol.* **26,** 216–221.

36. van Gulik, W., de Laat, W., Vinke, J., and Heijnen, J. (2000) Application of metabolic flux analysis for the identification of metabolic bottlenecks in the biosynthesis of penicillin-G. *Biotechnol. Bioeng.* **68,** 602–618.

37. Sauer, U., Lasko, D., Fiaux, J., Hochuli, M., Glaser, R., Szyperski, T., et al. (1999) Metabolic flux ratio analysis of genetic and environmental modulations of *Escherichia coli* central carbon metabolism. *J. Bacteriol.* **181,** 6679–6688.

38. Stanbury, P. and Whitaker, A. (1984) *Principles of Fermentation Technology.* Pergamon Press, Oxford, UK.

39. Wang, D., Cooney, C., Demain, A., Dunnhill, P., Humphrey, A., and Lilly, M. (1979) *Fermentation and Enzyme Technology.* John Wiley, New York, NY.

40. Yegneswaran, P., Gray, M., and Thompson, B. (1991) Experimental simulation of dissolved oxygen fluctuations in large fermentors: effect on *Streptomyces clavuligerus. Biotechnol. Bioeng.* **38,** 1203–1209.

41. Funahashi, H., Harada, H., Taguhi, H., and Yoshida, T. (1987) Circulation time distribution and volume mixing regions in highly viscous xanthan gum solution in a stirred vessel. *J. Ferment. Technol.* **20,** 277–282.

42. Bajpai, R. and Reuss, M. (1982) Coupling of mixing and microbial kinetics for evaluating the performance of bioreactors. *Can. J. Chem. Eng.* **60,** 384–392.

43. Enfors, S., Jahic, M., Rozkov, A., Xu, B., Hecker, M., Jurgen, B., et al. (2001) Physiological responses to mixing in large scale bioreactors. *J. Biotechnol.* **85,** 175–185.

44. Sweere, A., Luyben, K., and Kossen, N. (1987) Regime analysis and scale-down: tools to investigate the performance of bioreactors. *Enzyme Microb. Technol.* **9,** 386–398.

45. Geraats, S. (1994) Scaling-up of a lipase fermentation process: a practical approach, in *Advances in Bioprocess Engineering* (Galindo, E. and Ramirez, O., eds.), Kluwer Academic Publishers, Dordrecht, pp. 41–46.

46. Vardar, F. and Lilly, M. (1982) Effect of cycling dissolved oxygen concentrations on product formation in penicillin fermentations. *Eur. J. Appl. Microbiol. Biotechnol.* **14,** 203–211.

47. Larsson, G. and Enfors, S. (1988) Studies of insufficient mixing in bioreactors: effects of limiting oxygen concentrations and short term oxygen starvation on *Penicillium chrysogenum. Bioproc. Eng.* **3,** 123–127.

48. George, S., Larsson, G., and Enfors, S. (1993) A scale-down two compartment reactor with controlled substrate oscillations: metabolic response of *Saccharomyces cerevisiae. Bioproc. Eng.* **9,** 249–257.

49. Lashkari, D., DeRisi, J., McCusker, J., Namath, A., Gentile, C., Hwang, S., et al. (1997) Yeast microarrays for genome wide parallel genetic and gene expression analysis. *Proc. Natl. Acad. Sci. USA* **94,** 13,057–13,062.

50. Tao, H., Bausch, C., Richmond, C., Blattner, F., and Conway, T. (1999) Functional genomics: expression analysis of *Escherichia coli* growing on minimal and rich media. *J. Bacteriol.* **181,** 6425–6440.

51. Neidhardt, F. (1996) The enteric bacterial cell and the age of bacteria, in *Escherichia coli* and *Salmonella: Cellular and Molecular Biology,* 2nd ed. (Neidhardt, F. and Curtis., R., III, Ingraham, J., Lin, E., Low, K., Magasanik, B., et al., eds.), ASM Press, Washington, DC, pp. 1–3.

52. Schweder, T., Kruger, E., Xu, B., Jurgen, B., Blomsten, G., Enfors, S.-O., et al. (1999) Monitoring of genes that respond to process-related stress in large scale bioprocesses. *Biotechnol. Bioeng.* **65,** 151–159.

53. Tseng, G., Oh, M., Rohlin, L., Liao, J., and Wong, W. (2001) Issues in cDNA microarray analysis: quality filtering, channel normalizaion, models of variations and assessment of gene effects. *Nucleic Acids Res.* **29,** 2549–2557.

54. Eisen, M., Spellman, P., Brown, P., and Botstein, D. (1998) Cluster analysis and display of gneome-wide expression patterns. *Proc. Natl. Acad. Sci. USA* **95,** 14,863–14,868.

55. Wen, X., Fuhrman, S., Michaels, G., Carr, D., Smith, S., Barker, J., et al. (1998) Large-scale temporal gene expression mapping of central nervous system development. *Proc. Natl. Acad. Sci. USA* **95,** 334–339.

56. Tamayo, P., Slomin, D., Mesirov, J., Zhu, Q., Kitareewan, S., Dmitrovsky, E., et al. (1999) Interpreting patterns of gene expression with self-organizing maps: methods and applications to hematopoietic differentiation. *Proc. Natl. Acad. Sci. USA* **96,** 2907–2912.
57. Rigoutsos, I., Floratos, A., Parida, L., Gao, Y., and Platt, D. (2000) The emergence of pattern discovery techniques in computational biology. *Metab. Eng.* **2,** 159–177.

III

PLANT CELL CULTURE

8

Functional Genomics for Plant Trait Discovery

Sam Reddy, Ignacio M. Larrinua, Max O. Ruegger,
Vipula K. Shukla, and Yuejin Sun

1. Introduction

Functional genomics, as the name implies, approaches gene discovery on a genome-wide scale. Our understanding of the complex genetic interrelationships underlying plant traits such as yield or stress tolerance is being shaped by new technologies that facilitate the analysis of thousands of genes in a single experiment.

Most plant traits are determined by multiple genes. The contribution of an individual gene may be controlled at any number of levels including transcription, splicing, transcript degradation, translation, post-translational modification, protein targeting, and protein degradation. Such complexity has long presented a major challenge to scientists in their attempts to identify and understand all of the genes underlying a given trait. The traditional tools of molecular biology and biochemistry, such as RNA gel-blotting and protein immunoblotting, are generally limited to the study of one or a few genes per experiment. Without a more comprehensive, genome-scale approach, these studies can produce incomplete or even misleading results.

The advent of functional genomics technologies has revolutionized the potential for the discovery and characterization of the genes that provide the basis for plant traits. New high-throughput transcript profiling technology allows the expression levels of thousands of genes to be analyzed in a single experiment. Proteomics technology, combining the powers of two-dimensional electrophoresis, ESI-MS/MS (electrospray ionization-tandem mass spectrometry) and MALDI-TOF (matrix-assisted laser deposition-ionization-time of flight), provides the capacity to separate and compare thousands of proteins on a single gel and then determine the identity of individual proteins. Transformation technologies have also been developed that allow large-scale knockout or upregulation of endogenous plant genes as well as efficient, *in planta* analysis of heterologous genes.

Some of the key approaches to plant gene discovery using functional genomics strategies and tools are summarized in this chapter. For a more detailed understanding of functional genomics tools and plant traits, we encourage readers to consult the references provided at the end of this chapter.

From: *Handbook of Industrial Cell Culture: Mammalian, Microbial, and Plant Cells*
Edited by: V. A. Vinci and S. R. Parekh © Humana Press Inc., Totowa, NJ

2. Sequence Analysis: An *In Silico* Strategy

The availability of thousands of Expressed Sequence Tags (ESTs) and the complete genome sequence of *Arabidopsis* have provided new opportunities and challenges to modern biologists. Thorough knowledge-based analysis of both DNA and protein sequences may offer a unique potential to discover candidate genes for plant traits.

2.1. Databases

The most important DNA database for plants as well as for most other organisms is the National Center for Biotechnology Information (NCBI) Entrez nucleotide database (http://www.ncbi.nlm.nih.gov). This database, which is updated daily, contains an almost complete set of the publicly available DNA sequences. It exchanges daily information with the DNA Database of Japan (DDBJ, http://www.ddbj.nig.ac.jp) and the European Molecular Biology Laboratory (EMBL, http://www.ebi.ac.uk/). Other important databases include the Swiss-Protein database, a highly curated protein database, and the Protein Data Bank (http://www.rcsb.org/pdb), an X-ray crystallographic database. Finally, a number of sites maintain Pfam, a database of protein-domain families.

2.2. DNA Sequence Information

In the last few years, the amount of DNA sequences available for plants has experienced an exponential increase similar to that of other organisms. *Arabidopsis thaliana* was the first plant sequenced; and this wealth of information is easily accessible at http://www.arabidopsis.org. Currently, other more commercially important plants are also being completely sequenced. Of these, the most commercially important one is *Oryza sativa* (rice) because it is the most important food plant in the world, and it has one of the smallest cereal genomes, which makes it an important model crop. As of May 2001, over 100,000 non-EST plant-related entries were recorded in GenBank.

ESTs represent a second source of DNA information, separate from genomic sequencing. Defined as single-pass 5' sequencing of cDNA clones, ESTs provide a way of generating sequence information for only that part of the genome that is expressed, and presumably functional. As of mid-2001, some 8.5 million EST sequences were in the public domain. In the EST database at NCBI, 11 of the 20 top organisms are plants if one excludes human and mouse ESTs, which compose about 70% of the total number. Plant ESTs comprise about 50% of the rest, with well over one million ESTs. *See* Table 1 for a summary of the top plants in GenBank.

2.3. DNA and Protein Analysis Tools

DNA and protein-analysis tools can be divided into many categories, but a useful way of surveying the panoply of available tools is to classify them in two different but complementary ways. One possible classification divides publicly available tools provided directly by government institutions such as NCBI or by universities and independent researchers working under the GNU license from tool packages provided by commercial enterprises. The second way to categorize these tools is to separate stand-alone packages residing in individual workstations from enterprise-wide programs residing on larger computers and accessible to a large number of users simultaneously via a Web interface, Xwindows interface, or command line.

Table 1
Summary by Organism—May 4, 2001
GenBank Entries

Species	ESTs
Glycine max (soybean)	166,233
Medicago truncatula (barrel medic)	122,365
Lycopersicon esculentum (tomato)	114,999
Arabidopsis thaliana (thale cress)	113,000
Zea mays (maize)	89,125
Oryza sativa (rice)	75,057
Hordeum vulgare (barley)	68,480
Sorghum bicolor (sorghum)	65,040
Chlamydomonas reinhardtii (an alga)	64,973
Triticum aestivum (wheat)	60,022
Solanum tuberosum (potato)	38,074

Stand-alone publicly available tools can be found on the World Wide Web. For example, at the Sanger Centre (http://www.sanger.ac.uk/), one can download Artemis, a sequence visualization Java program. Otherwise, researchers can build their own programs with the help of publicly available tools. The most important of these is the NCBI toolbox (ftp://ncbi.nlm.nih.gov/toolbox/ncbi_tools/) available from NCBI, as well as more basic modules available at various central sites devoted to the use of the Perl, Java, or Python programming languages in bioinformatics (i.e., www.bioperl.org, www.java.org, www.python.org, www.bioxml.org, www.biocorba.org, www.bioinformatics.org).

Stand-alone commercial tools include such products as Sequencher (GeneCodes Corporation), Gene Construction Kit (Textco) and MacVector (Accelrys Inc.). These tools generally specialize in one function. For example, Sequencher is a useful program for a small sequencing project, since it does a good job of both allowing the reading of sequencing gels and putting together these reads into larger assemblies. In the same vein, Gene Construction Kit allows the user to construct and display their DNA constructs in a facile way. Typically, these products are only useful for small projects—usually by individual scientists or laboratories—and these tools are not suited for large bioinformatic projects.

These latter projects are best handled by larger and more complicated tools that usually reside on larger, more powerful computers accessed by a pool of users, although as the cost of computing power continues to drop this distinction is becoming increasingly artificial. For the interpretation of DNA sequence reads as well as putting these reads into larger assemblies and visualizing the result, the package of Phred/Phrap (University of Washington) is free to academic scientists and reasonably priced for industry. These programs are run from the UNIX command line with no graphical interface, thus limiting their use to users who are comfortable with the UNIX operating system. Consed (University of Washington) does allow the graphical visualization of results. These programs are a necessary adjunct to almost any large genomic sequencing project. Second, the NCBI makes its alignment tool BLAST (Basic Local Alignment Search Tool) available for downloading either as a fully functional package accessible via the command line or as a complete Web-based installation. Either installation will allow the user to run many sequence searches in batches. The BLAST command-line version is ideal for parsing the data into relational databases for latter retrieval. Finally, one can download a complete Web package that allows the user to perform Pfam-domain searches from Washington University (St. Louis). Pfam-domain searches are useful for looking at more advanced relationships among protein sets than can be achieved via BLAST, but they are much more computationally intensive and require the use of parallel computing solutions for batch searches.

Several companies provide many of these capabilities wrapped into one package. The Wisconsin Package (Accelrys, Inc.) usually abbreviated as GCG for historical reasons, is clearly the oldest of these, and offers a large number of individual functionalities at a modest price. It can be used as a command-line tool, as an Xwindows emulation (SeqLab), and as a Web-based tool (SeqWeb). Both the command line and Xwindows versions of this program are tied to the UNIX operating system, and they depend on internal nonstandard sequence files that do not allow for transparent import and export from the now standard FASTA format.

2.4. Information Storage

These tools generate large volumes of data, which must be effectively managed. The data can go into a notebook or can be stored as flat files (e.g., original BLAST reports) but the storage vehicle of choice is a relational database. The three most popular UNIX-based solutions are: Oracle (http://www.oracle.com), MySQL (http://www.MySql.com/) and PostgreSQL (http://www.postgresql.org). All three are relational databases that use a reasonably standard version of SQL (Structured Query Language). Oracle is a commercial high-end product. MySQL is free under a GNU license, and a commercial version can be purchased. By contrast, PostgreSQL was developed under a GNU license and is freely available to everyone.

2.5. Utility

To better conceptualize how the components described in **Subheading 2.4.** might work together, we will examine a hypothetical but realistic example. Let us assume that in a tagging experiment, a DNA sequence of interest has been isolated and sequenced with the help of our analysis tools. We still have many unanswered questions. Does it code for a protein? What is the function of this protein? Does it have homologs in other organisms? How do we begin to answer some of these questions?

As a first step, we might use NCBI's ORFFinder or GCG's Translate program to reveal the presence of any open reading frames (ORF). In the absence of these, a BLAST alignment using the appropriate databases may find homologies to known genes, and that might indicate that the sequencing contains errors. Once a protein-coding sequence has been identified, it can be run against a number of programs to try to extract information about its function:

1. BLAST searches of appropriate databases will reveal whether any known proteins share a homology with the gene in question.
2. More distant relationships can be distinguished with NCBI's PSI BLAST, which is more sensitive than normal BLAST.
3. Searching against Pfam families may uncover functional blocks, and programs such as GCG's Motifs can search for signature sequences of functional significance.

4. If a number of similar sequences are discovered, programs such as ClustalW or PAUP will allow the user to build possible phylogenetic trees and place the gene of interest in an evolutionary framework.

5. If a very close sequence match is found, a search for this match among those proteins with known crystallographic structures can be performed (http://130.14.22.106/entrez/query.fcgi?db=Structure). If a structure for the match can be found, programs exist to thread the sequence through the backbone of its homolog and provide a first approximation to its three-dimensional structure.

6. Database searches can also be used to identify homologs of the protein of interest in other organisms.

Using these tools or others, many or most proteins will yield information about their function and evolutionary history. Clearly, depending on the results, one may obtain a very clear picture of the gene involved, including an idea of its three-dimensional structure—or one may get only hints about its possible roles and functions. Regardless of how much information is found, a starting point will have been provided for laboratory experiments.

Finally, let us imagine that instead of one sequence we have thousands (or tens of thousands) of sequences. Although the logical experimental progression is very similar, the actual manipulation is not. The facile manipulation of large volumes of data in a realistic time frame dictates that the work will be done on a computer with an operating system that is a version of UNIX (SunOS, Linux). Furthermore, almost all of the manipulation will be done at the command line instead of from Web-based graphical user interfaces (GUIs), and it will require the help of text-manipulation programs or scripts written in languages such as Perl. For example, BPLite, a Perl BLAST parser module originally written by Ian Korf, could be used to parse thousands of BLAST reports into a tab delimited format suitable for import into a relational database.

All of these manipulations will be done by the computer without human supervision. It is therefore imperative that the experimental design receive all the attention needed to ensure that the scripts written will yield the results one expects and no other under any possible set of real-life circumstances. Furthermore, it is essential that once in place, the protocols are thoroughly tested before actual use with real data, and that the results of the analysis are scanned for anomalies. This process should not be put into production until one is certain that the process in place will deliver the correct results to the individual scientists who will use the data.

3. Expression Analysis: A Bio-Chip and *In Planta*-Based Approach

3.1. Transcript Profiling with Glass Slide-Based Microarrays

The DNA microarray refers to an orderly arrangement of DNA spots, of less than 300 μm in diameter, on a solid matrix. In general, DNA microarrays include high-density oligonucleotide arrays directly synthesized on the solid surface and arrays generated by spotting DNA on chemically coated microscopic glass slides. The utilities of the glass-slide-based microarray in plant science were first demonstrated by Schena et al. in 1995 using 45 *Arabidopsis* ESTs (1). Since then, plant scientists have explored a number of ways to apply this technology to plant-trait gene discovery. The *Arabidopsis* and maize genome projects founded by the National Science Foundation have played important roles in promoting this technology in academic research institutes. Although

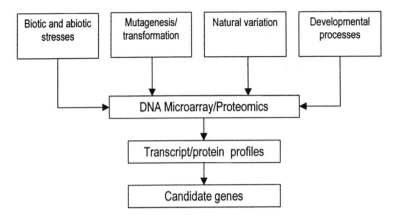

Fig. 1. Diagram illustrating the applications of DNA microarray and proteomics in plant-trait gene discovery. Transcript or protein profiles can be obtained from plants that are challenged with biotic or abiotic stresses, mutagenized or transformed through genetic manipulations, or altered during developmental processes. These profiles provide information that helps to identify candidate genes that encode desired traits.

the use of microarray technology in plant-trait discovery is still in its infancy, preliminary results have shown its promise. Since a large number of genes can be spotted on a single slide, microarrays are a powerful tool to identify novel genes associated with a trait of interest. In addition, microarrays depict the overall pattern of gene expression, allowing the comparison of patterns from multiple genotypes at multiple time-points, and identification of common components that respond to multiple environmental and developmental conditions.

Plant scientists have explored a number of major applications for DNA microarrays in the past few years (Fig. 1). First, microarrays have proven to be a powerful tool to investigate the responses of plants to biotic and abiotic stresses. During the past 6 yr, this technology has been used to analyze plant genes that respond to drought, cold, wounding, insect feeding, fungal infection, nitrate, herbicides, and herbivore-induced volatiles. Abiotic stresses, such as drought and cold, exert a pleiotropic effect on gene expression in higher plants, affecting a number of biochemical, molecular, and cellular processes. Microarray analysis with 1300 *Arabidopsis* genes identified 44 genes inducible by drought and 19 genes inducible by cold. This approach also allows the grouping of genes into subsets based on whether they respond to both drought or cold, or solely to one treatment or the other *(2)*. Microarrays have been used to identify defense-response genes. Upon infection with *Alternaria brassicicola* or treatment with salicylic acid, methyl jasmonate, and ethylene, 705 genes were affected among the 2375 *Arabidopsis* genes on the microarray slide. A coordinated defense response was identified by showing that 126 genes responded to all four treatments, accounting for approx 5% of the genes on the slide *(3)*. In another experiment, Maleck et al. spotted 10,000 ESTs on microarray slides and imposed 14 systematic-acquired-resistance-inducing or -repressing conditions to *Arabidopsis*. The experiment led to the identification of a common promoter element that binds to members of plant-specific transcription-factor families *(4)*.

Microarrays have been used to understand the molecular makeup of natural variations. A recent DNA microarray survey showed that in Pokkali, a salt-tolerance rice line, approx 10% of the transcripts were up- or downregulated within 1 h of salt treatment. This striking change in transcript levels became less obvious as the plant adapted to the stress condition a few hours after the treatment. In contrast, in the salt-sensitive IR29 rice line, the initial response was delayed and an overall downregulation of the transcripts was triggered, resulting in cell death *(5)*. These results provided valuable information about natural variation in salt tolerance.

In addition to natural variation, scientists have also attempted to create desirable plant traits with transformation or mutagenesis. The combination of transformation and mutagenesis with microarray technology has yielded even more in-depth knowledge of the downstream events that are regulated by the gene of interest. For example, T-DNA insertion in a MAP kinase (MPK4) caused a constitutive systematic acquired resistance in *Arabidopsis*. Microarray analysis showed that among 9861 tested genes, 16 were affected by the mutation with a fivefold or greater induction. Among the 16 affected genes, 8 encoded well-known pathogen resistance or wounding inducible genes *(6)*. These results further confirmed the role of MPK4 in disease resistance and demonstrated the downstream pathways that were affected by the kinase.

As illustrated in Fig. 1, comprehensive analysis of the developmental program with microarrays may lead to the identification of candidate genes that encode economically important traits. An excellent example in this regard is the identification of a novel strawberry alcohol acetyltransferase (SAAT) that is involved in flavor biogenesis *(7)*. Microarray experiments revealed that the level of SAAT transcript correlated with the maturation process of strawberry fruit, with a 16-fold induction at the ripening stage. Biochemical experiments showed that the SAAT had high substrate specificity to hexanol and octanol, and that its corresponding esters were the most important components of strawberry flavor formed during the ripening process.

Because of space limitations, it is not possible to list all the applications of DNA microarrays in plant trait gene discovery. Although the advantages of DNA microarray technology should be fully acknowledged, it is important to realize its pitfalls and limitations. First, a high level of variation is expected for low-abundance transcripts. Statistical analysis should be an integrated part of microarray data processing. Secondly, microarrays only establish a correlative relationship between the traits of interest and the candidate genes. The functions of a identified gene should be further tested with other molecular, biochemical, genetic, and cellular tools. Third, transcript profiling determines the steady-state level of messenger RNA of the expressed genes. It is well accepted that the abundance of transcript does not always reflect the level of gene expression. Translational regulation, post-translational modification and the turnover of proteins are among the additional steps in the gene expression regulatory pathway.

3.2. Proteomics

Proteomics is defined as "the systematic analysis of the proteins (the proteome) in a tissue, cell, organelle, or membrane" *(8)*. The "proteome" reflects gene expression at steady-state protein levels and is complementary to the "transcriptome" that provides information at the transcript level *(9)*. In brief, proteomic technology combines two-dimensional polyacrylamide gel electrophoresis (PAGE) with MALDI-TOF, a high-throughput and high-precision technology that determines the mass of peptides derived from endoproteoase digestion. When a protein sequence database of the species under investigation is available, peptide fingerprinting allows rapid identification of a protein spot on the polyacrylamide gel. When the protein-sequence data are not available, ESI-MS/MS provides sequence information of the endoprotease-digested peptides. Although the proteomic technologies are relatively new, applications of these technologies in plant-science research are increasing rapidly.

Similar to transcript-profiling technologies, protein-profiling tools have been widely used to understand the responses of higher plants to environmental cues, to identify proteins that are affected by mutations, and to investigate the developmental process of higher plants (Fig. 1). Although numerous experiments have used two-dimensional polyacrylamide gels to obtain protein profiles from plant tissues that were treated with different types of abiotic and biotic stresses, only a few of the investigations carried out a comprehensive identification of the proteins separated by the two-dimensional gel. Riccardi et al. compared the protein-expression profiles of drought-tolerant and drought-sensitive maize lines grown under water-deficit conditions, and found thirty-eight proteins that were differentially expressed between the two lines *(10)*. Microsequencing revealed that 16 of those proteins were well-known water stress-response proteins, and the rest were enzymes involved in glycolysis, Krebs cycle, and secondary metabolism. In another experiment, Chang et al. identified 48 proteins that changed their abundance during hypoxic acclimation *(11)*. These examples clearly show that the pleiotropic effect of environmental stress acts beyond the level of transcription to affect protein-expression levels as well.

Using proteomic technologies to obtain the protein profile of mutant lines has led to a better understanding of mutations *(12)*. The *opaque2* mutation, which increases the lysine content of corn seed, has long been of interest to the breeders and molecular biologists. Damerval and le Guilloux compared the protein profile from the seeds of *opaque2* mutant with the one from wild-type, and discovered that 36 polypeptides were affected by the mutation. In addition to six zeins that were known to be affected by *opaque2*, a variety of proteins, including glyceraldehyde 3-phosphate dehydrogenase, aspartic proteinase precursor, and sorbitol dehydrogenase, were found to be affected by the mutation. This finding suggests that the *opaque2* gene plays a central role in regulating multiple metabolic pathways *(13)*.

Recently, proteomic tools have also been used to analyze the developmental program during seed germination *(14)* and the response of *Arabidopsis* to fungal and bacterial elicitors *(15)*. The applications of proteomics in plant-gene discovery is rapidly increasing. When proteomic data are combined with transcript profiling and metabolite profiling data, we gain a more complete picture of the genetic makeup of a plant trait.

3.3. In Planta *Gene Expression Strategies*

Transcript profiling and proteomic experiments correlate the levels of gene expression with a plant trait. The function of a candidate gene identified by these technologies must be further tested *in planta*. Since the stable transformation of crops, such as maize and soybean, at high-throughput rates is prohibitively expensive, Arabidopsis, with a small size and relatively short life cycle, has proven to be a good model plant for functional testing of large numbers of candidate genes. A number of transformation methods are available to obtain stably transformed Arabidopsis *(16)*. *In planta* Agrobacterium-mediated transformation by vacuum infiltration has been widely used by the *Arabidopsis* research community. A high frequency of transformation was obtained through infiltration of adult *Arabidopsis*, with *Agrobacterium* harboring desired genes *(17)*. This method modified the previous *Arabidopsis* transformation methods of seed imbibition *(18)* and inoculation of young inflorescences with *Agrobacterium (11)*. Infiltration of adult plants has significantly increased the transformation efficiency. More than 50% of transformed plants carry a T-DNA insertion at single Mendelian locus. Therefore, it is conceivable that transformation of *Arabidopsis* with a population of cDNAs followed by screening for desired traits, such as herbicide and disease resistances, will allow the identification of large numbers of leads. However, to date, stable transformation of *Arabidopsis* has only been used as a measure to confirm the hits and leads generated from high-throughput screening.

4. Insertional Mutagenesis: A Reverse Genetics Approach

The ability to isolate and engineer genes has led to major advances in the agricultural biotechnology sector. Because plants are amenable to genetic manipulation, they represent a valuable opportunity to apply molecular genetic tools to crop improvement. This section examines one such tool, insertional mutagenesis, and how it is utilized in a reverse genetics approach aimed at discovering new genes/traits for plant improvement.

4.1. Reverse Genetics

Reverse genetics is a sequence-based approach to the isolation of genes. Reverse genetics first targets the genomic complement of an organism, using molecular/genetic tools to disrupt normal gene function, and then generates a population of mutagenized individuals. This disruption "marks" the gene at the molecular level, typically through an insertional event. Using polymerase chain reaction (PCR)-based methods, pooled DNA from these populations can be screened to identify and isolate mutants in genes of interest. Once identified, these mutant lines are characterized at the phenotypic level, and the function of the genes is determined. Reverse genetics is based on the concept that genes lead to function.

In contrast, forward genetics is a phenotypic approach to the isolation of genes. Like classical plant-breeding programs, it is based on the paradigm of mutation and selection. In this model, genetic mutations (natural or induced) lead to a desirable phenotype. The individual who carries this mutation is then selected and bred in order to propagate the trait or mutated gene. Through careful analysis and correlation with the observed phenotype, the responsible gene is eventually localized (mapped), physically isolated, and cloned. Contrary to reverse genetics, forward genetics is based on the concept that function leads to genes.

Both forward and reverse genetic approaches have advantages and limitations. Historically, forward genetics has been carried out in plant species such as corn, for which large collections of well-characterized mutants exist or can be easily created *(19)*. The analysis of these mutants allows researchers to select relevant phenotypes in the context of desirable germplasm and to correlate these phenotypes with genetic (breeding) data. A rate-limiting step in forward genetics is the isolation of gene(s) that give rise to phenotypes, since these genes are not "marked" and therefore are not easily identified or isolated.

Reverse genetics has been carried out primarily in model species. These species (e.g., *Arabidopsis thaliana, Zea mays*) have been extensively studied at the genetic level. Detailed physical maps and complete or near-complete genomic-sequence data *(20)* coupled with relative ease of manipulation allow the reverse genetic approach to rapidly lead to gene isolation. However, the range and classes of phenotypes observed in the existing mutant collections are limited, and may require extensive characterization under different conditions to extrapolate gene function *(21)*. Furthermore, when gene function is determined in a model species, that function does not necessarily predict activity in crop species grown under field conditions.

4.2. Insertional Mutagenesis

Reverse genetics approaches rely on a set of tools designed to disrupt normal gene function *in planta*. These disruptions are carried out primarily through the process of insertional mutagenesis, which introduces DNA elements into the genome at random positions. Insertional mutants, containing gene disruptions, fall into two distinct categories: loss-of-function (LOF) and gain-of-function (GOF) mutants.

To date, the vast majority of insertional mutants characterized are LOF mutants. Random insertion of DNA elements into a plant's genome results in many of these insertions targeting coding or regulatory regions associated with genes (the actual frequency depends on genome density). Interruption of these regions may lead to a complete (knockout) or partial (knockdown) disruption in the accumulation or activity of the proteins encoded by the affected genes. Using conventional genetic methods to isolate "clean" lines of insertional mutants that contain insertions in one and only one gene, the LOF phenotype is correlated with the disrupted gene, leading to functional predictions. However, a caveat of this approach lies in the organizational structure of genomes. It is increasingly evident that many, if not most, plant genes are members of multigene families *(20,22)*. This redundancy can be both structural (at the sequence level) and functional (at the activity level); therefore, a knockout in a single gene may not lead to an obvious phenotype because other homologs can compensate for essential functions. However, if a gene product is absolutely essential for viability, a LOF disruption will lead to lethality, thereby preventing in-depth analysis of phenotypes. Despite these limitations, LOF insertional mutants have proven to be very useful in defining gene function for certain types of gene families.

A second class of insertional mutants are the GOF mutants. GOF mutants are characterized by gene expression and/or gene-product accumulation at levels above normal, at different or de-regulated developmental stages, or in tissues where it normally does not occur (ectopic expression). GOF insertional mutants can be generated in one of two ways. First, an insertional element may target a negative-regulatory region of a gene. In this scenario, disruption takes place in a repressor element, thereby lifting repression and allowing gene expression to occur under conditions in which it normally does not. GOF mutants can also be generated using particular types of insertional elements known as "activation tags" *(23)* that hyperactivate the gene's own regulatory system, which results in supranormal expression and subsequent gene-product accumulation and/or activity. An advantage of GOF mutants is that they allow characterization of single members of multigene families, overcoming the limitations of redundancy. Additionally, genes for which a LOF mutation may prove lethal are amenable to GOF studies. At this time, the applicability of current GOF insertional tools appears to be limited to certain gene classes *(23)*. Since this technology is very new, further enhancements and development may overcome this hurdle.

Several collections of both LOF and GOF mutants have been generated and are available in *Arabidopsis thaliana*, the dicot model of choice (*see* http://www.arabidopsis.org). Furthermore, with the recent completion of the Arabidopsis genome sequencing project *(20)*, this model is particularly well-suited for the PCR-based screening methods employed in a reverse genetics approach. For monocots, *Zea mays* collections have been developed and characterized, and have proven very useful for the identification of new genes and traits *(24)*. Although the maize genome is significantly larger and more complex than that of Arabidopsis, detailed physical maps and some sequence data are available to facilitate discovery efforts at ZmDB, a maize genome database (*see* http://www.ZmDB.iastate.edu).

There are several types of insertional elements that give rise to GOF and LOF mutants. Historically, transposons provided the first evidence that DNA fragments could mobilize within the genome *(25)*; practically, this discovery has been exploited to determine gene function through the generation of LOF mutants. Transposons such as the *Ac/Ds* system in maize *(26)* insert at intervals throughout the genome, generating collections of individual mutants with knockout or knockdown phenotypes. One of the advantages of transposon-based insertional mutagenesis is that these elements excise and re-insert at relatively short genetic intervals, often generating multiple mutant alleles at the same locus. This allows dissection of gene function at a fine level when mutants are phenotypically analyzed.

The most widely used mechanism for the introduction of exogenous DNA insertional elements is via *Agrobacterium*-mediated transformation, either in whole plants or cell culture *(27)*. *Agrobacterium tumefaciens*, the causal agent of crown gall disease, inserts a transfer-DNA element (T-DNA) into the genome of plants that it infects *(28)*. This feature has also been engineered to allow introduction of heterologous sequences within the T-DNA. However, because not all plant species are susceptible to *Agrobacterium* infections, the current collections of mutants are limited to model plants such as *Arabidopsis thaliana*. Using T-DNA-based technology, several types of insertional mutagenic strategies have been created.

The most straightforward type of insertional element is a simple T-DNA. This element, which contains bacterial components necessary for transfer and integration into the host genome *(28)*, interrupts the coding or regulatory sequences of the gene it inserts into, leading to a LOF mutant. T-DNA marks the site of insertion at the molecular level, and can be readily identified using PCR methods. Recently, the University of Wisconsin, Madison, established a national Arabidopsis Knockout Facility to analyze and distribute materials from a collection of >60,000 tagged mutant lines *(29)*.

T-DNAs can also be engineered to introduce exogenous elements into a plant genome. Insertion of multimeric, virally derived enhancer elements (4X CMV-35S) has been used to activate ectopic expression of proximal genes, leading to GOF mutants *(23,30)*. Activation tagged lines allow the identification of mutants in genes that were previously inaccessible using LOF approaches *(23)*. However, insertion of T-DNAs containing enhancer elements can also lead to gene disruption if the insertion targets an open reading frame (ORF), resulting in a knockout instead of an activation *(31)*.

Finally, T-DNAs designed to deliver reporter genes at random sites within the genome have been used in an approach called Gene Trapping *(32)*. The basis for this approach is that reporter genes such as *gusA* (β-glucuronidase) form gene fusions (transcriptional and/or translational) in the chromosome that can be monitored to study the expression patterns of the targeted gene. Three basic types of Gene Traps exist: enhancer traps, promoter traps, and gene traps, and each is able to respond to cis-acting regulatory sequences at the site of insertion *(32)*. The advantage of this approach is that genes are identified based on reporter-gene expression, alleviating the need for a mutant phenotype; however, a gene trap insertion may disrupt gene function, resulting in a LOF. Gene-trap collections are available in Arabidopsis, for which several classes of expression patterns have been identified.

4.3. Use and Utility

Historically, mutagenic approaches have been used to determine gene function(s). Understanding gene functions leads to a better comprehension of plant systems and enables breeders to devise strategies for successful crop improvement. Now, with reverse genetics, the biotechnology sector has a powerful tool for pinpointing genes useful in genetic engineering strategies. The reverse genetics approach for plant-trait identification has successfully identified genes with obvious utility in the agricultural sphere. A T-DNA-tagged LOF mutant in the MADS-box gene *SHATTERPROOF1* (*SHP1*), when combined with a previously identified *shp2* mutant, leads to loss of fruit dehiscence and pod shatter in Arabidopsis *(33)*. This trait is especially relevant in species such as canola, in which pod shatter has a significant impact on crop yield. It is interesting to note that in this case, the functional redundancy between closely related members of a gene family means that only the double mutant *shp1shp2* displays the shatterproof phenotype *(33)*.

Examples of GOF traits identified by insertional mutagenesis include flowering time and lodging-resistance. The *FLOWERING TIME* (*FT*) gene was the first to be identified and isolated in plants using activation tagging *(30)*. Overexpression of this gene leads to early flowering in Arabidopsis, and this trait has usefulness in breeding programs of long-lived species such as trees, for which generation time may be dramatically shortened. Another trait identified by activation tagging, which is encoded by the *STURDY* gene, allows plants to be more resistant to lodging *(34)*. This trait is being developed for use in field crops, for which lodging leads to significant yield and quality losses.

In order to fully exploit the usefulness of reverse genetics in the industrial setting, biotechnology companies are in the process of modifying the molecular tools and systems for high-throughput, focused discovery programs. As we have described, most reverse-genetics efforts to date have been carried out using model systems such as Arabidopsis and maize, which are used to predict gene function. However, in order to efficiently adapt reverse genetics to agronomically important crops, it will be necessary to devise highly efficient transformation systems for those crops, to allow tagged populations to be generated in desirable genetic backgrounds and germplasm. Additionally, screening methodologies will need to automate and scale-up to account for the larger and more complex genomes of non-model species that may require screening of hundreds of thousands of lines to find the appropriate mutant.

If one assumes that technological hurdles such as these can be overcome, then the reverse genetics approach is very amenable to an industrial discovery environment. Identification of particular gene-family targets geared toward product concepts, automation of many tasks, and relatively short time frames to achieve results, coupled with availability of a variety of molecular tools, make reverse genetics a good strategy to include in the discovery toolbox.

5. Physical Mutagenesis: A Forward Genetics Approach

Mutations are "building blocks" for the production of new varieties as well as tools for gene discovery and characterization.

5.1. A Fresh Look at "Classical" Mutagenesis

Although DNA replication and repair activities occur with an amazing degree of accuracy, occasional errors or "spontaneous mutations" are inherent to these processes. Although the vast majority of these changes are either neutral or detrimental to the fitness of an organism, occasionally a DNA mutation leads to a trait favored by selective pressures. Spontaneous mutations underlie much of the remarkable variety of life on this planet, both in terms of the number of species and in the genetic diversity within species. Some 5,000 to 10,000 years ago, humans began exploiting spontaneous mutations and genetic diversity to select plants and animals with improved food and fiber traits *(35)*. Over time, human selection has led to domesticated species that often bear little resemblance to their wild ancestors.

In the early part of the twentieth century, Hermann Muller, seeking to improve the slow rate of spontaneous mutation in his fruit fly populations, reported the mutagenic effect of X-rays *(36)*. In successive decades, scientists utilized a variety of physical and chemical methods to induce mutations in a number of plant and animal species. Although most early mutagenesis experiments were conducted for basic research, a substantial interest in mutagenesis for applied crop breeding had arisen by the end of the 1960s *(37)*. By the mid 1990s, nearly 1800 mutant cultivars in more than 150 plant species were listed in the FAO/IAEA Mutant Varieties Database *(38)*. A current examination of that database (http://www-infocris.iaea.org/MVD/) identifies over 2200 entries and numerous agronomic improvements from mutagenesis, including disease resistance, lodging resistance, oil content, and yield. It is important to note that the vast majority of the contributions made by mutation breeding occurred without the benefit of recent discoveries in genomics or the development of highly automated, high-throughput analytical methods. When combined with these new and rapidly expanding capabilities, it is likely that what is now considered "classical" mutagenesis will lead to many novel improvements in crop traits. A particularly valuable aspect of this approach is that it does not rely upon recombinant DNA technology and that will avoid incurring the high costs associated with registration of genetically modified (GM) crops.

5.2. Selection of a Mutagen

One of the first steps in a mutagenesis program is the selection of a suitable mutagen. Apart from insertion elements, mutagens are usually grouped into two classes. The first class, physical mutagens, includes electromagnetic radiation such as ultraviolet light and X-rays as well as alpha particles and neutrons. The second class, chemical mutagens, includes a large number of compounds such as alkylating agents, base analogs, and intercalating agents (for reviews, *see 37,39*). Although many physical and chemical mutagens are known, a relatively small number are in routine use in plants. The physical mutagens, fast neutrons and gamma rays, typically cause DNA deletions that range in size from a few bases to many kilobases *(40,41)*, although chromosomal rearrangements can also occur (van Harten, *1998*). Deletions usually eliminate gene function by removing coding and regulatory regions to create null or "knockout" mutations. Large deletions may even remove multiple genes *(41,42)*. The chemical mutagen, ethylmethane sulfonate (EMS), an alkylating agent, is widely used to induce mutations in plants. By far the most common change observed in EMS-mutagenized DNA is C/G to T/A transitions *(43)*. EMS treatment may result in a null mutation, for example, by introducing a stop codon near the N terminus of a protein or by creating a missense mutation in a codon for an essential residue. EMS can also induce partial LOF (leaky) mutations, when important but nonessential residues are substituted, and it can induce temperature-sensitive mutations, where the gene product is inactive at high or low ("restrictive") temperature *(44,45)*. In some instances, EMS can induce GOF mutations like the single amino-acid change that blocks targeted degradation of an auxin-responsive protein in Arabidopsis to create a dominant auxin-resistant phenotype *(46)*.

Historically, both chemical and physical mutagens have been applied in forward genetics strategies, where unknown genes acting in a biological process are identified by mutant phenotypes. EMS mutagenesis has been favored in many experiments because it is convenient to use and it induces a large number and wide spectrum of mutations. A primary disadvantage of EMS mutagenesis is that the point mutations typically produced are generally the most difficult class of DNA mutations to clone. However, continuing progress in map-based cloning technology and detection of single-base changes makes the task of isolating genes identified by point mutations relatively straightforward, particularly in Arabidopsis *(47)*. The deletions produced by physical mutagens may be somewhat easier to identify in a map-based cloning effort because they typically result in a polymorphism that helps to identify the target gene. Genomic subtraction, a more direct method of cloning genes identified by deletion, has also met with some success *(48–50)*. Forward mutation projects often involve the screening of multiple populations, each treated with a different class of mutagen (including insertion elements), with the objective of isolating multiple mutant alleles for each target gene. Gene cloning then proceeds using the most favorable mutation type.

Until now reverse genetics, in which gene function is studied by altering the expression of sequenced but otherwise poorly characterized genes, has depended almost exclusively on transgenes or mobile elements as "mutagens" (discussed in **Subheading 4.2.**). For this reason, reverse-genetics studies have been limited to a relatively few species with well-developed transformation and/or transposon systems. However, recent innovations in detecting the deletions and point mutations induced by the classical mutagens promise to make these mutagenesis tools available for reverse-genetics studies in almost any species for which a gene sequence is known. PCR-based methods have been developed to facilitate high-throughput identification of deletions in known genes *(42,51)*. Since this technology should be transferable to crop species, the route from gene-function studies in *Arabidopsis* to modification of crops may be shortened substantially. The ability to detect larger deletions may even play an important role in functional genomics in *Arabidopsis*. The recent completion of the Arabidopsis Genome Sequencing Initiative has revealed a large number of gene families arrayed as tandem duplications of two or more members *(20)*. The scarcity of obvious phenotypes in single-gene insertion lines *(21)* suggests that two or more members of a gene family may contribute to gene function "overlap" in Arabidopsis, and by inference, in all plants. It is likely that the isolation of lines carrying deletions that eliminate two or more members within these multigene family clusters will reveal mutant phenotypes by substantially reducing or completely eliminating a particular function. Another recent development, also PCR based, allows the rapid identification of EMS-induced point mutations in known genes *(43)*. This technology should allow the isolation of a spectrum, or "allelic series" of mutations for a given gene—an approach that could be particularly helpful when a null mutation is relatively uninformative, as is the case when a mutation results in lethality in early embryogenesis. In such a case, a collection of leaky or temperature-sensitive alleles could provide the investigator with mutants that escape embryo lethality to progress to later stages of development, thereby providing the basis for more extensive studies of the gene's function. Many of the more than 200 embryo lethal genes in Arabidopsis that have been cataloged by the Meinke lab *(52;* http://mutant.lse.okstate.edu/embryopage/embryopage.html) and others could be investigated by this route. Potential applications of these nontransgenic approaches in agricultural biotechnology may include the reduction, modification, or elimination of undesirable components of crop products—for example, fiber and sinapine in canola or allergens in wheat and other crops.

The most common method of mutagenizing plants is to treat seed (i.e., embryos) with a mutagen, either by bombarding with radiation or by soaking in a chemical solution. Virtually all of the mutations generated in the embryo and subsequent M_1 plant will be heterozygous because the probability of an independent mutation in both alleles at a particular locus (in a diploid) is extremely low. Forward-mutant screening strategies typically plan for the recovery of recessive mutations by phenotype, expressed only in the M_2 generation following M_1 self-fertilization. The M_1 plants resulting from seed mutagenesis are chimeras—they are composed of multiple mutant sectors that each descended from independently mutagenized embryonic cells. Only a small subset of embryonic cells in the M_1 give rise to the reproductive structure and thus to the next generation. The number of these cells, referred to as the genetically effective cell number (GECN), varies by species. In Arabidopsis, the GECN is approx 2 *(39)*. Assuming the sectors produce a similar amount of seed and the mutation does not affect the transmission of gametes, a recessive mutation will segregate at approx 1:7 in the M_2 descendents of a mutagenized seed instead of the 1:3 ratio observed in a typical segregating F_2 population. In barley, the GECN is estimated to be about 4.6 *(53)*, and in maize, it is estimated to be between 3 and 6 *(54)*. Seed mutagenesis of maize is particularly problematic because the male and female flowers arise from different embryonic cells. Thus, virtually all mutations are heterozygous in the M_2 generation, and recessive mutant phenotypes do not appear until the M_3 generation after self-pollination of the M_2. For this reason, pollen mutagenesis is often used in maize to create unique, non-sectored M_1 seed that, as self-pollinated adult plants, will produce homozygous mutations in the

M_2 generation *(54)*. Although there is no need to screen homozygous individuals for DNA mutations in reverse-genetics screens, chimerism is an important consideration when deciding how to sample tissue for DNA extraction or when to pool DNA from multiple lines for screening.

5.3. Screening

The expected frequency of recovery for a mutation is an important consideration in the design of a screeing strategy. Unfortunately, the calculation of mutation frequency is not standardized, so it is sometimes difficult to compare the various reports in the literature *(53)*. One of the more widely accepted definitions is the frequency of detectable mutation events per haploid genome *(39)*. Mutation frequency depends upon a number of factors, including the species, the particular locus, the ability to detect the mutation, and the type and dose of mutagen *(55)*. Certain modifying factors may come into play for seed mutagenesis, including seed moisture content and condition and duration of post-treatment storage *(37)*. Koorneef et al. *(55)* recovered mutations at 15 individual loci in Arabidopsis populations mutagenized with X-rays, fast neutrons, or EMS, and reported that overall, mutation frequencies for fast neutrons and EMS were similar (approx 2×10^{-4} per locus per haploid genome) and about 5X greater than the mutation frequency for X-rays. Bird and Neuffer *(54)* reported the frequency of recessive mutations in one EMS-pollen mutagenesis study to be approximately 1×10^{-3} per locus per haploid genome. Koornneef et al. *(55)* compared their results with published mutation frequencies for barley, with a genome size approx 25 times that of Arabidopsis, and reported the mutation frequencies for the three mutagens to be within the same order of magnitude in both species, leading them to suggest that mutation frequency may be relatively independent of genome size.

Once a mutation frequency is determined (or assumed), the next step is to decide on the sizes of the M_1 and M_2 generations *(37,39,53)*. A relatively straightforward method is provided by van Harten *(37)*:

$$n = \log (1 - P)/\log (1 - f)$$

where n = number of "mutagenized cells" (equal number of M_1 if GECN = 1), P = level of confidence, and f = frequency of mutation. For example, for a confidence of 99% and a mutation frequency of 2×10^{-4},

$$n = \log (1 - 0.99)/\log (1 - 0.0002) = 23,024$$

This calculation assumes that all M_1 survive and produce a large number of M_2 progeny. To adjust for the GECN, divide the result by the GECN. For a GECN of 2, the formula predicts that 11,512 M_1 should be screened. Corrections for germination should be made. In practice, the number of M_1 plants may be increased as much as 10-fold in order to increase the opportunity for recovering multiple alleles or rarely occurring mutations *(35)*.

The balance between M_1 and M_2 numbers is important to the design of an efficient screen. Rédei and Koncz *(39)* and Bogyo *(53)* provide an in-depth treatment of this issue. As the number of screened M_2 progeny per mutagenized M_1 increases, the probability of detecting a mutation generated in an M_1 increases as well. However, with a fixed screening capacity, the increase in M_2 screened per M_1 would result in a reduction of M_1 (and independent mutation events) evaluated by the screen. The net result is that less independent mutation events will be recovered as the M_2 per M_1 number increases. The relative cost of producing M_1 vs M_2 should also be considered for efficient screen design. The GECN must also be considered. For example, a diploid M_1 with a GECN of 2 will produce M_2 seed that reflects the mutagenesis of four haploid genomes, and an M_1 plant derived from pollen mutagenesis will produce M_2 seed that reflects the mutagenesis of a single haploid genome. In general, the maximum efficiency for isolation of independent mutations is obtained from screening a small number of M_2 per M_1.

A mutation-screening strategy must consider how the mutants will be identified and retained. Forward-genetics screens typically examine M_2 plants individually for recessive (as well as dominant) phenotypes, whereas reverse genetics studies—because they often utilize PCR amplification to detect rare events—typically evaluate pooled DNA from M_2 or M_3 seeds or plants. In forward genetics screens, viable M_2 mutant candidates are simply saved for further characterization. Nonviable or sterile homozygous recessive M_2 lines can be recovered as heterozygous lines. This process is relatively straightforward when small M_2 seed populations derived from one or a few M_1 parents are created in advance. Although they require additional effort to construct, these small M_2 populations also make it easy to confirm that allelic mutations, when they come from separate populations, are derived from unique mutation events. For reverse genetics, a typical strategy is the creation of arrayed seed and DNA libraries. The DNA library can be screened any number of times for various genes; the seed library serves as the source for mutant lines carrying the gene disruptions identified in the DNA library.

5.4. Prospects for Application

Major advances in genome sequencing, bioinformatics, and the profiling of transcripts, proteins, and metabolites have laid the foundation for the new era of functional genomics *(56,57)*. Central to this new approach is the large-scale collection of information about the plant as it exists in two or more distinct "states"; for example, drought-stressed vs non-drought-stressed growth conditions or mutant versus wild-type. Using bioinformatics software, the large data sets from these studies are analyzed to predict a subset of genes that act in the process under investigation. Studies of this kind are likely to make rapid progress in Arabidopsis, given the number of tools and the amount of attention focused on this model plant *(58)*. However, the transfer of knowledge gained in Arabidopsis to practical improvements of crops is likely to present significant challenges, considering the greatly reduced sequence information, reduced transformation and transposon capabilities, increased generation time, and other issues associated with making modifications to crop species. Classical mutagenesis techniques coupled with the new methods for reverse genetics will certainly play an important role in translating knowledge gained in a model system into valuable products in crops.

Potentially, the past technology of mutation breeding may experience a revival when coupled with newer methods of high-throughput analyses. The use of sensitive assays on plants grown under uniform conditions, may make it possible to use forward genetics in crops to identify valuable phenotypes that were missed by past generations of plant breeders. When small changes are found in an important trait, combining two or more non-allelic mutations into the same background may result in additive or even synergistic improvements of the trait.

References

1. Schena, M., Shalon, D., Davis, R. W., and Brown, P. O. (1995) Quantitative monitoring of gene expression patterns with a complementary DNA microarray. *Science* **270,** 467–470.
2. Seki, M., Narusaka, M., Abe, H., Kasuga, M., Yamaguchi-Shinozaki, K., Carninci, P., et al. (2001) Monitoring the expression pattern of 1300 *Arabidopsis* genes under drought and cold stresses by using a full-length cDNA microarray. *Plant Cell* **13,** 61–72.
3. Schenk, P. M., Kazan, K., Wilson, I., Anderson, J. P., Richmond, T., Somerville, S. C., et al. (2000) Coordinated plant defense responses in *Arabidopsis* revealed by microarray analysis. *Proc. Natl. Acad. Sci. USA* **97,** 11,655–11,660.
4. Maleck, K., Levine, A., Eulgem, T., Morgan, A., Schmid, J., Lawton, K. A., et al. (2000) The transcriptome of *Arabidopsis thaliana* during systemic acquired resistance. *Nat. Genet.* **26,** 403–410.
5. Kawasaki, S., Borchert, C., Deyholos, M., Wang, H., Brazille, S., Kawai, K., et al. (2001) Gene expression profiles during the initial phase of salt stress in rice. *Plant Cell* **13,** 889–906.
6. Petersen, M., Brodersen, P., Naested, H., Andreasson, E., Lindhart, U., Johansen, B., et al. (2000) *Arabidopsis* map kinase 4 negatively regulates systemic acquired resistance. *Cell* **103,** 1111–1120.
7. Aharoni, A., Keizer, L. C., Bouwmeester, H. J., Sun, Z., Alvarez-Huerta, M., Verhoeven, H.A., et al. (2000) Identification of the SAAT gene involved in strawberry flavor biogenesis by use of DNA microarrays. *Plant Cell* **12,** 647–662.
8. van Wijk, K. J. (2000) Proteomics of the chloroplast: experimentation and predication. *Trends Plant Sci.* 5, 420–425.
9. Williams, K. L. and Hochstrasser, D. E. (1997) in *Proteome Research: New Frontiers in Functional Genomics.* Springer-Verlag, Berlin, Germany, pp. 1–12.
10. Riccardi, F., Gazeau, P., de Vienne, D., and Zivy, M. (1998) Protein changes in response to progressive water deficit in maize. Quantitative variation and polypeptide identification. *Plant Physiol.* **117,** 1253–1263.
11. Chang, S. S., Park, S. K., and Nam, H. G. (1994) Transformation of *Arabidopsis* by *Agrobacterium* inoculation on wounds. *Plant J.* **5,** 551–558.
12. Ziv, M. and de Vienne, D. (2000) Proteomics: a link between genomics, genetics and physiology. *Plant Mol. Biol.* **44,** 575–580.
13. Damerval, C. and Le Guilloux, M. (1998). Characterization of novel proteins affected by the o2 mutation and expressed during maize endosperm development. *Mol. Gen. Genet.* **257,** 354–361.
14. Gallardo, K., Job, C., Groot, S. P., Puype, M., Demol, H., Vandekerckhove, J., et al. (2001) Proteomic analysis of *Arabidopsis* seed germination and priming. *Plant Physiol.* **126,** 835–848.
15. Peck, S. C., Nuhse, T. S., Hess, D., Iglesias, A., Meins, F., and Boller, T. (2001) Directed proteomics identifies a plant-specific protein rapidly phosphorylated in response to bacterial and fungal elicitors. *Plant Cell* **13,** 1467–1475.
16. Martinez-Zapater, J. M. and Salinas, J. (1998) in, *Arabidopsis Protocols,* Humana Press, Totowa, NJ, pp. 209–267.
17. Bechtold, N., Ellis, J., and Pelletier, D. (1993) In planta Agrobaterium mediated gene transfer by infiltration of adult *Arabidopsis thaliana* plants. C. R. Acad. Sci. Paris, *Life Sci.* **316,** 1194–1199.
18. Feldmann, K. A. and Marks, M. D. (1987) Agrobacterium-mediated transformation of germinating seeds of *Arabidopsis thaliana*: a non-tissue culture approach. *Mol. Gen. Genet.* **208,** 1–9.
19. Neuffer, M. G., Coe, E. H. and Wessler, S. R. (1997) *Mutants of Maize,* Cold Spring Harbor Laboratory Press, Cold Spring Harbor, NY.
20. The Arabidopsis Genome Initiative (2000) Sequence and analysis of the flowering plant *Arabidopsis thaliana. Nature* **408,** 796–815.
21. Bouché, N. and Bouchez, D. (2001) *Arabidopsis* gene knockout: phenotypes wanted. *Curr. Opin. Plant Biol.* **4,** 111–117.

22. Bevan, M., Mayer, K., White, O., Eisen, J. A., Preuss, D., Bureau, T., et al. (2001) Sequence and analysis of the *Arabidopsis* genome. *Curr. Opin. Plant Biol.* **4,** 105–110.

23. Weigel, D., Ahn, J. H., Blazquez, M. A., Borevitz, J. O., Christensen, S. K., Fankhauser, C., Ferrandiz, C., et al. (2000) Activation Tagging in *Arabidopsis. Plant Physiol.* **122,** 1003–1013.

24. Walbot, V. (2000) Saturation mutagenesis using maize transposons. *Curr. Opin. Plant Biol.* **3,** 103–107.

25. McClintock, B. (1992) in, *The Dynamic Genome: Barbara McClintock's Ideas in the Century of Genetics*. (Federoff, N., and Botstein, D., eds.), Cold Spring Harbor Laboratory Press, Cold Spring Harbor, NY.

26. Federoff, N. (1983) in, *Mobile Genetic Elements*. (Shapiro, J., ed.), Academic Press, NY, pp. 1–63.

27. Bechtold, N. and Pelletier, G. (1998) In planta *Agrobacterium*-mediated transformation of adult *Arabidopsis thaliana* plants by vacuum infiltration. Methods Mol. Biol. 82, 259–266.

28. Zupan, J., Muth, T. R., Draper, O., and Zambryski, P. (2000) The transfer of DNA from *Agrobacterium tumefaciens* into plants: a feast of fundamental insights. *Plant J.* **23,** 11–28.

29. Krysan, P. J., Young, J. C., and Sussman, M. R. (1999) T-DNA as an insertional mutagen in *Arabidopsis. Plant Cell* **11,** 2283–2290.

30. Kardailsky, I., Shukla, V. K., Ahn, J. H., Dagenais, N., Christensen, S. K., Nguyen, J. T., et al. (1999) Activation tagging of the floral inducer FT. *Science* **286,** 1962–1965.

31. Christensen, S. K., Dagenais, N., Chory, J., and Weigel, D. (2000) Regulation of auxin response by the protein kinase PINOID. *Cell* **100,** 469–478.

32. Springer, P. S. (2000) Gene traps: tools for plant development and genomics. *Plant Cell* **12,** 1007–1020.

33. Liljegren, S. J., Ditta, G. S., Eshed, Y., Savidge, B., Bowman, J. L., and Yanofsky, M. F. (2000) SHATTERPROOF MADS-box genes control seed dispersal in *Arabidopsis. Nature* **404,** 766–770.

34. Huang, S., Cerny, R. E., Bhat, D. S., and Brown, S. M. (2001) Cloning of an *Arabidopsis patatin*-like gene, STURDY, by activation T-DNA tagging. *Plant Physiol.* **125,** 573–584.

35. Smith, B. D. (1994) *The Emergence of Agriculture*, W.H. Freeman and Company, New York, NY.

36. Muller, H. L. (1927) Artificial transmutations of the gene. *Science* **66,** 84–87.

37. van Harten, A. M. (1998) *Mutation Breeding, Theory and Practical Applications*, Cambridge University Press, Cambridge, UK, pp. 47–63.

38. Maluszynski, M., van Zanten, L., Ashri, A., Brunner, H., Ahloowalia, B., Zapata, F.P., et al. (1995) Mutation techniques in plant breeding, in *Induced Mutations and Molecular Techniques for Crop Improvement*, Proceedings of the FAO/IAEA Symposium, Vienna, IAEA, Vienna, pp. 489–504.

39. Rédei, G.P. and Koncz, C. (1992) Classical mutagenesis, in *Methods in Arabidopsis Research* (Koncz, C., Chua, N.-H., and Schell, J., eds.), J. World Scientific, Singapore, pp. 16–82.

40. Bruggemann, E., Handwerger, K., Essex, C., and Storz, G. (1996) Analysis of fast neutron-generated mutants at the *Arabidopsis thaliana* HY4 locus. *Plant J.* **10,** 755–760.

41. Cecchini, E., Mulligan, B. J., Covey, S. N., and Milner, J. J. (1998) Characterization of gamma irradiation-induced deletion mutations at a selectable locus in *Arabidopsis. Mutat. Res.* **401,** 199–206.

42. Li, X., Song, Y., Century, K., Straight, S., Ronald, P., Dong, X., et al. (2001) A fast neutron deletion mutagenesis-based reverse genetics system for plants. *Plant J.* **27,** 235–242.

43. McCallum, C. M., Comai, L., Greene, E. A., and Henikoff, S. (2000) Targeted screening for induced mutations. *Nat. Biotechnol.* **18,** 455–457.

44. Sablowski, R. W. and Meyerowitz, E. M. (1998) Temperature-sensitive splicing in the floral homeotic mutant apetala3-1. *Plant Cell* **10,** 1453–1463.

45. Whittington, A. T,. Vugrek, O., Wei, K. J., Hasenbein, N. G., Sugimoto, K., Rashbrooke, M. C., et al. (2001) MOR1 is essential for organizing cortical microtubules in plants. *Nature* **411,** 610–613.

46. Ouellet, F., Overvoorde, P. J., and Theologis, A. (2001) IAA17/AXR3. Biochemical insight into an auxin mutant phenotype. *Plant Cell* **13,** 829–842.

47. Spiegelman, J. I., Mindrinos, M. N., Fankhauser, C., Richards, D., Lutes, J., Chory, J., et al. (2000) Cloning of the Arabidopsis RSF1 gene by using a mapping strategy based on high-density DNA arrays and denaturing high-performance liquid chromatography. *Plant Cell* **12,** 2485–2498.

48. Straus, D. and Ausubel, F. M. (1990) Genomic subtraction for cloning DNA corresponding to deletion mutations. *Proc. Natl. Acad. Sci. USA* **87,** 1889–1893.

49. Sun, T. P., Straus, D., and Ausubel, F. M. (1992) Cloning *Arabidopsis* genes by genomic subtraction, in *Methods in Arabidopsis Research* (Koncz, C., Chua, N.-H., and Schell, J., eds.), World Scientific, Singapore, pp. 331–341.

50. Silverstone, A. L., Ciampaglio, C. N., and Sun, T. (1998) The *Arabidopsis* RGA gene encodes a transcriptional regulator repressing the gibberellin signal transduction pathway. *Plant Cell* **10,** 155–169.

51. Liu, L. X., Spoerke, J. M., Mulligan, E. L., Chen, J., Reardon, B., Westlund, B., et al. (1999) High-throughput isolation of *Caenorhabditis elegans* deletion mutants. *Genome Res.* **9,** 859–867.

52. Castle, L. A., Errampalli, D., Atherton, T. L., Franzmann, L. H., Yoon, E. S., and Meinke, D. W. (1993) Genetic and molecular characterization of embryonic mutants identified following seed transformation in *Arabidopsis. Mol. Gen. Genet.* **241,** 504–514.

53. Bogyo, T. P. (1991) Numerical aspects of mutation breeding programs, in *Induced Mutations and Molecular Techniques for Crop Improvement*, Proceedings of the FAO/IAEA Symposium, Vienna, IAEA, Vienna, pp. 489–504.

54. Bird, R. M. and Neuffer, M. G. (1987) Induced mutations in maize, in *Plant Breeding Reviews*, Vol. 5 (Janick, J., ed.), Timber Press, Portland, OR, pp. 139–180.

55. Koornneef, M., Dellaert, L. W., and van der Veen, J. H. (1982) EMS- and radiation-induced mutation frequencies at individual loci in *Arabidopsis thaliana* (L.) Heynh. *Mutat. Res.* **93,** 109–123.

56. Fiehn, O., Kloska, S., and Altmann, T. (2001) Integrated studies on plant biology using multiparallel techniques. *Curr. Opin. Biotechnol.* **12,** 82–86.

57. Somerville, C. and Somerville, S. (1999) Plant functional genomics. *Science* **285,** 380–383.

58. Somerville, C. and Dangl, J. (2000) Genomics. Plant biology in 2010. *Science* **290,** 2077–2078.

9

Molecular Tools for Engineering Plant Cells

Donald J. Merlo

1. Introduction

The introduction, establishment, and expression of foreign genes in the nuclei of plant cells involves three basic steps: the introduction of DNA into the cell, the identification and propagation of the transformed cell, and the evaluation of expression levels of the gene or genes of interest. This chapter examines the molecular tools required for the successful completion of plant transformation and desired gene expression. The goal is to illustrate some of the ideas and selection criteria that enter into the decision of which of the many choices available may best fit the object of the exercise. Particular components are mentioned to illustrate a specific point, because, as the field advances, the number of options steadily increases. A wide range of published information is available for more in-depth pursuit of a specific topic.

Chapter 10 covers the subject of plant-cell tissue culture in the development of transformation technologies. That chapter discusses some of the molecular components that will be mentioned in greater detail here. It should also be noted that substantial research has been directed toward transformation of plants through the integration of foreign genes into the chloroplast genome, but that topic is discussed separately (Chapter 11).

The categorization of transforming genes is based on two functions. The first category includes genes that produce proteins that confer upon the transformed cell a phenotype that is scored physiologically or visually (selectable or screenable markers, respectively). The second category is represented by genes that produce proteins (or in some instances, RNAs), which confer commercially or scientifically important phenotypes (for example, insect resistance), or have intrinsic value. In other uses, the genes encode enzymes that participate in biochemical steps that lead to commercially relevant compounds.

2. Selectable Marker Genes

Selectable marker genes confer a growth advantage to transformed cells under conditions in which nontransformed cells will not grow. These genes are generally necessary because current plant transformation methods are extremely ineffective when considered at the cell level. Without selection, the identification and isolation of a small population of transformed cells from among an overwhelming and growing population

From: *Handbook of Industrial Cell Culture: Mammalian, Microbial, and Plant Cells*
Edited by: V. A. Vinci and S. R. Parekh © Humana Press Inc., Totowa, NJ

of nontransformed material is nearly impossible. Although modern high-throughput methods may allow screening of large numbers of samples to identify the rare events, these methods are time-consuming and expensive. Furthermore, plants regenerated from nonselected tissues are likely to be chimeric (composed of both transformed and nontransformed tissues), and a substantial amount of genetic follow-up work is required to identify purely transgenic progeny.

2.1. Hormone Independence

The earliest plant transformation experiments exploited the ability of pathogenic *Agrobacterium tumefaciens* cells to transfer T-DNA via the Ti plasmid to the plant cell, with eventual integration into the genome (discussed in Chapter 10). Encoded within the T-DNA are genes that encode enzymes which are biochemically responsible for the synthesis of two types of plant hormone—an auxin and a cytokinin *(1)*. Cells of some plant species, when transformed with the wild-type T-DNA, acquire the ability to grow in vitro on defined growth media that lack plant hormones. Nontransformed cells are unable to grow without the exogenous hormones. Although it allows for selection of transformed cells and tissues, the acquisition through the T-DNA of the uncontrolled hormone biosynthesis genes disrupts the ability of the transformed cells to organize into differentiated growth, and the result is the formation of tumor-like callus or teratomas *(2)*. Therefore, selection for hormone-independent growth, although useful in the early days of developing and understanding plant transformation, has been replaced by dominant selectable marker genes that offer the advantage of positive selection of transformed cells on media that has all the components needed to induce regeneration of fertile whole plants.

2.2. Antibiotic Resistance

Bacterially derived antibiotic resistance genes were the first to be used successfully to select for totipotent transformed plant cells. The native bacterial promoters of these genes are not functional in plant cells, so the coding regions of the resistance genes must be cloned under the control of plant expression elements. In most instances, the antibiotic is a protein-synthesis inhibitor, and resistance is mediated through metabolic detoxification of the drug, usually by phosphorylation or acetylation. Genes that encode aminoglycoside phosphotransferases (*aph* genes) have seen the most widespread use for selection of transformed plant cells in the presence of normal inhibitory concentrations of aminoglycoside antibiotics such as kanamycin or G418 *(3)*. The neomycin phosphotransferase II (NptII) coding region (encoded by *aphII*) from bacterial transposon Tn*5* has been used broadly in both commercial and research applications, after being placed under the control of appropriate plant gene-expression elements *(3)*. At least 36 plant genera have been genetically modified and shown to express the NptII protein at levels sufficiently high to confer drug resistance *(4)*.

The coding regions of the *aphI* genes from Tn*601* and Tn*903*, both of which encode neomycin phosphotransferase I (NptI), have also been used to a limited extent *(5)*. Other drug resistance genes have been used as plant selectable marker genes, primarily in experimental settings. Examples are genes that encode chloramphenicol acetyltransferase (from Tn*9*) *(5)*, bleomycin resistance (from Tn*5*) *(6)*, streptomycin phosphotransferase (from Tn*5*) *(7)*, and hygromycin phosphotransferase (*hpt*, from an *Escherichia coli* plasmid) *(8)*. Methotrexate resistance is mediated not by a detoxification mechanism but through overexpression of a bacterial (from plasmid R67) *(3)* or mouse *(9)* methotrexate-insensitive dihydrofolate reductase. A mouse gene encoding adenosine deaminase has been expressed in transgenic maize, and confers resistance to 2'deoxyadenosine *(10)*.

Emerging public concerns about the issue of widespread release of transgenic plants harboring antibiotic resistance genes *(4)* has led to the general discontinuance of these genes as plant selectable markers for commercial crops. The basis for these objections primarily revolves around the perceived possibility that the genes will escape into the microbial environment, thereby initiating eventual transfer to human or animal pathogens *(11,12)*. It is a matter of open debate whether this concern is scientifically justified. The risk of an antibiotic-resistance gene escaping from a corn plant to a soil microbe seems inconsequential when compared to the proven interspecific transfer of antibiotic resistance via multidrug-resistance conjugal plasmids that are ubiquitous in bacteria-rich environments such as sewage treatment plants *(13)*.

2.3. Herbicide Resistance

Genes that confer plant resistance to herbicidal compounds have three advantages in the production of commercial transgenic crops. First, the genes are often used for in vitro selection of transformed plant cells during the first stages of transformation and regeneration. Later, during the trait introgression process (i.e., plant breeding) in which transgenes are introgressed into commercially important varieties, the herbicide resistance trait is often used to identify transformants and to cull nontransformed or weakly expressing plants from the segregating population. Finally, the herbicide resistance trait is commercially valuable, as it allows growers to spray a crop with a broad-spectrum chemical to kill susceptible weeds, and causes no damage to the crop plants that express engineered resistance to the chemical.

For many commercial herbicides, naturally selected or mutation-induced resistant plants have been identified. Many published reports have described the identification of the target-site enzyme and elucidation of the modes of resistance to particular compounds or classes of herbicides. Plant herbicide resistance mechanisms generally fall into only a few categories *(14,15)*. Exclusion of the herbicide molecule from the site in the plant where it normally induces the toxic effect is accomplished by inhibiting the uptake/translocation of the compound, by compartmentalization of the compound, or by metabolism of the compound. In other cases, growth-sustaining levels of the target enzyme activity are attained by overproduction of the target enzyme, or by accumulation of site mutations that render the target enzyme resistant to inhibition by the herbicide molecule. A third category of resistance—one that has had substantial commercial success—involves production of metabolic detoxification enzymes. In a few instances, a single gene introduced transgenically into plant cells has been sufficient for both transgenic selection and field-level resistance.

2.3.1. Acetolactate Synthase (ALS) Inhibitors

ALS (sometimes called AHAS; acetohydroxyacid synthase), catalyzes the first committed step in the synthesis of the branched-chain amino acids *(14)*. ALS-inhibitor herbicides comprise four chemical families: the sulfonylureas (products of DuPont Crop Protection), the imidazolinones (products of American Cyanamid/BASF), the triazolopyrimidine sulfonanilides (products of Dow AgroSciences) and the pyrimidyl-oxy benzoates (products of Kumiai Chemical Industry Company). Naturally selected resistance mutants of several plant species are known. The ALS-encoding genes of several of these plants have been cloned and sequenced, and resistance to particular classes of compounds has been traced to point mutations in the ALS protein. At least 10 mutation sites have been identified—some of which, alone or in combination, confer cross-resistance to more than one compound class. Engineered resistance through overproduction of the mutant ALS enzyme has been successful in several plant species, including crops. This type of resistance is problematic because of the presence of a basal level of susceptible endogenous enzymes in the cell. Generally, this limitation to attaining herbicide resistance is more serious when high levels of endogenous enzymes must be maintained.

From a commercial herbicide safety standpoint, chemical inhibitors of ALS are desirable because mammals lack the target enzyme. Currently, multiple agricultural chemical companies produce ALS inhibitors; many of which have been developed to be crop-selective. Maximum value for herbicides and herbicide-resistant crops is derived from the combination sale of the chemical and the seeds of the resistant crops. Since many of the ALS mutations confer cross-resistance to multiple chemical classes, the deployment of crops that are resistant to one compound class may prompt sales of a herbicide sold by a competitor who has not incurred the expense of developing the resistant crop. If crop-selective ALS inhibitors already exist to control weeds in the crop, there is probably no advantage to be gained by incurring the expenses of producing and registering a herbicide-resistant crop that harbors a mutant ALS gene.

2.3.2. Glutamine Synthetase (GS) Inhibitors

Glutamine synthetase is involved in nitrogen metabolism through the conversion of L-glutamate to L-glutamine. A bacterially derived compound, phosphinothricin (an analog of L-glutamate), and its synthetic version, glufosinate, are both highly effective GS inhibitors *(14)*. A third compound, bialaphos, a bacterial tripeptide precursor to phosphinothricin, is also a GS inhibitor via conversion in plants to the highly toxic phosphinothricin. All three compounds are useful as selective agents for plant transformation. The fermentation product bialaphos, and particularly glufosinate, are successful field-use herbicides. Two *Streptomyces* genes that detoxify these compounds by acetylation are routinely used to confer plant resistance to these compounds. The bar (bialaphos resistance) gene was cloned from *S. hygroscopicus*, the source of bialaphos and hygromycin *(16)*, and the pat (phosphinothricin acetyltransferase) gene was discovered in *S. viridochromogenes (17)*. When placed under the control of the appropriate plant expression control elements, both of these genes have been useful as selectable markers in a wide variety of plant species, including about 20 crops *(14)*. To date, no attempts to attain herbicide resistance through overproduction of herbicide-resistant GS have been reported, although phosphinothricin resistance has been observed in plants in which the native GS copy number has been naturally amplified *(18)*.

2.3.3. EPSP Synthase Inhibition

Glyphosate is a low-residual broad-spectrum herbicide that is a potent inhibitor of 5-enolpyruvylshikimate-3-phosphate synthase (EPSPS). This enzyme is involved in the synthesis of many aromatic compounds in plants, including the aromatic amino acids

(14), but is not found in animals. Selection of transformed cells and plant-level resistance to glyphosate has been accomplished via transgenic production of various enzymes. Overexpression of chimeric glyphosate-resistant EPSPS-encoding *aroA* genes from bacteria *(Salmonella* or *E. coli) (19,20)* requires targeting of the enzyme for transport into the chloroplast for high-level glyphosate resistance *(20)*. Expression of glyphosate-resistant EPSPS-encoding plant genes provides moderate to high levels of resistance *(21)*. In many cases, the resistant EPSPS is less enzymatically efficient than the wild-type, and thus high-level expression is required to support normal plant metabolism. A gene for a glyphosate-resistant EPSPS with relatively high enzymatic efficiency has been isolated from an *Agrobacterium* strain, and alleviates some of the necessity for overproduction *(21)*. In addition, production in plants of glyphosate-degrading enzymes from bacteria such as *Achromobacter* has been an effective method of establishing the glyphosate-resistant trait *(21,22)*. The combination sale of glyphosate-resistant crop seeds and the formulated herbicide has been commercially successful.

2.3.4. Photosystem II Inhibitors

Bromoxynil and the triazines (e.g., atrazine) inhibit electron transport in the photosystem II reaction center in plant chloroplast thylakoid membranes *(14)*. Dicot crops are particularly sensitive to bromoxynil, and resistance to this herbicide could be a valuable trait in crops such as cotton. Although some weed resistance to bromoxynil has emerged, the mechanism of the resistance, and any genes associated with it, have yet to be exploited. So far, engineered resistance to bromoxynil has been accomplished through expression of a chimeric gene for bromoxynil nitrilase isolated from *Klebsiella ozaenae (23)*. In cotton, transgenic lines carrying this gene are resistant to up to 10-fold the field-rate recommendation of formulated bromoxynil *(24)*.

Many plant transformation protocols require incubation of early-stage processed materials in the dark prior to beginning the whole-plant regeneration steps. Because bromoxynil is a photosynthesis inhibitor, it is not as useful as certain other compounds in the selection of early-stage transformants. Selection must be applied later on after photosynthetically competent plants are produced.

The emergence of weeds resistant to atrazine in the field was first reported in 1970 *(25)*, and is usually found to be the result of a single amino-acid mutation in a chloroplast gene that encodes a thylakoid protein of about 32,000 Daltons *(26)*. Engineered atrazine resistance has not been used either as a selectable marker or as a commercial trait.

2.4. Metabolic Inhibitors

Two recent developments in selectable marker systems employ genes that confer resistance to an amino-acid analog or enable sugar utilization. In the first example, *Arabidopsis* plants grown from seeds treated with a mutagen were screened for the ability to grow in inhibitory concentrations of O-methyl-threonine. This trait was named *omr1* (for O-methyl-threonine resistance) and through genetic and molecular studies the *omr1* gene was found to encode a mutated threonine deaminase enzyme. Threonine deaminase (also called threonine dehydratase) catalyzes the first committed step in the biosynthesis of isoleucine. Although it is useful for selection of transformed tissues of some plant species *(27)*, the omr1 gene is not generally as valuable as some other genes.

The second example utilizes the observation that mannose cannot be converted into metabolizable sugars by plant cells, and inhibits seed germination in *Arabidopsis* and respiration in wheat and tomato *(28)*. Mannose that enters plant cells is converted to mannose-6-phosphate, which accumulates and is believed to be responsible for the growth-inhibitory effects. However, when transformed with a modified *E. coli manA* gene that encodes phosphomannose isomerase, the transformed cells are able to thrive on medium with mannose as the primary carbon source, and thus outgrow nontransformed tissues *(28)*.

3. Screenable Marker Genes

Screenable markers are proteins that do not confer a growth advantage to transformed cells, but are recognized by their ability to cause physical or chemical changes in the transformed cell. In the ideal case, such changes can be detected via a nondestructive, high-throughput assay on putatively transformed samples. Less useful are assays that consume a sample removed from a precious candidate tissue that is often available in limited amounts or only for a short time.

3.1. Opine Synthesis

Early plant-cell transformation protocols utilized oncogenic, teratogenic, or disabled Ti plasmid T-DNA transfers via *Agrobacterium tumefaciens* or *A. rhizogenes*. The T-DNA in these early experiments transferred a small number of genes that produced, among other compounds, modified amino acids (opines) in the transformed plant cells. Transformation by non-oncogenic T-DNAs, which allowed regeneration of whole fertile plants, was detected by paper electrophoretic analysis of opines—(e.g., octopine, nopaline, or mannopine) *(29)* in plant-cell extracts from putatively transformed tissues. **Note:** The octopine synthase (lysopinedehydrogenase) enzyme has been used as a selectable marker to overcome plant-cell toxicity of L-homoarginine. Because homoarginine is not a potent inhibitor of plant cell growth, this selection regime is not routinely useful *(30,31)*.

3.2. Chloramphenicol Acetyltransferase (CAT)

The CAT protein is encoded by the *cmr* (chloramphenicol resistance) gene of bacterial transposon Tn9. After modification for plant expression, the CAT coding region is useful as a screenable marker *(32)*. The assays are quite cumbersome, and involve thin-layer chromatography or ethyl acetate extraction to determine the transfer of radiolabeled acetate to chloramphenicol *(33,34)*. Total amounts of the CAT protein can be measured by enzyme-linked immunosorbent assay (ELISA) *(32)*. Because plant-cell growth is not particularly sensitive to inhibition by chloramphenicol, the CAT protein has limited use as a selectable marker *(5)*.

3.3. β-Glucuronidase (GUS)

GUS is encoded by the *E. coli uidA* gene *(35)*. When placed under control of plant-expression elements, production of the GUS protein has proven extremely useful for a variety of experiments beyond the simple identification of transformed tissues. The most commonly used substrates are XGLUC (5-bromo-4-chloro-3-indolyl glucuronide for histochemical assays) and MUG (4-methyl umbelliferyl glucuronide for quantitative fluorescence assays) *(36)*. GUS acts upon the XGLUC substrate by cleaving the glucuronide bond to release an indoxyl derivative, which dimerizes in the presence of oxygen to form a highly colored indigo blue dye precipitate. This property makes the GUS protein highly useful in the analysis of promoter-expression patterns, as the cells in which the promoter is active are stained blue. The GUS protein is relatively stable, and thus it is possible to overestimate promoter activity because of the accumulation of GUS and the resulting intense blue staining in histochemical assays. This artifact makes GUS expression somewhat less useful than markers such as luciferase or green fluorescent protein (GFP) in comparative analysis of promoter-expression levels (*see* below).

Sensitive fluorometric assay of the level of GUS enzymatic activity in plant-cell extracts is possible through the use of the MUG substrate. MUG is not fluorescent until GUS activity releases 4-methyl umbelliferone, which is fluorescent at pH8-9. Standard curves established with commercially available GUS enzyme allow precise quantitation of GUS activity in multiple samples; therefore, quantitative determination and comparison of chimeric gene expression at various time-points is possible.

3.4. Luciferases

The gene that encodes the luciferase enzyme responsible for light emission from firefly lanterns (*Photinus pyralis*) has been cloned, and the coding region has proven quite useful as a marker gene for promoter expression studies in plants *(37)*. The addition of ATP and luciferol substrate are sufficient to induce light emission that can be captured quantitatively by photon-capture devices in vitro, or qualitatively by standard photography.

It is relatively common in transgenic promoter expression studies to compare the expression level or tissue specificity of a test promoter to a standard reference promoter that is co-transformed with the test gene construct *(38)*. In these cases, the reference promoter may be used to drive expression of the GUS gene, for example, while the test promoter is used to drive expression of the luciferase gene. The relative expression levels of the two promoters can be assayed in vitro in a single extract, and the test promoter activity is normalized to the activity of the reference promoter, which serves as an internal standard. Samples taken from different tissues or on different days can thus be compared. It is not an understatement that these reporter genes, GUS and luciferase, have been the two most valuable enzymes utilized in the advancement of our understanding of plant gene expression.

Another substrate-dependent light-emitting enzymatic activity that has been used to a limited extent in plant gene-expression studies involves the products of the *Vibrio harveyi luxAB* genes *(39)*. The active enzyme is normally comprised of two subunits, but they can be genetically joined into a single protein that retains activity *(40)*. In addition, a gene that encodes the luciferase from *Renilla reniformis* (sea pansy) is functional in plants *(41,42)* and is useful because it emits light of a wavelength different from that of the firefly luciferase. This permits both the firefly and *Renilla* reporter genes to be used in the same experiment *(43)*. Engineered genes for the firefly and *Renilla* luciferases, as well as reagent kits, are commercially available (Promega, Madison, WI).

3.5. Fluorescent Proteins

The first gene that encodes a non-substrate-dependent light-emitting (fluorescent) protein that was used as a reporter gene in plants *(44,45)* was cloned from the jellyfish *Aequoria victoria (46)*. This protein, called the green fluorescent protein (GFP) emits fluorescent light with a peak wavelength of 508–509 nm when excited with ultraviolet light

(360–400 nm) or blue light (440–480 nm) *(45)*. No external substrate is required; the fluorescence is produced from an active chromophore formed by cyclization and oxidation of three amino acids (serine-65, tyrosine-66, and glycine 67) within the peptide. At first, plant expression of the native gene was problematic because of native sequence similarity with known plant introns. Improved expression resulted from re-engineering the gene to reduce the AU (adenosine + uracil) content of the mRNA and the addition of a targeting signal peptide to direct the protein to the endoplasmic reticulum (ER) *(47)*. With these improvements, production of this protein in transgenic plants provides the opportunity to assay gene function in a nondestructive manner. Newer versions of the protein have been engineered to emit more intense green light *(48)*, and other versions have emission peaks at wavelengths different from the native GFP, including cyan, yellow, and red *(49,50)*. These differences in emission spectra allow microscopic imaging of the amounts of the respective proteins *(51)*. Genes that encode fluorescent proteins from other sources are also available—for example, the red-fluorescing DsRed protein from *Discosoma* spp. *(52)*.

3.6. β-*Galactosidase*

The *lacZ* gene of *E. coli* encodes β-galactosidase and is widely used as a reporter gene in bacterial studies. In contrast, the production of β-galactosidase in plant cells as a screenable marker has had only limited use *(53–55)*.

4. Expression Control Elements

The success of any plant transformation effort, particularly one whose object is the commercialization of crops with an engineered trait, ultimately relies on how well the phenotype produced matches the needs of the market. High level production of a protein that is toxic to corn rootworms, for example, is not likely to confer a market advantage to corn plants that have an average yield that is substantially below competitive varieties ("yield drag"). A market-viable transgenic must produce adequate levels of the insect toxin in the appropriate tissues (the roots), at the appropriate time (essentially the first two-thirds of the growing season), and result in no detrimental side effects such as yield drag, low harvestability, or lowered feed value. Achievement of these goals requires insightful choices of gene-expression control elements and appropriate combinations of the chosen elements. These considerations include: the choice of host plant, which promoter to use, the sequence of the 5' untranslated region, whether or not introns will be included, how the coding region for the desired protein will be configured, and which sequences will be used for the 3' untranslated region that encodes the transcription termination and polyadenylation signals. These decisions for individual genes must be made in the larger context of the element composition of the other transgenes that will be co-transformed (e.g., the selectable marker gene). A brief description of some of the factors to be considered is presented here. This discussion is necessarily superficial because the entire field of plant gene-expression control is advancing so rapidly.

4.1. *Promoters*

Most transgenic plants destined for commercial production employ promoters derived from plants or plant viruses to control transgene expression. Although some research has examined the plant functionality of promoters from animals or other sources, regulatory approval of commercial transgenics is facilitated when the promoter of the transgene is of plant (or plant virus) origin, particularly if the promoter was originally derived from the host plant. Examples of commonly used promoters are the maize polyubiquitin1 (*ubi*1) promoter *(56)* and the 35S promoter derived from the Cauliflower Mosaic Virus *(57)*.

Both of these promoters are considered to be constitutive in their expression patterns. Thus, they are usually expressed at some level throughout the tissues of the plants during the lifetime of the plant. However, this designation is not absolute, as the expression level and pattern of an individual promoter is markedly affected by the position into which it integrates in the chromosome during the transformation process. This "position effect" results in expression patterns in individual, independently transformed plants in which some degree of tissue specificity, developmental preference, or inactivity usually occurs *(32,58)*.

Constitutive promoters are useful to drive expression of genes when tissue specificity is not a prerequisite—for example, selectable marker genes. As mentioned previously, herbicide-resistance selectable marker genes often have a triple role. They are used early in the transformation process to select for transformed cells, in the breeding process to follow transgenes in whole plant populations, and finally as a commercially valuable trait in the field to provide crop resistance to weed-control chemicals. A constitutive promoter may also be a good choice to control the expression of an insecticidal protein when the target pest is one that feeds on stems, leaves, and other plant parts.

In contrast to constitutive expression characteristic of cellular housekeeping genes, for example, many plant promoters are highly regulated in terms of when, where, and to what level they are functional. Regulation can take many forms, and plant promoters have been characterized that respond to internal chemistries such as plant hormone levels or osmotic potential, or to external stimuli such as heat, light, (or the absence of light), touch, drought, or stress induced by insect feeding or pathogen attack *(59,60)*. In addition, there are genes controlled by promoters that function only at limited times during development (e.g., seed formation) or only in a very limited number of cells comprising a single cell layer (e.g., seed aleurone). Other promoters are specifically functional in certain tissues that exist throughout the life of the plant, for example, the root or leaf meristems *(59)*. The choice of such a tightly regulated promoter may be mandatory for the commercial success of some transgenic traits such as the modification of oil levels or composition in oil seed crops, in which the oil bodies are present only in the embryo of the seed.

Chimeric promoters have been useful to combine the regulated aspects of one promoter with the high expression characteristics of another. For example, a seed-specific sequence element can be combined with a constitutive, strong promoter (CaMV 35S) to confer seed-specific expression *(61)*. Many other response elements that respond to light, hormone, stress, or other stimuli have been characterized, and continue to be exploited.

In some instances, it would be desirable, or even necessary, that the expression of a particular trait or production of a particular protein be completely under the control of the grower. It may be advantageous that high-level production of a high-value protein be inducible by chemical treatment of the transgenic plant. Chemically regulated plant promoters have been available for some time *(62,63)*, but to date none of the published systems has been developed to the point that it is commercially useful in the field. One primary barrier to deployment of such systems is the governmental regulatory requirement that the chemical must be registered for environmental release on large acreages. Borrowing from the bacterial world, a tetracycline-controlled promoter has proven functional in plants, and expression (either induction or repression) is controlled by external application of the antibiotic tetracycline *(65)*. This system has obvious limitations for field-level application. Another inducible promoter system employs a mammalian glucocorticoid receptor and is controlled by the application of the steroid dexamethasone *(66)*. An analogous system that employs an estrogen receptor yielded higher constitutive expression *(66)*. Again, it is obvious that spraying large acreages with a steroid hormone poses many problems.

The most useful chemically controlled system is one based on a compound that is already registered as a pest control agent. Two such systems are under development, a safener-induced system and a system based on insect ecdysone receptors. In the first case, promoters were identified by screening for genes that were activated following treatment of plants with chemicals added to a commercial herbicide formulation to help prevent damage to the non-target crop plant (a "safener"). Safeners comprise a group of structurally diverse compounds, and induce expression of genes such as glutathione-S-transferases and cytochrome P450 mixed-function oxygenases *(62,63)*. The inducing chemicals activate a number of stress or defense genes, so these inducible promoter systems can be used only in applications in which the pleiotropic effects can be tolerated.

In the second case, a receptor for the insect molting hormone ecdysone *(67)* was modified and exploited for its ability to respond to synthetic ecdysone agonists, which are registered for use as insecticides *(68)*. Despite the commercial nature of the inducing compounds, there are no reports of the use of either of these systems to control gene expression in the field.

To summarize, one should begin the development of any commercial transgenic project with the end in mind. It is essential to fully understand which phenotype the final transgenic plant will have and then work backwards to the point where one can design the entire gene cassette so that it incorporates as many features as possible to increase the likelihood of attaining the trait. Often, this requires an up-front extensive biochemical and molecular biological characterization of the host plant, to determine which factors must be considered. After laying the groundwork with a full understanding of what the object of the effort is, one can then make the proper choice of which promoter to use.

4.2. Untranslated Flanking Regions and Introns

The rapidly advancing and impressive array of molecular biology tools and genomics technologies now available (*see* Chapter 8) allow the isolation of virtually every promoter in a plant, if one is willing to expend the effort. This process will be even easier in the future, as the entire genome sequences of more plants are determined *(69)*. But the promoter is only one element in the list of choices to be made in assembling a successful gene construct.

Much work has been published regarding the interactions of coding region flanking sequences on gene expression in plants *(70,71)*. It is known that the sources of the 5' and 3' untranslated regions (UTR) can play an important role in the level of gene expression. Particular sequence features of both of these elements are important. The 5' UTR must be able to efficiently assemble ribosomes and present the start codon in an appropriate configuration. It is not uncommon to utilize the native 5' UTR associated with a particular plant promoter, since this combination of promoter/ 5' UTR has evolved to function as a unit. However, there are instances in which substitution of a different 5' UTR—for example, from a plant virus gene—has been found to increase expression levels *(72)*. The 5' UTRs from the genomic RNA of tobacco mosaic virus *(73)* and the alfalfa mosaic virus coat protein gene *(74)* have been shown to be particularly effective in enhancing transgene expression. This effect may be the result of the relatively low degree of secondary structure provided by these sequences, which may facilitate easy passage of the initiating ribosomes *(73)*. From a practical standpoint, one may expect that any gene with a product that is required in relatively large quantities such as a viral coat protein may possess sequence elements that have evolved for high expression.

The 3' UTR provides mRNA stability and facilitates polyadenylation of the mRNA. Many gene constructs have utilized the 3' UTR captured from the nopaline synthase (nos) gene of *Agrobacterium* Ti plasmids (C58 or T37) *(75)*. This small unit (about 260 bp) functions well in both dicots and monocots, and is probably the most widely used 3' UTR today. Experiments with advanced constructs, however, have shown that the nos 3' UTR is fairly weak in terminating transcription, and transcriptional readthrough is sometimes seen *(76,77)*. Utilization of plant-derived 3' UTRs often facilitates better chimeric gene expression than seen with the T-DNA derived 3' UTRS *(77,78)*, but some combinations of 3' UTRs and other gene components work better than other combinations *(71,77,79)*. As with the 5' UTR, it can be advantageous to utilize the 3' UTR associated with the plant gene from which the promoter was isolated, but some testing with alternative sources may reveal a better combination of elements.

Although some plant genes (and all plant virus genes) lack introns, the addition of a plant intron to a construct can have a dramatic positive effect on expression level, particularly in monocots *(76,80)*. Intron processing is a stringent process in the plant kingdom, and the recognition/splicing mechanisms are loosely conserved between monocots and dicots. However, the phylogenetic origin of a plant intron does not completely dictate its ability to be spliced in a heterologous plant system *(71)*. Some monocot introns are spliced inefficiently or not at all in dicots, and other monocot introns are spliced in dicots with reasonable efficiency. As a general rule, monocots are able to efficiently process both monocot and dicot introns, but some dicot introns are not spliced in other dicots *(71)*.

When added to a construct, placement of the intron can affect the expression level attained *(76,81)*. The most difficult task is to introduce an intron directly into a protein-coding region. If a heterologous intron sequence is cloned as a restriction fragment into a compatible site within a coding region, the splicing reaction will sometimes leave several bases behind. The effects of these "footprint" sequences range from the addition of a few new amino acids to the protein, which may or may not affect its function, to a reading-frame shift that terminates translation or results in production of a new chimeric protein. Thus, the engineering must be carefully planned, and the spliced product must be tested to determine that no loss in activity or stability of the protein results from the addition of the new bases. If the genomic version of a coding region contains an intron, the simplest solution is to add a native intron to the cDNA, in the same position that it occurs in the genomic copy. Poly-merase chain reaction (PCR) techniques now allow the addition of an intron sequence to occur so that splicing does not leave extraneous bases *(82)*. Looking outside the coding region, it is not uncommon to find native plant genes that have an intron in the 5' UTR *(77,79)*. These introns can be quite long; the maize ubiquitin1 gene has a 1010-basepair (bp) intron in the 5' UTR sequence *(56)*, and the 5' leader/intron of the potato Sus4 gene is 1612 bp *(77)*. It has been well-demonstrated that engineering an intron into the 5' UTR can be beneficial to expression levels *(76,80)*. The degree of expression enhancement is not predictable, because it depends on the plant system used (dicot or monocot), the type of study (stable or transient expression), the source of the intron and promoter, the intron placement, and the assay method (mRNA level/processing efficiency or encoded protein levels) *(71)*.

As previously stated, certain 5' and 3' sequences appear to work best in combination with others, and the effect may be different when the same elements are combined with different coding regions and promoters *(77, 79)*. Thus, it is often necessary to examine the expression activity of several combinations of components to determine the combi-nation that is best suited to the ultimate goal. This is no easy task, although some short-cuts are available. A preliminary understanding of how a particular combination will function can be gained through transient expression systems such as electroporated protoplasts or bombarded tissues. Usually, these calibration experiments substitute the coding region of an easily assayed reporter protein for the ultimate protein of choice. Such experiments require substantial calibration, and must include well-characterized internal control genes, against which expression levels of the test genes can be normal-ized *(38)*. Because of large variations between individual experiments, many replicates are needed to establish a statistically sound answer. Once a particular combination of promoter and other elements is identified as providing good expression, the experi-ments can be repeated using the coding region for the desired protein.

Because protoplast experiments do not involve differentiated tissues, these systems are only useful for examinations of element combinations driven by constitutive pro-moters. Sometimes this is adequate to identify appropriate UTR/coding region combi-nations, but ultimately the gene must be tested in whole plants.

Transgenic *Arabidopsis* plants are frequently used when large numbers of indepen-dently transformed plants are needed *(83)*. The *Agrobacterium*-mediated transforma-tion method results in the production of transformed seeds directly from the treated plant. Seeds are germinated on selective medium, and several hundreds of independent events can be quickly generated. The short seed-to-seed generation time of *Arabidopsis* allows studies of gene expression in multiple tissues and generations, and this system is extremely useful as a next step up from the in vitro systems. However, a growing body of evidence indicates that results obtained with a particular gene in *Arabidopsis* are not necessarily predictive of those seen in other dicots (such as tobacco) *(84)*, and are certainly not always predictive of expression in monocots such as corn and rice. Thus, these model systems are useful in eliminating advancement of genes that have severe expression problems (frameshift errors), but for the final analysis, the gene must be tested in the crop of interest. This is usually an expensive, time-consuming, resource-intensive process that can extend over several years for crops such as corn and cotton. Because the results obtained from a particular combination of elements are often unpredictable, it is desirable to have an extensive library of promoters and other expression components available, so that the best combination can be assembled.

A recent publication *(85)* nicely illustrates the points that: i) plant virus promoters can be exploited for transgenic expression (banana bunchy-top virus promoters BT1 to 5), ii) promoters from the same virus differ in strength (BT4 and BT5 are stronger than BT1, BT2, or BT3), iii) some coding regions work better with certain promoters than other coding regions (GFP is better than GUS), iv) transgenic promoters can promote tissue-preferential expression (mostly vascular-associated), v) introns increase expression (maize polyubiquitin1 intron), and vi) relatively exotic crops are transformable (banana).

4.3. Other Expression Controls

In addition to the gene components mentioned in **Subheading 4.2.**, other constraints may limit or even prevent the production of acceptable levels of a transgenic protein.

4.3.1. mRNA Stability and Translation

During early attempts at plant genetic engineering, it was frequently observed that only miniscule amounts of the foreign proteins were accumulated. This was particularly true in efforts to produce insect-resistant plants through the introduction of genes that encoded insecticidal crystal proteins of *Bacillus thuringiensis* (Bt). Ultimately, the low levels of expression were found to be the results of messenger RNA instability *(86–88)*. Careful analysis of the sequences of the introduced genes revealed that several sequence motifs associated with plant mRNA instability or processing were present in the native Bt toxin gene sequence. For example, the 3537-bp coding region of the HD73 Cry1Ac delta-endotoxin gene contains 10 motifs similar to plant 5' or 3' consensus intron splice sites, and 32 motifs associated with plant mRNA polyadenylation sites *(89)*. The result of the presence of these extraneous motifs within the toxin-coding mRNA is that the full-length message is highly unstable in plant cells, and thus very little translation of intact protein occurs. These motifs have no effect on expression of the gene in the bacteria, and they arise as a result of the relatively high adenosine-plus-thymidine (AT) composition (and correspondingly low guanosine-plus-cytosine content; GC) of *Bacillus* genomes (approx 60% AT). Many eukaryotic gene-control sequence elements are rich in AT content, so the presence of sequence homologs to the eukaryotic motifs in a high-AT-content genome is favored.

A second consequence of the high-AT genome composition is that average codon usage for the amino acids encoded by redundant codons is different between plants and many bacteria. A comparative analysis of the codon bias statistics of maize genes and the Cry1Ac coding region shows that for amino acids specified by two or more codons, the bacterial gene is comprised of codons that rarely occur in maize genes *(89)*. Furthermore, whereas maize genes usually employ codons with G or C in the third position, the Cry1Ac coding region has a majority of A's or T's in the third codon position. This difference in codon bias, coupled with the presence of sequences deleterious to mRNA stability in plant cells, renders it extremely difficult to attain commercial levels of the Bt protein toxin from the native coding region.

The solution to this problem is to completely synthesize the protein toxin-coding region in vitro, using a sequence design that substitutes plant-preferred codons for rarely used ones *(90–92)*. Since the gene design is *de novo*, all the deleterious sequences can be eliminated from the coding region, and other desirable features such as strategically placed restriction-enzyme recognition sites can be incorporated into the sequence. This gene-rebuilding process is now a routine step in transgenic plant programs. The abundance of gene sequences from many different crops allows computation of codon usage tables that facilitate the design of genes tailored specifically for expression in a particular crop (e.g., corn vs cotton), or class of plants (dicots vs monocots) if desired.

4.3.2. Position Effects

None of the plant transformation systems available today is capable of directing the integration of transgene DNA into a predetermined site in the genome. Many experiments have attempted to exploit endogenous plant homologous recombination mechanisms by flanking the transgene with large pieces of native plant DNA. All have failed to produce targeted events at high frequency. Consequently, transgenes integrate into the chromosomes at random positions, and in one to many partial or full-length copies. The result is a wide distribution of expression activity, partially a function of the nature of the flanking genomic DNA. Sorting through a population of transformed individuals for the few that exhibit desired expression patterns and heritability is a costly, time-consuming process. Studies on nuclear matrix structure have shown that regions of the chromosome that are populated by highly expressed genes are flanked by sequences designated as matrix attachment regions (MARs) or structural attachment regions (SARs). Substantial work is in progress to examine the effects that the use of MARs or SARs sequences to flank transgenes will have on transgenic expression. It is possible that certain MARs will provide more predictable, and perhaps higher overall—expression levels and patterns (93,94) with some genes or transformation systems. It is apparent that different types of MARs, SARs, and other flanking elements exist, and that they can have different effects on gene expression in different systems (95). In one study (94), the use of chicken lysozyme A elements did not reduce variability of NPTII expression in tobacco, but the elements did cause a highly significant reduction in variation in GUS expression. As with other expression-control elements, prediction of the expression characteristics imbued by a new combination of MARs elements is somewhat risky, and experimental determination of the answer often yields surprises and new insight on the control of plant gene expression.

4.3.3. Gene Silencing

In addition to unpredictable expression patterns, another result of the random integration of transgenes into the chromosome is the phenomenon of gene silencing. Broadly defined, this is seen as the diminution and eventual elimination of transgene expression in the first or successive plant generations. Several gene-silencing patterns have been discovered, and more than one mechanism exists (96,97). Transcriptional silencing results in the loss of mRNA synthesis and is mediated at least in part by hypermethylation of the promoter and/or coding region of the transgene. Triggering of this DNA modification may be induced by some mechanism in the nucleus that scans the chromosome for foreign DNA that has, for example, a DNA GC content that is not typical of that particular chromosomal region. A variation of this type of gene silencing is apparently driven to some degree by sequence homology between the transgene and native plant sequences, or between multiple copies of introduced transgenes *(98)*. Although silencing of single-copy transgenes does occur *(96)* it may be less likely than silencing of multiple copies.

This homology association is not absolute, as demonstrated by two examples. First, the viral 35S promoter seems particularly susceptible to transcriptional silencing in monocots, although no known sequence homologs in monocot genomes are known *(99)*. Others have shown that unlinked sequences as small as 300 bp can detect and inactivate each other in a genome of $>10^9$ bp *(100)*, so unknown monocot homologies responsible for the silencing effect may exist. Second, plant promoters re-introduced back into the native host plant are not always silenced. If large numbers of independently transformed lines are examined, it is usually possible to find suitable candidates for exploitation.

Some studies have shown that the inclusion of flanking MARs or SARs sequences in transgene constructs may insulate the integrated transgene from silencing mechanisms, but more studies are needed to confirm the general utility of this approach *(93,94)*.

Post-transcriptional silencing does not affect the transcription level of nascent mRNAs, but results in the rapid degradation of the targeted mRNA. What induces this process is unknown, but a partial trigger may be the absolute level of a particular mRNA present in the cell. Homology to endogenous RNA is also implicated in some mechanisms (cosuppression) *(97)* and can be used to deliberately downregulate endogenous genes to engineer new traits *(101)*.

Although they are detrimental to the outcome of some research programs, gene downregulation mechanisms can be turned to an advantage. One of the first commercial transgenic products was developed by means of anti-sense RNA mediated downregulation of a tomato polygalacturonase gene, which encodes an enzyme that plays an important role in fruit softening *(102,103)*. Production of this anti-sense mRNA induced silencing of the endogenous gene, resulting in tomatoes that had a longer shelf life. In addition to cosuppression and anti-sense RNA production, other approaches to gene downregulation such as ribozymes *(104)* have been examined in attempts to produce commercial products, or used as fundamental tools to study gene function.

4.3.4. Polycistronic mRNA

Polycistronic mRNAs, which contain the coding regions for more than one protein, are common in bacteria and plant chloroplasts. In contrast, the genes in eukaryotic nuclei are generally expressed as mRNAs that contain the information for only a single protein (for the purposes of this discussion mechanisms such as alternative splicing are not considered). Translation processes in plants are very similar to those in other eukaryotic organisms, and can generally be explained with the scanning model, wherein the ribosomes assemble on 5'-end-capped RNAs and scan down the 5' leader to begin translation at the first ATG start codon *(105)*. Particularly among plant viruses, (e.g., caulimoviruses, badnaviruses, crucifer tobacco mosaic virus) *(106)* unconventional mRNAs are frequent and use modulated translation processes for their expression. Examples include leaky ribosome scanning, translational-stop-codon readthrough or frameshifting, and transactivation by virus-encoded proteins that are used to translate polycistronic mRNAs. In some cases, 5' leader and 3' trailer sequences confer efficient, cap-independent ribosome binding. Although this usually happens via an end-dependent mechanism, true internal ribosome entry may also occur to initiate translation at internal start codons *(106)*. Translation in plant cells is known to be regulated under conditions of stress and during development, but the underlying molecular mechanisms of regulation are unknown. Thus far, only a small number of nonviral plant mRNAs have been discovered with a structure that suggests that they may require some unusual translation mechanisms *(105)*.

The use of polycistronic mRNAs in plant engineering would simplify the alteration of traits such as pest resistance, which may involve multiple proteins. In a hypothetical case, these proteins would be produced through expression from a single nuclear promoter that drives transcription of the polycistronic mRNA. This would obviate the need for coordinated expression of individual promoters driving genes that encode the individual proteins. An alternative to nuclear transformation is the use of chloroplast transformation (*see* Chapter 11), or infection by engineered viruses *(106)*, but these systems have complications of their own.

The expression of polycistronic messages of caulimovirus (a type of plant pararetrovirus) employs a highly unusual mechanism to express the multiple cistrons of their pregenomic RNA. It involves translation of the polycistronic mRNA utilizing *cis*-acting viral RNA sequences and a transacting virus-encoded protein (P6) *(107)*. In addition to its role in polycistronic translation, the translational trans-activator protein P6 also activates its own expression from a monocistronic subgenomic RNA. The efficient expression of polycistronic and monocistronic caulimovirus mRNAs in plant cells thus requires compatible interactions between the P6 translational trans-activator and its cognate *cis*-element at the 3' end of the mRNA. Exploitation of this type of translational mechanism to express an engineered polycistronic mRNA will probably require co-expression of the P6 protein or an analog.

A variation on the theme of producing multiple proteins from a single mRNA is also seen in some plant viruses, in which a polypeptide primary translation product is self-processed to produce multiple proteins *(108)*.

4.3.5. mRNA Targeting

In addition to protein-targeting mechanisms mediated by specific amino acid sequences, proteins may also be targeted to specific subcellular compartments by localization of their mRNAs *(109,110)*. Studies of mRNA localization are a relatively new area in plant science, and results thus far reveal that mRNA localization is involved in many aspects of expression and cell structure. mRNA localization has been implicated in the assembly of macromolecular structures within the cell, in the formation of endoplasmic reticulum subdomains, in facilitating protein localization in the endomembrane system, and in the control of gene expression. In addition, mRNA localization in plants may be part of a mechanism for controlling intracellular communication, not only between adjacent cells but between cells separated by relatively long distances. Although mRNAs can be localized by various mechanisms, evidence gathered to date implicates the role of a translation initiation codon along with cytoskeletal elements, microfilaments, and microtubules that function in the transport and anchoring of RNAs to specific subcellular locations within the cell *(110)*.

5. Heterologous Recombination Systems

Given the unpredictable outcomes of transformation methods that integrate transgenes randomly into the plant genome, various approaches have been used in an attempt to lessen the variability of transgene expression, or to control or alter the result of transgene integration. The inclusion of MARS sequences in the transgene construct was mentioned previously. Site-directed integration through homologous recombination between the plant chromosome and plant DNA included on the transgene construct has met with limited success *(111)*. Another approach that holds some promise is the use of sequence-specific recombination systems derived from bacteria or yeast. These systems have two features in common, a recombinase that promotes DNA recombination and a set of specific DNA sequences that comprise 20–30 bp and serve as the recombination substrate for the recombinase *(112)*. Each recombinase recognizes a pair of the specific sequences and catalyzes recombination between them, with a result that depends on the initial physical relationship of the members of the substrate pair. For example, if the individual members of the substrate sequences are on separate plasmids, the recombinant result is a co-integrated plasmid comprised of both starting plasmids, with a copy of the substrate sequence at each junction of the plasmid sequences. These two intramolecular copies of the substrate sequences can act as substrates for a second recombination event that resolves the cointegrant plasmid into the starting molecules. Thus, the outcome an intramolecular event can be predetermined by controlling the orientation of the two substrate sequences relative to one another. The outcome of recombination between directly repeated substrate sequences is deletion of the DNA located between the substrates (plasmid resolution in the previous example). In contrast, the outcome of recombination between substrates in inverted orientation relative to one another is inversion of the intervening DNA.

Several such sequence-specific systems have been engineered to work in plants (reviewed in ref. *111*). The first was the cre/loxP system derived for the bacteriophage P1. By including copies of the substrate sequence (loxP, locus of crossing-over) in the transgene construct, various types of recombination events can be driven in planta by expression of the recombinase causes recombination (cre) protein. Examples include gene activation or inactivation, deletion of antibiotic selectable marker genes, and non-reversible chromosomal rearrangements mediated by mutant lox sites *(113)*.

The development of other recombination systems for plant use has followed the examples set with the cre/lox system. These include the FLP/FRT system from yeast (*Saccharomyces cereviseae*) *(114,115)*, the R/RS system (from *Zygosaccharomyces rouxii*) *(116)*, the Gin/gix system (from *E. coli* bacteriophage Mu) *(117)* and the attP system (from the *E. coli* lambda phage) *(118)*. The combination of an estrogen-induced transactivator system and the cre/lox recombination system allows chemical-inducible, site-specific DNA excision in *Arabidopsis (119)*.

6. Intellectual Property Issues

The scientific field of plant genetic engineering has existed for about twenty years. Although it is not my intent to diminish in any way the huge public-domain contributions made by scientists in academic institutions, it is important to understand that many of the basic enabling technologies in the field were developed and patented by agricultural research companies, or the intellectual property rights to the technologies were licensed or were purchased by them. It is prudent for any research program initiated with an intent to commercialize the products of its research to conduct an analysis of the intellectual property landscape surrounding all aspects of the technology. Many of the initial pioneering patents were filed in the 1980s and early 1990s. Competing parties sometimes filed patents within a few days or weeks of one another. As a result, many of the patent applications are still being examined in the patent offices, are in interference proceedings, or have been issued but are being challenged in the courts. As these cases are resolved, the intellectual property ownership will sometimes shift from one party to another, with the result that a license to use a technology from one party may be rendered in question by the issuance of a new patent or court decision. Thus, a certain amount of risk is incurred by the decision to use a particular promoter or transformation method in the development of a commercial product. What may seem to be "free and clear" at the inception of the program may become encumbered later on as the intellectual landscape evolves.

In academic or other public institutions, it is common to find that exciting new genes being discovered or technologies being developed are "contaminated" through the use of components that must be licensed for commercial use. Although many industrial research companies will grant free licenses to some of their intellectual property to academic scientists for research purposes only, the line between what constitutes basic research and commercial development research is becoming increasingly blurred. The researcher may begin with good intentions and obtain readily available promoters or selectable marker genes in order to get his or her research program off to a fast start. As the years pass and excitement builds over the progress of the research, these components become so ubiquitous in all the constructs that it becomes extremely time-consuming to go back and re-engineer with different, unencumbered components. Eventually the program may reach a point when an invention is patentable, and the institution's intellectual property licensing office tries to find a royalty-paying commercial partner that will acquire a license, sell a product, and turn money back to the institution. In some cases, this further commercial development is hindered or made impossible because of high licensing and royalty costs owed to third-party companies that own particular pieces of intellectual property essential to practicing the new invention. As more public institutions become more aggressive in filing patents to protect the inventions of their faculty, it might be advisable for them to institute intellectual property workshops or other types of training for their research community. This training could emphasize the importance of decisions made early in the initiation of a research program, and how those decisions may affect future deployment of the products from that research. The results may benefit both the institution and the prospective commercialization partner through the availability of a "cleaner" invention with higher value.

A current example is provided by the development of transgenic "golden rice" *(120)*, which was engineered with the complement of genes needed to create a biosynthesis pathway for beta-carotene, which humans convert into vitamin A. A potential benefit of golden rice (so named because the beta-carotene imparts a yellow color to the grain) is that it could help alleviate a widespread public health problem, vitamin A deficiency. Globally, about 400 million people suffer from this affliction, which can lead to vision impairment and increased disease susceptibility. Although the beta-carotene content of the first sets of transgenic plants is not sufficient to supply the total dietary requirement of vitamin A, widespread inclusion in human diets could provide some benefits. However, two studies have found that between 25 and 70 proprietary techniques and materials were involved in the gene transfers *(121)*, and agreements must be reached with the affected parties before the transgenic plants can be passed on to third parties, such as rice-breeding institutes.

A second example is provided by the events following the discovery of an ALS gene that contained two point mutations that together confer resistance to all four chemical classes of ALS inhibitors (*see* **Subheading 2.3.1.**) *(122)*. One mutation conferred resistance to sulfonylureas and/or triazolopyrimidine sulfonanilides, and the second mutation mediated resistance to the imidazolinones and pyrimidyl-oxy benzoates. If expressed in a transgenic crop, such a four-way gene may allow the application of a mixture of two different herbicides, one from each group corresponding to the resistances conferred by the individual mutations. It was proposed that these combinations of chemistries could thus delay or eliminate the appearance of weeds resistant to the ALS inhibitors *(123)*, since it seemed extremely unlikely that both point mutations would occur in the ALS genes of a weed in the same generation. When approached by an enthusiastic seeds industry group about commercializing the gene, however, the company that held patents on the ALS gene responded negatively, and ultimately the public research on the gene was terminated.

The lesson to be emphasized is that, as public institutions become more involved in research that lends itself to development of a commercial product, these institutions must gain a much better understanding of the scope and nature of business decisions. United States patent laws, combined with the potential for large profits in the agricultural biotechnology industry, are profoundly changing the nature of "public domain" research.

References

1. Weiler, E. W. and Schröder, J. (1987) Hormone genes and crown gall disease. *Trends Biol. Sci.* **12**, 271–278.
2. Pengelly, W. L., Vijayaraghavan, S. J., and Sciaky, D. (1986) Neoplastic progression in crown gall in tobacco without elevated auxin levels. *Planta* **169**, 454–461.
3. Herrera-Estrella, L., De Block, M., Messens, E., Hernalsteens, J.-P., Van Montagu, M., and Schell, J. (1983) Chimeric genes as dominant selectable markers in plant cells. *EMBO J.* **2**, 987–995.
4. Flavell, R. B., Dart, E., Fuchs, R. L., and Fraley, R. T. (1991) Selectable marker genes: safe for plants? *Bio-Technology* **10**, 141–144.
5. Pietrzak, M., Shillito, R. D., Hohn, T., and Potrykus, I. (1986) Expression in plants of two bacterial antibiotic resistance genes after protoplast transformation with a new plant expression vector. *Nucleic Acids Res.* **14**, 5857–5868.
6. Hille, J., Verheggen, F., Roelvink, P., Franssen, H., van Kammen, A., and Zabel, P. (1986) Bleomycin resistance: a new dominant selectable marker for plant cell transformation. *Plant Mol. Biol.* **7**, 171–176.
7. Jones, J. D. G., Svab, Z., Harper, E. C., Hurwitz, C. D., and Maliga, P. (1987) A dominant nuclear streptomycin resistance marker for plant cell transformation. *Mol. Gen. Genet.* **210**, 86–91.
8. Waldron, C., Murphy, E. B., Roberts, J. L., Gustafson, G. D., Armour, S. L., and Malcolm, S. K. (1985) Resistance to hygromycin B. *Plant Mol. Biol.* **5**, 103–108.
9. Eichholtz, D. A., Rogers, S. G., Horsch, R. B., Klee, H. J., Hayford, M., Hoffmann, N. L., et al. (1987) Expression of a mouse dihydrofolate reductase gene confers methotrexate resistance in transgenic petunia plants. *Somatic Cell Mol. Genet.* *13,* 67–76.

10. Petolino, J. F., Young, S., Hopkins, N., Sukhapinda, K., Woosley, A., Hayes, C., et al. (2000) Expression of murine adenosine deaminase (ADA) in transgenic maize. *Transgenic Res.* **9**, 1–9.

11. Bertolla, F., Kay, E., and Simonet, P. (2000) Potential dissemination of antibiotic resistance genes from transgenic plants to microorganisms. *Infect. Control Hosp. Epidemiol.* **21**, 390–393.

12. Simonet, P. (2000) Évaluation des potentialités de transfert de l'ADN des plantes transgéniques vers les bactéries du sol. Ol., *Corps Gras, Lipides* **7**, 320–323.

13. Lorenz, M. G. and Wackernagel, W. (1994) Bacterial gene transfer by natural genetic transformation in the environment. *Microbiol. Rev.* **58**, 563–602.

14. Dekker, J. and Duke, S. O. (1995) Herbicide-resistant field crops. *Advances in Agron.* **54**, 69–116.

15. Devine, M. D. and Eberlein, C. V. (1997) Physiological, biochemical and molecular aspects of herbicide resistance based on altered target sites, in *Herbicide Activity: Toxicity, Biochemistry and Molecular Biology* (Roe, R. M., et al., eds.), IOS Press, Amsterdam, The Netherlands, pp. 159–185.

16. White, J., Chang, S.-Y. P., Bibb, M. J., and Bibb, M. J. (1990) A cassette containing the bar gene of *Streptomyces hygroscopicus*: a selectable marker for plant transformation. *Nucleic Acids Res.* **18**, 1062.

17. Wohlleben, W., Arnold, W., Broer, I., Hillemann, D., Strauch, E., and Puehler, A. (1988) Nucleotide sequence of the phosphinothricin N-acetyltransferase gene from *Streptomyces viridochromogenes* Tue494 and its expression in *Nicotiana tabacum*. *Gene* **70**, 25–37.

18. Donn, G., Tischer, E., Smith, J. A., and Goodman, H. M. J. (1984) Herbicide-resistant alfalfa cells: an example of gene amplification in plants. *Mol. Appl Genet.* **2**, 621–635.

19. Comai, L., Facciotti, D., Hiatt, W. R., Thompson, G., Rose, R., and Stalker, D. (1985) Expression in plants of a mutant aroA gene from *Salmonella typhimurium* confers tolerance to glyphosate. *Nature* **317**, 741–744.

20. Della-Cioppa, G., Bauer, S. C., Taylor, L. M., Rochester, D. E., Klein, B. K., Shah, D. M., et al. (1987) Targeting a herbicide-resistant enzyme from *Escherichia coli* to chloroplasts of higher plants. *Bio-Technology* **5**, 579–584.

21. Padgette, S. R., Re, D. B., Barry, G. F., Eichholtz, D. E., Delanney, X., Fuchs, R. L., et al. (1996) New weed control opportunities: development of soybeans with a Roundup Ready gene, in *Herbicide-Resistant Crops: Agricultural, Environmental, Economic, Regulatory, and Technical Aspects* (Duke, S. O., ed.), CRC Press, Boca Raton, FL, pp. 53–84

22. Barry, G., Kishore, G. M., Padgette, S., Taylor, M., Kolacz, K., Welson, M., et al. (1992) Inhibitors of amino acid biosynthesis: strategies for imparting glyphosate tolerance to crop plants, in *Biosynthesis and Molecular Regulation of Amino Acids in Plants* (Singh, B. K., et al., eds.), American Society of Plant Physiologists, Rockville, MD, pp. 139–145.

23. Stalker, D. M., McBride, K. E., and Malyj, L. D. (1988) Herbicide resistance in transgenic plants expressing a bacterial detoxification gene. *Science* **242**, 419–423.

24. Stalker, D. M., Kiser, J. A., Baldwin, G., Coulombe, B., and Houck, C. M. (1996) Cotton weed control using the BXNTM system, in *Herbicide-Resistant Crops: Agricultural, Environmental, Economic, Regulatory, and Technical Aspects* (Duke, S. O., ed.), CRC Press, Boca Raton, FL, pp. 93–105.

25. Ryan, R. F. (1970) Resistance of common groundsel to simazine and atrazine. *Weed Sci.* **18**, 614–616.

26. Hirschberg, J., and McIntosh, L. (1983) Molecular basis of herbicide resistance in *Amaranthus hybrides*. *Science* **222**, 1346–1349.

27. Mourad, G S. (1999) Feedback-insensitive threonine dehydratase/deaminase from an *Arabidopsis thaliana* mutant and its use in genetic engineering of plants and microorganisms. *PCT Int. Appl.* **WO 9902656**, pp. 122.

28. Negrotto, D., Jolley, M., Beer, S., Wenck, A. R., and Hansen, G. (2000) The use of phosphomannose-isomerase as a selectable marker to recover transgenic maize plants (*Zea mays* L.) via *Agrobacterium* transformation. *Plant Cell Rep.* **19**, 798–803.

29. Otten, L. A. B. M. and Schilperoort, R. A. (1978) A rapid micro scale method for the detection of lysopine and nopaline dehydrogenase. *Biochim. Biophys. Acta* **527,** 497–500.

30. Dahl, G. A., and Tempé, J. (1983) Studies on the use of toxic precursor analogs of opines to select transformed plant cells. *Theor. Appl. Genet.* **66,** 233–239.

31. Van Slogteren, G. M. S., Hooykaas, P. J. J., Planqué, K., and De Groot, B. (1982) The lysopinedehydrogenase gene used as a marker for the selection of octopine crown gall cells. *Plant Mol. Biol.* **1,** 133–142.

32. Gendloff, E; H., Bowen, B., and Buchholz, W. G. (1990) Quantitation of chloramphenicol acetyltransferase in transgenic tobacco plants by ELISA and correlation with gene copy number. *Plant Mol. Biol.* **14,** 575–583.

33. Davey, M. R., Blackhall, N. W., and Power, J. B. (1995) Chloramphenicol acetyltransferase assay. *Meth. Mol. Biol.* **49,** 143–148.

34. Sleigh, M. J. (1986) A nonchromatographic assay for expression of chloramphenicol acetyltransferase gene in eukaryotic cells. *Anal. Biochem.* **156,** 251–256.

35. Jefferson, R. A., Kavanagh, T. A., and Bevan, M. W. (1987) GUS fusions: β-glucuronidase as a sensitive and versatile gene fusion marker in higher plants. *EMBO J.* **6,** 3901–3907.

36. Jefferson, R. A. (1987) Assaying chimeric genes in plants: the GUS gene fusion system. *Plant Mol. Biol. Rep.* **5,** 387–406.

37. Ow, D. W., Wood, K. V., DeLuca, M., De Wet, J. R., Helinski, D. R., and Howell, S. H. (1986) Transient and stable expression of the firefly luciferase gene in plant cells and transgenic plants. *Science* **234,** 856–859.

38. Lepetit, M., Ehling, M., Gigot, C., and Hahne, G. (1991) An internal standard improves the reliability of transient expression studies in plant protoplasts. *Plant Cell Rep.* **10,** 401–405.

39. Langridge, W. H. R., and Szalay, A. A. (1998) Bacterial and coelenterate luciferases as reporter genes in plant cells. *Meth. Mol. Biol.* **82,** 385–396.

40. Kirchner, G., Roberts, J. L., Gustafson, G. D., and Ingolia, T. D. (1989) Active bacterial luciferase from a fused gene: expression of a *Vibrio harveyi luxAB* translational fusion in bacteria, yeast, and plant cells. *Gene* **81,** 349–354.

41. Lorenz, W. W., McCann, R. O., Longiaru, M., and Cormier, M. J. (1991) Isolation and expression of a cDNA encoding *Renilla reniformis* luciferase. *Proc. Natl. Acad. Sci. USA* **88,** 4438–4442.

42. Mayerhofer, R., Langridge, W. H. R., Cormier, M. J., Szalay, A. A. (1995) Expression of recombinant *Renilla luciferase* in transgenic plants results in high levels of light emission. *Plant J.* **7,** 1031–1038.

43. Park, J. B. (2001) Concurrent measurement of promoter activity and transfection efficiency using a new reporter vector containing both *Photinus pyralis* and *Renilla reniformis* luciferase genes. *Anal. Biochem.* **291,** 162–166.

44. Niedtz, R. P., Sussman, M. R., and Satterlee, J. S. (1995) Green fluorescent protein: an *in vivo* reporter of plant gene expression *Plant Cell Rep.* **14,** 2403–2406.

45. Leffel, S. M., Mabon, S. A., and Stewart, C. N., Jr. (1997) Applications of green fluorescent protein in plants. *BioTechniques* **23,** 912–918.

46. Chalfie, M., Tu, Y., Euskirchen, G., Ward, W. W., and Prasher, D. C. (1994) Green fluorescent protein as a marker for gene expression. *Science* **263,** 725–888.

47. Haseloff, J., Siemering, K. R., Prasher, D. C., and Hodge, S. (1997) Removal of a cryptic intron and subcellular localization of green fluorescent protein are required to mark transgenic *Arabidopsis* plants brightly. *Proc. Natl. Acad. Sci. USA* **94,** 2122–2127.

48. Reichel, C., Mathur, J., Eckes, P., Langenkemper, K., Koncz, C., Schell, J., et al. (1996) Enhanced green fluorescence by the expression of an *Aequorea victoris* green fluorescent protein mutant in mono-and dicotyledonous plant cells. *Proc. Natl. Acad. Sci. USA* **93,** 5888–5893.

49. CLONTECH Laboratories Product Catalog, Palo Alto, CA.

50. Delagrave, S., Hawtin, R. E., Silva, C. M., Yang, M. M., and Youvan, D. C. (1995) Red-shifted excitation mutants of the green fluorescent protein. *Bio-Technology* **13**, 151–154.

51. Yang, T.-T., Kain, S. R., Kitts, P., Kondepudi, A., Yang, M. M., and Youvan, D. C. (1996) Dual color microscopic imagery of cells expressing the green fluorescent protein and a red-shifted variant. *Gene* **173**, 19–23.

52. Matz, M. V., Fradkov, A. F., Labas, Y. A., Savitsky, A. P., Zaraisky, A. G., Markelov, M. L., et al. (1999) Fluorescent proteins from nonbioluminescent *Anthozoa* species. *Nat. Biotechnol.* **17**, 969–973.

53. Teeri, T. H., Lehväslaiho, H., Franck, M., Uotila, J., Heino, P., Palva, E. T., et al. (1989) Gene fusions to lacZ reveal new expression patterns of chimeric genes in transgenic plants. EMBO J. 8, 343–350.

54. Helmer, G., Casadaban, M., Bevan, M., Kayes, L., and Chilton, M,-D. (1984) A new chimeric gene as a marker for plant transformation: the expression of *Escherichia coli* β-galactosidase in sunflower and tobacco cells. Bio-Technology (June), 520–527.

55. Matsumoto, S., Takebe, I., and Machida, Y. (1988) *Escherichia coli lacZ* gene as a biochemical and histochemical marker in plant cells. *Gene* **66**, 19–29.

56. Christensen, A. H. and Quail, R. H. (1996) Ubiquitin promoter-based vectors for high-level expression of selectable and/or screenable marker genes in monocotyledonous plants. *Transgenic Res.* **5**, 231–218.

57. Odell, J. T., Nagy, F., and Chua, N.-H. (1985) Identification of DNA sequences required for activity of the cauliflower mosaic virus 35S promoter. *Nature* **313**, 810–812.

58. Peach, C. and Velten, J. (1991) Transgene expression variability (position effect) of CAT and GUS reporter genes driven by linked divergent T-DNA promoters. *Plant Mol. Biol.* **17**, 49–60.

59. Datla, R., Anderson, J. W., and Selvaraj, G. (1997) Plant promoters for transgene expression. *Biotech. Annu. Rev.* **3**, 269–296.

60. Braam, J. and Davis, R. (1990) Rain-, wind-, and touch-induced expression of calmodulin and calmodulin-related genes in *Arabidopsis. Cell* **60**, 357–364.

61. Chen, Z.-L., Pan, N.-S., and Beachy, R. N. (1988) A DNA sequence element that confers seed-specific enhancement to a constitutive promoter. *EMBO J.* **7**, 297–302.

62. Ward, E. R., Ryals, J. A., and Miflin, B. J. (1993) Chemical regulation of transgene expression in plants. *Plant Mol. Biol.* **22**, 361–366.

63. Gatz, C. (1997) Chemical control of gene expression. *Annu. Rev. Plant Physiol. Plant Mol. Biol.* **48**, 89–108.

64. Zuo, J. and Chua, N. H. (2000) Chemical-inducible systems for regulated expression of plant genes. *Curr. Opin. Biotechnol.* **1**, 146–151.

65. Weinmann, P., Gossen, M., Hillen, W., Bujard, H., and Gatz, C. (1994) A chimeric transactivator allows tetracycline-responsive gene expression in whole plants. *Plant J.* **5**, 559–569.

66. Lloyd, A. M., Schena, M., Walbot, V., and Davis, R. W. (1994) Epidermal cell fate determination in *Arabidopsis*: patterns defined by a steroid-inducible regulator. *Science* **266**, 436–439.

67. No, D., Yao, T.-P., and Evans, R. M. (1996) Ecdysone-inducible gene expression in mammalian cells and transgenic mice. *Proc. Natl. Acad. Sci. USA* **93**, 3346–3351.

68. Padidam, M. and Cao, Y. (2001) Elimination of transcriptional interference between tandem genes in plant cells. *Bio-Techniques* **31**, 328–334.

69. The *Arabidopsis* Genome Initiative. (2000) Analysis of the genome sequence of the flowering plant *Arabidopsis thaliana. Nature* **408**, 796–815.

70. *Post-Transcriptional Control of Gene Expression in Plants* (1996) (Filipowicz, W. and Hohn, T., eds.), *Plant Mol. Biol.* **32**, Nos. 1 & 2. p. 414.

71. *A Look Beyond Transcription: Mechanisms Determining mRNA Stability and Translation in Plants* (1998) (Bailey-Serres, J. and Gallie, D. R., eds.), American Society of Plant Physiologists, Rockville, MD, p. 183.

72. Gallie, D. R., Sleat, D. E., Watts. J. W., Turner, P. C., and Wilson, T. M. A. (1987) The 5' leader of tobacco mosaic virus RNA enhances the expression of foreign gene transcripts *in vitro* and *in vivo*. *Nucleic Acids Res.* **15**, 3257–3273.

73. Dowson Day, M. J., Ashurst, J. L., Mathias, S. F., Watts, J. W., Wilson, T. M. A., and Dixon, R. A. (1993) Plant viral leaders influence expression of a reporter gene in tobacco. *Plant Mol. Biol.* **23**, 97–109.

74. Jobling, S. A. and Gehrke, L. (1987) Enhanced translation of chimaeric messenger RNAs containing a plant viral untranslated leader sequence. *Nature* **325**, 622–625.

75. Depicker, A., Stachel, S., Dhaese, P., Zambryski, P., and Goodman, H. M. (1982) Nopaline synthase, transcript mapping and DNA sequence. *J. Mol. Appl. Genet.* **1**, 561–573.

76. Rose, A. B. and Last, R. L. (1997) Introns act post-transcriptionally to increase expression of the *Arabidopsis thaliana* tryptophan pathway gene PAT1. *Plant J.* **11**, 455–464.

77. Fu, H., Kim, S. Y., and Park, W. D. (1995) High-level tuber expression and sucrose inducibility of a potato *Sus4* sucrose synthase gene require 5' and 3' flanking sequences and the leader intron. *Plant Cell* **7**, 1387–1394.

78. An, G., Mitra, A., Hong, K. C., Costa, M. A., An, K., Thornburg, R. W., et al. (1989) Functional analysis of the 3' control region of the potato wound-inducible proteinase inhibitor II gene. *Plant Cell* **1**, 115–122.

79. Fu, H., Kim, S. Y., and Park, W. D. (1995) A potato sucrose synthase gene contains a context-dependent 3' element and a leader intron with both positive and negative tissue-specific effects. *Plant Cell* **7**, 1395–1403.

80. Luehrsen, K. R. and Walbot, V. (1991) Intron enhancement of gene expression and the splicing efficiency of introns in maize cells. *Mol. Gen. Genet.* **225**, 81–93.

81. Snowden, K. C., Buchholz, W. G., and Hall, T. C. (1996) Intron position affects expression from the tpi promoter in rice. *Plant Mol. Biol.* **31**, 689–692.

82. Pachuk, C. J., Samuel, M., Zurawski, J. A., Snyder, L., Phillips, P., and Satishchandran, C. (2000) Chain reaction cloning: a one-step method for directional ligation of multiple DNA fragments. *Gene* **243**, 19–25.

83. The entire December, 2000, issue of *Plant Physiology* (Vol. 124, No. 4, pp.1449–1865) is devoted to *Arabidopsis* studies.

84. De Rocher, E. J., Vargo-Gogola, T. C., Diehn, S. H., and Green, P. J. (1998) Direct evidence for rapid degradation of *Bacillus thuringiensis* toxin mRNA as a cause of poor expression in plants. *Plant Physiol.* **117**, 1445–1461.

85. Dugdale, B., Becker, D. K., Beetham, P. R., Harding, R. M., and Dale, J. L. (2000) Promoters derived from banana bunchy top virus DNA-1 to –5 direct vascular-associated expression in transgenic banana (*Musa* spp.). *Plant Cell Rep.* **19**, 810–814.

86. Diehn, S. H., Chiu, W-L., De Rocher, E. J., and Green, P. J. (1998) Premature polyadenylation at multiple sites within a *Bacillus thuringiensis* toxin gene-coding region. *Plant Physiol.* **117**, 1433–1443.

87. Ohme-Takagi, M., Taylor, C. B., Newman, T. C., and Green, P. J. (1993) The effect of sequences with high AU content on mRNA stability in tobacco. *Proc. Natl. Acad. Sci. USA* **90**, 11,811–11,815.

88. Diehn, S. H., De Rocher, E. J., and Green, P. J. (1996) Problems that can limit the expression of foreign genes, in plants: lessons to be learned from B. t. toxin genes, in *Genetic Engineering*, Vol. 18, (Setlow, J. K., ed.), Plenum Press, New York, NY, pp. 83–99.

89. D. Merlo, unpublished

90. Campbell, C. H. and Gowri, G. (1990) Codon usage in higher plants, green algae, and cyanobacteria. *Plant Physiol.* **92**, 1–11.

91. Koziel, M. G., Carozzi, N. B., and Desai, N. (1996) Optimizing expression of transgenes with an emphasis on post-transcriptional events. *Plant Mol. Biol.* **32**, 393–405.

92. Adang, M. J., Brody, M. S., Cardineau, G., Eagan, N., Roush, R. T., et al. (1993) The reconstruction and expression of a *Bacillus thuringiensis CryIIIA* gene in protoplasts and potato plants. *Plant Mol. Biol.* **21,** 1131–1145.

93. Ülker, B., Allen, G. C., Thompson, W. F., Spiker, S., and Weissinger, A. K. (1999) A tobacco matrix attachment region reduces the loss of transgene expression in the progeny of transgenic tobacco plants. *Plant J.* **18,** 253–263.

94. Mlynárová, L., Jansen, R. C., Conner, A. J., Stiekema. W. J., and Nap, J.-P. (1995) The MAR-mediated reduction in position effect can be uncoupled from copy number-dependent expression in transgenic plants. *Plant Cell* **7,** 599–609.

95. Lewin, B. (1994) Chromatin and gene expression: constant questions, but changing answers. *Cell* **79,** 397–406.

96. Kumpatla, S. P., Chandrasekharan, M. B., Iyer, L. M., Li, G., and Hall, T. C. (1998) Genome intruder scanning and modulation systems and transgene silencing. *Trends Plant Sci.* **3,** 97–104.

97. Baulcombe, D. C. (1996) RNA as a target and an initiator of post-transcriptional gene silencing in transgenic plants. *Plant Mol. Biol.* **32,** 79–88.

98. Flavell, R. B. (1994) Inactivation of gene expression in plants as a consequence of specific sequence duplication. *Proc. Natl. Acad. Sci. USA* **91,** 3490–3496.

99. Kumpatla, S. P., Teng, W., Buchholz, W. G., and Hall, T. C. (1997) Epigenetic transcriptional silencing and 5-azacytidine-mediated reactivation of a complex transgene in rice. *Plant Physiol.* **115,** 361–373.

100. Bestor, T. H. and Tycko, B. (1996) Creation of genomic methylation patterns. *Nat. Genet.* **12,** 363–367.

101. Seymour, G. B., Fray, R. G., Hill, P., and Tucker, G. A. (1993) Down-regulation of two non-homologous endogenous tomato genes with a single chimaeric sense gene construct. *Plant Mol. Biol.* **23,** 1–9.

102. Shewmaker, C. K., Kridl, J. C., Hiatt, W.R., Knauf, V. (1995) Antisense regulation of gene expression in plant cells. U.S. Patent Appl., Cont.-in-part of U.S. 5,107,065. 16 pp.

103. Smith, C. J. S., Watson, C. F., Ray, J., Bird, C. R., Morris, P. C., Schuch, W., et al. (1988) Antisense RNA inhibition of polygalacturonase gene expression in transgenic tomatoes. *Nature* **334,** 724–726.

104. Owens Merlo, A., Cowen, N., Delate, T., Edington, B., Folkerts, O., Hopkins, N., et al. (1998) Ribozymes targeted to stearoyl-ACP D9 desaturase mRNA produce heritable increases of stearic acid in transgenic maize leaves. *Plant Cell* **10,** 1601–1621.

105. Fuetterer, J. and Hohn, T. (1996) Translation in plants - rules and exceptions. *Plant Mol. Biol.* **32,** 159–189.

106. Hohn, T., Corsten, S., Hemmings-Miesczak, M., Hyun-Sook, P., Poogin, M., Ryabova, L., et al. (2000) Polycistronic translation in plants. What can we learn from viruses? *Dev. Plant Genet. Breed.* **5,** 126–129.

107. Edskes H. K., Kiernan J. M., and Shepherd R. J. (1996) Efficient translation of distal cistrons of a polycistronic mRNA of a plant pararetrovirus requires a compatible interaction between the mRNA 3′ end and the proteinaceous trans-activator. *Virology* **224,** 564–567.

108. Marcos, J. F. and Beachy, R. N. (1994) In vitro characterization of a cassette to accumulate multiple proteins through synthesis of a self-processing polypeptide. *Plant Mol. Biol.* **24,** 495–503.

109. Okita, T. W., Choi, S.-B., Ito, H., Muench, D. G., Y. Wu, and Zhang, F. (1998) Entry into the secretory system—the role of mRNA localization. *J. Exper. Bot.* **49,** 1081–1090.

110. Choi, S. B.; Wang, C., Muench, D. G., Ozawa, K., Franceschi, V. R., Wu, Y., et al. (2000) Messenger RNA targeting of rice seed storage proteins to specific ER subdomains. *Nature* **407,** 765–767.

111. *Homologous Recombination and Gene Silencing in Plants.* (1994) (Paszkowski, J., ed.), Kluwer Academic Publishers, Dordrecht, The Netherlands.

112. Odell, J. T. and Russell, S. H. (1994) Use of site-specific recombination systems in plants, in *Homologous Recombination and Gene Silencing in Plants* (Paszkowski, J., ed.), Kluwer Academic Publishers, Dordrecht, The Netherlands. pp. 219–270.

113. Ow, D. W. and Medberry, S. L. (1995) Genome manipulation through site-specific recombination. *Crit. Rev. Plant Sci.* **14,** 239–261.

114. Lyznik, L. A., Mitchell, J. C., Hirayama, L., and Hodges, T. K. (1993) Activity of yeast FLP recombinase in maize and rice protoplasts. *Nucleic Acids Res.* **21,** 969–975.

115. Kilby, J., Davies, G. J., Snaith, M. R., and Murray, J. A. H. (1995) FLP recombinase in transgenic plants: constitutive activity in stably transformed tobacco and generation of marked cell clones in *Arabidopsis. Plant J.* **8,** 637–652.

116. Sugita, K., Kasahara, T., Matsunaga, E., and Ebinuma, H. (2000) A transformation vector for the production of marker-free transgenic plants containing a single copy transgene at high frequency. *Plant J.* **22,** 461–469.

117. Maeser, S. and Kahmann, R. (1991) The Gin recombinase of phage Mu can catalyze site-specific recombination in plant protoplasts. *Mol. Gen. Genet.* **230,** 170–176.

118. Zubko, E., Scutt, C., and Meyer, P. (2000) Intrachromosomal recombination between attP regions as a tool to remove selectable marker genes from tobacco transgenes. *Nat. Biotechnol.* **18,** 442–445.

119. Zuo, J., Niu, Q.-W., MØller, S. G., and Chua, N.-H. (2001) Chemical-regulated, site-specific DNA excision in transgenic plants. *Nat. Biotechnol.* **19,** 157–161.

120. Ye, X., Al-Babili, S., Klöti, A., Zhang, J., Lucca, P., Beyer, P., et al. (2000) Engineering the provitamin A (-Carotene) biosynthetic pathway into (carotenoid-free) rice endosperm. *Science* **287,** 303–305.

121. Normile, D. (2000) Monsanto donates its share of golden rice. *Science* **289,** 843–845.

122. Mourad, G., Haughn, G., and King, J. (1994) Intragenic recombination in the CSR1 locus of Arabidopsis. *Mol. Gen. Genet.* **243,** 178–184.

123. Knudsen, N. S. (1998) Discovery of gene with 4-way herbicide resistance—the end of innocence for public domain researchers? *Seed and Crops Digest* **49,** 6–10.

10

Plant Cell Culture

A Critical Tool for Agricultural Biotechnology

Joseph F. Petolino, Jean L. Roberts, and Ponsamuel Jayakumar

1. Introduction

Agricultural biotechnology was born during the 1980s, when the first published reports of the successful delivery, integration, and expression of foreign genes in plants began to appear *(1,2)*. Since that time, exceptionally rapid progress in extending gene-transfer capabilities to economically important crop plants has been made. Today, representatives from virtually all the major families of crop plants have been successfully transformed *(3)*. Improved methods of DNA delivery, development of effective selectable marker genes, and availability of potent gene-expression signals are among the most important factors that contribute to the production of transgenic plants over the last decade (*see* Chapter 9). However, the foundation for all agricultural biotechnology has been the establishment of methods for culturing plant cells and tissues in vitro and subsequently regenerating fertile plants. This chapter focuses on plant cell and tissue culture as it relates to transgenic plant production.

2. Somatic Cell Genetics: "Plant Breeding" in a Test Tube

2.1. Totipotency: From Cell to Fertile Plant

Schleiden and Schwann proposed the concept of totipotency, which in essence is the ability of undifferentiated cells to develop into whole organisms or organs when exposed to appropriate environmental stimuli *(4,5)*. In 1898, the botanist Gottlieb Haberlandt was the first to attempt the systematic in vitro culture of single plant cells *(6)*. His purpose was to study the mutual influences of cells as "living units" within a multicellular body. Haberlandt visualized the theoretical potential of the culture approach in experimental plant morphology and physiology; however, nearly half a century elapsed before his far-reaching ideas were realized.

"There has been, so far as I know, up to present, no planned attempt to cultivate the vegetative cells of higher plants in suitable nutrients. Yet the results of such attempts should cast many interesting sidelights on the peculiarities and capacities which the cell, as an elementary organism, possesses: they should make possible conclusions as to the interrelations and reciprocal influences to which the cell is subjected within the multicellular organism. Without permitting

From: *Handbook of Industrial Cell Culture: Mammalian, Microbial, and Plant Cells*
Edited by: V. A. Vinci and S. R. Parekh © Humana Press Inc., Totowa, NJ

myself to pose further questions, I believe, in conclusion, that I am not making too bold a prediction if I point to the possibility that, in this way, one could successfully cultivate artificial embryos from vegetative cells *(6)*."

Cells within some plant tissues (such as undifferentiated parenchyma, meristematic, and embryonic), under certain environmental conditions (such as wounding or infection), are capable of switching to alternative pathways of development. Such undifferentiated cells can rapidly proliferate to produce amorphous cell masses known as callus. Callus cells appear to exhibit a high degree of plasticity in their response to physical and environmental stimuli. By formulating appropriate culture conditions (medium, light, or temperature) it is possible to induce callus formation in aseptically dismembered plant parts (buds, roots, stems, leaves, seeds, or anthers).

Many tissue explants can be induced to form callus by the application of plant growth regulators. Modifying the ratio of the auxin and cytokinin concentration has proven to be a very effective means of inducing alternative developmental states *(7)*. Different tissues have specific requirements, which can only be determined empirically. Once established, callus cultures can be maintained by regular subculture. Under certain nutritional and/or hormonal conditions, some callus cultures can be induced to develop bipolar embryo-like structures. This phenomenon, known as somatic embryogenesis, follows an ontogenetic sequence through pro-embryo, globular, and torpedo stages that is similar to zygotic embryo development *(8)*. The resulting cultures are said to be embryogenic. The synthetic auxin 2,4-D is routinely used to induce somatic embryogenesis from tissue explants. The embryogenic cultures are then transferred to a 2,4-D-free medium to induce plant regeneration.

The general approach of using plant-growth regulators to induce alternative developmental pathways in totipotent plant cells has been extremely successful in establishing in vitro capabilities for a broad spectrum of plant species, including most major crops. This capability represents the foundation on which virtually all of the current transgenic plant production systems are based (*see* **Subheading 4.**).

2.2. Clonal Propagation: Reproduction Without Sex

Plants have two natural modes of multiplication, sexual and vegetative propagation. Vegetative propagation is a process that involves the production of axillary buds and adventitious roots. This type of multiplication occurs regardless of flowering and sexual reproduction, and results in clonal propagation. Clonal propagation is routinely used for multiplying tuber-bearing plants such as potatoes. Moreover, fruit trees, ornamental plants, and many flowers are propagated from cuttings or scions. Although clonal propagation has long been used for plant multiplication, the advent of in vitro culture techniques has accelerated this process and allowed for the production of a wide range of identical plants. The process of in vitro clonal propagation has been referred to as micropropagation.

Micropropagation is routinely used to generate large numbers of high-quality clonal plants, including ornamental and vegetable species, as well as some plantation crops and fruit trees *(9)*. Micropropagation has significant advantages over traditional clonal propagation techniques. These include the potential of combining rapid, large-scale propagation of new genotypes, the use of small amounts of original germplasm (particularly at the early breeding and/or transgenic plant stage, when only a few individuals are available), and the generation of pathogen-free propagules.

In recent years, micropropagation has taken on a new dimension with the use of large-scale bioreactors *(10)*, automation facilities *(11)*, and synthetic seeds *(12)* that have the potential to further increase efficiency. Continued improvement is needed to ensure consistent reproducibility and quality of the micropropagated plants.

2.3. Somaclonal Variation: Culture-Induced Variability

Theoretically, plants derived from tissue explants should represent clones of the parent material from which the explant was derived—i.e., identical in genotype and phenotype. Although this has been observed in the vast majority of cases, it has long been known that sporadic abnormalities attributed to the tissue-culture process can lead to variant plants *(13)*. Somaclonal variation is the variability observed among plants regenerated from in vitro culture *(14)*. This variability is heritable because it is transmitted through meiosis, and is usually irreversible *(15)*. Indeed, the appearance of variants in tissue culture is now known to be a general occurrence for certain plant species and/or specific explant sources.

Somaclonal variation has been observed at several levels, including morphological and physiological modification, changes in chromosome number and/or structure, and altered protein and DNA composition. This phenomenon is particularly apparent in plants regenerated from long-term cultures. Another type of variation observed in plants regenerated from tissue culture is that of epigenetic origin. Epigenetics refers to modifications in gene expression brought about by heritable, but potentially reversible, changes in chromatin structure, and/or DNA methylation *(16)*. Although somewhat of a nuisance when trying to maintain genetic fidelity, the existence of culture-induced variability can also provide a source of potentially beneficial variation for use in new cultivar development *(17,18)*.

3. Gene Transfer Into Cultured Plant Cells and Tissues

3.1. Protoplasts: "Naked" Plant Cells

A protoplast is the component of an individual plant cell that is delimited by the plasma membrane. Although usually surrounded by individual cell walls, the development of technology to enzymatically remove this structural component allows the manipulation of isolated protoplasts in the absence of the physical and chemical barrier of the cell wall *(19)*. In many cases, isolated protoplasts can be induced to resynthesize cell walls and reinitiate developmental programs resulting in plant regeneration. A wide variety of methods developed to manipulate protoplasts representing a broad spectrum of plant species has spawned a whole new field of somatic-cell genetics.

Somatic hybridization combines complete genomes (including cytoplasmic genomes) by fusing protoplasts of different species and regenerating plants from the resulting fusion product. The diploid genome contributed by each cell parent usually results in a tetraploid product. As with sexual interspecific hybridization, reproductive fertility barriers await the regenerated plant material. Introgression of single characters into a cultivated crop can be facilitated by X-ray fragmentation of the donor genome, creating asymmetric hybrids, plus the ability to efficiently select the transferred trait. Somatic hybridization offers the unique possibility of generating new combinations of cytoplasmic genomes or traits. As a route to new cultivars, somatic hybridization presents the breeder with a high multiplicity of possible outcomes that are difficult to predict. Nonetheless, this method has been successfully used to introgress virus resistance from Solanum brevidens into the cultivated potato *(20)*.

Exogenous DNA containing cloned genes can be introduced directly into plant protoplasts. Polyethylene glycol can be used to permeabilize protoplasts, thereby allowing DNA uptake *(21)*. Plastid as well as nuclear transformation has been achieved by this method (*see* Chapter 11). Alternatively, DNA can be electroporated into the cells. The application of high-voltage DC pulses induces the formation of transient pores in cell membranes. Experimental protocols must be empirically optimized for the system, as electrical parameters including field strength and pulse duration appear to be highly specific to each cell type.

The transient expression of genes transferred to protoplasts has been used to establish optimal conditions for DNA uptake and to test vector components, including promoters, and coding regions *(22)*. Stable transformation generally results from the same conditions as transient expression, but at markedly lower frequencies, 10^{-3} to 10^{-5} per treated cell *(23)*. Although attempts have been made to improve the efficiency of stable transformation, protoplast-based methods remain highly variable in their effectiveness between labs as well as between species *(24)*.

The regeneration of fertile plants from isolated protoplasts remains a major barrier to the effective implementation of these methods in several groups of important crops. Although the technique has been particularly effective within the *Brassicaceae* and *Solanaceae (20)*, protoplast regeneration for most major crops such as legumes and grasses has proven to be technically challenging. There appears to be as much "art" as there is "science" associated with the development of successful methods for any crop species. Nonetheless, for species whose protoplasts are readily manipulated, this powerful technology can be exploited to transfer very large segments of DNA such as chromosomes and cytoplasmically inherited organelle DNA, as well as bacterial and yeast artificial chromosomes. However, even when available, methods to regenerate plants from isolated protoplasts are quite laborious.

3.2. Agrobacterium: The "Taming" of a Pathogen

Since it was first demonstrated 25 years ago that crown gall disease involves the transfer of a specific segment of DNA (T-DNA) from a bacterial plasmid to the plant nucleus *(25)*, a significant effort has been made to in convert the disease-causing organism, *Agrobacterium tumefaciens*, into a useful vector for plant genetic transformation. Within the *Agrobacterium* cell, a chemically responsive promoter system directs the expression of over 20 virulence genes that encode proteins needed to mobilize T-DNA. In addition, some of the virulence gene products are DNA-binding proteins containing plant nuclear-localization signals *(26)*. In this way, T-DNA is transferred to the plant cell's nucleus, where it can become integrated into the genome.

The conversion of the pathogen into a plant vector depended on modification of the virulence-causing plasmid system of *Agrobacterium*. Early dissection of the *Agrobacterium*-plant interaction showed that part of the disease process involved the transfer and subsequent expression of genes associated with the biosynthesis of plant-growth regulators, resulting in the characteristic disorganized crown-gall growth. Elimination of these genes from the vector DNA ("disarmed" vectors) was a first step to harnessing the T-DNA for gene transfer into plants. Further modification of the very large virulence plasmids provided ease of molecular manipulation, so that the gene of interest could be placed onto a smaller plasmid that was separate from the virulence functions ("binary" plasmid), or on a plasmid that subsequently recombined with the resident virulence plasmid ("cointegrate" plasmid).

The limited host range of naturally occurring *Agrobacterium* isolates was another early barrier to the broad use of this vector system. Screening of natural isolates for broader host range characteristics, plus manipulation of virulence determinants and plasmid configurations, led to the derivation of today's commonly used strains. Although initially restricted to dicotyledonous plants, the important monocot crops can now be transformed with Agrobacterium *(27,28)*. Recent studies have shown that *Agrobacterium* can transfer T-DNAs to yeast and filamentous fungi *(13,29,30)*, as well as to cultured human cells *(31)*.

The integration of the T-DNA into the host genome appears to be the result of illegitimate, non-homologous recombination, although there appears to be a preference for insertion into transcriptionally active chromosomal regions *(32,33)*. This process generates a range of DNA insertions with varying expression patterns, so that sorting of transgenic events for optimal transgene expression is a common step required in plant transformation protocols. T-DNA mediated transformation, with its protein-coated linear DNA intermediate, usually generates simpler insertion patterns with fewer rearrangements than other DNA delivery methods, and as such, has become the technology of choice for transgenic plant production.

As with most plant transformation protocols available today, Agrobacterium-mediated transformation typically depends on the ability to regenerate plants from a tissue culture that is capable of being infected and subsequently transformed. This means that the process relies on the identification of developmental pathways in which plant cells can be transformed at a practical frequency and identified by selection or screening. *Agrobacterium* must then be removed from the tissue by antibiotic treatment so that fertile transgenic plants can be recovered.

3.3. Direct DNA Delivery: "Shrapnel" Biology

Methods that allow DNA to be delivered into intact cells and tissues avoid the need to regenerate fertile plants from isolated protoplasts, and also effectively circumvent any host-range limitations that accompany the use of Agrobacterium. Microparticle bombardment and WHISKERS™ are examples of physical methods that are routinely used to stably transform intact plant cells and tissues *(34,35)*. Although different from each other in terms of the physical aspects of DNA delivery, both methods function by causing some degree of cellular injury, so that DNA enters cells and ultimately becomes integrated into the plant's genome.

Microparticle bombardment involves the acceleration to penetrating velocities of high-density, micron-sized particles coated with DNA *(35)*. These DNA-coated particles are sufficiently small (1–2 μm in diameter) to allow them to penetrate and be retained by plant cells without killing them. WHISKERS™ are silicon carbide microfibers, 10–80 μm long and 0.6 μm in diameter, capable of penetrating cell walls upon vigorous agitation *(34)*. Collisions between these sharp microfibers and plant cells cause cell-wall damage, thereby allowing DNA entry through either active or passive means. The challenge with both of these methods is to effectively deliver DNA to a large number of cells without causing enough damage to negatively impact survival.

The availability of receptive tissue that displays particular characteristics relative to morphology, resistance to physical stress, and proliferation capacity is a prerequisite to transgenic production via these direct DNA delivery methods. Microparticle bombardment and WHISKERS™ effectively deliver DNA only to surface cell layers *(36,37)*. Conditions that allow deeper penetration are usually not conducive to cell survival; therefore, tissue cultures with surface cells that are competent for DNA uptake and resistant to physical stress are a prerequisite. In addition, since integrative transformation is a rare event, transformed cells must be competent to proliferate in the presence of a selection agent to enable transgenic cultures to be established. Highly embryogenic callus and suspension cultures have been the most effective targets for direct DNA delivery because they contain cells that are accessible, selectable, and totipotent *(38)*.

4. Transgenic Crop Production

Tissue-culture-free transformation systems may one day be available for all crop species (*see* **Subheading 6.**). Until then, the most important factor in the determination of a plant species' relative transformability is the ease with which one can initiate and propagate cell cultures and subsequently recover fertile plants. The first successful work with transformation of plants relied upon the use of species that were members of a single family, the Solanaceae, which includes tobacco as well as the tomato and potato *(39)*. This was a function of the fact that tissue cultures could be readily established from these species and, perhaps more importantly, plants could be easily recovered from such cultures. In addition, these species were particularly susceptible to *Agrobacterium* infection. These two characteristics were combined to create an easy-to-use method for producing large numbers of transgenic plants.

Once beyond a few model species, the production of transgenic plants is a highly specialized and resource-intensive pursuit. Most commercially significant crops belong to families that are much more difficult to transform because tissue-culture initiation and plant regeneration are not trivial. Although no plant species should be considered a priori to fall outside the range of those amenable to transformation, the development and subsequent optimization of transgenic production protocols for a given crop are usually empirical activities, and as such, require a sustained commitment or resources. It is no coincidence that as recalcitrance to in vitro manipulation and plant regeneration from in vitro cultures breaks down, so do barriers to successful transgenic production.

5. Industrial Applications

5.1. Insect Resistance: Bacillus thuringiensis

Crop resistance to insect damage is an essential component of any integrated pest-management strategy. Breeding for insect resistance requires the identification of resistance genes that can be effectively crossed into commercial varieties. Although highly effective, conventional approaches are limited to genes that exist in the gene pools of crop plants and their wild relatives. The ability to produce transgenic plants has led to alternative approaches to the development of insect-resistant crop plants.

Genes from bacteria such as *Bacillus thuringiensis* (Bt) are among the most promising for use in the genetic manipulation of crops for insect resistance *(40,41)*. The first Bt insecticidal gene was isolated in 1981 *(42)*. Since then, several crop species have been genetically engineered to control target insect pests including maize, cotton, potato, tobacco, rice, broccoli, lettuce, walnuts, apples, alfalfa, and soybeans *(43–45)*.

Bt genes, referred to as "cry" genes after the crystalline nature of the encoded protein toxin, are classified according to insect specificity and sequence homology *(46,47)*. These crystalline proteins are inactive until solubilized by insect-gut proteases, at which time they become toxic to midgut function *(41,48,49)*. In the Bt δ-endotoxins family of proteins, 140 genes have been described that are toxic to *Lepidoptera*, *Coleoptera*, and *Diptera* *(50)*.

The first transgenic plants expressing Bt involved a full-length cry gene *(51–53)*. The expression was rather weak in these plants, resulting in only 20% mortality of the tobacco hornworm (*Manduca sexta*) larvae. Later, truncated Bt genes, which encoded for the toxic N-terminal fragment, provided better protection. Subsequently rebuilding Bt genes to have more favorable codon bias, the removal of mRNA instability sequences, and the use of strong genetic regulatory elements has resulted in the accumulation of higher levels of cry protein and the generation of highly efficacious transgenic plants.

Cotton plants transformed with Bt displayed total protection against *Trichoplusia ni*, *Spodoptera exigua*, and *Heliothis zea* *(42)*. Bt protein accumulated to 0.1% of the total soluble protein in these plants. Transgenic rice expressed Bt to nearly 0.05% of the total soluble leaf protein, and conferred resistance to the rice-leaf folder (*Cnaphalocrosis medinalis*) and yellow stem borer (*Chilo suppressalis*) *(54)*. Transgenic maize-expressing cry proteins have shown that they are highly effective against the European corn borer (*Ostrinia nubilalis*) and can withstand up to 50 larvae per plant at the whorl leaf stage and about 300 larvae at the anthesis stage *(55)*. Resistance to corn root worm has also been achieved by expressing two separate cry genes *(56)*. Transgenic sugarcane plants showed significant activity against neonate larvae of sugarcane borer (*D. saccharalis*), despite low expression of cry protein. Transformed eggplants showed significant activity against fruit borer (*Leucinodes orbonalis*) larvae *(57)*. Synthetic cry genes have been expressed in tobacco and potato plants for the control of Colorado potato beetle (*Leptinotarsa decemlineata*) *(58)* and broccoli (*Brassica oleracea* ssp. *italica*) to control the diamond back moth (*Plutella xylostella*) *(42)*.

Bt has been an extremely rich source of insect-resistance genes. Transgenic plants that express cry genes will no doubt figure prominently in future pest-management practices. However, further studies are needed on the environmental impact of large-scale production of transgenic crops, particularly as it relates to nontarget insects.

5.2. Herbicide Resistance: Glyphosate Tolerance

Glyphosate is a top-selling, nonselective herbicide. Unlike most herbicides, it has very limited selectivity to crops and very short residual activity. Glyphosate kills plants by inhibiting 5-enolpyruvyl-shikimate-3-phosphate synthase (EPSPS), an early enzyme in aromatic amino-acid biosynthesis. This biosynthetic pathway is present in bacteria, fungi, and plants, but not in animals or insects. The comparative biochemistry of its mode of action thus confers an inherent safety margin.

In early work directed at the development of glyphosate-tolerant transgenic plants, mutant EPSP synthases with decreased glyphosate sensitivity were overproduced and directed to the plastid with targeting sequences *(59,60)*. When this approach yielded insufficient resistance, bacterial collections were surveyed in an attempt to find the most resistant naturally occurring variant EPSPS. The enzyme with the lowest K_m for the true substrate, phosphoenolpyruvate, and the highest K_i for the inhibitor, glyphosate, would display the most enzyme activity, yet bind the herbicide most poorly. EPSPS from an *Agrobacterium* strain performed most reliably in transgenic plants. When EPSPS was targeted to the plastid by a transit peptide sequence, transgenic soybean lines were produced that proved resistant to field treatments with the herbicide *(61)*.

Approval to market glyphosate-tolerant transgenic soybeans required extensive safety testing *(62)*. Compositional analysis showed the new variety to be essentially equivalent to the nontransgenic progenitor. Feeding studies on four species of animals demonstrated feed equivalence. Soybean allergens and desirable phytoestrogens were not altered in the transgenic derivative. Methods were developed to detect the genetically engineered seeds in mixtures. Yields were shown to be equivalent or better than nontransgenic soybeans, with traditional weed control in most trials *(63–66)*.

Until the introduction of genetically engineered glyphosate-tolerant soybeans, application of this herbicide in crop production was limited to preplant "burn-down" treatments for no-till crops. Since their introduction in 1995, glyphosate-tolerant soybeans have quickly taken a dominant position in the market, reaching an estimated 50% of the total US soybean acreage in 1999 *(64)*. The market for glyphosate-tolerant crops continues to expand. Glyphosate-tolerant corn and cotton are making significant market inroads, and the economics of these cropping systems appear to be generally favorable. Sugarbeet and canola have also been transformed to be glyphosate-tolerant. As the economic value of the new cropping systems utilizing herbicide-tolerant seeds becomes generally recognized and accepted, new genes for glyphosate tolerance continue to be identified from microbial sources *(67)*.

5.3. Nutritional Enhancement: Golden Rice

The development of crops with enhanced nutritional characteristics can have a positive impact on human health, particularly in areas suffering from extreme poverty. Milled rice is naturally low in the orange-yellow pigment β-carotene, which mammals convert into the essential nutrient, vitamin A. Vitamin A deficiency is a major problem in areas in which rice is the main food. In Southeast Asia, 70% of children under the age of five suffer from vitamin A deficiency, which can lead to vision impairment and increased susceptibility to disease *(68)*. Providing the β-carotene in the grain would be an effective approach to alleviating this problem.

β-carotene is an accessory pigment that is essential to photosynthesis. It is synthesized from the isoprenoid pathway in all green tissue. Although rice seed does not contain much β-carotene, a precursor, geranylgeranyl diphosphate (GGPP) is present. To synthesize β-carotene in rice seed, transgenic plants were generated that express two additional genes in the endosperm *(69)*. A phytoene desaturase isolated from daffodil and a bacterial phytoene desaturase from *Erwinia uredovora* under the control of an endosperm-specific glutelin and cauliflower mosaic virus 35S promoter, respectively, were transformed into rice-tissue cultures. Plants regenerated from these cultures produced golden-colored seed that accumulated up to 1.6 mg/g β-carotene *(94)*. These transgenic materials should provide the foundation for the development of highly nutritious rice varieties.

5.4. High-Value Protein: Avidin

Avidin is an antibacterial protein purified from chicken egg whites. This 66-Kd tetrameric, glycosylated protein binds the vitamin/coenzyme biotin with very high affinity. The ultra-strong binding of biotin by avidin has been exploited in numerous biochemical purification protocols and diagnostic tests. The market for avidin as a biochemical reagent has led to its transgenic expression in maize for commercial purification and sale *(70–72)*.

Codon-adapted, synthetic versions of the barley α-amylase-leader sequence and chicken egg-white avidin-coding sequence, each driven by the maize ubiquitin-1 promoter, were used to generate transgenic maize plants. The barley α-amylase leader sequence was included to direct expressed avidin to the endoplasmic reticulum (ER) for glycosylation, and from there to the extracellular spaces, to avoid potential negative phenotypic effects from accumulation in the cytoplasm *(70)*. The ubiquitin promoter provided strong expression in vegetative tissues. High-expressing lines selected for leaf expression were tested for seed expression. Segregating seeds on the ear were found to express high levels of avidin—2–5% of the aqueous extracted protein.

Avidin is isolated from transgenic seed by grinding or flaking, aqueous extraction, and affinity purification on biotinylated columns *(72)*. Avidin produced by this method is equivalent to avidin purified from chicken eggs *(70,71)*. The N-terminal sequence is the same, showing that the α-amylase leader sequence was correctly processed in the transgenic maize. Chicken- and maize-produced avidin demonstrate very similar biotin-affinity constants. The glycosylation of avidin from these two sources is different, however, resulting in a lower apparent mol wt for the maize homotetramer. Although correct post-translational glycosylation is cited by many as a potential limiting factor in plant expression of biopharmaceutical agents *(73)*, it is apparently not a major contributor to the biochemical functionality of avidin.

Production of avidin in maize offers several benefits that apply to the production of many other transgenic proteins in maize seed. First, maize seed as a production vehicle provides well-established cultivation and harvest practices. The 300% increase in seed mass per growing season facilitates rapid scale-up of selected transgenic lines. Processing and purification of transgenic avidin from maize seed benefits from the low protease load in the seed as well as the differential solubility of most maize storage proteins (the majority of protein stored in the endosperm is not extracted in aqueous buffers). When the ease of extraction is coupled to the relative stability of avidin, and the opportunity for affinity-chromatography purification, a relatively simple purification process can be designed.

Using transgenic maize, a kilogram of avidin could potentially be purified from the seed yield of an acre crop. Avidin from transgenic seed would be produced at about 1/50th the cost, and would be from about 1/50th the mass of chicken eggs that would have been required to produce the equivalent amount of native avidin. The higher titer of avidin in transgenic maize seed may offer a less expensive purification protocol. Although these cost estimates are highly favorable, avidin as a biochemical reagent may have a limited total market size. Transgenic plant-derived avidin demonstrates the potential for high quantitative yield of feedstocks, and provides the first example of commercial sales of biochemicals produced in transgenic plants.

6. The Future: Tissue-Culture-Free Systems

6.1. Arabidopsis: The "Perfect" Subject

This diminutive weed from the *Brassicaceae* was proposed as a subject for genetic studies because of its small genome size and its low proportion of repetitive DNA *(74–76)*. In the fifteen years since its adoption by plant molecular biologists, its contribution to plant genetic research has been strongly enhanced by the recent development of methods for *in planta* transformation. This convenient transformation protocol is doubly powerful, now that it is coupled with the recently published *Arabidopsis* genome sequence *(77)*.

Tissue-culture methods for this species were never particularly efficient or powerful. By the late 1980s, transformation of *Arabidopsis* could be accomplished by *Agrobacterium* co-cultivation of leaf discs *(78)* or root pieces *(79)*. A protocol also became available for direct DNA delivery to *Arabidopsis* protoplasts *(80)*. However, to achieve the most powerful genetic analysis, new methods were needed to clone *Arabidopsis* genes based on phenotype (reverse genetics), to mutagenize a given DNA and observe its phenotype (forward genetics), and to confirm these assignments by complementation with transformed genes *(75,81)*. These approaches became possible with the advent of effective in planta *Arabidopsis* transformation.

The first reported *in planta* transformation used a seed or seedling treatment with *Agrobacterium (82)*. Now a highly convenient method is available using a floral dip into an *Agrobacterium* culture *(81,83,84)*. Approximately 1% of seed collected following such a treatment is transformed. The mechanism of transformation with this protocol apparently involves *Agrobacterium* colonization of the *Arabidopsis* influorescence prior to closure of individual ovules *(81)*. The female gametophyte is transformed, and multiple independent individual transformation events are produced *(83)*. Using this method, the time required to produce transgenic plants is shorter, as no regeneration from tissue culture is required. The range of techniques currently employed in *Arabidopsis* now includes: testing genes of interest *in planta*; testing of regulatory elements; map-based cloning of mutants and their validation by complementation analysis; insertional mutagenesis programs generating collections of sequence-tagged insertion mutants; and activation tagging. Following the demonstration in *Arabidopsis* of *in planta* transformation, similar methods have been found to be effective in other species, such as *Pakchoi (85)* and *Medicago truncatula (86)*. Although this method represents an almost ideal transgenic production system, only the future will show how far it can be extended.

6.2. The Pollen-Tube Pathway

Although the floral dip method is attractive, it may not be applicable to all species. A related method that shows great promise for tissue-culture-free transgenic production is the pollen-tube pathway. Attempts to exploit the pollination-fertilization process as a means of plant genetic engineering started in the 1970s *(87)*. The underlying concept for this work was based on the assumption that pollen could be either transformed per se or carry DNA to the embryo sac for subsequent transformation. DNA was either mixed directly with pollen prior to fertilization *(88,89)* or applied directly to the floral structures before or after pollination *(90)*. It is hard to imagine an easier, more rapid transformation method.

The earliest "successes" with this method involved "transformation" with exogenous genomic DNA extracted from a donor species and observed "variant" offspring in recipient species *(88–90)*. These early studies lacked unequivocal transformation, confirmed by integration of clearly defined DNA segments. Once plasmid vectors containing clearly identifiable reporter genes were available, integrative transformation (as measured by Southern analysis) was reported for cotton *(91)*, wheat *(92)*, and rice; however, independent confirmation of integrative transformation, including segregation of the introduced transgene beyond the first generation, could not be obtained *(93)*.

Although the pollen-tube pathway method has been studied for over 15 years, it still is not generally accepted as a viable means of transgenic production outside of China. Moreover, little systematic research has been conducted in the areas of procedural optimization and/or mechanisms of DNA entry *(94)*. However, in light of the potential for overcoming some of the most significant barriers to transgenic production (such as avoidance of tissue culture, genotype independence, and speed), full evaluation of the pollen-tube pathway should be pursued with a high degree of rigor.

7. Conclusion

Agricultural biotechnology depends on the ability to generate transgenic plants. A prerequisite to the production of transgenic plants in most species, regardless of the DNA delivery method used, is the ability to initiate and maintain cell and tissue cultures that are capable of regenerating fertile plants. Experimental totipotency and the subsequent optimization of in vitro methodologies for a given species are usually empirical activities that require a sustained commitment to acquire the necessary "feeling for the organism." This has been accomplished, at least in some genotypes, for most major crop species.

Most of the difficulties with generating large numbers of transgenic events in a short period of time for gene testing and product development in crop plants relate to the dependence on in vitro cultures. It is anticipated that future transgenic production methods will be less dependent on cell and tissue cultures. The Arabidopsis floral dip method provides a glimpse of what might be possible. It is not a coincidence that most of the cutting-edge research in plant developmental biology and functional genomics is being performed with this species.

References

1. Fraley, R. B., Rogers, S. G., Horsch, R. B., Sanders, P. R., Flick, J. S., Adams, S. P., et al. (1983) Expression of bacterial genes in plant cells. *Proc. Natl. Acad. Sci. USA* **80**, 4803–4807.
2. Zambryski, P., Joss, H., Genetello, C., Leemans, J., Van Montagu, M., and Schell, J. (1983) Ti plamid vector for the introduction of DNA into plant cells without alteration of their normal regeneration capacity. *EMBO J.* **2**, 2143–2150.
3. McElroy, D. (1996) The industrialization of plant transformation. *Nat. Biotechnol.* **14**, 715,716.

4. Schleiden M. J. (1838) Beitrage zur Phytogenesis. *Archiv fuer Anatomie und Physiologie in Verbindung mit mehreren gelehrten herausgegeben von Johannes Mueller,* Verlag von veit ET COMP. Berlin, pp. 137– 176.

5. Schwann T. (1839) Mikroskopishe Untersuchungen uber die Ubereinstimmung in der Struktur unddem Wachstun der Tiere und Pflanzen, (Smith, H., transl.) London, Sydenham Society.

6. Haberlandt, G, (1902) Culturversuche mit isolierten Pflanzenzellen. *Sitzungsber. Kais. Akad. Wiss.—Math. Naturw. Klasse, Wien Bd.* **CXI**, 69–92.

7. Dodds, J. H. and Roberts, L. W. (1987) *Experiments in Plant Tissue Culture,* 2nd ed., Cambridge University Press, New York, NY, pp. 101–113.

8. Bhojwani, S. S. and Razdan, M. K. (1999) Cellular totipotency, in *Plant Tissue Culture: Theory and Practice,* Elsevier, NY, pp. 95–123.

9. Fehr, W. R. (1987) Principles of Culture Development, MacMillan Publishing Co., New York, pp. 533–576.

10. Takayama, S. and Akita, M. (1994) The type of bioreactors used for shoots and embryos. *Plant Cell Tissue Organ Cult.* **39,** 147–156.

11. Vasil, I. K. (1994) Automation of plant propagation. *Plant Cell Tissue Organ Cult.* **39,** 105–108.

12. Sudha Vani, A. K. and Reddy, G. M. (1995) Micropropagation of banana through synseed technology. *In Vitro Cell. Dev. Biol.* **31,** 57A.

13. Xavier, J. L., Karine L, and Branchard, M. (2000) Plant genomic instability detected by microsatellite-primers. Vol. 3 No. 2, Issue of August 15, 2000. *EJB Electronic Journal of Biotechnology.* http://www.ejb.org/content/vol3/issue2/full/2/bip/

14. Larkin, P. J. and Scowcroft, W. R. (1981) Somaclonal variation - a novel source of variability from cell cultures. *Theor. App. Genet.* **67,** 197–201.

15. Phillips, R. L., Kaeppler, S. M., and Olhoft, P. (1994) Genetic instability of plant tissue cultures: breakdown of normal controls. *Proc. Natl. Acad. Sci. USA* **91,** 5222–5226.

16. Henikoff, S. and Matzke, M. A. . (1997) Exploring and explaining epigenetic effects. *Trends Genet.* **13,** 293–295.

17. Evans, D. A. (1989) Somaclonal variation: Genetic basis and breeding applications. *Trends Genet.* **5,** 46–50.

18. Reichert, N. A. and Baldwin, B. S. (1996) Potential for kenaf improvement via somaclonal variation, in *Progress in New Crops,* (Janick, J., ed.), ASHS Press, Arlington, VA, pp. 408–411.

19. Cocking, E. C. (1960) A method for the isolation of plant protoplasts and vacuoles. *Nature* **187,** 962,963.

20. Waara, S. and Glimelius, K. (1995) The potential of somatic hybridization in crop breeding. *Euphytica* **85,** 217–233.

21. Hinchee, M. A., Corbin, D. R., Armstrong, C. L., Fry, J. E., Sato, S. S., DeBoer, D. L., et al. (1994) Plant Transformation, in *Plant Cell and Tissue Culture,* (Vasil, I. K. and Thorpe, T. A., eds.), Kluwer, The Netherlands, pp. 231–270.

22. Dehneke, J., Gossele, V., Bottermann, J., and Cornelissen, M. (1989) Quantitative analysis of transiently expressed genes in plant cells. *Methods Mol. Cell. Biol.* **7,** 725–737.

23. Lyznik, L. A., Ryan, R. D., Ritchie, S. W., and Hodges, T. K. (1989) Stable co-transformation of maize protoplasts with gusA and neo genes. *Plant Mol. Biol.* **13,** 151–161.

24. Benediktsson, I., Spampinato, C., and Schieder, O. (1995) Studies of the mechanism of transgene integration into plant protoplasts: improvement of the transformation rate. *Euphytica* **85,** 53–61.

25. Chilton, M. D., Drummond, D. M., Merlo, D. J., Sciaky, D., Montoya, A. L., Gordon, M. P., et al. (1977) Stable incorporation of plasmid DNA into higher plant cells: the molecular basis of crown gall tumorigenesis. *Cell* **11,** 263.

26. Gelvin, S. (2000) Agrobacterium and plant genes involved in T-DNA transfer and integration. *Annu. Rev. Plant Physiol. Plant Molec. Biol.* **51,** 223–256.

27. Hiei, Y., Ohta S., Komari, T., and T. Kumashiro. (1994) Efficient transformation of rice (Oryza sativa L.) mediated by *Agrobacterium* and sequence analysis of the boundaries of the T-DNA. *Plant J.* **6,** 271–282.

28. Ishida, Y., Saito, H., Ohta, S., Hiei, Y., Komari, T., and Kumashiro, T. (1996) High efficiency transformation of maize (*Zea mays L.*) mediated by *Agrobacterium tumefaciens. Nat. Biotechnol.* **14,** 745–750.

29. Bundock, P. and Hooykaas, P. (1996) Integration of *Agrobacterium tumefaciens* T-DNA in the *Saccharomyces cerevisiae* genome by illegitimate recombination. *Proc. Natl. Acad. Sci. USA* **93,** 15,272–15,275.

30. de Groot, M. et al. (1998) *Agrobacterium tumefaciens*-mediated transformation of filamentous fungi. *Nat. Biotechnol.* **16,** 839–842.

31. Kunik, T., Tzfira, T., Kapulnik, Y., Gafni, Y., Dingwall, C., and Citovsky, V. (2001) Genetic transformation of HeLa cells by *Agrobacterium. Proc. Natl. Acad. Sci. USA* **98,** 1871–1876.

32. Koncz, C., Martini, N., Mayerhofer, R., Koncz-Kalman, Z., Korber, H., Redei, G., et al. (1989) High frequency T-DNA mediated gene tagging in plants. *Proc. Natl. Acad. Sci. USA* **86,** 8467–8471.

33. Koncz, C., Nemeth, N., Redei, G. P., and Schell, J. (1992) T-DNA insertional mutagenesis in *Arabidopsis. Plant Mol. Biol.* **20,** 963–976.

34. Kaeppler, H. F., Gu, W., Somers, D. A., Rines, H. W., and Cockburn, A. F. (1990) Silicon carbide fiber-mediated DNA delivery into plant cells. *Plant Cell Rep.* **8,** 415–418.

35. Sanford, J. (1988) The biolistic process. *Trends Biotechnol.* **6,** 299–302.

36. Pareddy, D., Petolino, J., Skokut, T., Hopkins, N., Miller, M., Welter, M., et al. (1997) Maize transformation via helium blasting. *Maydica* **42,** 143–154.

37. Serik, O., Ainur, I., Murat, K., Tetsuo, M., and Masaki, I. (1996) Silicon carbide fiber-mediated DNA delivery into cells of wheat (*Triticum aestivum L.*) mature embryos. *Plant Cell Rep.* **16,** 133–136.

38. Petolino, J. F. (2001) Direct DNA delivery into intact cells and tissues, in *Transgenic Plants and Crops*, (Hui, Y. H., Khachatourians, G. G., McHughen, A., Nip, W. K., and Scorza, R., eds.), Marcel Dekker, Inc., New York, NY, in press.

39. Uchimiya, H., Handa, T., and Brar, D. S. (1989) Transgenic plants. *J. Biotechnol.* **12,** 1–20.

40. Charles, J. F., Nielsen-LeRoux, C., and Delecluse, A. (1996) *Bacillus sphaericus* toxins: molecular biology and mode of action. *Annu. Rev. Entomol.* **41,** 451–472.

41. Gill, S. S., Cowles, E. A., and Pietrantonio, F. V. (1992) The mode of action of *Bacillus thuringiensis* endotoxins. *Ann. Rev. Entomol.* **37,** 615–636.

42. Sharma, C. H., Sharma, K. K., Seetharama, N., and Ortiz, R. (2000) Prospects for using transgenic resistance to insects in crop improvement. *EJB Electronic Journal of Biotechnology.* Vol. 3 No. 2, Issue of August 15. http://www.ejb.org/content/vol3/issue2/full/3/index.html

43. Bennett, J. (1994) DNA-based techniques for control of rice insects and diseases: transformation, gene tagging and DNA fingerprinting, in *Rice Pest Science and Management*, (Teng, P. S., Heong, K. L., and Moody, K., eds.), International Rice Research Institute, Los Banos, Philippines, pp. 147–172.

44. Federici, B. A. (1998) Broad-scale leaf pest-killing plants to be true test. *Calif. Agri.* **52,**14–20.

45. Griffiths, W. (1998) Will genetically modified crops replace agrochemicals in modern agriculture? *Pesticide Outlook* **9,** 6–8.

46. Hoftey, H. and Whiteley, H. R. (1989) Insecticidal crystal proteins of *Bacillus thuringiensis. Microbiology Rev.* **53,** 242–255.

47. Tailor, R., Tippett, J., Gibb, G., Pells, S., Pike, D., Jordan, L., et al. (1992) Identification and characterisation of a novel *Bacillus thuringiensis*-endotoxin entomocidal to coleopteran and lepidopteran larvae. *Mol. Microbiol.* **7,** 1211–1217.

48. Milne, R. and Kaplan, H. (1993) Purification and characterisation of a trypsin like digestive enzyme from spruce budworm (*Christoneura fumiferana*) responsible for the activation of d-endotoxin from *Bacillus thuringiensis. Insect Biochem. Mol. Biol.* **23,** 663–673.

49. Tojo, A. and Aizawa, K. (1983) Dissolution and degradation of endotoxin by gut juice protease of silkworm, Bombyx mori. *Appl. Environ. Microbiol.* **45,** 576–580.

50. Crickmore, N., Ziegler, D. R., Fietelson, J., Schnepf, E., Van Rie, J., Lereclus, D., et al. (1998) Revision of the nomenclature for *Bacillus thuringiensis* pesticidal crystal proteins. *Microbiol. Mol. Biol. Rev.* **62,** 807–813.

51. Barton, K., Whitely, H., and Yang, N. S. (1987) *Bacillus thuringiensis* d-endotoxin in transgenic Nicotiana tabacum provides resistance to lepidopteran insects. *Plant Physiol.* **85,** 1103–1109.

52. Fischhoff, D. A., Bowdish, K. S., Perlak, F. J., Marrone, P. G., McCormick, S. M., Niedermeyer, J. G., et al. (1993) Insect resistant rice generated by a modified delta endotoxin genes of *Bacillus thuringiensis. BioTechnology* **11,** 1151–1155.

53. Vaeck, M., Reynaerts, A., Hoftey, H., Jansens, S., DeBeuckleer, M., Dean, C., et al. (1987) Transgenic plants protected from insect attack. *Nature* **327,** 33–37.

54. Fujimoto, H., Itoh, K., Yamamoto, M., Kayozuka, J., and Shimamoto, K. (1993) Insect resistant rice generated by a modified delta endotoxin genes of *Bacillus thuringiensis. BioTechnology* **11,** 1151–1155.

55. Armstrong, C. L., Parker, G. B., Pershing., J. C., Brown, S. M., Sanders, P. R., Duncan, D. R., et al. (1995. Field evaluation of European corn borer control in progeny of 173 transgenic corn events expressing an insecticidal protein from *Bacillus thuringiensis. Crop Sci.* **35,** 550–557.

56. Moellenbeck, D. J., Peters, M. L., Bing, J. W., Rouse, J. R., Higgins, L. S., Sims, L., et al. (2001) Insecticidal proteins from *Bacillus thuringiensis* protect corn from corn rootworms. *Nat. Biotechnol.* **19,** 668–672.

57. Kumar, P. A., Mandaokar, A., Sreenivasu, K., Chakrabarti, S. K., Bisaria, S., Sharma, S. R., et al. (1998) Insect-resistant transgenic brinjal plants. *Mol. Breed.* **4,** 33–37.

58. Perlak, F. J., Stone, T. B., Muskopf, Y. N., Petersen, L. J., Parker, G. B., McPherson, S. A., et al. (1993) Genetically improved potatoes: protection from damage by Colorado potato beetles. *Plant Mol. Biol.* **22,** 313–321.

59. Della-Cioppa, G., Bauer, S. C., Taylor, M. L., Rochester, D. E., Klein, B. K., Shah, D. M., et al. (1987) Targeting a herbicide-resistant enzyme from *Escherichia coli* to chloroplasts of higher plants. *BioTechnology* **5,** 597–584.

60. Thompson, G. A., Hiatt, W. R., Facciotti, D., Stalker, D. M., and Comai, L. (1987) Expression in plants of a bacterial gene coding for glyphosate resistance. *Weed Sci.* **351,** 19–23.

61. Re, D., Padgette, S., Delannay, X., LaVallee, B., Eichholtz, D., Barry, G., et al. (1992) Characterization of EPSPS enzymes and their use in the production of Roundup™ tolerant soybean. *Miami Short Rep.* **2,** 77.

62. Fuchs, R. L., Re, D. B., Rogers, S. G., Hammond, B. G., and Padgette, S. R. (1994) Commercialization of soybeans with the Roundup Ready™ gene, in *The Biosafety Results of Field Tests of Genetically Modified Plants and Microorganisms*, (Jones, D. D., ed.), Proc. 3rd Intl. Symp., Monterey, CA, Division of Agriculture and Natural Resources, University of California, pp. 233–244.

63. Ateh, C. and Harvey, R.. (1999) Annual weed control by glyphosate in glyphosate-resistant soybean. *Weed Technol.* **13,** 394–398.

64. Baldwin, F. (1999) The value and exploitation of herbicide-tolerant crops in the US. *Brighton Crop Protection Conference: Weeds. Proc. Intl. Conf.* Brighton, UK **2,** 653–660.

65. McKinley, T. L., Roberts, R. K., Hayes, R. M., and English, B. C. (1999) Economic comparison of herbicides for johnsongrass (*Sorghum halepense*) control in glyphosate-tolerant soybean (Glycine max) *Weed Technol.* **13,** 30–36.

66. Webster, E. P., Bryant, K. J., and Earnest, L. D. (1999) Weed control and economics in nontransgenic and glyphosate-resistant soybean (*Glycine max*) *Weed Technol.* **13,** 586–593.

67. Penaloza-Vazquez, A., Mena, G. L., Oropeza, A., and Bailey, A. M. (1997) The genes involved in glyphosate utilization by *Pseudomonas pseudomallei* and the tolerance conferred to plants. *Dev. Plant Pathol.* **9,** 417–423.

68. Guerinot, M. L. (2000) The green revolution strikes gold. *Science* **287,** 241–243.

69. Ye, X., Al-Babili, S., Kloti, A., Zhang, J., Lucca, P., Beyer, P., et al. (2000) Engineering the provitamin A (β-carotene) biosynthetic pathway into (carotenoid-free) rice endosperm. *Science* **287,** 303–305.

70. Hood, E. E., Kusnadi, A., Nikolov, Z., and Howard, J. A. (1999) Molecular farming of industrial proteins from transgenic maize. *Adv. Exp. Med. Biol.* **464,** 127–147.

71. Hood, E. E., Witcher, D. R., Maddock, S., Meyer, T., Baszczynski, C., Bailey, M., et al. (1997) Commercial production of avidin from transgenic maize: characterization of transformant, production, processing, extraction and purification. *Mol. Breed.* **3,** 291–306.

72. Kusnadi, A., Hood, E., Witcher, D., Howard, J., and Nikolov, Z. . (1998) Production and purification of two recombinant proteins from transgenic corn. *Biotechnol. Progr.* **14,** 149–155.

73. Bakker, H., Bardo, M., Moldhoff, J., Gomord, V., Elbers, I., Stevens, L. H., et al. (2001) Galactose-extended glycans of antibodies produced by transgenic plants. *Proc. Nat. Acad. Sci. USA* **98,** 2899–2904.

74. Meyerowitz, E. (1987) *Arabidopsis thaliana. Annu. Rev. Genet.* **21,** 93–111.

75. Meyerowitz, E. and Pruitt, R. (1985) *Arabidopsis thaliana* and plant molecular genetics. *Science* **229,** 1214–1218.

76. Pang, P. and Meyerowitz, E. (1987) *Arabidopsis thaliana*: a model system for plant molecular biology. *BioTechnology* **5,** 1177–1181.

77. Arabidopsis Genome Initiative. (2000) Analysis of the genome sequence of the flowering plant *Arabidopsis thaliana. Nature* **408,** 796–815.

78. Lloyd, A. M., Barnason, A. R., Rogers, S. G., Byrne, M. C., Fraley, R. T., and Horsch, R. B. (1986) Transformation of *Arabidopsis thaliana* with *Agrobacterium tumefaciens. Science* **234,** 464–466.

79. Valvekens, D., Van Montagu, M., and Van Lijsebettens, M. (1988) *Agrobacterium tumefaciens*-mediated transformation of *Arabidopsis thaliana* root explants by using kanamycin selection. *Proc. Natl. Acad. Sci. USA* **85,** 5536–5540.

80. Bilang, R. and Potrykus, I. (1993) Transformation in *Arabidopsis*, in *Biotechnology in Agriculture and Forestry*, Vol. 22, Plant Protoplasts and Genetic Engineering III, (Bajaj, Y. P. S., ed.), Springer-Verlag, Heidelberg, Germany, pp.123–134.

81. Bent, A. F. (2000) *Arabidopsis* in planta transformation: uses, mechanisms, and prospects for transformation of other species. *Plant Physiol.* **124,** 1540–1547.

82. Feldman, K. and Marks, M. (1987) *Agrobacterium*-mediated transformation of geminating seeds of *Arabidopsis thaliana*: a non-tissue culture approach. *Molec. Gen. Genet.* **208,** 1–9.

83. Bechtold, N., Jaudeau, B., Jolivet, S., Maba, B., Vezon, D., Voisin, R., et al. (2000) The maternal chromosome set is the target of the T-DNA in the in planta transformation of *Arabidopsis thaliana. Genetics* **155,** 1875–1887.

84. Clough, S. and Bent, A. (1998) Floral dip: a simplified method for *Agrobacterium*-mediated transformation of *Arabidopsis thaliana. Plant J.* **16,** 735–743.

85. Liu, F., Cao, M. Q., Yao, L., Li, Y., Robaglia, C., and Tourneur, C. (1998) In planta transformation of pakchoi (*Brassica campestris L.* ssp. *Chinensis*) by infiltration of adult plants with *Agrobacterium. Acta Hortic.* **467,** 187–192.

86. Trieu, A. T., Burleigh, S. H., Kardailsky, I. V., Maldonado-Mendoza, I. E., Versaw, W. K., Blaylock, L. A., et al. (2000) Transformation of *Medicago truncatula* via infiltration of seedlings or flowering plants with *Agrobacterium. Plant J.* **22,** 531–541.

87. Hess, D. (1987) Pollen-based techniques in genetic manipulation, in *International Review of Cytology*, (Giles, K. L. and Prakash, J., eds.), Academic Press, Orlando, FL, pp. 367–395.

88. DeWet, J. M. J., DeWet, A. E., Brink, D. E., Hepburn, A. G., and Woods, J. H. (1986) Gametophyte transformation in maize, in *Biotechnology and Ecology of Pollen* (Mulcahy, D. C., Bergamini-Mulcahy, G., and Ottaviano, E., eds.), Springer-Verlag, New York, NY, pp. 59–64.

89. Ohta, Y. (1986) High efficiency genetic transformation of maize by a mixture of pollen and exogenous DNA. *Proc. Natl. Acad. Sci. USA* **83,** 715–719.

90. Zhou, G. Y., Weng, J., Zeng, Y., Huang, J., Qiean, S., and Liu, G. (1983) Introduction of exogenous DNA into cotton embryos, in *Methods in Enzymology*, (Wu, R., Grossman, L., and Moldave, K., eds.), Academic Press, New York, NY, pp. 433–481.

91. Jian, W., Shen, W. F., Wang, Z. F., Chen, K. Q., Yang, W. X., Gohg, Z. Z., et al. (1984) A molecular demonstration of the introduction into cotton embryos of exogenous DNA. *Acta. Biochem. Biophys. Sin.* **16,** 325–327.

92. Picard, E., Jacquemin, J. M., Granier, F., Bobin, M., and Forgeois, P. (1988) Genetic transformation of wheat by plasmid DNA uptake during pollen tube germination. 7th International Wheat Genetics Symposium, Cambridge University Press, Cambridge, UK, pp. 779–787.

93. Langridge, P., Brettschneider, R., Lasseri, P., and Lorz, H. (1992) Transformation of cereals via Agrobacterium and the pollen pathway: a critical assessment. *Plant J.* **2,** 631–638.

94. Hu, C. Y. and Wang, L. (1999) In planta soybean transformation technologies developed in China: procedure, confirmation, and field performance. *In Vitro Cell. Dev. Biol.* **35,** 417–420.

11

Expression of Recombinant Proteins via the Plastid Genome

Jeffrey M. Staub

1. Introduction

Genetic engineering of plant cells was first accomplished more than 20 years ago. Since this time, nuclear transgenic plants have been used for everything from the study of gene function to the use of plants as production vehicles for industrial enzymes. However, plant cells contain two additional genetic compartments—plastids and mitochondria—for which transgenic technology could provide exciting alternative approaches over existing methods of nuclear transformation. Although mitochondrial transformation has not yet been achieved, the ability to incorporate transgenes of interest directly into the plastid genome of higher plants via transformation became a reality in the early 1990s *(1,2)*, following similar success in the unicellular green alga *Chlamydomonas reinhardtii (3)*. Compared to nuclear transformation, several potential advantages for protein engineering via plastid transformation may exist (Table 1). These include the ability to accumulate extraordinarily high levels of recombinant proteins, in some cases up to 30-40% of the total protein in a cell. In contrast to the random integration of transgenes observed in nuclear transformation, in plastid transformation, transgene integration into the plastid genome is exclusively directed by homologous recombination. Therefore, all events derived from a single transformation vector have predictable insert quality and uniform gene-expression characteristics. Also, in contrast to nuclear transformation, transgene silencing has not been observed in plastids, so subsequent transgenic plant generations maintain faithful expression of the genes of interest. Importantly, plastids are maternally inherited in most crop plants and only very rarely transmitted through pollen *(4,5)*, therefore, transgenes are restricted to the female parent and maternally-derived progeny.

Despite the numerous advantages, plastid transformation has only recently been used to test proof of concept for potentially commercial applications. Along the way, great progress has been made in understanding the rules of plastid gene expression, and in developing technologies that will aid in the eventual commercialization of products derived from plastid transformation. This chapter describes such progress, using examples of current approaches that utilize plastid transformation in the overproduction of recombinant proteins.

From: *Handbook of Industrial Cell Culture: Mammalian, Microbial, and Plant Cells*
Edited by: V. A. Vinci and S. R. Parekh © Humana Press Inc., Totowa, NJ

**Table 1
Comparison of Gene-Expression Technology
via Nuclear and Plastid Transformation**

Gene expression technology	via nuclear transformation[a]	via plastid transformation[b]
Integration of transforming DNA	Random insertion	Targeted by homologous recombination
Insertion of multiple genes	Using multiple expression cassettes	Using multiple expression cassettes or polycistronic operons
Transgene expression uniformity	No; variable in all events from a single construct	Yes; similar in all tested events from a single construct
Expression of unmodified prokaryotic genes	Usually no	Yes
Complex folding and disulfide linkages	Using secretory pathway	Yes
Protein accumulation levels	Usually less than ~3% tsp typical	As high as 5–40% tsp possible
Gene silencing	Possible	Not observed
Transgene inheritance	Mendelian	Uniparental (maternal in most crops)

[a]Data based on numerous plant species.
[b]Data based herein on tobacco references only.

2. The Structure of the Plastid Genome

In higher plants, plastids are abundant in all cell types, and arise from undifferentiated progenitor organelles termed proplastids in the meristematic cells of shoots and roots. Plastid development is closely linked to the cell type in which the plastid is located, and within one given cell type the plastid population is relatively uniform with regard to its differentiation state *(6-7)*. Plastids differentiate into specialized types with different functions—for example, chlorophyll production and photosynthetic reactions in leaf chloroplasts, starch accumulation in amyloplasts of roots or tubers, and pigment production in chromoplasts of fruits and flower petals. A unique feature of plastids is their ability to interconvert—they are capable of differentiating into any of the other plastid types given the appropriate environmental cues. Plastids do not arise *de novo*, but reproduce by binary fission of existing plastids.

The plastid genetic system is characterized by polyploidy, with numerous identical plastid genome (ptDNA) copies in each plastid, and numerous plastids in each cell. In general, photosynthetic cells have the highest number of plastids per cell, with the highest genome copy number. A typical dicot leaf cell may contain as many as 100 chloroplasts with up to 100 ptDNA molecules each, for a total of ~10,000 ptDNA copies per cell. The number can soar to ~50,000 ptDNA copies per leaf cell in monocots. At the low end are non-photosynthetic root cells—~50 plastids per cell each with ~10 genome copies for a total of ~500 ptDNAs per cell *(8)*. This extraordinarily high genome ploidy level is partly responsible for the ability to overexpress recombinant proteins to levels as high as 30–40% of total soluble protein (tsp).

To date, the plastid genome of about a dozen plant species has been completely sequenced, and an algal plastid genome sequencing effort currently exists (http://megasun.bch.umontreal.ca/ogmpproj.html). The plastid genome size ranges from 120 kb–160 kb, and some very large algal genomes or the very small genomes of non-photosynthetic plants are the outliers *(9)*. Most higher plant plastid genomes have the typical arrangement of an Inverted Repeat (IR) region separated by a Small Single Copy (SSC) and Large Single Copy (LSC) regions (Fig. 1); exceptions include some legumes and conifers that lack the IR region. The IR region is characterized by the presence of the ribosomal RNA genes and a variable number of additional genes depending on the plant species. A recombination mechanism termed "copy correction" maintains sequence identity between the two inverted repeats *(10)*.

The typical plant plastid genome has a coding capacity of ~120 genes. These include genes required for transcription and translation, such as an entire complement of ribosomal and transfer RNAs, as well as genes for RNA polymerase subunits, and ribosomal protein genes. In addition, some components of the photosynthetic apparatus and a respiratory-chain NADH dehydrogenase complex are encoded in the organelle genome *(9)*. In contrast to the nuclear genome, the plastid genome encodes mostly operons that produce polycistronic transcription units from a single promoter. This feature provides potential advantages for the simultaneous expression of multiple transgenes. The coding capacity of the organelle also includes introns in a small number of genes and the ability to direct RNA-editing *(11,12)*. To date, the involvement of these RNA maturation processes in the regulation of transgene expression have not been extensively studied.

3. The Process of Plastid Genetic Engineering

The process of plastid genetic engineering has recently been extensively reviewed *(13–19)*; thus only the salient features are described here. The process that was developed in tobacco leaves *(1,2)* (Fig. 2) has remained as the general guide for development of the technology in several additional plant species. This process requires a robust method to introduce DNA into plastids, tissue culture, and a selection scheme amenable to plastid transformation, and vectors that direct transgene integration by homologous recombination and allow subsequent expression of foreign genes in the plastid genome.

For the transforming DNA to penetrate the double membrane of the chloroplast, the particle bombardment (e.g., biolistic) process has been routinely used *(19)*, although PEG-mediated protoplast transformation has also been successful *(20)*. The transformation of tobacco leaves has been reported using either supercoiled or linear DNA, tungsten, or gold microparticles, and the original gunpowder gun or helium gun with similar efficiencies. This is in contrast to *Chlamydomonas*, in which linear DNA was up to 10-fold more efficient *(21)*, and particle size played a role *(22)*. Regardless, the events following particle penetration into the cell are unknown, and it is believed that only one or a very few chloroplasts are hit by the microprojectile after which the transforming DNA integrates by homologous recombination into one or only a very few plastid genomes.

Thus far, selection of plastid transformants has been based on resistance to antibiotics that inhibit plastid protein synthesis. Initial success was obtained with a spectinomycin-

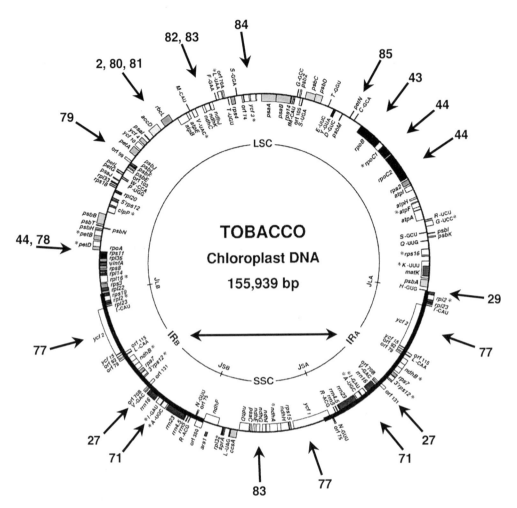

Fig. 1. Map of the tobacco plastid genome and sites of foreign gene insertions. The updated tobacco plastid genome map (gift of M. Sugiura) is shown with a partial listing of the genomic sites (arrows) where transgenes have been used to study gene function. The number for representative published studies are listed next to each site. Note that transgenes directed to the Inverted Repeat (IR) region will be present in two copies in the transplastomic genome as a result of "copy correction" gene conversion (horizontal arrow). LSC-Large Single Copy region. SSC-Small Single Copy region.

resistant allele of the plastid 16S ribosomal RNA gene. Transformation frequency using this recessive marker was about 1 transformant for every 100 bombardments (1,23). A stepwise process improvement was subsequently achieved using a bacterial *aadA* gene driven by plastid expression signals (2). This antibiotic-inactivating-enzyme is a dominant selectable marker that confers resistance to spectinomycin and streptomycin. Using spectinomycin selection, a 100-fold improvement in transformation efficiency was observed, to about 1 transformant per shot. Selection of plastid transformants using spectinomycin resistance markers has been considered a non-lethal selection scheme—although growth on the sucrose-containing tissue-culture medium continues, cells

A

Particle bombardment
of leaves

↓

1-2 day delay

↓

Leaf pieces on
spectinomycin 500 ug/mL

↓

4-8 weeks selection

↓

Primary shoots arise

↓

Second round regeneration

↓

Molecular confirmation
of homoplasmy

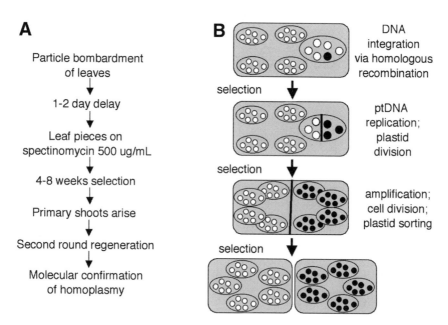

B

DNA
integration
via homologous
recombination

selection

ptDNA
replication;
plastid
division

selection

amplification;
cell division;
plastid sorting

selection

Fig. 2. The biolistic transformation process and selection of tobacco plastid transformants. (**A**) The biolistic process in tobacco. Transforming DNA is precipitated onto the surface of gold or tungsten microprojectiles. Particle bombardment is used to introduce the DNA-coated particles into tobacco-leaf cell chloroplasts. After a delay period, the leaf is cut into sections and placed on selective spectinomycin-containing medium. Over a period of several weeks, the leaf pieces bleach on the medium. Green shoots that arise from the bleached leaf sections are putative plastid transformants. Plastid transformants are typically distinguished by molecular identification of the aadA transgene, or by resistance to the second selective antibiotic, streptomycin. Plastid-transformed shoots are often heteroplasmic at the initial stage of identification; a second round of plant regeneration under selective conditions can be performed to obtain homoplasmic plants. (**B**) Selection and sorting of plastid transformed cells. Integration of transforming DNA into the recipient plastid genome occurs through homologous recombination. Integration probably occurs in only one plastid genome of a single chloroplast in the leaf cell (black circle). During the process of plastid DNA replication and plastid division, newly replicated plastids randomly receive transformed plastid genomes. Plastids that carry transformed genomes are able to express the antibiotic resistance marker and acquire phenotypic resistance. Amplification of transformants occurs first at the organellar level and subsequently at the cellular level. Cells with a predominant population of transformed chloroplasts have a selective advantage in the presence of spectinomycin selection and will go on to form green shoots.

bleach out and transformed cells are identified by greening. The nonlethal selection scheme may allow time for transformed plastid genomes to amplify until phenotypic resistance at the cellular level can be expressed *(19)*. Kanamycin selection using a chimeric *npt*II gene has also been reported, although at relatively low efficiency *(24)*, and more recently, a selection scheme based on the conversion of toxic betaine aldehyde to nontoxic glycine betaine using a chimeric betaine aldehyde dehydrogenase gene (BADH) has also been reported *(25)*.

Because of the high ploidy level of the plastid genome, initial plastid transformants are usually heteroplasmic. Thus, only a fraction of the plastid genomes in a cell contain the transforming insert DNA. Transformed cells must be maintained on selective medium to allow replication of transformed plastid genomes, partitioning of resistant genomes to daughter organelles, and subsequent sorting of homoplasmic organelles. The sensitive, wild-type plastids are lost during this process. Once homoplasmic plants are obtained, the transgene is stable through multiple generations *(19)*. Plastid transformed plants are termed "transplastomic" or "transplastidic" to distinguish these from nuclear transgenic plants.

In contrast to nuclear transformation in which the random integration of transforming DNA occurs, the rule for insertion of foreign DNA into the plastid genome is through homologous recombination. The enzymes involved in plastid recombination have not been identified, although a nuclear-encoded RecA homolog may be involved in at least intrachromosomal recombination in plastids *(26)*. Plastid transformation vectors, therefore, are designed with the selectable marker and genes of interest cloned into surrounding homologous ptDNA on either side of the insert; a convenient series of tobacco plastid transformation vectors have been described *(27)*. There appears to be no significant bias in insertion sites, and numerous regions of the plastid genome have been successfully targeted to study various gene functions (Fig. 1). Although the upper limit of insert size has not been tested, two genes (selectable marker and gene of interest) are routinely integrated, and insertion of at least ~7 kb has been reported *(28)*. Additionally, cotransformation—which targets inserts to two different locations within the same genome—is possible *(29)*, and may be useful in technological applications to increase the number of inserted transgenes.

A distinguishing feature of organelles are their cytoplasmic inheritance pattern; maternal inheritance is observed for plastids in most crop plants. Therefore, plastid transgenes are inherited only in maternally derived seed progeny (Fig. 3), and should not be transmitted via pollen. As a result, transplastomic plants may have simpler field management requirements than nuclear transgenic plants in some cases. Although maternal inheritance may be favorable for field management in some cases, the presence of an antibiotic-resistance selectable marker may still be a barrier to the widespread commercialization of transplastomic plants. For this reason, methods for selectable marker excision from plastids have been developed. The Cre/lox site-specific recombination system *(30,31)*, as well as the homologous recombination between transgene expression signals used to create deletion derivatives *(32)*, have recently shown to be efficient for marker removal from transplastomic tobacco plastids.

4. Development of Plastid Transformation in Food Crops

Plastid transformation in the model plant—tobacco—has been used to answer many specific biological questions. However, attempts to broaden the transformation technology to additional model plants *(33)* and crop species have thus far met with only partial success. To date, most technology development has focused on dicot species with tissue-culture systems that are amenable to leaf-based transformation and regeneration protocols. Although these examples take advantage of abundant chloroplasts in the target tissue, each has required a significant effort for success.

Potato (variety FL1607) *(34)* and tomato (L. esculentum variety IAC-Santa Clara) *(35)* were the first food crops in which plastid transformation technology was developed. Both were chosen because of their close relationship to tobacco as members of

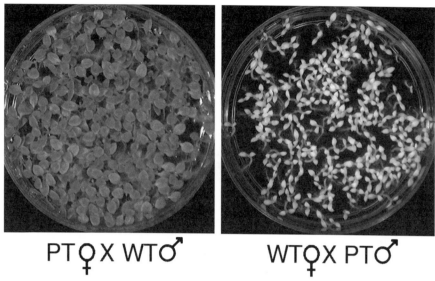

PT♀ X WT♂   WT♀ X PT♂

Fig. 3. Maternal inheritance of plastid transgenes. Homoplasmic plastid transformed tobacco plants were grown to maturity and used in crosses to confirm maternal inheritance. The plastid transformant (PT) is used as maternal recipient parent in a cross to wild-type (WT) plants used as pollen donor (left panel), or used as pollen donor to wild-type recipient plants (right panel). Seeds are sown on selective spectinomycin medium where bleaching indicates sensitivity to the antibiotic. Note that all maternally derived seedlings are resistant and green (dark in the photograph), and all seedlings from the cross with the plastid transformant as pollen donor are sensitive (white).

the *Solanaceae* family, with their plastid genomes that maintain a very high overall degree of sequence identity (~98%) and sharing gene order and content. This high degree of sequence identity allowed testing of plastid transformation vectors first in tobacco, with the prediction that transgene integration by homologous recombination would not be compromised and the genes of interest would be properly expressed. In both cases, tissue culture and transformation protocols were considered the rate-limiting steps, and required significant process improvements for success. In potato, sensitivity to spectinomycin and bleaching occurred at only 40–50 mg/L whereas tobacco required 500 mg/L for complete bleaching. Nuclear transformants in potato were obtained at a high frequency, using 40 mg/L spectinomycin for selection. Interestingly, plastid transformants could not be obtained at 40 mg/L spectinomycin, and required a much more stringent selection pressure of 300–400 mg/L spectinomycin for the successful recovery of transformants. In contrast, tobacco nuclear and plastid transformants are obtained at about the same frequency, using spectinomycin selection at 500 mg/L. In potato, plastid transformation frequency was only ~fivefold lower than nuclear transformation efficiency, suggesting that commercial applications may be feasible. On the other hand, tomato plastid transformants were obtained at very low frequency (~1 in 80–100 selection plates), and the tissue-culture regime required nearly 2 yr to obtain fruit-bearing plants.

In addition to the successes with the dicot plastid transformation systems described here, plastid transformation of the monocot—rice—has also recently been reported

(36). Significant differences in monocot and dicot tissue-culture systems exist: most monocots utilize non-green embryogenic cells for transformation, and are naturally resistant to spectinomycin but can be selected on streptomycin. Despite these differences, Khan and Maliga *(36)* were able to use an aadA-GFP fusion to select transformed cells on streptomycin medium and regenerate rice plants that carried transgenic chloroplasts. In rice, only a small fraction of chloroplasts in the regenerated plants expressed the transgene, and individual cells contained a mixed population of wild-type and transgenic chloroplasts. These results underscore the differences in plastid transformation efficiency and show the importance of developing plastid-specific transformation protocols even among closely related plant species, and between monocots and dicots.

5. Considerations for Foreign Gene Expression in Plastids

Plastids are derived from an ancient prokaryotic endosymbiont, and still share many similarities with present-day cyanobacteria and eubacteria, including similar gene-expression characteristics. However, the expression of plastid transgenes is complicated by differences in ptDNA copy number, and tissue-specific and developmental differences in plastid gene-expression patterns. Furthermore, relative to the field of nuclear transgenic technology, plastid transformation is in its infancy, and detailed knowledge about the *cis-* and *trans*-acting factors that govern transgene expression is only beginning to be revealed. Described here are the major controlling points to consider for high-level protein expression, and some examples of their successful use in transplastomic plants.

5.1. Transcription of Plastid Transgenes

The accumulation of plastid transcripts depends on transcription rate, mRNA stability, and relative ptDNA copy number. For most plastid genes, transcription rates and RNA accumulation are greatest during rapid chloroplast development in the light. In addition to the regulation of overall plastid RNA levels, there is also variation in the relative ratios of individual plastid RNAs during development and in different tissue-types. Despite the dynamic regulation of plastid RNA accumulation, plastid genes are typically constitutively transcribed, with strong promoters transcribed at higher rates relative to genes with weaker promoters (reviewed in refs. *7,37*).

Plastids transcription depends on at least two DNA-dependent RNA polymerase (RNAP) activities (reviewed in refs. *38,39*). The plastid genome encodes the subunits of the catalytic core—proteins that are highly homologous to the *E. coli* RNAP core subunits. The plastid-encoded core subunits assemble with nuclear-encoded sigma-like factors to form the functional holoenzyme, known as the plastid-encoded polymerase (PEP). PEP has been biochemically characterized and shown to exist in various forms in etioplasts and chloroplasts. Both forms carry a similar set of sigma factors that copurify with PEP *(40)*, but these may be functionally distinct based on different phosphorylation patterns *(41)*. The genes for several families of nuclear-encoded sigma factors have now been cloned from several plant species, and their gene products have been shown to be targeted to plastids. Recent evidence suggests that the accumulation of individual sigma factors may occur in response to light, plastid type, or developmental stage, and that their promoter specificities are being identified *(39)*.

Not surprisingly, PEP recognizes promoters that are similar in structure to the *E. coli* sigma^{70}-type, consisting of –35 and –10 consensus elements. The relative strengths of these promoters have been characterized in homologous plastid extracts and in vivo by transcriptional run-on assays (reviewed in ref. *37*). However, in some cases, transcription from PEP promoters also requires elements upstream of the core promoter *(42)*.

In addition to PEP, a nuclear-encoded plastid RNA polymerase (NEP) RNA activity has also been characterized in plastids. NEP activity is relatively weak, and has been revealed in plastids that lack PEP activity. Deletion of PEP subunits from the tobacco plastid genome *(43,44)* allowed the identification of a set of genes transcribed by NEP, and a loose 10-nucleotide consensus sequence around the transcriptional start site was derived *(45)*. The catalytic NEP core was recently cloned from several plant species and shown to resemble the T3/T7 RNAP class of enzymes (reviewed in ref. *46*). More recently, a putative third nuclear-encoded polymerase activity was identified (NEP-2), which may be specifically involved in transcription of the spinach *rrn* operon and requires interaction with a nuclear-encoded transcription factor (CDF2) *(47)*. The molecular identity of this enzyme is unknown, and a broader role in transcription of other genes is still unclear.

Plastid genes may be classified according to their promoters: those with PEP, NEP and PEP, and NEP promoters only *(45)*. Plastid genes that carry only PEP promoters include the photosynthetic genes such as *rbcL*, *psbA*, and *psbD* and are transcribed at high levels in leaves. Plastid genes that carry both PEP and NEP promoters include the *rrn16* operon, whose promoter is among the strongest promoters characterized in plastids. Interestingly, this class of plastid genes may have one or more NEP promoters, positioned upstream or downstream, or both, in relation to the PEP promoter. Plastid genes that carry NEP alone include *accD*, *ycf2*, and *rpoB*. These genes are transcribed at relatively low levels in leaves, but their activity may be upregulated in non-green tissues *(45)*.

As a practical matter, plastid transgenic constructs designed for protein overexpression have thus far attempted to utilize the controlling elements of the most abundantly expressed plastid genes. These have typically been PEP promoters from *rrn16* and *psbA*, the first promoters used to successfully drive transgene expression in tobacco to high levels *(2,48)*. The role of NEP promoters in plastid transgenes is largely untested, but may be important for expression in non-green tissues *(45)*. Future transplastomic studies are needed to fully understand their role.

5.2. Role of Plastid 3'-ends

The 3'-untranslated region (3'-UTR) of nearly all plastid genes contains an inverted repeat region that can form a stem-loop structure. This stem-loop structure is reminiscent of bacterial rho-independent transcriptional terminators. However, in plastids, these elements act as mRNA processing and stabilization elements, but are ineffective transcription terminators *(49,* reviewed in *50)*. RNA maturation involves readthrough past the 3' stem-loop structure, followed by processing to form a discrete mRNA sequence with a mature 3'-end that is immediately downstream of the stem-loop. Processing may occur either by a 3'-5' exonuclease activity or by endonucleolytic cleavage followed by exonuclease trimming *(50)*, and may be mediated by a high mol-wt protein complex involving nuclear-encoded RNA-binding proteins *(51)*. Interestingly, processing of 3'-ends may be an important determinant for translation. In *Chlamydomonas*, only processed *atpB* and *rbcL* mRNA was found to be associated with polysomes *(52)*. In addition to RNA maturation, the 3'-UTR stem-loop structure is important for mRNA stability. For example, deletions that remove the stem-loop from the *Chlamydomonas atpB* 3'-UTR result in an unstable RNA species and reduced protein accumulation with no effect on transcription rate, *(53)*, and replacement with the stem-loop of other plastid genes (*petD*, *petA*, *rbcL*) in chloroplast transformants resulted in a stable discrete-sized mRNA, but only when the sense orientation of the heterologous 3'-UTR was used *(54)*. Likewise, exchange of the tobacco *psbA* 3'-UTR for the *rps16* or *rbcL* 3'-UTR had no effect on transcript accumulation in tobacco plastid transformants *(55)*. Stem-loop structures involved in mRNA stability may also be found in the 5'-UTR of some plastid genes—for example, in the tobacco *rbcL* gene *(56)*.

Recently, a homolog of bacterial polynucleotide phosphorylase (PNPase) was purified from the RNA processing complex involved in 3'-end maturation. This PNPase was subsequently shown to be involved in both endonucleolytic cleavage and polyadenylation of chloroplast mRNAs *(57)*. Evidence suggests that polyadenylation occurs only on a very small subfraction of the total mRNA pool, and presumably marks these for RNA degradation. In *Chlamydomonas*, polyadenylation occurred mostly at or near the mature 3'-end, and in spinach chloroplasts polyadenylation occurred at endonucleolytic cleavage spots throughout the mRNA and only rarely at the processed 3'-end *(58)*.

5.3. The Critical Role of Translation Initiation in Plastids

Translation initiation is a major controlling point for expression of plastid-encoded photosynthetic proteins and transgene-encoded proteins in vivo *(48)*. The 5'-untranslated region (5'-UTR) is clearly involved in translation, but the cis-acting sequences required for control are usually unknown. In *E. coli*, the Shine-Dalgarno sequence (S-D) is a purine-rich sequence of 3–9 contiguous bases that form base pairs (bp) with a complementary sequence at the 3'-end of the 16S rRNA component of the small ribosomal subunit. The S-D is located optimally five nucleotides upstream from the initiation codon, but can function as far as 13 nucleotides from the AUG codon. A sequence downstream of the initiation codon called the "downstream box" sequence has been theorized to increase translation by increasing complementarity to anti-downstream box sequences in the penultimate stem of the 16S rRNA. However, the downstream box shows only patchy complementarity to 16S rRNA, and changes in this sequence also change the amino-acid sequence of the encoded protein, changing protein stability (reviewed in ref. *59*).

In the unicellular green alga *Chlamydomonas*, S-D sequences are rarely observed near the initiation codon of plastid mRNAs, and translation can even start from a leaderless RNA *(60)*. In many cases, a stem-loop structure in the 5'-UTR of *Chlamydomonas* plastid messages are bound by nuclear-encoded translational activator proteins required for translation *(61)*. More recently, the –1 triplet codon, upstream of the initiation codon, was shown to be involved in extended base pairing with the initiator tRNA anticodon *(62)*. In higher plants, the sequences in plastid 5'-UTRs that control translation are less understood. Bonham-Smith and Bourque *(63)* analyzed the sequences of plastid gene leaders and determined that at least two-thirds of these did not contain a recognizable consensus S-D sequence within the prescribed distance from the translation initiation site. In many cases, a putative S-D could be seen further upstream in the

5'-untranslated leader, and it was theorized that some RNA secondary structure was necessary to bring the S-D to within the appropriate distance of the AUG initiator.

Using the newly available tool of plastid transformation, early work showed that the 5'-UTR of the plastid *psbA* gene, encoding the D1 protein of Photosystem II, contained all of the sequences required for light-regulated and tissue-specific protein accumulation of a chimeric reporter gene *(55)*. The reporter gene encoded glucuronidase (GUS), which accumulated to ~2.5% of total soluble protein in leaves, but 100-fold less in dark-grown seedlings and undetectable levels in roots, similar to the D1 protein *(48)*. Another early plastid transformation vector utilized a chimeric promoter/leader combination consisting of the strong, constitutive *Prrn* promoter linked to a small synthetic-leader sequence patterned after the putative S-D region of the plastid *rbcL* gene. This chimeric expression signal was predicted to facilitate translation of transgenes in all tissues, and was first used to achieve a stepwise improvement in plastid transformation *(2)*.

The synthetic *rbcL* leader was subsequently shown to be unpredictable for transgene translation, working well for some genes and less well for others, despite its consensus *E. coli* S-D sequence. To test the influence of non-native ribosome-binding site regions on plastid translation, a systematic series of chimeric genes were tested, that differed only by their translational control regions (64). As a reporter, the coding region of 5-enolpyruvyl-shikimate-3-phosphate synthase (EPSPS) from *Agrobacterium* strain CP4 was used, for which antibodies, enzyme, and phenotypic glyphosate-resistance assays existed. To drive EPSPS translation, the leader from the bacteriophage *gene 10* (G10L), shown to significantly enhance translation of foreign genes in *E. coli (65)*, was compared to the *rbcL* synthetic leader. The *rbcL* and G10L ribosome-binding-site regions differ in their length, putative S-D sequences, and the spacing of the S-D to the translation start codon *(64)*. Additionally, the 14 N-terminal amino acids of the GFP protein, shown previously to be efficiently translated in potato plastids *(34)*, was N-terminally fused to the EPSPS to test the effects of sequences that were downstream of the initiator codon of translational efficiency. The series of chimeric genes were introduced into tobacco plastids linked to the selectable *aadA* marker gene, and transplastomic plants were analyzed for EPSPS protein accumulation.

A dramatic 10,000-fold difference in EPSPS protein accumulation among the various constructs in transplastomic plants was observed. Protein accumulation was barely detectable from the synthetic *rbcL* leader, with a 200-fold increase using the G10L and a further 50-fold increase with the GFP-EPSPS translational fusion. Total EPSPS protein was >10% of soluble protein in leaves of these plants. A parallel increase in EPSPS accumulation based on translation elements was also seen in non-photosynthetic tissues, although the overall protein levels were much lower as a result of fewer plastids and lower transcription/translation activity in these tissues. Interestingly, changing the codon usage of the *Agrobacterium* gene to more closely resemble that of plastid photosynthetic genes increased protein accumulation only moderately, as was also observed for a chimeric *bar* gene *(66)*.

The enhanced accumulation of the GFP-EPSPS translational fusion indicates that sequences downstream of the translational start site are also important determinants for plastid translation. However, the rules involved are not clear. Addition of the *E. coli* downstream box sequence along with the G10L increased NPTII protein accumulation to 16% of total soluble protein, and the addition of a plastid sequence modeled after the same positions in the tobacco plastid 16S rRNA resulted in only 0.16% NPTII *(67)*. Furthermore, when the plastid *atpB* gene 5'-UTR, which does not contain a recognizable S-D sequence, was used to drive transgene expression, the presence or absence of *atpB* sequences downstream of the initiation codon had little effect. However, when the full plastid *rbcL* 5'-UTR including its S-D was used to drive transgene expression, high-level protein accumulation unexpectedly required sequences downstream of the start codon *(68)*.

The data described here indicate that translation initiation is the major controlling point for plastid transgene expression, and that the sequences involved include the ribosome binding site region within the 5'-UTR and extend to sequences downstream of the translational start codon. These examples also illustrate the need for a more systematic mutational analysis of higher plant plastid translational control regions.

6. Heterologous Expression Systems for Plastids

A current limitation in plastid engineering is the inability to control tissue-specific or developmentally regulated transgene expression, or to limit the timing of high-level expression of transgenes to prevent potentially detrimental effects on agronomic traits. To address these issues, McBride et al. *(69)* developed a heterologous two-component activation system for plastids, capable of controlling transgene expression independent of the plastid gene-expression apparatus. The two-component system utilizes both a nuclear-encoded transgene, a bacteriophage T7 RNA polymerase targeted to plastids by a chloroplast transit peptide (CTP-RNAP), and a plastid-encoded transgene driven by a phage T7 promoter. The T7 RNA polymerase is a single-subunit enzyme with strong specificity for the promoter associated with the plastid transgene, thus making its expression dependent on the nuclear-encoded plastid-targeted polymerase.

The first step of the two-component system was to establish nuclear-transformed tobacco lines carrying the CTP-RNAP gene. In this example, the nuclear-encoded CTP-RNAP gene was driven by the 35S promoter for constitutive expression in all plant tissues. Nuclear-transformed lines were generated using standard *Agrobacterium*-mediated transformation and selection protocols, and subsequently characterized by in vitro RNA polymerase enzyme activity assays. Chosen lines were then bred to homozygosity to be used in crosses to plastid-transformed lines, or used directly as recipients for plastid transformation, in the second round of the two-step procedure.

The second step of the transactivation system was to generate plastid-transformed tobacco lines carrying a transgene linked to the selectable marker used for plastid transformation. In this case, a reporter *uidA* transgene was utilized to monitor the success of the two-component activation assay. The *uidA* transgene was preceded by the phage T7 *gene* 10 promoter and 5'-UTR, necessary for recognition by the T7 RNAP and translation of the encoded mRNA.

Using either the crossing or retransformation strategy, GUS expression was activated to very high levels in plants carrying the two-component expression system. GUS enzyme activity was measured in all tissues of the plant, and was shown to accumulate to an impressive 20–30% of total soluble protein in mature leaves as judged by Coomassie blue-stained SDS polyacrylamide gels. These results with the two-component transactivation system led to subsequent proof-of-concept work with potentially agronomically important proteins. In this case, overexpression to 3–5% of soluble protein in tobacco leaves of the protoxin form of the *Bacillus thuringiensis* (Bt) crystal protein gene cry1A(c) with nearly 100% effective insecticidal activity to target insects was observed *(70)*.

7. Expression of Engineered Polycistronic Operons in Plastids

The organization of most plastid genes into operons, and the prokaryotic characteristics of plastid gene expression, suggest that engineered or native bacterial polycistronic operons may be a practical way to express multiple transgenes in plastids from a single promoter. Expression of multiple coordinately regulated transgenes as a polycistron would allow introduction of metabolic pathways required to produce complex carbohydrates, fatty acids or oils, and production of complex products such as biodegradable plastics. Expression of polycistronic operons also eliminates the need for multiple engineered transcriptional promoter and terminator elements, easing cloning requirements and potential limits to insert size.

To test translation of promoter distal open reading frames (ORFs) in plastids, Staub and Maliga *(71)* engineered a promoterless reporter *uidA* transgene downstream of the endogenous tobacco plastid *rbcL* gene. Because of inefficient transcription termination at the *rbcL* 3'-end, polycistronic messages containing the promoterless *uidA* as the second cistron were formed. GUS protein accumulated to high levels in these plants although no monocistronic *uidA* was observed, indicating efficient translation of the second ORF. Interestingly, the presence of numerous stop codons in all three reading frames between the first (*rbcL*) and second (*uidA*) coding regions indicated that translational coupling between the two genes did not occur, and that internal ribosome loading was probably responsible for translation initiation of the downstream gene in these experiments.

More recently, De Cosa et al. *(72)* modified the three-gene Bt cry2Aa2 operon for expression in plastids. The protoxin gene is the third ORF in the natural Bt operon, preceded by the *orf2* gene that encodes a putative chaperonin that facilitates crystal protein formation and the *orf1* gene of unknown function. This operon was cloned downstream of the *aadA* gene used as selectable marker for tobacco-plastid transformation. To complete the engineered plastid operon, the four genes were then cloned downstream of the strong plastid *Prrn* promoter and a synthetic ribosome-binding site provided as the 5'-untranslated leader for translation in plastids.

After the selection of tobacco-plastid transformants, expression of the cry2Aa2 protein was monitored by enzyme-linked immunosorbent assay (ELISA) assay and by observation of crystal proteins via electron microscopy of leaf samples. Using the ELISA assay, accumulation of cry2Aa2 up to 46.1% total soluble protein in old leaves was claimed *(72)*. Cuboidal crystals that consume a large volume of the chloroplasts in leaves were also observed, consistent with high-level protein accumulation. Significantly lower levels of cry2Aa2 protein (only up to 0.36% total soluble protein) were reported when cry2Aa2 was expressed as a single gene.

Transcript accumulation patterns were not shown in the cry2Aa2 studies, so it is not clear if a stable polycistronic message was produced or if processing events may have resulted in translatable monocistronic mRNAs, as is seen in some native plastid operons *(73)*. Nevertheless, the results described above indicate that translation of ORFs within an engineered operon can occur.

8. Expression of Human Therapeutics in Tobacco Chloroplasts

The possible extraordinarily high protein accumulation observed for numerous bacterial proteins described here made the chloroplast transformation system a logical vehicle to test the "plant as protein factory" concept for much more complex eukaryotic proteins such as human therapeutics. As a first test for the chloroplast expression system, Staub et al. *(74)* chose a well-characterized, relatively simple human therapeutic protein with a long history of recombinant protein production—human somatotropin (hST). This pituitary protein was one of the first recombinant proteins produced in the biotechnology industry and has indications in hypopituitary dwarfism in children, treatment of Turner's syndrome, chronic renal failure, HIV wasting syndrome, and treatment of the elderly.

Despite the fact that hST was one of the first products of the biotechnology industry and is a relatively simple signaling molecule, expression in plastids poses several unique challenges. These included the need for controlled disulfide bond formation, as hST requires two disulfide-bonds for biological activity. Furthermore, to maintain human equivalency of the recombinant hST protein, expression in plastids required formation of a non-methionine N-terminus, as processing of the signal peptide from hST leaves a phenylalanine at the N-terminus during secretion of hST in the human pituitary gland. As normal translation in plastids initiates at methionine, a ubiquitin-hST fusion was designed to yield a phenylalanine N-terminus in the final hST product. Because ubiquitin is not present in plastids *(75)*, cleavage of ubiquitin from the recombinant protein by ubiquitin protease could occur during extraction and purification of the recombinant protein.

Constructs used to express hST and ubiquitin-hST have been introduced into chloroplasts and transplastomic plants have been generated. For comparison, nuclear transgenic plants that express chloroplast-targeted hST (CTP-hST) or hST targeted through the endoplasmic reticulum to the secretory pathway (ER-hST) were also created. Quantitative ELISA and Western Blot analysis of these plants indicated that hST was produced at up to >7% total soluble protein in transplastomic plants, an amount more than 300-fold greater than the accumulation of hST from the nuclear-encoded chloroplast-targeted protein (Table 2). This result confirmed that recombinant protein production in chloroplasts may have significant advantages for some proteins.

In addition to high-level protein accumulation from the chloroplast-expressed genes, the ubiquitin-hST fusion protein was cleaved, apparently during extraction and purification of the protein from leaf samples, proving the utility of this approach in generating a non-methionine N-terminus on recombinant proteins expressed in chloroplasts. The chloroplast-recombinant hST was also shown to have biological activity in an in vitro assay that utilizes a rat lymphoma cell line that proliferates in the presence of somatotropin. The response of the cell line to recombinant chloroplast somatotropin was equivalent to *E. coli* produced and refolded hST protein used as control. The formation of proper disulfide linkages in recombinant hST produced in chloroplasts, as indicated by the bioassay, was also shown by peptide-mass mapping of tryptic peptides. Surprisingly, no improperly paired cysteine residues were observed in the purified sample as compared to native human somatotropin *(74)*. The proper formation of disulfide bonds and biological activity of chloroplast-expressed somatotropin indicated that chloroplasts have the ability to process complex mammalian secretory proteins.

Table 2
Expression of Human Somatotropin Genes in Transgenic Plants

Plasmid	Gene location	Protein location	Protein	Expression[a]
wrg4747	Nucleus	Chloroplast	CTP-hST	ND - 0.025
wrg4776	Nucleus	ER	ER-hST	0.004, 0.008
wrg4838	Chloroplast	Chloroplast	hST	0.2
pMON38794	Chloroplast	Chloroplast	Ubiq-hST	7.0

[a]Protein is expressed as the percentage of total soluble protein and was quantitated by ELISA assay or Western Blot analysis. ND-None detected. CTP-chloroplast transit peptide. ER-secretory (endoplasmic reticulum) peptide. Ubiq-ubiquitin fusion. Table modified from ref. *(73)*.

Table 3
Partial List of Foreign Genes and Proteins Expressed in Plastids

Functional class	Gene or protein	References
Selectable markers	*aadA*, *npt*II, BADH, aadA-GFP, *cod*A	2,24,25,27,36,86
Reporter genes	GUS, GFP	34,48
Herbicide resistance	EPSPS, *bar*, *Bxn*, *crt*I	4,32,64,66, WO 00/03022[a]
Insect or microbe resistance	cry1A, cry2Aa2, MSI-99	70,72,87
Metabolism	Anthranilate synthase, Rubisco large subunit	81, 88, 89
Enzymes	Aprotinin, cellulase	WO 00/03012[a] WO 98/11235[a]
Therapeutic proteins	Human somatotropin, cholera toxin B subunit, tetanus antigen, human serum albumin, monoclonal antibody	74, 76, WO 01/72959 A2[a]

[a]Published PCT patent applications.

Furthermore, in contrast to the hST expressed in bacterial systems, the plastid-expressed recombinant protein accumulated in soluble form in the stromal compartment of the chloroplast. These results suggest that the prokaryotic-like organelle may be capable of producing additional complex proteins destined for human therapeutics.

9. Future Prospects

As detailed in other chapters of this volume, the use of transgenic plants for industrial applications such as biomaterials and feedstocks has already begun. Using the approaches described in this chapter, a wide variety of recombinant proteins have now been expressed in plastids (Table 3), suggesting that this technique may provide an attractive alternative production approach in some cases. With examples of recombinant protein accumulation in leaves in some cases up to 40% tsp and the biological activity of complex eukaryotic proteins, transplastomic plants may also make an attractive alternative production vehicle for some pharmaceutical proteins destined for human or animal use. In fact, recent expression of candidate subunit vaccines for tetanus (Pal Maliga, Rutgers University, personal communication) and cholera toxin *(76)* in transplastomic tobacco plants confirms that this may be possible in the near future. In the case of the tomato *(35)*, recombinant protein accumulation in green or red-ripe fruits up to ~1% tsp also suggests that edible parts of transplastomic plants may someday find commercial applications as nutraceuticals or oral vaccines. Further understanding of transplastomic expression and transfer of these tools to additional crop species, however, is a prerequisite for widespread use in the industry.

Acknowledgments

The author gratefully acknowledges the expert contributions of members of the Plastid Transformation initiative at Monsanto, and the support of Drs. Ganesh Kishore, Mike Montague, Steve Padgette, and Ken Barton. The author also thanks Professor Masahiro Sugiura, of Nagoya City University, for the generous gift of the tobacco plastid genome map.

References

1. Svab, Z., Hajdukiewicz, P., and Maliga, P. (1990) Stable transformation of plastids in higher plants. *Proc. Natl. Acad. Sci. USA.* **87,** 8526–8530.
2. Svab, Z. and Maliga, P. (1993) High-frequency plastid transformation in tobacco by selection for a chimeric *aadA* gene. *Proc. Natl. Acad. Sci. USA* **90,** 913–917.
3. Boynton, J. E., Gillham, N. W., Harris, E. H., Hosler, J. P., Johnson, A. M., Jones, A. R., et al. (1988) Chloroplast transformation in *Chlamydomonas* with high velocity microprojectiles. *Science* **240,** 1534–1538.
4. Daniell, H., Datta, R., Varma, S., Gray, S., and Lee, S. B. (1998) Containment of herbicide resistance through genetic engineering of the chloroplast genome. *Nat. Biotechnol.* **16,** 345–348.
5. Scott, S. E. and Wilkinson, M. J. (1999) Low probability of chloroplast movement from oilseed rape (*Brassica napus*) into wild *Brassica rapa. Nat. Biotechnol.* **17,** 390–392.
6. Link, G. (1991) Photoregulated development of chloroplasts. *Cell Cult. Somatic Cell Genet. Plants* **7B,** 365–394.
7. Mullet, J. E. (1988) Chloroplast development and gene expression. *Annu. Rev. Plant Physiol. Plant Mol. Biol.* **39,** 475–502.
8. Bendich, A. J. (1987) Why do chloroplasts and mitochondria contain so many copies of their genome? *BioEssays* **6,** 279–282.
9. Sugiura, M. (1995) The chloroplast genome. *Essays Biochem.* **30,** 49–57.
10. Palmer, J. D. (1985) Comparative organization of chloroplast genomes. *Annu. Rev. Genet.* **19,** 325–354.
11. Sugita, M. and Sugiura, M. (1996) Regulation of gene expression in chloroplasts of higher plants. *Plant Mol Biol.* **32,** 315–326.
12. Freyer, R., Kiefer-Meyer, M. C., and Kossel, H. (1997) Occurrence of plastid RNA editing in all major lineages of land plants. *Proc. Natl. Acad. Sci. USA* **94,** 6285–6290.
13. Boynton, J. E. and Gillham, N. W. (1993) Chloroplast transformation in *Chlamydomonas. Methods Enzymol.* **217,** 510–536.
14. Maliga, P., Carrer, H., Kanevski, I., Staub, J. M., and Svab, Z. (1993) Plastid engineering in land plants: a conservative genome is open to change. *Philos. Trans. R. Soc. Lond. B. Biol. Sci.* **342,** 203–208.

15. Daniell, H. (1993) Foreign gene expression in chloroplasts of higher plants mediated by tungsten particle bombardment. *Methods Enzymol.* **217,** 536–556.
16. Bock, R. (2001) Transgenic plastids in basic research and plant biotechnology. *J. Mol. Biol.* **312,** 425–438.
17. Hager, M. and Bock, R. (2000) Enslaved bacteria as new hope for plant biotechnologists. *Appl. Microbiol. Biotechnol.* **54,** 302–310.
18. Heifetz, P. (2000) Genetic engineering of the chloroplast. *Biochimie* **82,** 655–666.
19. Maliga, P. (1993) Towards plastid transformation in flowering plants. *TIBTECH* **11,** 101–106.
20. Golds, T., Maliga, P., and Koop, H.-U. (1993) Stable plastid transformation in PEG-treated protoplasts of *Nicotiana tabacum. BioTechnology* **11,** 95–97.
21. Blowers, A. D., Bogorad, L., Shark, K. B., and Sanford, J. C. (1989) Studies on *Chlamydomonas* chloroplast transformation: foreign DNA can be stably maintained in the chromosome. *The Plant Cell* **1,** 123–32.
22. Randolph-Anderson, B., Boynton, J. E., Dawson, J., Dunder, E., Eskes, R., Gillham, N. W., et al. (1995) Sub-micron gold particles are superior to larger particles for efficient biolistic transformation of organelles and some cell types. *BioRad Technical Bulletin 2015.*
23. Staub, J. M. and Maliga, P. (1992) Long regions of homologous DNA are incorporated into the tobacco plastid genome by transformation. *The Plant Cell* **4,** 39–45.
24. Carrer, H., Hockenberry, T. N., Svab, Z., and Maliga, P. (1993) Kanamycin resistance as a selectable marker for plastid transformation in tobacco. *Mol. Gen. Genet.* **241,** 49–56.
25. Daniell, H., Muthukumar, B., and Lee S.-B. (2001) Marker free transgenic plants: engineering the chloroplast genome without the use of antibiotic selection. *Curr. Genet.* **39,** 109–116.
26. Cerutti, H., Johnson, A. M., Boynton, J. E, and Gillham, N. W. (1995) Inhibition of chloroplast DNA recombination and repair by dominant negative mutants of *Escherichia coli RecA. Mol. Cell. Biol.* **15,** 3003–3011.
27. Zoubenko, O. V., Allison, L. A., Svab, Z., and Maliga, P. (1994) Efficient targeting of foreign genes into the tobacco plastid genome. *Nucleic Acids Res.* **22,** 3819–3824.
28. Staub, J. M. and Maliga, P. (1995) Marker rescue from the Nicotiana tabacum plastid genome using a plastid/*Escherichia coli* shuttle vector. *Mol. Gen. Genet.* **249,** 37–42.
29. Carrer, H. and Maliga, P. (1995) Targeted insertion of foreign genes into the tobacco plastid genome without physical linkage to the selectable marker gene. *BioTechnology.* **13,** 791–794.
30. Hajdukiewicz, P. T. J., Gilbertson, L., and Staub, J. M. (2001) Multiple pathways for Cre/lox-mediated recombination in plastids. *The Plant J.* **27,** 161–170.
31. Corneille, S., Lutz, K., Svab, Z., and Maliga, P. (2001) Efficient elimination of selectable marker genes from the plastid genome by the Cre-lox site-specific recombination system. *The Plant J.* **27,** 171–178.
32. Iamtham, S. and Day, A. (2000) Removal of antibiotic resistance genes from transgenic tobacco plastids. *Nat. Biotechnol.* **18,** 1172–1176.
33. Sikdar, S. R., Serino, G., Chaudhuri, S., and Maliga, P. (1998) Plastid transformation in *Arabidopsis thaliana. Plant Cell Rep.* **18,** 20–24.
34. Sidorov, V. A., Kasten, D., Pang, S.-Z., Hajdukiewicz, P. T. J., Staub, J. M., and Nehra, N. (1999) Stable chloroplast transformation in potato: use of green fluorescent protein as a plastid marker. *The Plant J.* **19,** 209–216.
35. Ruf, S., Herrmann, M., Berger, I. J., Carrer, H., and Bock, R. (2001) Stable genetic transformation of tomato plastids and expression of a foreign protein in fruit. *Nat. Biotechnol.* **19,** 870–875.
36. Khan, M. S. and Maliga, P. (1999) Fluorescent antibiotic resistance marker for tracking plastid transformation in higher plants. *Nat. Biotechnol.* **17,** 910–915.
37. Gruissem, W. and Tonkyn, J. C. (1993) Control mechanisms of plastid gene expression. *Crit. Rev. Plant Sci.* **12,** 19–55.
38. Maliga, P. (1998) Two plastid RNA polymerases of higher plants: an evolving story. *Trends in Plant Science* **3,** 4–6.

39. Allison, L. A. (2000) The role of sigma factors in plastid transcription. *Biochimie* **82,** 537–548.

40. Tiller, K. and Link, G. (1993) Sigma-like transcription factors from mustard (*Sinapis alba L.*) etioplast are similar in size to, but functionally distinct from, their chloroplast counterparts. *Plant Mol. Biol.* **21,** 503–513.

41. Tiller, K. and Link, G. (1993) Phosphorylation and dephosphorylation affect functional characteristics of chloroplast and etioplast transcription systems from mustard (*Sinapis alba L.*). *EMBO J.* **12,** 1745–1753.

42. Allison, L. A. and Maliga, P. (1995) Light-responsive and transcription-enhancing elements regulate the plastid psbD core promoter. *EMBO J.* **14,** 3721–3730.

43. Allison, L. A., Simon, L. D., and Maliga, P. (1996) Deletion of rpoB reveals a second distinct transcription system in plastids of higher plants. *EMBO J.* **15,** 2802–2809.

44. Serino, G. and Maliga, P. (1998) RNA polymerase subunits encoded by the plastid rpo genes are not shared with the nucleus-encoded enzyme. *Plant Physiol.* **117,** 1165–1170.

45. Hajdukiewicz, P. T. J., Allison, L. A., and Maliga, P. (1997) The two RNA polymerases encoded by the nuclear and the plastid compartments transcribe distinct groups of genes in tobacco plastids. *EMBO J.* **16,** 4041–4048.

46. Cahoon, A. B. and Stern, D. (2001) Plastid transcription: a menage a trois? *Trends in Plant Science* **6,** 45–46.

47. Bligny, M., Coutois, F., Thaminy, S., Chang, C. C., Lagrange, T., Baruah-Wolff, J., et al. (2000) Regulation of plastid rDNA transcription by interaction of CDF2 with two different RNA polymerases. *EMBO J.* **19,** 1851–1860.

48. Staub, J. M., Maliga, P. (1993) Accumulation of D1 polypeptide in tobacco plastids is regulated via the untranslated region of the *psbA* mRNA. *EMBO J.* **12,** 601–606.

49. Stern, D. and Gruissem, W. (1987) Control of plastid gene expression: 3' inverted repeats act as mRNA processing and stabilization elements, but don not terminate transcription. *Cell* **51,** 1145–1157.

50. Monde, R. A., Schuster, G., and Stern, D. B. (2000) Processing and degradation of chloroplast mRNA. *Biochimie.* **82,** 573–582.

51. Hayes, R., Kudla, J., Schuster, G., Gabay, L., Maliga, P., and Gruissem, W. (1996) Chloroplast mRNA 3'-end processing by a high molecular weight protein complex is regulated by nuclear encoded RNA binding proteins. *EMBO J.* **15,** 1132–1141.

52. Rott, R., Levy, H., Drager, R. G., Stern, D. B., and Schuster, G. (1998) 3'-processed mRNA is preferentially translated in *Chlamydomonas reinhardtii* chloroplasts. *Mol. Cell. Biol.* **18,** 4605–4611.

53. Stern, D. B., Radwanski, E. R., and Kindle, K. L. (1991) A 3' stem/loop structure of the *Chlamydomonas* chloroplast *atpB* gene regulates mRNA accumulation in vivo. *The Plant Cell* **3,** 285–297.

54. Rott, R., Liveanu, V., Drager, R. G., Stern, D. B., and Schuster, G. (1998) The sequence and structure of the 3'-untranslated regions of chloroplast transcripts are important determinants of mRNA accumulation and stability. *Plant Mol Biol.* **36,** 307–314.

55. Staub, J. M. and Maliga, P. (1994) Translation of *psbA* mRNA is regulated by light via the 5'-untranslated region in tobacco plastids. *The Plant J.* **6,** p. 547–553.

56. Shiina, T., Allison, L. A., and Maliga, P. (1998) rbcL transcript levels in tobacco plastids are independent of light: reduced dark transcription rate is compensated by increased mRNA stability. *The Plant Cell* **10,** 1713–1722.

57. Yehudai-Resheff, S., Hirsh, M., and Schuster G. (2001) Polynucleotide phosphorylase functions as both an exonuclease and a poly(A) polymerase in spinach chloroplasts. *Mol. Cell. Biol.* **21,** 5408–5416.

58. Schuster, G., Lisitsky, I., and Klaff, P. (1999) Polyadenylation and degradation of mRNA in the chloroplast. *Plant Physiol.* 937–944.

59. Kozak, M. (1999) Initiation of translation in prokaryotes and eukaryotes. *Gene* **234**, 187–208.

60. Fargo, D. C., Zhang, M., Gillham, N. W., and Boynton, J. E. (1998) Shine-Dalgarno-like sequences are not required for translation of chloroplast mRNAs in *Chlamydomonas reinhardtii* chloroplasts or in *Escherichia coli. Mol. Gen. Genet.* **257**, 271–282.

61. Rochaix, J.-D. (1996) Post-transcriptional regulation of chloroplast gene expression in *Chlamydomonas reinhardtii. Plant Mol. Biol.* **32**, 327–341.

62. Esposito, D., Hicks, A. J., and Stern, D. B. (2001) A role for initiation codon context in chloroplast translation. *The Plant Cell.* **13**, 2373–2384.

63. Bonham-Smith, P. C. and Bourque, D. (1989) Translation of chloroplast-encoded mRNA: potential initiation and termination signals. *Nucleic Acids Res.* **17**, 2057–2080.

64. Ye, G.-N., Hajdukiewicz, P. T. J., Broyles, D., Rodriguez, D., Xu, C. W., Nehra, N., et al. (2001) Plastid-expressed 5-enolpyruvylshikimate-3-phosphate synthase genes provide high level glyphosate tolerance in tobacco. *The Plant J.* **25**, 261–270.

65. Olins, P. O., Devine, C., Rangwala, S. H., and Kavka, K. S. (1988) The T7 phage gene 10 leader RNA, a ribosome binding site that dramatically enhances the expression of foreign genes in *Escherichia coli. Gene* **73**, 227–235.

66. Lutz, K. A., Knapp, J. E., and Maliga, P. (2001) Expression of bar in the plastid genome confers herbicide resistance. *Plant Physiol.* **125**, 1585–1590.

67. Kuroda, H. and Maliga, P. (2001) Complimentarity of the 16S rRNA penultimate stem with sequences downstream of the AUG destabilizes the plastid mRNAs. *Nucleic Acids Res.* **29**, 970–975.

68. Kuroda, H. and Maliga, P. (2001) Sequences downstream of the translation initiation codon are important determinants of translation efficiency in chloroplasts. *Plant Physiol.* **125**, 430–436.

69. McBride, K., Schaaf, D. J., Daley, M., and Stalker, D. M. (1994) Controlled expression of plastid transgenes in plants based on a nuclear DNA-encoded and plastid-targeted T7 RNA polymerase. *Proc. Natl. Acad. Sci. USA* **91**, 7301–7305.

70. McBride, K., Svab, Z., Schaaf, D. J., Hogan, P. S., Stalker, D. M., and Maliga, P. (1995) Amplification of a chimeric *Bacillus* gene in chloroplasts leads to an extraordinary level of an insecticidal protein in tobacco. *BioTechnology (NY)* **13**, 362–365.

71. Staub, J. M. and Maliga, P. (1995) Expression of a chimeric *uidA* gene indicates that polycistronic mRNAs are efficiently translated in tobacco plastids. *Plant J.* **7**, 845–848.

72. De Cosa, B., Moar, W., Lee, S.-B., Miller, M., and Daniell, H. (2001) Overexpression of the Bt cry2Aa2 operon in chloroplasts leads to formation of insecticidal crystals. *Nat. Biotechnol.* **19**, 71–74.

73. Barkan, A., Walker, M., Nolasco, M., and Johnson, D. (1994) A nuclear mutation in maize blocks the processing and translation of several chloroplast mRNAs and provides evidence for the differential translation of alternative mRNA forms. *EMBO J.* **13**, 3170–3181.

74. Staub, J. M., Garcia, B., Graves, J., Hajdukiewicz, P. T., Hunter, P., Nehra, N., et al. (2000) High-yield production of a human therapeutic protein in tobacco chloroplasts. *Nat. Biotechnol.* **18**, 333–338.

75. Vierstra, R. D. (1996) Proteolysis in plants: mechanisms and functions. *Plant Mol. Biol.* **32**, 275–302.

76. Daniell, H., Lee, S.-B., Panchal, T., and Wiebe, P. O. (2001) Expression of the native cholera toxin B subunit gene and assembly of functional oligomers in transgenic tobacco chloroplasts. *J. Mol. Biol.* **311**, 1001–1009.

77. Drescher, A., Ruf, S., Calsa Jr., T., Carrer, H., and Bock, R. (2000) The two largest chloroplast genome-encoded open reading frames of higher plants are essential genes. *The Plant J.* **22**, 97–104.

78. Suzuki, J. Y. and Maliga, P. (2000) Engineering of the *rpl23* gene cluster to replace the plastid RNA polymerase a subunit with the *Escherichia coli* homologue. *Curr. Genet.* **38**, 218–225.

79. Bock, R., Kossel, H., and Maliga, P. (1994) Introduction of a heterologous editing site into the tobacco plastid genome: the lack of RNA editing leads to a mutant phenotype. *EMBO J.* **13**, 4623–4628.

80. Kanevski, I. and Maliga, P. (1994) Relocation of the plastid *rbcL* gene to the nucleus yields functional ribulose-1,5-bisphosphate carboxylase in tobacco chloroplasts. *Proc. Natl. Acad. Sci. USA* **91**, 1969–1973.

81. Whitney, S. M. and Andrews, T. J. (2001) Plastome-encoded bacterial ribulose-1,5-bisphosphate carboxylase/oxygenase (RubisCO) supports photosynthesis and growth in tobacco. *Proc. Natl. Acad. Sci. USA* **98**, 14,738–14,743.

82. Burrows, P. A., Sazanov, L. A., Svab, Z., Maliga, P., and Nixon, P. J. (1998) Identification of a functional respiratory complex in chloroplasts through analysis of tobacco mutants containing disrupted plastid *ndh* genes. *EMBO J.* **17**, 868–876.

83. Kofer, W., Koop, H.-U., Wanner, G., and Steinmuller, K. (1998) Mutagenesis of the genes encoding subunits A, C, H, I, J and K of the plastid NAD(P)H-plastoquinone-oxidoreductase in tobacco by polyethylene glycol-mediated plastome transformation. *Mol. Gen. Genet.* **258**, 166–173.

84. Ruf, S., Kossel, H., and Bock, R. (1997) Targeted inactivation of a tobacco introns-containing open reading frame reveals a novel chloroplast-encoded Photosystem I-related gene. *J. Cell Biol.* **139**, 95–102.

85. Hager, M., Biehler, K., Illerhaus, J., Ruf, S., and Bock, R. (1999) Targeted inactivation of the smallest plastid genome-encoded open reading frame reveals a novel and essential subunit of the cytochrome b6f complex. *EMBO J.* **18**, 5834–5842.

86. Serino, G. and Maliga, P. (1997) A negative selection scheme based on the expression of cytosine deaminase in plastids. *The Plant J.* **12**, 697–701.

87. DeGray, G., Rajasekaran, K., Smith, F., Sanford, J., and Daniell, H. (2001) Expression of an antimicrobial peptide via the chloroplast genome to control phytopathogenic bacteria and fungi. *Plant Physiol.* **127**, 852–862.

88. Zhang, X.-H., Brotherton, J. E., Widholm, J. M., and Portis Jr., A. R. (2001) Targeting a nuclear anthranilate synthase a-subunit gene to the tobacco plastid genome results in enhanced tryptophan biosynthesis. Return of a gene to its pre-endosymbiotic origin. *Plant Physiol.* **127**, 131–141.

89. Kanevski, I., Maliga, P., Rhoades, D. F., and Gutteridge, S. (1999) Plastome engineering of ribulose-1,5-bisphosphate carboxylase/oxygenase in tobacco to form a sunflower large subunit and tobacco small subunit hybrid. *Plant Physiol.* **119**, 133–141.

12

Oleosin Partitioning Technology for Production of Recombinant Proteins in Oil Seeds

Maurice M. Moloney

1. Introduction

Plant-based expression systems for the production of recombinant proteins have been the subject of numerous studies in recent years. It is now clear that plant-based systems offer a number of substantial advantages over conventional fermentation and cell-culture systems, including the exclusion of mammalian pathogens or bacterial endotoxins, improved scale-up economics, and reduced cost of goods. Plants have also proven to be capable of the assembly of complex multimeric proteins such as secretory IgAs, which are difficult to produce in other systems. However, even among plant systems, there are numerous configurations and modalities for the expression of a particular protein. For example, the actual choice of host species may be affected by a number of considerations, including biochemistry, extraction, and downstream processing or regulatory parameters. Similarly, within a plant, the deposition of the protein in certain tissues , organs, or even subcellular compartments may have a profound effect on the efficiency and economics of production. Although it has proven possible to express recombinant proteins in vegetative cells such as leaves, tubers and roots, there are some singular advantages to the use of seeds as the site of recombinant protein deposition.

In their natural state, seeds utilize sophisticated and versatile storage mechanisms for a variety of products. Seeds store lipids or carbohydrates as a carbon source, protein for carbon and nitrogen, and certain inorganics such as phosphate, which is sequestered in inositol hexaphosphate (phytate). Seeds are adapted to storing these products for extended periods of time, often under extreme conditions of cold, heat, and water stress. In seeds, the deposition and storage of proteins occurs in a variety of different tissues, including cotyledons, perisperm, and endosperm.

The fact that the seed is the site of extensive protein deposition in plants raises the question of whether other (recombinant) proteins could be stored in seeds of transgenic plants with the same efficiency. If so, with the use of gene-transfer techniques it should be possible to modify the types of proteins in seeds to help them accumulate those that are useful for therapeutic or industrial purposes.

From: *Handbook of Industrial Cell Culture: Mammalian, Microbial, and Plant Cells*
Edited by: V. A. Vinci and S. R. Parekh © Humana Press Inc., Totowa, NJ

2. The Rationale for Using Oil Seeds as a Host

Plant seeds typically store part of the energy needed for germination in organelles known as oil bodies or oleosomes. Oil bodies are spherical structures, made up of oil droplets of neutral lipid (most frequently triacylglycerides) surrounded by a half-unit phospholipid membrane. A micrograph of typical seed cells containing oil bodies is shown in Fig. 1. The entire surface of these oil bodies covered by a unique class of protein called oleosins. These proteins have an extremely hydrophobic core and appear to be the most lipophilic proteins reported thus far in the protein and DNA databases. Despite an intensely hydrophobic core, their N- and C- termini are hydrophilic or amphipathic *(1)*. Oleosins become associated with nascent oil bodies during their bio-genesis on the ER by a cotranslation mechanism *(2,3)*. Oleosins in many oil seeds accumulate at relatively high levels. It has been reported for the *Brassica* species, for example, oleosins may comprise as much as 8–20% of total seed protein *(4)*. This level of accumulation indicates that oleosin genes are strongly transcribed during seed development, and probably undergo minimal turnover.

Traditionally, industrial oil seeds are extracted under harsh conditions to manufacture vegetable oil. These conditions may include high pressure and shear forces, elevated temperatures (60–80°C) and extraction with organic solvents such as hexane. However, when oil seeds are extracted in aqueous media, they form a three-phase mixture of insoluble material, aqueous extract, and a very stable emulsion of oil bodies, which can be resolved by low-speed centrifugation, resulting in the flotation of the oil bodies accompanied by their protein complement. Successive aqueous washes of this oil body fraction remove any extraneous proteins bound peripherally to oil bodies. The oleosins, conversely, remain tightly associated with the oil bodies because of their highly lipophilic core, which acts as an transmembrane anchor.

Protein analysis of oil body preparations that have undergone flotation centrifugation and washing reveals that virtually all the other seed proteins are absent—they have partitioned into the aqueous phase. In such oil body preparations, the oleosin fraction is highly enriched, as shown in Fig. 2.

These observations suggest that oleosins could serve as vehicles or carriers for heterologous proteins expressed in plant seeds. This would require that the desired proteins were bound covalently or noncovalently to oil bodies. If this association was strong enough to survive extraction and centrifugation, it would facilitate the recovery and purification of recombinant proteins in plants. If successful, such a process would have an enormous impact on production economics, because in most protein production systems, a significant proportion of the cost-of-goods is accounted for by the cost of downstream purification of the protein.

3. Oleosin Subcellular Targeting and its Relevance to Recombinant Protein Production

Oleosins are seed proteins that are present in all common oil seeds. They are also constituents of many other seeds that are not cultivated primarily for their oil, including corn and cotton. In oil seeds such as sunflower (*Helianthus annuus*), safflower (*Carthamus tinctorius*), canola (*Brassica napus*), and flax (*Linum usitatissimum*), triacylglycerol may comprise as much as 40% of the dry weight of the seed, and there-

Typical oilseed cells:

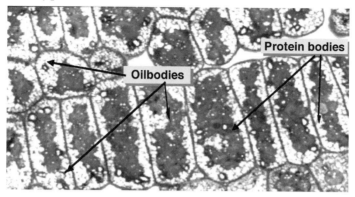

Fig. 1. Transverse section of hypocotyl cells of developing *Brassica napus* seeds, indicating the protein bodies (dark staining), oil bodies (lighter-stained spheres), and apoplastic space. These subcellular locations have each been shown to act as suitable sites for recombinant protein deposition in seeds.

Coomassie Blue Stained Gel

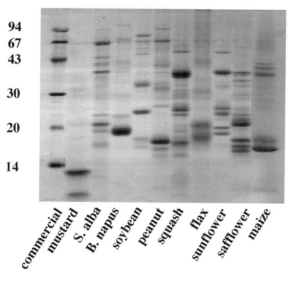

Fig. 2. Coomassie blue-stained protein gel of oil body fractions from a variety of common oil seeds. "Commercial mustard" refers to an extract of a comestible mustard preparation ("Grey Poupon") commonly consumed in North America and indicative of the ubiquity of oleosins in many human food products. The protein patterns of oil body fractions are very simple compared to whole-cell extracts.

fore, up to 50% of seed cellular volume may be occupied by oil bodies. Oil bodies in seeds comprise three components: a neutral lipid lumen, a phospholipid half-unit membrane derived from the endoplasmic reticulum (ER), and a protein "shell" comprising one or more isoforms of the oleosins *(1)*. Oil bodies in seeds occur in relatively large numbers (often hundreds per cell), and are typically 0.5–2 µm in diameter. Other proteins may adhere to oil bodies in vivo —and certainly upon extraction from seeds, other proteins not usually resident on oil bodies can adhere peripherally *(5)*. However, it is clear that oleosins associate with oil bodies in a manner that is very distinct from other cellular proteins. By using relatively harsh treatments such as washing with 8 *M* urea or 0.5 *M* sodium carbonate, it is possible to remove all surface-adhering proteins except oleosins from the oil bodies of most species. It is now clear that the topology of oleosins in vivo is such that their highly lipophilic core is embedded in the lumen of the oil body, while the N- and C-termini are cytoplasmically oriented *(1,6)*. It appears that oleosins effectively form a "shell" around oil bodies, as suggested by treatments with phospholipase C, which has no effect on the disruption of oil bodies unless they are pretreated with a protease such as trypsin, which presumably breaches this proteinaceous "shell." Each seed cell contains a large number of small oil bodies, presenting a large surface area that is densely packed with oleosins. As a result, the oleosin content of many seeds may approach those normally associated with storage proteins. The fundamental role of oleosins appears to be in encapsulating oil bodies and probably controlling oil-body surface-to-volume ratio, a property, that presumably facilitates lipolysis during germination. It has also been suggested that oleosins help to prevent coalescence of oil bodies, particularly during the early phases of germination *(7)*.

3.1. Oleosin Targeting to Oil Bodies

The targeting of oleosins to the oil body has been a subject of several studies. Unlike seed-storage proteins, oleosins undergo subcellular targeting to oil bodies without any structural modifications such as cleavage of N- or C-termini *(8,9)*. Oleosins appear to use a signal-anchor sequence, which is unaffected by ER lumen proteases. The nascent oleosin polypeptide appears to associate with the ER in a signal-recognition particle (SRP)-dependent pathway *(10,11)*. The ER-associated oleosin polypeptide is mobilized to the developing oil bodies by a mechanism that is not yet known. However, it is known that certain structural features such as the presence of a motif known as a proline knot, are essential for this migration, or at least for the maintenance of oleosin stability on the oil body after transfer *(6)*. This feature comprises three prolines distributed in a 12 amino-acid stretch, which results in a hairpin bend in the polypeptide. It has been suggested that this structural feature also permits the formation of an antiparallel interaction between the flanking amino acids of the hydrophobic core, which may impart structural stability to oleosins *(1)*.

3.2. Subcellular Targeting of Recombinant Oleosins

Although it is now clear that native oleosins are targeted with high avidity to oleosomes in vivo, it was an open question whether the creation of recombinant oleosins by addition of polypeptide "tails at the N-or C-termini would be deleterious to targeting. Such aberrant targeting has been encountered previously with modified storage proteins such as recombinant phaseolin *(12)* or 2S albumin *(13)*. Aberrant targeting generally results in protein instability and turnover. Consequently, experiments were performed to test the idea that modifications of oleosins at either the N- or C- termini end may affect overall targeting to oil bodies *(8,9)*. Those studies showed that both N-

and C- terminal translational fusions of oleosin with b-glucuronidase (GUS), did not significantly impair the basic targeting mechanism. Furthermore, the C-terminal oleosin-GUS fusion remained enzymatically active. This activity was followed in oil body extracts to test whether the oleosin-GUS protein was attached to the oil body with similar avidity to that of native oleosin. It was found that oleosin fusions associate with oil bodies with similar avidity to native oleosins.

The domains that determine the targeting of oleosins were investigated by van Rooijen and Moloney *(8)*. In these experiments, it was shown that removal of the highly variable C-terminal of the oleosin protein had no significant impact on its subcellular targeting. This was subsequently verified using an in vitro analytical method, which showed that ER association of the oleosin nascent polypeptide is unaffected by removal of the C-terminus of the protein *(6)*.

These experiments have elucidated a number of important properties of oleosins. First, they can be extended at either end and will still undergo oil body targeting. Second, long polypeptide extensions do not seem to pose a problem (GUS has a mW of ~67 kDa). Furthermore, these experiments were performed using a β-glucuronidase known to be susceptible to N-glycosylation at position 358. It has been shown that if such a β-glucuronidase is exposed to the ER lumen, GUS is inactivated as a result of glycosylation *(14)*. In the case of the oleosin C-terminal extensions, GUS was fully functional, indicating that it was not glycosylated. This finding establishes that fusions at the C-terminal of the oleosin are not exposed to the ER lumen and remain on the cytoplasmic side. More recent studies *(10)* unequivocally established that under normal conditions, the N- and C- termini of oleosins are always displayed on the cytoplasmic side of the ER.

These fundamental properties of oleosins and their targeting suggest that a recombinant protein expressed as an oleosin translational fusion would be targeted much as a native oleosin, and would not undergo post-translational modifications associated with the secretory pathway. Furthermore, they would predict that an oleosin-recombinant protein fusion would associate exclusively with the oil body fraction. This prediction has been confirmed with several recombinant proteins and is illustrated in Fig. 3, where various subcellular fractions are analyzed. As seen in this figure, an oleosin-fused polypeptide associates predominantly with the oil body, and is not substantially associated with other cellular compartments. Upon flotation cetrifugation, the desired protein-fusion is the only protein found in the oil body fraction other than the native oleosin complement.

This subcellular targeting approach can be further refined by interposing specific labile cleavage sites between the oleosin and recombinant protein domain. In the early experiments, these were four amino-acid proteolytic sites hydrolyzed by enzymes such as thrombin or factor Xa. Such a configuration should permit the recombinant polypeptide domain to be cleaved from the surface of the oil bodies (*see* Fig. 4). As seen in Fig. 4, this cleavage can indeed be effected using oil bodies suspended in cleavage buffer and subjected to a specific protease treatment. The resulting product in the aqueous "undernatant" phase is predominantly the desired protein or polypeptide. Factors that affect the choice of cleavage site and cleavage agent are considered in **Subheading 4.**

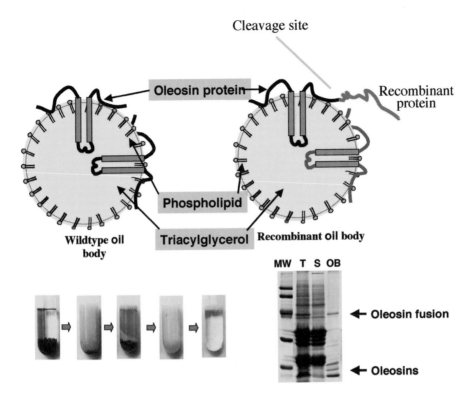

Fig. 3. (**A**) Configurations of native and recombinant oleosins on oil bodies. Oleosins comprise a lipophilic central domain, which anchors the oleosin to the oil body. The N- and C-termini of the oleosin are always displayed on the cytoplasmic or aqueous side of the oil body. (**B**) Diagram illustrating the purification of oil bodies by flotation centrifugation and the extent of protein enrichment by this process shown in a polyacrylamide gel. MW = mol wt marker, T = total cell protein, S = soluble protein, OB = oil body-associated protein.

4. Production and Purification of the Anti-Coagulant, Hirudin, as an Oleosin Fusion

Although the experiments detailed in the previous section demonstrate the basic principles of oil body-based recombinant protein production, it is intriguing to apply it in cases where the recombinant protein is of great value and difficult to manufacture economically. Accordingly, we created a translational fusion between an *Arabidopsis* oleosin-coding sequence and the coding sequence of the mature form of the blood anticoagulant, Hirudin. Hirudin is a naturally occurring thrombin inhibitor that is produced and secreted in the salivary glands of the medicinal leech, *Hirudo medicinalis*. Hirudin is a powerful blood anticoagulant with a number of very desirable properties, including stoichiometric inhibition of thrombin, short clearing time from the blood, and low immunogenicity. The unit cost of production in leeches would be prohibitive for extensive therapeutic applications. Hirudin has been made in a variety of microorganisms including *E. coli* and yeast, but these entail significant fixed costs associated with fermentation. Therefore, our objective was to test the oleosin expression and purification system as an alternative and potentially inexpensive source of hirudin.

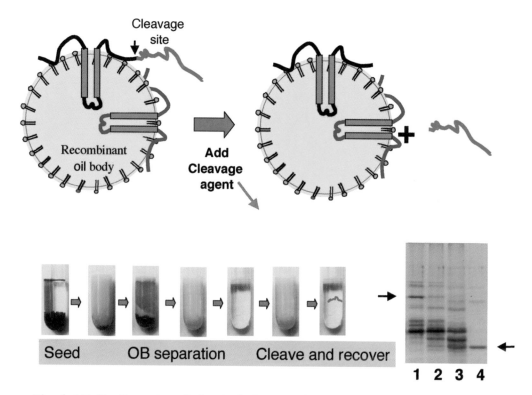

Fig. 4. (**A**) Configuration of oleosin fusion proteins and their cleavage from oil bodies. (**B**) Depiction of the purification and cleavage process and the analysis of the cleavage reaction. 1. Oilbody preparation of recombinant oil bodies, 2. Treatment of oil bodies with 1:1000 cleavage enzyme to fusion protein, 3. Treatment of oil bodies with 1:100 cleavage enzyme to fusion protein, 4. Aqueous "undernatant" fraction after cleavage.

Constructs containing the translational fusion of oleosin-hirudin under the transcriptional control of an oleosin promoter from *Arabidopsis* were introduced into *Brassica napus* and *Brasica carinata* using *Agrobacterium*-mediated transformation *(15)*. The resulting transgenic plants yielded seed in which a protein of approx 25 kDa could be detected. This is shown in Fig. 5, where the fusion protein is visualized both with an anti-hirudin antibody and an anti-oleosin antibody. The fusion protein crossreacted efficiently with a monoclonal antibody (MAb) raised against hirudin. This protein proved to be associated tightly with the oil bodies and could not be removed from the oil bodies by successive washings. Attempts to determine whether the oleosin-hirudin fusion protein had anti-thrombin activity suggested that the fusion protein was completely inactive. As part of the construction of the translational fusion, we incorporated four additional codons that specified a Factor Xa cleavage site (I-E-G-R). This configuration was designed to permit release of the hirudin polypeptide sequence from the oil body by proteolytic cleavage. When the washed oil bodies from these seeds were treated with Factor Xa, hirudin polypeptide was released into the aqueous phase. As can be seen in Fig. 6, only the cleaved hirudin product showed this anti-thrombin activity. No refolding steps were required to reveal this activity.

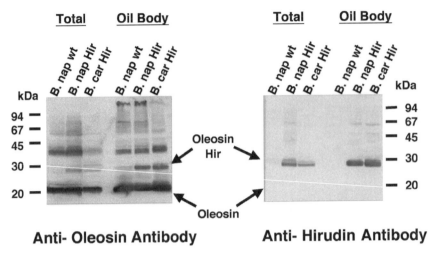

Fig. 5. Western blot analysis of oleosin-hirudin fusions in seeds of *Brassica napus* (B. nap) and *Brassica carinata* (B. car). Left-hand panel was visualized using anti-oleosin antibodies, right-hand panel was visualized using anti-hirudin antibodies.

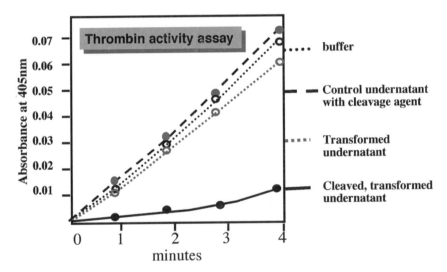

Fig. 6. Enzymatic assay of thrombin in the presence of oil body extracts with and without hirudin. Increases in the absorbance at 405 nm indicate cleavage of substrate by thrombin. In this assay, only cleaved oleosin-hirudin gives rise to an active anti-thrombin product.

Authentic hirudin has three disulfide bridges. These are essential to its activity. Thus, if the plant were to make hirudin polypeptide, but did not allow its correct folding, no thrombin inhibition would be detected unless a refolding treatment was applied. In fact, after Factor Xa treatment, the aqueous phase showed strong antithrombin activity, demonstrating that biologically active hirudin was released. The specific activity of the inhibitor was similar to that of recombinant hirudin secreted from yeast cells, indicating that the majority of the hirudin released was correctly folded and that disulfide bridges were appropriately configured. The released hirudin was then subjected to ion-exchange chromatography. The fractions showing anti-thrombin activity were concentrated and loaded onto a reversed-phase analytical high-performance liquid chromatography (HPLC) column. This showed that the hirudin was substantially pure after the ion-exchange step *(16,17)*. The HPLC trace showed two peaks close to the expected retention volume of hirudin. Mass spectrometric analyses of these two peaks using MALDI-TOF revealed that the two peaks corresponding to full-length hirudin and a truncated product from which two C-terminal amino acids were missing. A similar truncated form was also found when Hirudin was expressed in yeast *(18)*. Both forms of the molecule are potent thrombin inhibitors.

5. Oilbody Fusions for the Production of Immobilized Enzymes

The potential for using oleosin fusions as an adaptable means of recombinant protein expression in seeds is under investigation. It is interesting to note that the production of oleosin-polypeptide fusions appears to function for rather short peptides such as IL-1β and Hirudin, but equally well for much longer polypeptides such as β-glucuronidase. When larger polypeptides are produced, there is strong evidence that proper folding occurs. This is supported by the fact that in a number of cases including β-GUS and xylanase, the fusion protein retains its enzymatic activity. This finding has also led to experiments that illustrate the utility of oil bodies as immobilization matrices *(9,19)*. Using β-glucuronidase as a model, it was shown that a dispersion of oil bodies carrying β-glucuronidase on their surface would hydrolyze glucuronide substrates for GUS with catalytic properties indistinguishable from soluble GUS *(19)*. Furthermore, it was shown that virtually all the enzymatic activity could be recovered and recycled by the use of flotation centrifugation to obtain the oil bodies. The enzyme in dry seed remains fully active for several years. Once extracted onto oil bodies, the half-life of the enzyme was about 4 wk *(9)*.

The expression of the enzyme, xylanase, on oil bodies illustrates several ways in which the basic technology may be used *(20)*. The enzyme accumulated on seed oil bodies in active form. From here it can be cleaved to release soluble xylanase into solution. In these experiments, it was shown that the immobilized xylanase had essentially the same kinetic properties as soluble xylanase from the same source (*Neocallimastix patriciarum*). Such a preparation could be useful in such processes as de-inking of recycled paper. Alternatively, without significant purification, the oil bodies could be mixed with insoluble substrate such as crude wood pulp to help break down the xylan crosslinks. This could be helpful in paper-making. Cost would be minimized because the enzyme itself could be recovered and re-used. In addition to these formulations, the enzyme could be delivered in seed meal to animals without purification. In this application, the enzyme would function in the stomach of a monogastric animal to improve digestibility of fiber. Clearly, this latter formulation would be created in conjunction with other cellulytic enzymes. The result, however, would be to provide monogastrics with enhanced feed efficiency and reduced biological waste.

6. Expression of Somatotropins as Oleosin Fusions— Oil Bodies as Delivery Vehicles

As demonstrated here, oleosin fusions may remain active without cleavage from the oil body. This may result in a novel formulation for the delivery of a biologically active molecule. This is exemplified by the expression of a cyprinid (carp) somatotropin as an N-terminal oleosin fusion *(21)*. It has proven difficult to express somatotropins in plants at economically viable levels *(22)*. Oleosins appear to offer several advantages for the accumulation and use of this class of protein. In this experiment, the coding sequence of growth hormone from the common carp was fused in frame with an oleosin coding sequence. This was used to create transgenic *Brassica napus* plants, which expressed the fusion in a seed-specific manner. The resulting fusion also contained a cleavage site so that it could be released from the oil body by endoproteolysis. The resulting polypeptide was of the correct mol wt and showed biological activity when injected into goldfish, as determined by the stimulation of insulin-like growth factor-1 (IGF-1) transcription. Fish somatotropins are also known to be orally active *(23)*; however, it was not clear whether the fusion polypeptide would show oral activity. To test this, formulations of feed pellets were prepared into which were incorporated oil body preparations containing the oleosin-somatotropin fusion protein. It was shown with rainbow trout that such feeding experiments resulted in substantial, and statistically significant increases in growth rates over an 8-wk period. Results from one of these feeding trials are shown in Fig. 7. These results further demonstrate the use of oil bodies not only for protein expression and recovery, but also for their use as delivery vehicles and as media for formulation of orally delivered biologically active molecules.

7. Storage and Recovery of Recombinant Proteins in Seeds

Seeds lend themselves readily to the accumulation, recovery, and purification of the transgene product. Seeds themselves are natural storage tissues in which proteins must remain intact for long periods, until germination of the plant. Thus, it is interesting to question whether a recombinant protein would also remain stable in seeds for extended periods. Using the example of hirudin, Fig. 8 shows that an oleosin fusion protein remains stable for an extended period (> 2.5 yr). Extractions after this period result in the recovery of fully functional proteins. This observation is important because it demonstrates that expression in seeds completely decouples the availability of recombinant protein from the growing season. Growth of redundant amounts of seed—which can then be stored—is very inexpensive, and contributes only a small percentage of the cost of goods. This inexpensive raw material also provides a low-capital intensive inventory, and allows for flexibility in responding to demand.

To date, there have been only a few studies performed on the recovery of recombinant proteins from seeds *(17,24–26)*. One significant advantage of seeds as hosts for recombinant proteins is that seed processing is a very sophisticated industry. As a result, a wide range of existing processing equipment and technology is available.

8. Aqueous Extraction Systems

When no purification is required, a technique as simple as dry milling may be adequate for the production of a recombinant protein formulation. This was the approach used by Pen et al. *(27)* for phytase. However, it is usually essential to extract and at least partially purify the desired protein. Where seeds contain substantial quantities of oil, it is customary to employ hexane extraction to recover maximal amounts of the oil (*see* Fig. 9a). In most instances involving recombinant proteins, this step is not

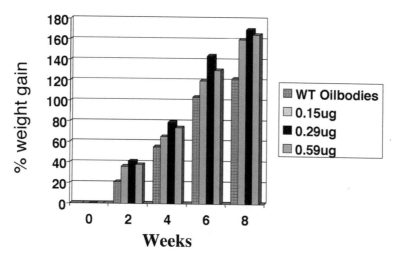

Fig. 7. Effect of orally delivered oleosin-carp somatotropin as a fusion protein on oil bodies, which were incorporated directly into feed. The figures 0.15, 0.29, 0.59 µg are incorporation rates in µg per gram of body wt of the fish at the start of the experiment.

Recombinant proteins are stable for years in seeds

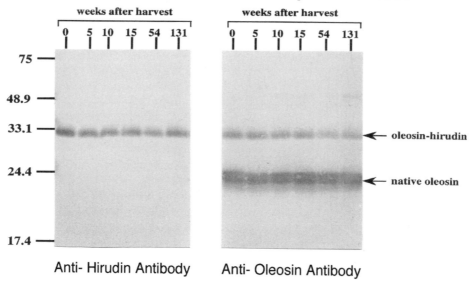

Anti- Hirudin Antibody   Anti- Oleosin Antibody

Fig. 8. Stability of recombinant proteins in oil seeds as oleosin fusions. Oleosin-hirudin fusion proteins visualized on western blots using an anti-hirudin antibody (left-hand panel) or an anti-oleosin antibody (right-hand panel). Times are weeks after harvest. Note that the fusion protein remains stable over the test period compared to the native oleosin. Storage conditions were at room temperature (23°C) at about 25–30% relative humidity.

possible because of the denaturant property of hexane and many organic solvents. Therefore, the extraction and purification of recombinant seeds usually requires aqueous extraction (*see* Fig. 9b). This process has been performed successfully, at scale, by

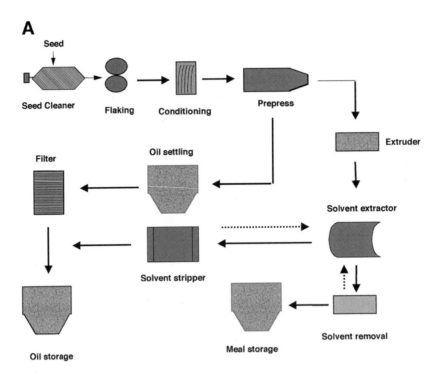

Fig. 9A. Process diagram depicting traditional extraction of oleaginous seeds, which typically involves a heating step (conditioning) and hexane extraction. Without significant modification, it would be difficult to extract biologically active proteins using this traditional procedure.

Kusnadi et al. *(25,26)* for the extraction of avidin and β-glucuronidase (GUS) from corn seed. They noted that despite the use of a constitutive (i.e., non-tissue-specific) promoter up to 98% of the GUS activity was deposited in the germ (embryo). This observation led to an advantageous extraction scheme in which the kernel separated into the endosperm and the embryo. The full-fat germ was then extracted using aqueous buffers to obtain a protein fraction, or alternatively, the germ was first defatted with hexane at 60°C. Surprisingly, the GUS enzyme was not inactivated by this solvent treatment, although this may be a property specific to GUS that is generally considered to be a very stable enzyme. The aqueous protein-rich fraction was subjected to four rounds of chromatography, including two ion exchanges, one hydrophobic interaction, and one size-exclusion step. The overall yield of purified protein was 10%. This purification scheme would be quite costly on a large scale and with such a low recovery would only be economical for high-value proteins. Yet, the scheme is simple, and lends itself readily to optimization of each individual step of the process.

9. Two-Phase Extraction Systems Using Oil Bodies

The alternative extraction scheme for separation and purification to homogeneity of a protein is based on oil body partitioning *(17)*. In this example, the host-seed that was used is canola. Thus, apart from dehulling, little pre-fractionation was possible. However, aqueous extraction of the whole seed followed by centrifugation to separate oil

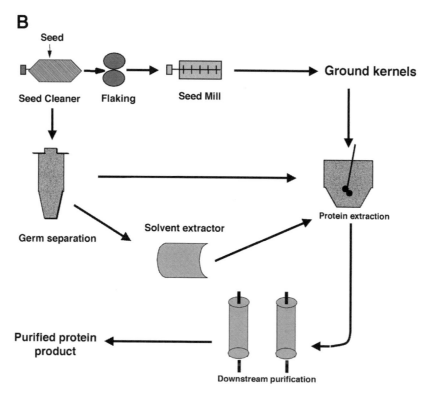

Fig. 9B. Process diagram for a modified procedure to permit extraction of recombinant proteins from corn while recovering some traditional byproducts (after Kusnadi et al., 1998 a, b).

bodies has proven to be a major enrichment step. This alternative process is depicted in Fig. 10. Once the oil bodies are washed, only small amounts of other seed proteins remain. For example, in the case of hirudin, the cleavage of recombinant oleosin-fusion protein from oil bodies provided a significant enrichment step. This extract was subjected to anion and reverse-phase chromatography and yielded an extraction efficiency of purified hirudin of about 40%, with a purity greater than 99% *(17)*. Minimization of chromatography steps and early-stage enrichment of the desired protein both assist greatly in improving overall rates of recovery.

The recovery and purification of expressed recombinant proteins from transgenic plants is probably the most critical factor in establishing plants as a practical alternative system for protein production. Although a variety of schemes may be envisioned, it is essential to minimize the number of processing steps and to carry out each step at much higher efficiencies than have thus far been reported. However, this area has only recently received attention and it is clear that greater efficiencies will be forthcoming.

10. Cleavage of Recombinant Proteins from Oil Bodies

In the initial work resulting the use of oleosins as carriers of recombinant proteins, the desired protein was expressed as a translational fusion. This poses the problem of cleavage of the recombinant protein from the oil body. Although there are many examples of cleavage enzymes that can be used at a laboratory scale—such as Factor

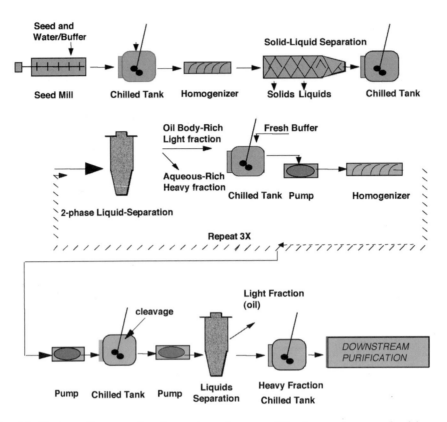

Fig. 10. Process diagram showing a two-phase partitioning system used with aqueous extraction to recover recombinant proteins from oil seeds, particularly when the desired protein is associated with oil bodies or oleosins.

Xa, thrombin, and enterokinase—it is generally accepted that these enzymes are too expensive for larger-scale use, and often do not provide a high efficiency of cleavage. These observations have led us to consider alternative approaches to cleavage reactions, using examples of high-efficiency cleavage in nature. Many proteins undergo specific cleavage reactions as part of their post-translational processing. This includes the cleavage of signal sequences during subcellular targeting and the maturation of a number of proteases. The cleavage reactions of some of the self-maturing proteases have been widely studied. However, it appears that little work has been performed on understanding their ability to function *in trans* with a heterologous polypeptide. Therefore, we have investigated this, using the self-processing of the milk-clotting enzyme chymosin as a model maturation reaction. In this study, we examined the potential of chymosin to function as a cleavage enzyme of a recombinant fusion protein which comprised the chymosin A propeptide leader that was fused translationally to a number of heterologous proteins. Using this configuration, it was shown that chymosin can act as a cleavage enzyme for a heterologous polypeptide associated with the chymosin pro-sequence. In fact, the cleavage reaction appears to be highly efficient, and can result in almost stoichiometric reactions *(28)*.

This approach offers a broad applications not only with fusion proteins associated with oil bodies, but also with fusion proteins produced by more conventional routes such as fermentation. Clearly, given the efficiency of the reaction and the relative cost of chymosin rather than Factor Xa, this process holds much promise for commercial-scale preparative procedures. A full report of this cleavage system and its uses will be published elsewhere *(29)*.

11. Use of Oil Bodies as Affinity Matrices: Noncovalent Polypeptide Attachment

In the original view of oleosins as carriers of recombinant proteins, the desired polypeptide was expressed as a covalently associated fusion protein, as described in **Subheading 10.** Nevertheless, it is clear that there are many polypeptides, which may require exposure to a subcellular compartment other than the cytoplasm, in which oleosins are effectively displayed. For example, it may be necessary for certain proteins to be exposed to the lumen of the ER, where appropriate disulfide bridging or other post-translational modifications can occur. Therefore, it was reasoned that proteins deposited into other cellular compartments may be recoverable by the association of the polypeptide with oil bodies using noncovalent binding. An example of this would be the inclusion of an "affinity tag" on the recombinant polypeptide—when it is exposed to the oil body during processing it preferentially associates with the oil body fraction. This would permit the recovery of such proteins using the same process of flotation centrifugation that is currently used for covalently attached polypeptides. An example of this is the use of a single-chain antibody (scFv) against oleosin, which has a high affinity for the oil bodies. This scFv can be expressed as a fusion protein with other polypeptides. In the presence of oil bodies carrying the recognized oleosin, the scFv adheres with high avidity to the oil bodies and thus allows for oil body washing and protein recovery. This method *(30)* has broad applicability, and allows for the recovery of proteins deposited in cellular compartments other than the oil body itself. Thus, with the use of a single manufacturing system, it is possible to recover and purify proteins that are covalently bound to oleosins, and any protein targeted to other cellular compartments including the secretory pathway. A detailed description of this work will appear elsewhere.

12. Conclusions and Future Prospects

The use of plant oil bodies and their associated proteins—oleosins—as vehicles for recombinant protein product has been exemplified with several proteins. The major advantage of using oil bodies as carriers is the ease with which proteins can be recovered and purified. Oleosin targeting does not seem to be impaired even by very long polypeptide extensions to the N- or C-termini of the oleosin. This greatly enhances the versatility of this system in contrast to alternative approaches for recombinant protein production in plants. Separation of oil bodies from seed extracts is amenable to scale-up using equipment typical to dairy operations such as cream-separators *(31)*. Thus, it seems likely that this system could be used for the production of a wide range of proteins of therapeutic, industrial, and feed or food use. We have investigated the production of a wide range of commercially attractive polypeptides in this system and have developed scaled-up extraction and purification systems, which could be applied to a variety of different oil seeds engineered for production of such oleosin-polypeptide fusions.

References

1. Huang, A. H. C. (1992) Oil bodies and oleosins in seeds. *Annu. Rev. Plant Physiol. Plant Mol. Biol.* **43,** 177-200.
2. Hills, M. J., Watson, M. D., and Murphy, D. J. (1993) Targeting of oleosins to the oil bodies of oil seed rape (*Brassica napus L.*). *Planta* **189,** 24–29.
3. Loer, D. and Herman, E. M. (1993) Cotranslational integration of soybean (Glycine max) oil body membrane protein oleosin into microsomal membranes. *Plant Physiol.* **101,** 993–998.
4. Huang, A. H. C. (1996) Oleosins and oil bodies in seeds and other organs. *Plant Physiol.* **110,** 1055–1061.
5. Kalinski, A., Melroy, D. L., Dwivedi, R. S., and Herman, E. M. (1992). A soybean vacuolar protein (p34) related to thiol proteases is synthesized as a glycoprotein precursor during seed maturation. *J. Biol. Chem.* **267,** 12,068–12,076.
6. Abell, B. M., Holbrook, L. A., Abenes, M., Murphy, D. J., Hills, M. J., and Moloney, M. M. (1997) Role of the proline knot motif in oleosin endoplasmic reticulum topology and oil body targeting. *Plant Cell* **9,** 148–193.
7. Leprince, O., van Aelst, A. C., Pritchard, H. W., and Murphy, D. J. (1998) Oleosins pre-vent oil-body coalescence during seed imbibition as suggested by a low-temperature scanning electron microscope study of desiccation-tolerant and -sensitive oil seeds. *Planta* **204,** 109–119.
8. van Rooijen, G. J. H. and Moloney, M. M. (1995a) Structural requirements of oleosin domains for subcellular targeting to the oil body. *Plant Physiol.* **109,** 1353–1361.
9. van Rooijen, G. J. H. and Moloney, M. M. (1995b) Plant seed oil-bodies as carriers for foreign proteins. *BioTechnology* **13,** 72–77.
10. Abell, B. M., High, S., and Moloney, M. M. (2002) Membrane protein topology of oleosin is constrained by its long hydrophobic domain. *J. Biol. Chem.* **277,** 8602–8610.
11. Beaudoin, F., Wilkinson, B. M., Stirling, C. J., and Napier, J. A. (2000) In vivo targeting of a sunflower oil body protein in yeast secretory (sec) mutants. *Plant J.* **23,** 159–170.
12. Hoffman, L. M., Donaldson, D. D., and Herman, E. M. (1988) A modified storage protein is synthesized, processed and degraded in the seeds of transgenic plants. *Plant Mol. Biol.* **11,** 717–729.
13. Krebbers, E. and Vandekerckhove, J. (1990) Production of peptides in plant seeds. *TIBTECH* **8,** 1–3.
14. Iturriaga, G., Jefferson, R. A., and Bevan, M. W. (1990) Endoplasmic reticulum targeting and glycosylation of hybrid proteins in transgenic tobacco. *Plant Cell* **1,** 381–390.
15. Moloney, M. M., Walker, J. M., and Sharma, K. K. (1989) High efficiency transformation of *Brassica napus* using *Agrobacterium* vectors. *Plant Cell Rep.* **8,** 238–242.
16. Parmenter, D. L., Boothe, J. G., van Rooijen, G. J. H., Yeung, E. C., and Moloney, M. M. (1995) Production of biologically active hirudin in plant seeds using oleosin partitioning. *Plant Mol. Biol.* **29,** 1167–1180.
17. Parmenter, D. L., Boothe, J. G., and Moloney, M. M. (1996) Production and purification of recombinant hirudin from plant seeds, in *Transgenic Plants: A Production System for Industrial and Pharmaceutical Proteins*, (Owen, M. R. L. and Pen, J., eds.), John Wiley and Sons Ltd., pp. 261–280.
18. Heim, J., Takabayashi, K., Meyhack, B., Maerki, W., and Pohlig, G. (1994) C-terminal proteolytic degradation of recombinant desulfato-hirudin and its mutants in the yeast *Saccharomyces cerevisiae*. *Eur. J. Biochem.* **226,** 341–353.

19. Kühnel, B., Holbrook, L. A., Moloney, M. M., and Van Rooijen, G. J. H. (1996) Oil bodies of transgenic *Brassica napus* as a source of immobilized β-glucuronidase. *J.A.O.C.S.* **73,** 1533–1538.

20. Liu, J.-H., Selinger, L. B., Cheng, K.-J., Beauchemin, K. A., and Moloney, M. M. (1997) Plant seed oil-bodies as an immobilization matrix for a recombinant xylanase from the rumen fungus *Neocallimastix patriciarum. Mol. Breed.* **3,** 463–470.

21. Mahmoud, S. (1999) Production of recombinant carp growth hormone in the seed of *Brassica napus*. Ph. D. Dissertation, University of Calgary, Canada.

22. Bosch, D., Smal, J., and Krebbers, E. (1994) A trout growth hormone is expressed, correctly folded and partially glycosylated in the leaves but not the seeds of transgenic plants. *Transgenic Res.* **3,** 304–310.

23. Jeh, H. S., Kim, C. H., Lee, H. K., and Han, K. (1998) Recombinant flounder growth hormone from *Escherichia coli*, overexpression, efficient recovery, and growth-promoting effect on juvenile flounder by oral administration. *J Biotechnol.* **60,** 183–193.

24. Hood, E. E., Witcher, D. R., Maddock, S., Meyer, T., Baszczynski, C., Bailey, M., et al. (1997). Commercial production of avidin from transgenic maize, characterization of transformant, production, processing, extraction and purification. *Mol. Breed.* **3,** 291–306.

25. Kusnadi, A. R., Hood, E. E., Witcher, D. R., Howard, J. A., and Nikolov, Z. L. (1998a) Production and purification of two recombinant proteins from transgenic corn. *Biotechnol. Prog.* **14,** 149–155.

26. Kusnadi, A. R., Evangelista, R. L., Hood, E. E., Howard, J. A., and Nikolov, Z. L. (1998b) Processing of transgenic corn seed and its effect on the recovery of recombinant beta-glucuronidase. *Biotechnol. Bioeng.* **60,** 44–52.

27. Pen, J., Verwoerd, T. C., van Paridon, P. A., Beudeker, R. F., van den Elzen, P. J. M., et al. (1993) Phytase-containing transgenic seeds as a novel feed additive for improved phosphorus utilization. *BioTechnology* **11,** 811–814.

28. Moloney, M. M., Alcantara, J., and van Rooijen, G. J. H. (1998) Method for cleavage of fusion proteins PCT. Patent Application # WO 98/49326.

29. Kühnel, B., Alcantara, J., Boothe, J., van Rooijen, G., and Moloney, M. (2002) A novel method for cleavage of recombinant fusion proteins. Submitted.

30. Moloney, M., van Rooijen, G., and Boothe, J. (1999) Oil bodies and associated proteins as affinity matrices. United States Patent No. 5,856,452.

31. Jacks, T. J., Hensarling, T. P., Neucere, J. N., Yatsu, L. Y., and Barker, R. H. (1990) Isolation and physiocochemical characterization of the half-unit membranes of oil seed lipid bodies. *J.A.O.C.S.* **67,** 353–361.

IV

CRITICAL TOOLS FOR BIOTECHNOLOGY

13

Genome Sequencing and Genomic Technologies in Drug Discovery and Development

Lawrence M. Gelbert

1. Introduction

Drug discovery and development is rooted in empiricism, exemplified by testing of compounds in animal models of disease, and on the identification and optimization of the active ingredients in traditional folk medicines. Over the past century, this has radically changed with the introduction of new scientific paradigms and cutting-edge technologies into the pharmaceutical R&D process *(1,2)* (Fig. 1). These innovations include the automation of compound screening *(3)* and improvements in medicinal chemistry, along with the development of combinatorial chemistry *(4)*, and advances in molecular biology. Most of these technologies have focused on increased throughput through one portion of the drug discovery and development pipeline and the result has been to relocate—rather than eliminate—the bottlenecks and failures in the overall process. The attrition rate of the drug discovery and development pipeline continues to be a major problem, as highlighted by the large number of compounds that fail in clinical trials because of poor pharmacology, toxicity, and/or lack of efficacy *(5,6)*. At the same time, economic factors apply, increasing pressure to reduce this attrition *(7)*.

Genomics (also known as functional genomics) is the comprehensive analysis of the entire genetic content of an organism, and is part of a newly defined discipline known as systems biology. Systems biology can be defined as the analysis of all components of a biological system before and after a genetic or chemical insult *(8–10)*. The concept of systems biology has been previously discussed and alternatively called modular biology *(11)* and "omic" biology *(12)*. However it is defined, genomics holds the promise of reducing the attrition currently seen in drug discovery and development, unlike previous innovations *(13,14)*. Genomics provides comprehensive views of biological pathways representing disease pathophysiology and drug activity that were previously unavailable in contemporary pharmaceutical development *(9)*. Such comprehensive views should improve the quality of drug target identification/validation and other steps in the drug discovery process, resulting in higher-quality compounds entering clinical development. Because genomic data is digital, it will foster the integration of biological information from all phases of drug

From: *Handbook of Industrial Cell Culture: Mammalian, Microbial, and Plant Cells*
Edited by: V. A. Vinci and S. R. Parekh © Humana Press Inc., Totowa, NJ

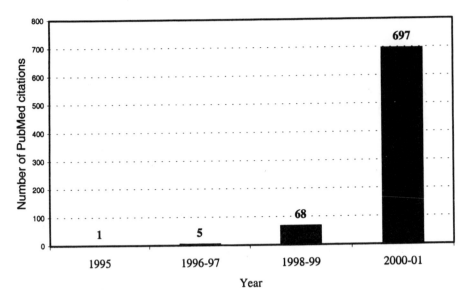

Fig. 1. Relative increase in publications describing the use of microarrays. The data presented was generated by searching the PubMed database at NCBI (http://www.NCBI.NLM.NIH.gov/entrez) with the terms "gene expression" and "microarray," and does not include the use of microarrays for non-expression experiments such as DNA resequencing and genotyping.

discovery and development and alter the traditional pharmaceutical research loop *(15)*. Thus, through increased knowledge of the relevant biological process (signal transduction, physiology, pharmacology, and toxicology) and integration of biological information from various scientific disciplines, genomics should supplement the contemporary drug discovery process and reduce attrition.

1.1. From Genetics to Genomics

Most of the common diseases that are the focus of drug discovery are multifactorial, resulting from a combination of genetic and environmental factors. The biological result of these factors is called a phenotype *(16)*. An example of such a multifactoral phenotype is environmental carcinogenesis, in which cancer susceptibility after exposure to a carcinogen is modulated by individual genetic variation *(17,18)*. This interaction of genetic and environmental factors is also characteristic of other biological phenotypes, including disease susceptibility and response to therapeutic intervention. Until recently, it has been at best difficult to identify these genetic and environmental factors. The promise of genomics is that it holds the potential to identify the genetic factors associated with these phenotypes and illuminate the associated biological pathways.

Until recently, genomics has been synonymous with molecular biology and positional cloning. Positional cloning allows for the identification of disease genes through inheritance. After an initial chromosomal region is identified through genetic linkage analysis in families that carry a mutant copy of the disease gene, the DNA in the region is physically cloned, all the genes in the region are identified, and the gene responsible for the disease is determined through identification of mutations in one of these genes in all affected family members *(19)*. Positional cloning has become an area of great interest, and there have been some spectacular successes in this area *(20)*, but its use in drug discovery has been limited for several reasons. Positional cloning has been successful for the identification of disease genes with major affects (monogenic or Mendelian disorders), but less successful for identifying the genes responsible for the complex traits that are the targets for drug discovery *(21)*. Positional cloning requires the collection of large families or other family designs (such as sib-pairs). Such pedigrees are not easily collected, and are not used for clinical drug development. Finally, positional cloning efforts are very expensive and labor-intensive, and do not fit within the time constraints for drug discovery programs. Positional cloning has provided insight into the underlying molecular pathophysiology of complex human diseases, yet such efforts have had relatively little impact on the discovery of new drugs.

Although positional cloning makes it possible to identify individual genes, genomics is focused on the systematic identification and analysis of all the genes for an organism. A genome sequence can be considered the equivalent of the periodic table to a chemist, and refocuses the role of the scientist from gene identification to the analysis of biological pathways and functions *(22)*. The human genome project has not only provided new global views of human biology, it has established the technical infrastructure that allows for the sequencing of other genomes of biological, medical, and agricultural significance *(23,24)*. It is too early to determine whether genomics will fulfill the promise of a Kuhnian revolution in biology and drug discovery *(25)*, but it is now possible to conclude that genomic approaches will have a positive impact and are considered an essential technology for all major pharmaceutical and biotechnology companies *(26,27)*. The focus of this chapter is to describe the sequencing of the human genome and the genomes of several important model systems, to provide an overview of several new genomic technologies and paradigms that use genomic information, and to provide several examples of the application of genomics to drug discovery and development.

2. Genome Sequencing and Model Systems

A genome is the complete DNA sequence for an organism *(28)*. Genome sequencing provides all the genetic information encoded in an organism's genome, as well as the foundation for all genomics tools, including bioinformatics, microarrays, high-throughput model systems, and pharmacogenomics. The genomes for approximately 800 organisms have been or are being sequenced, and include sequences from all three domains of living organisms and several eukaryotes *(29,30)*. A complete list of these genomes can be found on the National Center for Biotechnology Information (NCBI) website *(31)*. The size of these completed sequences range from 0.58 megabases (Mb, one million basepairs [bp] of DNA) for *Mycoplasma gentitalium* to 3300 Mb for *Homo sapiens*. Genome sequencing provides a wealth of biological information in addition to a complete catalog of genes. Comparison of genome sequences (comparative genomics) identifies conserved sequences and defines essential genes and biological pathways *(32)*. For example, the comparison of positionally cloned human disease genes in yeast, the nematode *Caenorhabditis elegans* (*C. elegans*), and several bacterial genomes identified orthologs with significant homology that are phylogenetically conserved *(33)*. The strong conservation of these genes through evolution and their association with human disease highlights how comparative genomics can be used to identify potential drug discovery targets, and validates the use of lower eukaryotic model systems in drug discovery *(34,35)*. Comparative genomics can be also be used to compare pathogenic microbial genomes to identify new antibiotic targets *(34,36)*.

Whole-genome sequences allow the comprehensive analysis of noncoding sequences for regulatory regions such as transcriptional promoters, which when mutated can cause phenotypic changes as important as those seen in the protein-coding regions *(37–39)*. Genome sequences provide the information needed to understand biology and enable drug discovery and development, and can be analzyed (also called mining) to identify proteins that can be used directly as biopharmaceuticals, in vaccine development and antibiotics, and for use as industrial enzymes.

The sequencing of genomes for several lower eukaryotes has now led to their widespread use as tools for target validation and drug screening *(40–42)*. These model systems are defined by short lifespans, and are amenable to genetic manipulation and automated handling *(43,44)*. Four model systems currently fit these criteria: yeast (*Saccharomyces cerevisiae*), nematode (*Caenorhabditis elegans*), fly (*Drosophila melanogaster*), and zebrafish (*Danio rerio*). Details of these model systems are provided in **Subheadings 2.1-2.4.**

2.1. The Human Genome Sequence

The most widely recognized milestone in genome sequencing has been the publication of the first drafts of the human sequence in February 2001 *(45,46)*. When first proposed, the sequencing of the human genome was controversial. There were several concerns: that the project would divert money from other research programs, that sequencing the noncoding regions of the genome was wasteful, and that sequencing the human genome was technically impossible. After much debate, funding was secured and a initial set of goals was established for a joint National Institutes of Health/Department of Energy Human Genome Project (HGP) *(47)*. The HGP plan was to generate a genetic map of the human genome, use the map to enable ordered physical cloning, and perform DNA sequencing. The genetic map and ordered cloning would then act as a foundation to align the complete genome sequence with known chromosomal and genetic markers. The HGP goals also included the sequencing of several eukaryotic model organisms. The experience gained from sequencing these genomes of increasing size and complexity would allow for the development of the needed automation and sequencing technology, and provide information necessary to analyze the human sequence *(48)*.

By 1998 these original goals were achieved or exceeded, and a new set of objectives were defined *(49)*. These objectives included the complete sequence of the human genome by the year 2003. Also in 1998, Celera Genomics, a private biotechnology company, announced a competing effort to complete the sequence the human genome two years ahead of the public HGP. Celera's announcement was controversial, not only in attempting to complete the sequence more quickly, but because Celera would use an alternative sequencing approach called shotgun-sequencing. Shotgun-gun sequencing relies on computer analysis to assemble a large number of small overlapping sequences into larger assemblies, and ultimately, a complete genome. The resulting competition between the HGP and Celera greatly accelerated the completion of the human genome, and separate draft sequences were published in February 2001 *(50,51)*.

Although not identical, the HGP and Celera draft sequences are surprisingly similar in size and information content, and several important observations have been made from their initial analysis. The human genome consists of approx 3 billion bp of DNA. The biggest surprise and source of debate comes from the number of predicted genes, which was approx 32,000. This was far fewer than the expected 70,000–100,000 genes predicted from previous studies *(45,46)*. When compared to the number of genes found in lower eukaryotes—14,000 in the fly and 19,000 in the worm—it is difficult to envision how 32,000 genes can encode the information necessary for the more biologically complex humans. The lower number of predicted genes in humans certainly arises in part from the initial quality of the two genome sequences and the computer programs used for gene prediction *(52)*. Work continues on filling the gaps in the draft sequence and refining the analysis. Recently, an independent analysis of the draft sequences has predicted 65,000–75,000 genes *(53)*. Others have suggested that the complexity encoded in the human genome can be explained via alternative splicing, post-translational modification of proteins, and other mechanisms. However, although the catalog of human genes will not be complete for several years, the information derived from the current versions of the sequence still provides the most comprehensive view of human biology available *(54)*. Additional observations from the draft sequences have provided clues about the organization and evolution of the human genome. These include estimates for the level of gene duplication, location and distribution of genes across each chromosome, genome content and distribution of repetitive sequences and recombination sites, and the level of genetic variation *(55)*. These observations provide important clues to the function and control of the information content of the genome and the genes associated with cancer *(56,57)*, immunology *(58)*, neurobiology *(59,60)*, evolution *(61,62)*, and ultimately the practice of medicine *(63–66)*.

2.2. Nonmammalian Eukaryote Model Systems

The first eukaryotic genome to be sequenced was for the yeast *Saccharomyces cerevisiae (67,68)*. The yeast genome and was completed in 1996 and consists of 12.1 Mb, which was 6 times the size of the previously largest genome to be sequenced at that time (*Haemophilus influenzae* at 1.8 Mb). The initial analysis of the yeast genome identified 6200 genes, and a mammalian homolog was identified for approx 30% of these genes *(69)*. The similarity to mammalian genes was an important observation because it indicates that there has been significant conservation of biological function between a simple single-cell eukaryote and more complex mammals. In contrast to this observation, for more than 25% of the predicted genes, no homology or function could be determined through comparison to known genes from other organisms *(68,70)*. One important lesson learned from yeast is the complexity of whole-genome analysis. A recent study identified and confirmed 137 new genes in the yeast, which represents 2% of the previously known gene content *(71,72)*. These genes were identified 5 yr after the completion and initial analysis of the yeast sequence. Considering the larger size and complexity of the human genome, it is important to realize that the completion of a genome sequence actually represents the beginning, not the end, of the genetic analysis.

The next major eukaryote genome to be sequenced was the nematode *Caenorhabditis elegans*. The *C. elegans* sequence represented a significant milestone because it was the first genome for a multicellular animal. As a model system, *C. elegans* has several unique properties. It contains only 959 cells, but includes many cell types found in higher eukaryotes, including a 300-cell nervous system, as well as muscle, and reproductive and digestive systems *(73)*. Analysis of the 97-Mb sequence predicted approx 19,000 potential genes *(74)*. The *C. elegans* sequence was immediately analyzed to more fully examine many of the features that make it such a useful model system. *C. elegans* has a central nervous system, and has long been used as model for mammalian neurobiology. Comparison of the *C. elegans* sequence to known genes controlling nervous-system functions (including ion channels, neuropeptides, and G-protein-coupled receptors) provided the first complete inventory of such gene families *(75)*. Likewise, a similar survey was performed to identify signal transduction and transcription-factor genes that regulated cellular lineage and development *(76)*. These surveys identified novel family members, thus providing a more comprehensive view of the genes and the biology controlling these processes in a multicellular organism.

The last major model system genome to be sequenced as a prequel to the human genome was for the fruit fly *Drosophila melanogaster (77)*. *D. melanogaster* has been used as a genetic model system since 1910, and continues to be a major eukaryotic genetic model system for the study of biology *(78,79)*. Approximately 14,000 genes have been identified in the 180-Mb *D. melanogaster* genome. The sequencing of the *Drosophila* genome was partly a technical milestone, validating the shotgun-sequencing approach for use in sequencing the human genome *(80)*. And at the time, *Drosophila* provided a third genome, which improved the resolution of eukaryotic comparative genomics *(81)*.

2.3. Microbial Genomes

Sequencing the genome of pathogenic organisms provides a unique opportunity to study pathogenicity, host-pathogen interactions, and virulence *(36)*. Genomic analysis of pathogenic microorganisms will enhance the development of new antibiotics by improving microbial target identification in several ways. It is expected to accelerate the shift from whole-cell bacterial screening to target-based screening, which is easier and faster than whole-cell screening and facilitates rational drug design *(34)*. The comparison of several strains of a bacterial pathogen can identify biological pathways associated with pathogenicity and/or drug resistance. This has been done for *Helicobacter pylori*, a Gram-negative bacterium found in the stomach of humans that causes gastric ulcers *(82)*. The 1.7-Mb genome sequence was completed in 1997 *(83)* and comparison of several clinical isolates and laboratory strains has identified genes associated with pathogenicity *(84–87)*. Similarly, parallel analysis of both the pathogen and host genomes will be used for the exploration of host-pathogen intereactions that affect pathogenicity and host response *(35)*. For example, microarray analysis of human monocytes infected with *Salmonella typhimurium* has identified several host-cell genes that may play a role in macrophage death caused by salmonellosis *(88)*.

A comprehensive list is maintained on The Institute for Genomic Research Comprehensive Microbial Resource (TIGR-CMR), which can be accessed on the internet (http://www.tigr.org/). Currently, there are complete sequences for over 48 different pathogens, including many that are associated with major infectious diseases. Bacterial pathogens include those for tuberculosis (*Mycobacterium tuberculosis*) and leprosy (*Mycobacterium leprae*) *(89–91)*. Nonbacterial pathogens have also been sequenced, including the malaria parasite (*Plasmodium falciparum*) *(92–94)*. In addition to pathogenic species, the genomes of several other nonpathogenic microbes have been sequenced. These include the K12 laboratory strain of *Escherichia coli (95,96)* and the genome of *Agrobacterium tumefaciens*, which is widely used as a gene-transfer system in plant biology *(97–99)*.

2.4. Future Directions in Genome Sequencing

One benefit of the HGP was the development of a large public infrastructure of genome centers. Combined, this network has the capacity to generate 172 Mb of DNA sequence per day and are responsible for many of the tools associated with analysis of the genomic data *(100)*. Currently, these centers are still focused on completing (or refining) the human genome, which still contains many gaps and errors. However, attention has already turned toward the next focus for new projects by the public genome resources *(101)*. Researchers are actively sequencing the genome of the next eukaryote model systems. These projects include widely used laboratory models such as the zebrafish (*Danio rerio*), mouse, and rat. These organisms have been selected for both scientific and practical reasons. The Zebrafish has all the features of other high-throughput models described here, but is unique because it is a vertebrate that permits whole-genome mutational screens *(102–106)*. The sequencing of the mouse and rat genomes is important for comparative genomic studies of the human genome *(24,107)*, and will direct existing transgenic and gene-knockout technologies to develop models for drug discovery and development *(108,109)*. Several pilot programs also exist to decipher the genomes of other animal species relevant to biomedical research, including primates *(24,110)*.

Genome sequencing is also being performed on plants, as highlighted by the recent publication of the 100-Mb genome sequence for the plant model *Arabidopsis thaliana (111)*. Analysis of the sequence predicts approx 25,000 genes, and this information has been applied to the study of plant biology in areas such as response to light and water movement in roots *(112–114)*. Recent studies are now sequencing the genome of several important crop plants including corn and rice. A summary of current plant genome projects can be found on the Plant Genome Database: http://www.hgmp.mrc.ac.uk/GenomeWeb/plant-gen-db.html.

2.5. DNA Sequence Variation

Large-scale resequencing of the human genome is now being used to measure naturally occurring genetic variation in the human population. These genetic variants are known as polymorphisms, and are important for several reasons. First, because of their stability, polymorphisms can be used as markers for a specific regions of a chromosome. Such genetic markers are used to construct genetic maps of a genome and for the analysis of inheritance by genetic linkage or association studies *(115,116)*. If a genetic polymorphism occurs in a functional region of the genome such as a gene, it can have a dramatic impact on genome function. These polymorphisms are of great interest to the pharmaceutical industry because it has been shown that linkage and/or association between these genetic polymorphisms (a genotype) can be performed from samples in clinical trials to explain individual patient-drug response or adverse events (the phenotype). The analysis of genetic factors that influence response to drug response is known as pharmacogenetics. Until recently, pharmacogenetics focused on the analysis of drug-metabolizing enzymes (DMEs) *(117)*. Previous studies have shown that genetic polymorphisms exist for many DME genes (such as the cytochrome P450 gene family), and there are common variants in these genes that affect the metabolism of a significant number of prescription drugs *(118–120)*. Similarly, since DMEs are involved in the metabolism of environmental xenobiotics, DME polymorphisms are also associated with increased sensitivity to carcinogens *(17,121,122)*. The tremendous growth of genomic information from genome-sequencing efforts and technical advances in genotyping has resulted in the expanded definition of pharmacogenetics known as pharmacogenomics. Beyond drug metabolism, pharmacogenomics is the analysis of genetic factors that affect drug efficacy, toxicity, and disease susceptibility *(123,124)*. One of the technologies driving pharmacogenomics is the identification and analysis of single-nucleotide polymorphisms, or SNPs (pronounced snips).

A SNP is a specific position in a stretch of DNA in which there is a single nucleotide substitution. Each alternate nucleotide is called an allele. SNPs are the most abundant type of polymorphism in the human genome *(125)* and have several advantages over previous genetic markers: the large number of SNPs and distribution across the genome provides a higher level of resolution for genetic studies than was previously possible, their power can be further enhanced by analyzing clusters of closely spaced SNPs called haplotypes, and because they are diallelic, their identification and analysis (genotyping) can be automated. These advantages allow for genetic marker/phenotype studies that were previously impossible. Preliminary studies resulting in the first SNP-based genetic map of the human genome *(126)* confirmed that SNPs could be used to study genome evolution, genome organization, and population diversity *(127–129)*. Several studies are now underway to identify and develop a high-resolution SNP map of the human genome. These include the SNP Consortium (TSC), an organization of private companies and academic genome centers (details this effort can be found on the TSC website, http://snp.cshl.org/). So many human SNPs have been identified that a public database has been established on the NCBI website to collate SNP data (dbSNP, http://www.ncbi.nlm.nih.gov/SNP/). The result of these efforts was a SNP map containing 1.42 million SNPs, which was published along with the HGP genome sequence *(130)*.

Considerable efforts are now focused on construction of higher-resolution SNP and haplotype maps of the human genome, and the development of genotyping technology that will allow for genome-wide genetic association studies *(126,131–136)*. The intensity of these efforts demonstrates that the analysis of genetic diversity and its role in biology and human disease will be a major outcome of the sequencing of the human genome *(137)*. Recent studies have shown the utility of these new approaches, and several examples are described later in this chapter.

3. Transcript Profiling

Currently, the most mature and widely used genomic technology is transcript profiling. Transcript profiling, also called expression profiling, can be defined as the analysis of gene expression for all or of a large portion of a genome. Temporal expression of genes, the level of expression, and the processing of RNA all act to regulate the information encoded in genomic DNA. Analysis of RNA expression patterns provides valuable information about the underlying molecular basis of cellular development and differentiation, allows for the identification of signal-transduction pathways, can be used to identify the function of novel genes, provides insight into the pathophysiology of diseases, and to study drug mechanism of action (MOA) *(22,138)*. Before the advent of microarray technology, several approaches were used to monitor gene expression. These included Northern blotting *(139)*, EST sequencing *(140,141)*, differential display (RADE) *(142–144)*, and serial analysis of gene expression (SAGE) *(145,146)*. Although all these approaches allowed semi-quantitative measurement of gene expression, they were hampered by high cost and low throughput, and were too labor-intensive to be broadly applied.

Microarrays are manufactured and automated tools that allow for the analysis of a large number of nucleic-acid sequences on a small, gridded and solid support. Current microarray technology is based on the technology first described by Ed Southern in 1975 for the immobilization of nucleic acids on a solid support *(147,148)*. The immobilized nucleic acid (probes) are then hybridized to a labeled nucleic acid sample (the target). After allowing sufficient time for hybridization, the microarray is washed with a series of buffers that remove target molecules that are not hybridized to any of the probes on the array, and the information about the presence and quantity of thousands of individual genes are then determined through analysis of the hybridization pattern on the microarray. There are currently two predominant types of microarrays, cDNA and oligonucleotide microarrays. cDNA microarrays were first described in 1995 *(149)* and there are a large selection of custom and commercial cDNA microarray systems *(150–152)*. These systems use glass microscope slides that have been chemically treated to facilitate the adherence of DNA fragments or clones to the surface. The DNA templates to be arrayed can come from a variety of sources, including clones from cDNA libraries or PCR fragments. Next, a robot is used to spot the nucleic acid onto the glass surface. The amount of material spotted varies from a volume of 0.26 nL (0.13 ng of probe) up to approx 32 nL (16 ng probe). These volumes generate individual spots with a radius between 50 and 250 μ, allowing for approx 6000 spots to be gridded on a standard 1" × 3" microscope slide. The RNA targets to be analyzed are labeled through enzymatic incorporation of fluorescent nucleotides. The use of two fluorescent labels usually allows for two different targets to be hybridized to one microarray. After hybridization and washing the array, the data is collected using a laser confocal scanner that excites the individual dyes in the targets to emit light at a specific wavelength. The resulting data are captured on a computer for subsequent analysis.

Oligonucleotide arrays are similar to cDNA arrays, except that they use single-stranded oligonucleotides (25–60 nucleotides) as probes instead of double-stranded DNA molecules. The oligonucleotides are synthesized directly on a solid support either using photolithography *(153)* or ink-jet technology *(154)*. These techniques allow for the synthesis of over 400,000 different oligonucleotides on a 1.28 × 1.28 cm surface *(153)*. Procedures for the labeling of target RNA, hybridization, and scanning are similar to those used for cDNA microarrays.

Although the techniques for the manufacturing of arrays are now well-established, the protocols for the analysis of array data are less mature. Initial analysis approaches focused on pattern recognition and clustering *(155)*, which continues to be the predominant method of analyzing microarray data *(156)*. However, statistical approaches are gaining favor because of the large amount of variation observed in biological systems and with array techniques *(157–161)*. A major problem that emerges from the growing amount of published array data is a lack of standards for experimental design, microarrays, and data analysis. Several groups are attempting to establish a uniform set of standards for arrays experiments *(162,163)*, but currently no standards exist. This makes it impossible to compare published-array data generated by different groups. Until such standard protocols are established, it will not be possible to establish large open databases of array data, ultimately limiting the full potential of the data.

4. Applications of Genomics in Drug Discovery and Development

4.1. Model Systems

Because the yeast sequence was the first complete eukaryotic genome, and with existing methods to manipulate the yeast genome, the first examples of using high-throughput models in drug discovery and development have been in yeast. The availability of the genome sequence allows for a complete inventory of genes representing a specific class of proteins. This gene inventory enables drug-target identification and validation through the analysis of whole classes of targets such as G-protein-coupled receptors (GPCRs), kinases, and proteases. The first such inventory for a eukaryote was performed in yeast for ABC transport proteins *(164)*. Analysis of such target platform information is now considered critical in drug-target identification and validation for several reasons. Comparison across species for homology identifies the members of a gene family that are conserved through evolution, suggesting an essential role for those genes *(165)*. Such conservation is also a characteristic of many disease genes, and this information can be used when selecting new drug discovery targets *(13,33)*.

The first studies to illustrate the value of whole-genome drug screening were performed in yeast *(166,167)*. These studies used whole-genome expression profiling with microarrays of samples from yeast treated with a series of compounds acting on the same molecular target *(168,169)*. The studies showed that similar to in vitro biochemical assays, whole-genome screening in yeast could distinguish active from inactive compounds. However, unique to the profiling drugs in yeast, distinct biological activities in structurally similar compounds were seen that were not identified by in vitro biochemical assays. The testing of drugs in model systems is further enhanced by the ease and scale with which the genome of these organisms can be manipulated. For example, expression profiling of drugs was performed on strains of haploid yeast for which the gene for the known molecular target of the drug was deleted *(168,169)*. This allowed for testing of drugs in living cells that completely lack the known target and when combined with whole-genome expression profiling, defines the gene-expression profile resulting from activity independent of the known molecular target. This approach identifies pathways and molecular targets that were not previously associated with a compound. These "off-target" affects are important because they can define secondary biological properties for a compound that is not possible with conventional biochemical assays, and can explain both novel efficacy or adverse events *(166)*.

The complete sequence of the yeast genome has enabled a program to systematically develop an entire set of yeast strains where each of the open reading frames (ORFs) for all the proteins encoded in the genome have been deleted. The high rate and accuracy of genetic recombination, and the ability to grow yeast as both haploid and diploid cells has facilitated this program. A high-throughput system was developed for which individual ORFs are simultaneously mutated and insertion of a unique 20-bp tag (or bar-code) into the genome *(41)*. The bar-code allows for the identification of each strain and for multiplex experiments analyzing a large number of different strains. Approximately 93% of the predicted genes in the yeast genome have been bar-coded, and details of this program can be found at the *Saccharomyces* Genome Deletion Project (http://www-sequence.stanford.edu/group/yeast_deletion_project/ deletions3.html). For genes essential to cell viability, the mutants are maintained as diploid heterozygous cells with one wild-type and one mutant allele. The development of the complete panel of yeast mutants will further enhance the use of yeast as a drug discovery tool and similar approaches are now being developed for other model systems with defined genomes.

4.2. Genetic Diversity in Human Disease and Drug Response

The following examples illustrate the potential of pharmacogenomics to define genetic factors associated with common diseases and drug response. This information can then be used to identify and validate new drug discovery targets and more accurately predict how patients will respond in clinical trials.

- Apolipoprotein E (APOE) is a ligand for the low-density lipoprotein receptor and is involved in the regulation of blood cholesterol. There are three common alleles for the gene, APOE2, -E3, and -E4. In two initial studies, an increased frequency of the APOE4 allele was associated with both sporadic and familial Alzheimer's disease *(170,171)*. Many additional studies have confirmed the association of the APOE4 allele with increased risk of Alzheimer's disease. In a separate study, the increased frequency of the -E4 allele was found to be associated with a decreased response of Alzheimer's disease patients taking tacrine, a cholinesterase inhibitor used to improve cognitive function *(172)*. These studies illustrate how a genetic marker can be used to identify patients who were at risk for developing a disease and to predict drug response, which can be used to enhance the selection of patients in clinical drug trials. Such patient stratification results in smaller and more cost effective clinical trials for drug development.
- Peroxisome proliferator-activated receptor-gamma (PPARG) is a nuclear-hormone receptor that regulates adipocyte gene expression and differentiation. Thiazolidinediones are agonists for PPARG and are used for the treatment of type 2 diabetes (non-insulin-dependent diabetes mellitus, NIDDM, or adult-onset diabetes) by increasing insulin sensitivity. There are many conflicting reports on the role of genetic variants in the PPAG gene that increase risk of developing type 2 diabetes. Many factors can affect the results of genetic-association studies including patient phenotyping and sample size. Using new genotyping technology all 16 SNPs in PPARG previously reported to be associated with type 2 diabetes were reanalyzed in a large population *(173)*. Through the analysis of over 3000 samples to increase power, the authors were able to confirm one of the 16 previous associations. This retrospective study illustrates how genomics can be used for drug-target identification, and how new SNP genotyping technology allows for large clinical genetic studies that were previously not possible.
- Two strong genetic associations have been made that predict response to anti-asthma treatments. Leukotrienes play important roles in inflammation, immediate hypersensitivity and asthma. 5-lipoxygenase (ALOX5) catalyzes the initial steps in the synthesis of leukotrienes, and compounds that inhibit ALOX5 are one class of drugs used to treat asthmatics. The ALOX5 promoter contains a variable tandem repeat sequence in its promoter. Studies have shown that specific variants in the repeat sequence reduce the

expression of ALOX5. By genotyping the repeat sequence in patients, a pharmaco-genomic association was found between specific variants and a subset (6%) of asthmatics that do not respond to ALOX inhibitors *(174)*. A similar observation between response and a promoter polymorphism was discovered for another class of asthma drugs. The β2-adrenergic receptor modulates bronchodilation, and agonists for this receptor are used to treat asthma. Analysis of the β2-adrenergic-receptor promoter identified specific haplotypes (but not the individual SNPs) that are associated with response to the receptor agonist albuterol, and highlights the increased resolution of haplotype analysis *(39,175)*.

4.3. Microarrays and Transcript Profiling

Since the publication of the first microarray paper in 1995, there has been a five- to 10-fold increase every 2 yr in publications describing the use of microarrays for transcript profiling (Fig. 1). Considering the strong commitment to the technology by the biotechnology and pharmaceutical industry, where many experiments go unpublished, this is certainly an underestimate of the actual acceptance and use of microarray technology. Transcript profiling holds great promise in making drug discovery more efficient through the analysis of drug (MOA) *(176)*, both for small-molecule drugs and bioproducts. Two of the first expression profiling experiments on microarrays were the analysis of cyclin-dependent kinase (cdk) inhibitors and the immunosuppressant FK506 in yeast *(168,169)*. FK506 has now been reanalyzed by transcript profiling in mouse B cells, where novel signal-transduction pathways through which the drug acts were identified *(177,178)*. The MOA of many anticancer compounds have now been analyzed, and transcript profiling is becoming a primary paradigm in the analysis of anticancer MOA *(179–181)*. Transcript profiling has been used to analyze how cells become resistant to the thymidylate synthase (TS) inhibitor 5-fluorouracil (5-FU). 5-FU is used to treat most solid tumors and acquired chemoresistance is a major clinical problem. To explore the molecular mechanisms of chemoresistance, a panel of 5-FU-sensitive and resistance tumor cells have been analyzed *(182)*. One known resistance mechanism is to overexpress the drug target and increased TS expression was observed in some of the resistant tumors. In 80% of the 5-FU-resistant tumors expression of another gene, YES1, was also seen to be overexpressed. YES1 is a kinase and a proto-oncogene, and the data suggests that this may be a novel biological pathway used by tumors to overcome the action of 5-FU. Another example is the analysis of the MOA for flavopiridol, a cdk inhibitor currently in cancer clinical trials. Although some of the molecular targets of flavopiridol are known, there is evidence that it also acts through additional pathways. Transcript profiling in tumor cells treated with flavopiridol show a significant decrease in expression of 5% of the genes analyzed *(183)*. A dose-response experiment showed that the decrease in gene expression correlated strongly with the cytotoxicity. These results suggest that the cytotoxicity of flavopiridol may partly result from the broad destabilization of mRNAs. These examples illustrate how transcript profiling provides information on drug MOA that cannot otherwise be found using conventional biochemical assays and animal testing.

Transcript profiling has also been used to analyze a number of therapeutic proteins. The interferons (IFNs) are a class of cytokines with therapeutically relevant biological properties, including anti-viral, anti-tumor, and immunomodulation. The biological activity of the interferons (IFNs) results from the induction of specific groups of genes known as interferon (IFN)-simulated genes (ISG). Over the years, many ISGs have been identified, but function of many of these ISGs were unknown. Der et al. used transcript profiling to further elucidate the MOA of the interferons identified a large number of novel ISG that provide new insights into biology of the interferons, and may suggest additional therapeutic applications *(184)*.

Another bioproduct that has been analyzed with transcript profiling is human activated protein C (rhAPC), which has recently been approved for the treatment of severe sepsis. Severe sepsis is a systemic inflammatory disorder resulting from a complex host response to infection. The mortality rate in severe sepsis is between 30 and 50%, and is a major concern for a variety of conditions, including AIDs patients, cancer patients, and burn victims. Severe sepsis is the 11th leading cause of death in the United States, and has been resistant to therapeutic intervention *(185)*. The pathophysiology of severe sepsis includes endothelial-cell damage, abnormal coagulation, uncontrolled inflammation, multi-organ failure, shock, and death *(186)*. rhAPC is natural anticoagulant factor and the was first compound approved to treat servere sepsis. Endothelial dysfunction is a central feature in severe sepsis, and transcript profiling has been used to investigate the direct effects of rhAPC on endothelial function. The model for severe sepsis used was primary cultured human endothelium (HUVEC) cells, which were treated with the inflammatory cytokine TNF-α, rhAPC, or both compounds. The results provided evidence for direct modulation of genes in pro-inflammatory and apoptotic pathways *(187)*. The results of this study showed that rhAPC acts directly on endothelial cells as an anti-inflammatory, anti-apoptotic, and cell-survival factor. Taken together, these results suggest novel MOA for rhAPC that may explain its clinical efficacy in treating severe sepsis *(185,188)*.

The studies described here explore just a fraction of the novel biology that has been identified through transcript profiling. Exciting results have also been achieved through profiling to perform molecular classification of human tumors to predict patient response *(189,190)*, to analyze the effect of the cellular microenvironment on cell cultures *(191,192)*, and to study cellular response to metabolic changes *(193)*.

5. Conclusion

Over the past decade, a dramatic transformation has occurred in the science of genetics. The HGP has established a technical infrastructure allowing for the sequencing of the genomes of several species, including the human genome. At the same time, new technologies such as microarrays, SNPs and genotyping, and bioinformatics have been developed, which allow for the analysis of the information content in genome sequences that were previously not possible. This industrialization of genetics has resulted in the formation of a new scientific discipline known as genomics. Genomic experiments are now providing global views of biology, and are defining the genetic factors and biological pathways associated with phenotypes of medical importance, including common and complex human diseases such as cancer, diabetes, asthma, and severe sepsis. As more genomes are sequenced and genomic technologies become ubiquitous, it should be expected that genomic approaches are expected to impact all areas of biology, from microbiology and medicine to plant biology.

The pharmaceutical and biotechnology industry has embraced genomics, and it is believed that genomics will influence every phase of drug discovery and development. Genome sequencing, comparative genomics, and transcript profiling are now being used to identify important biological pathways to be targeted for drug intervention, and promise the identification of new, robust drug targets that can be pharmacologically modulated. The defined genomes of model organisms such as yeast, the fruit fly, and zebrafish will allow their use to perform high-throughput screening of compounds in whole organisms, a more physiologically relevant setting and pharmacogenomics promises to have a profound effect on the clinical development and marketing of new drugs. The analysis of genetic variation in a patient population should result in the stratification of patients for a more customized drug-therapy regimen. Pharmacogenomics will also result in the customization of the development of new drugs, and will offer novel methods for the differentiation of drugs with a similar MOA. Finally, pharmcogenomics should result in the development and widespread use of diagnostic genetic tests in drug development and in the marketing of drugs.

It is now possible to conclude that genomics will have a profound effect on biomedical research, and in particular, drug discovery and development. Genomics is now viewed as an essential drug discovery and development technology, and is currently being integrated into the larger drug discovery process. The ultimate success of genomics in biomedical research has not yet been determined, but its future seems bright.

Acknowledgments

I would like to thank my colleagues at Lilly Research Laboratories for their helpful comments and review, and Randee, Rachel, and Teegan at home for their patience while I wrote this chapter.

References

1. Drews, J. (1999) In Quest of Tomorrows Medicines. Springer, New York, NY, p. 272.
2. Drews, J. (2000) Drug discovery, a historical perspective. *Science* **287,** 1960–1964.
3. Broach, J. R. and Thorner, J. (1996) High-throughput screening for drug discovery. *Nature* **384(suppl),** 14–16.
4. Hogan, J. C., Jr. (1996) Directed combinatorial chemistry. *Nature* **384(suppl),** 17–19.
5. Tapolczay, D., Chorlton, A., and McCubbin, Q. (2000), Probing drug structure improves the odds. *Drug Discovery and Development* **2000,** 30–33.
6. Prentis, R. A., Lis, Y., and Walker, S. R. (1988) Pharmaceutical innovation by the seven UK-owned pharmaceutical companies (1964–1985). *Br. J. Clin. Pharmacol.* **25,** 387–396.
7. Drews, J. and Ryser, S. (1997) The role of innovation in drug development. *Nat. Biotechnol* **15,** 1318–1319.
8. Aebersold, R., Hood, L. E., and Watts, J. D. (2000) Equipping scientists for the new biology. *Nat. Biotechnol.* **18,** 359.
9. Brent, R. (2000) Genomic biology. *Cell* **100,** 169–183.
10. Vidal, M. (2001) A biological atlas of functional maps. *Cell* **104,** 333–339.
11. Hartwell, L. H., et al. (1999) From molecular to modular cell biology. *Nature* **402(suppl),** C47–52.
12. Evans, G. A. (2000) Designer science and the "omic" revolution. *Nat. Biotechnol.* **18,** 127.
13. Cockett, M., Dracopoli, N., and Sigal, E. (2000) Applied genomics, integration of the technology within pharmaceutical research and development. *Curr. Opin. Biotechnol.* **11,** 602–609.
14. Gelbert, L. M. and Gregg, R. E. (1997) Will genetics really revolutionize the drug discovery process? *Curr. Opin. Biotechnol.* **8,** 669–674.

15. Bumol, T. F. and Watanabe, A. M. (2001) Genetic information, genomic technologies, and the future of drug discovery. *JAMA* **285,** 551–555.

16. Lewin, B., (2000) Genes VII. Oxford University Press, Oxford, NY, p. 990.

17. Perera, F. (1997) Environment and cancer, who are susceptible? *Science* **278,** 1068–1073.

18. Fearon, E. R. (1997) Human cancer syndromes, clues to the origin and nature of cancer. *Science* **278,** 1043–1050.

19. Collins, F. S. (1995) Positional cloning moves from perditional to traditional. *Nat. Genet.* **9,** 347–350.

20. Cohen, J. (1997) The genomics gamble. *Science* **275,** 767–772.

21. Schafer, A. J. and Hawkins, J. R. (1998) DNA variation and the future of human genetics. *Nat. Biotechnol.* **16,** 33–39.

22. Lander, E. S. (1996) The new genomics, global views of biology. *Science* **274,** 536–539.

23. Gewolb, J. (2001) Genome research. DNA sequencers to go bananas? *Science* **293,** 585–586.

24. O'Brien, S. J., Eizirik, E., and Murphy, W. J. (2001), GENOMICS: On choosing mammalian genomes for sequencing. *Science* **292,** 2264–2266.

25. Strohman, R. C. (1997) The coming Kuhnian revolution in biology. *Nat. Biotechnol.* **15,** 194–200.

26. Drews, J. (2000) Drug discovery today—and tomorrow. *Drug Discov Today* **5,** 2–4.

27. Drews, J. (1997) Strategic choices facing the pharmaceutical industry, a case for innovation. *Drug Discov. Today,* 72–78.

28. Bork, and Copley, R. (2001) Genome speak. *Nature* **409,** 815.

29. Morell, V. (1996) Life's last domain. *Science* **273,** 1043–1045.

30. Bult, C. J., et al. (1996) Complete genome sequence of the methanogenic archaeon, *Methanococcus jannaschii*. *Science* **273,** 1058–1073.

31. NCBI, National Center for Biotechnology Information. (http://www.NCBI.NLM.NIH.gov)

32. Dacks, J. B. and Doolittle, W. F. (2001) Reconstructing/Deconstructing the earliest eukaryotes. How comparative genomics can help. *Cell* **107,** 419–425.

33. Mushegian, A. R., et al. (1997) Positionally cloned human disease genes, patterns of evolutionary conservation and functional motifs. *Proc. Natl. Acad. Sci. USA* **94,** 5831–5836.

34. Rosamond, J. and Allsop, A. (2000) Harnessing the power of the genome in the search for new antibiotics. *Science* **287,** 1973–1976.

35. Broder, S. and Venter, J. C. (2000) Sequencing the entire genomes of free–living organisms, the foundation of pharmacology in the new millennium. *Annu. Rev. Pharmacol. Toxicol* **40,** 97–132.

36. Koonin, E. V. (1997) Big time for small genomes. *Genome Res* **7,** 418–421.

37. Davuluri, R. V., Grosse, I., and Zhang, M. Q. (2001) Computational identification of promoters and first exons in the human genome. *Nat. Genet.* **29,** 412–417.

38. Pilpel, Y., Sudarsanam, , and Church, G. M. (2001) Identifying regulatory networks by combinatorial analysis of promoter elements. *Nat. Genet.***29,** 153–159.

39. Drysdale, C. M., et al. (2000) Complex promoter and coding region beta 2-adrenergic receptor haplotypes alter receptor expression and predict in vivo responsiveness. *Proc. Natl. Acad. Sci. USA* **97,** 10,483–10,488.

40. Winzeler, E. A., et al. (1999) Functional characterization of the *S. cerevisiae* genome by gene deletion and parallel analysis. *Science* **285,** 901–906.

41. Shoemaker, D. D., et al. (1996) Quantitative phenotypic analysis of yeast deletion mutants using a highly parallel molecular bar–coding strategy. *Nat. Genet.* **14,** 450–456.

42. Hughes, T. R., et al. (2000) Functional discovery via a compendium of expression profiles. *Cell* **102,** 109–126.

43. Scangos, G. (1997) Drug discovery in the postgenomic era. *Nat. Biotechnol.* **15,** 1220–1221.

44. Matthews, D. J. and Kopczynski, J. (2001), Using model-system genetics for drugbased target discovery. *Drug Discov. Today* **6,** 141–149.

45. Baltimore, D. (2001) Our genome unveiled. *Nature* **409,** 814–816.

46. Pennisi, E. (2001) The human genome. *Science* **291,** 1177–1180.

47. Collins, F. and Galas, D. (1993) A new five-year plan for the U.S. Human Genome Project. *Science* **262,** 43–46.

48. Roberts, L. (2001) The human genome. Controversial from the start. *Science* **291,** 1182–1188.

49. Collins, F. S., et al. (1998) New Goals for the U.S. Human Genome Project, 1998–2003. *Science* **282,** 682–689.

50. Lander, E. S., et al. (2001) Initial sequencing and analysis of the human genome. *Nature* **409,** 860–921.

51. Venter, J. C., et al. (2001) The sequence of the human genome. *Science* **291,** 1304–1351.

52. Aach, J., et al. (2001) Computational comparison of two draft sequences of the human genome. *Nature* **409,** 856–859.

53. Wright, F. A., et al. (2001) A draft annotation and overview of the human genome. *Genome Biol.* **2,** 25.

54. Claverie, J.M. (2001) Gene number. What if there are only 30,000 human genes? *Science* **291,** 1255–1257.

55. Green, E. D. and Chakravarti, A. (2001) The human genome sequence expedition, views from the "base camp." *Genome Res.* **11,** 645–651.

56. Cheung, V. G., et al. (2001) Integration of cytogenetic landmarks into the draft sequence of the human genome. *Nature* **409,** 953–958.

57. Futreal, A., et al. (2001) Cancer and genomics. *Nature* **409,** 850–852.

58. Fahrer, A. M., et al. (2001) A genomic view of immunology. *Nature* **409,** 836–838.

59. Nestler, E. J. and Landsman, D. (2001) Learning about addiction from the genome. *Nature* **409,** 834–835.

60. Clayton, J. D., Kyriacou, C., and Reppert, S. M. (2001) Keeping time with the human genome. *Nature* **409,** 829–831.

61. Li, W. H., et al. (2001) Evolutionary analyses of the human genome. *Nature* **409,** 847–849.

62. Caron, H., et al. (2001) The human transcriptome map, clustering of highly expressed genes in chromosomal domains. *Science* **291,** 1289–1292.

63. Peltonen, L. and McKusick, V. A. (2001) Genomics and medicine, dissecting human disease in the postgenomic era. *Science* **291,** 1224–1229.

64. Collins, F. S. and McKusick, V. A. (2001) Implications of the Human Genome Project for medical science. *JAMA* **285,** 540–544.

65. McKusick, V. A. (2001) The anatomy of the human genome, a neo–Vesalian basis for medicine in the 21st century. *JAMA* **286,** 2289–2295.

66. Sander, C. (2000) Genomic medicine and the future of health care. *Science* **287,** 1977–1978.

67. Walsh, S. and Barrell, B. (1996) The *Saccharomyces cerevisiae* genome on the World Wide Web. *Trends Genet* **12,** 276–277.

68. Hieter, Bassett, D. E., and Valle, D. (1996) The yeast genome—a common currency. *Nat. Genet.* **13,** 253–255.

69. Botstein, D., Chervitz, S. A., and Cherry, J. M. (1997) Yeast as a model organism. *Science* **277,** 1259–1260.

70. Mewes, H. W., et al. (1997) Overview of the yeast genome. *Nature* **387(suppl),** 7–65.

71. Kumar, A., et al. (2002) An integrated approach for finding overlooked genes in yeast. *Nat. Biotechnol* **20,** 58–63.

72. Oliver, S. (2002) 'To-day, we have naming of parts.' *Nat. Biotechnol.* **20,** 27–28.

73. Pennisi, E. (1998) Worming secrets from the C. elegans genome. *Science* **282,** 1972–1974.

74. The *Caenorhabditis elegans* Sequencing Consortium, Genome Sequence of the *Nematode C. elegans*, a platform for investigating biology. *Science* **282**, 2012–2018.

75. Bargmann, C. I. (1998) Neurobiology of the *Caenorhabditis elegans* genome. *Science* **282**, 2028–2033.

76. Ruvkun, G. and Hobert, O. (1998) The taxonomy of developmental control in *Caenorhabditis elegans*. *Science* **282**, 2033–2041.

77. Adams, M. D., et al. (2000) The genome sequence of *Drosophila melanogaster*. *Science* **287**, 2185–2195.

78. Kornberg, T. B. and Krasnow, M. A. (2000) The *Drosophila* genome sequence, implications for biology and medicine. *Science* **287**, 2218–2220.

79. Pennisi, E. (2000) Ideas fly at gene-finding jamboree. *Science* **287**, 2182–2184.

80. Myers, E. W., et al. (2000) A whole-genome assembly of *Drosophila*. *Science* **287**, 2196–2204.

81. Rubin, G. M., et al. (2000) Comparative genomics of the eukaryotes. *Science* **287**, 2204–2215.

82. Doolittle, R. F. (1997) A bug with excess gastric avidity. *Nature* **388**, 515–516.

83. Tomb, J. F., et al., The complete genome sequence of the gastric pathogen *Helicobacter pylori*. *Nature* **388**, 539–547.

84. Covacci, A., et al. (1999) *Helicobacter pylori* virulence and genetic geography. *Science* **284**, 1328–1333.

85. Alm, R. A., et al. (1999) Genomic-sequence comparison of two unrelated isolates of the human gastric pathogen *Helicobacter pylori*. *Nature* **397**, 176–180.

86. Bjorkholm, B., et al. (2001) Comparison of genetic divergence and fitness between two subclones of *Helicobacter pylori*. *Infect. Immun.* **69**, 7832–7838.

87. Bjorkholm, B. M., et al. (2001) Genomics and proteomics converge on *Helicobacter pylori*. *Curr. Opin. Microbiol.* **4**, 237–245.

88. Detweiler, C. S., Cunanan, D. B., and Falkow, S. (2001) Host microarray analysis reveals a role for the *Salmonella* response regulator phoP in human macrophage cell death. *Proc. Natl. Acad. Sci. USA* **98**, 5850–5855.

89. Young, D. B. (1998) Blueprint for the white plague. *Nature* **393**, 515–516.

90. Cole, S. T., et al. (1998) Deciphering the biology of *Mycobacterium tuberculosis* from the complete genome sequence. *Nature* **393**, 537–544.

91. Cole, S. T., et al. (2001) Massive gene decay in the leprosy *Bacillus*. *Nature* **409**, 1007–1011.

92. Wahlgren, M. and Bejarano, M. T. (1999) A blueprint of 'bad air'. *Nature* **400**, 506–507.

93. Bowman, S., et al. (1999) The complete nucleotide sequence of chromosome 3 of *Plasmodium falciparum*. *Nature* **400**, 532–538.

94. Gardner, M. J., et al. (1998) Chromosome 2 sequence of the human malaria parasite *Plasmodium falciparum*. *Science* **282**, 1126–1132.

95. Blattner, F. R., et al. (1997) The complete genome sequence of *Escherichia coli* K–12. *Science* **277**, 1453–1474.

96. Pennisi, E., Laboratory workhorse decoded. *Science* **277**, 1432–1434.

97. Pennisi, E. (2001) Microbial genomes, new genome a boost to plant studies. *Science* **294**, 2266a.

98. Wood, D. W., et al. (2001) The genome of the natural genetic engineer *Agrobacterium tumefaciens* C58. *Science* **294**, 2317–2323.

99. Goodner, B., et al. (2001) Genome sequence of the plant pathogen and biotechnology agent *Agrobacterium tumefaciens* C58. *Science* **294**, 2323–2328.

100. Galas, D. J. (2001) Sequence interpretation, making sense of the sequence. *Science* **291**, 1257–1260.

101. Pennisi, E. (2001) What's next for the genome centers? *Science* **291**, 1204–1207.

102. Fishman, M. C. (2001) Genomics. Zebrafish—the canonical vertebrate. *Science* **294**, 1290–1291.

103. Vogel, G. (2000) Genetics, Zebrafish earns its stripes in genetic screens. *Science* **288,** 1160–1161.

104. Patton, E. E. and Zon, L. I. (2001) The art and design of genetic screens, zebrafish. *Nat. Rev. Genet.* **2,** 956–966.

105. Roush, W. (1997) Developmental biology, a Zebrafish genome project? *Science* **275,** 923.

106. Duyk, G. and Schmitt, K. (2001) Fish x 3. *Nat. Genet.* **27,** 8–9.

107. Nadeau, J. H., et al. (2001) Sequence interpretation, functional annotation of mouse genome sequences. *Science* **291,** 1251–1255.

108. Rulicke, T. (1996) Transgenic technology, an introduction. *Int. J. Exp. Pathol.* **77,** 243–245.

109. Dayan, A. D. (1996) Transgenic rodents in toxicology. *Int. J. Exp. Pathol.* **77,** 251–256.

110. Gewolb, J. (2001) Genomics, animals line up to be sequenced. *Science* **293,** 409–410.

111. The Arabidopsis Genome Initiative, Analysis of the genome sequence of the flowering plant *Arabidopsis thaliana. Nature* **408,** 796–815.

112. Maloof, J. N., et al. (2001) Natural variation in light sensitivity of *Arabidopsis. Nat. Genet.* **29,** 441–446.

113. Millar, A. J. (2001) Light responses of a plastic plant. *Nat. Genet.* **29,** 357–358.

114. Quigley, R. A., Rosenberg, J. M., Shachar-Hill, Y., Bohnert, H. J. (2001) From genome to function, the *Arabidopsis aquaporins.* Genome Biology http,//genomebiology.com/2001/3/1/ research/0001.1.

115. Botstein, D., et al. (1980) Construction of a genetic linkage map in man using restriction fragment length polymorphisms. *Am. J. Hum. Genet.* **32,** 314–331.

116. Risch, N. and Merikangas, K. (1996) The future of genetic studies of complex human diseases. *Science* **273,** 1516–1517.

117. Nebert, D. W. (1997) Polymorphisms in drug–metabolizing enzymes, what is their clinical relevance and why do they exist? *Am. J. Hum. Genet.* **60,** 265–271.

118. Puga, A., et al. (1997) Genetic polymorphisms in human drug–metabolizing enzymes, potential uses of reverse genetics to identify genes of toxicological relevance. *Crit. Rev. Toxicol.* **27,** 199–222.

119. Sachse, C., et al. (1997) Cytochrome P450 2D6 variants in a caucasian population, allele frequencies and phenotypic consequences. *Am. J. Hum. Genet.* **60,** 284–295.

120. Meyer, U. A. and Zanger, U. M. (1997) Molecular mechanisms of genetic polymorphisms of drug metabolism. *Annu. Rev. Pharmacol. Toxicol.* **37,** 269–296.

121. Nakajima, T., et al. (1995) Expression and polymorphism of glutathione S–transferase in human lungs, risk factors in smoking–related lung cancer. *Carcinogenesis* **16,** 707–711.

122. Yengi, L., et al. (1996) Polymorphism at the glutathione S-transferase locus GSTM3, interactions with cytochrome P450 and glutathione S-transferase genotypes as risk factors for multiple cutaneous basal cell carcinoma. *Cancer Res.* **56,** 1974–1977.

123. Norton, R. M. (2001) Clinical pharmacogenomics, applications in pharmaceutical R&D. *Drug Discov. Today* **6,** 180–185.

124. Roses, A. D. (2001) How will pharmacogenetics impact the future of research and development? *Drug Discov. Today* **6,** 59–60.

125. Nickerson, D. A., et al., Identification of clusters of biallelic polymorphic sequence–tagged sites (pSTSs) that generate highly informative and automatable markers for genetic linkage mapping. *Genomics* **12,** 377–387.

126. Wang, D. G., et al., Large-scale identification, mapping, and genotyping of single-nucleotide polymorphisms in the human genome. *Science* **280,** 1077–1082.

127. Chakravarti, A. It's raining SNPs, hallelujah? *Nat. Genet.* **19,** 216–217.

128. Clark, A. G., et al., Haplotype structure and population genetic inferences from nucleotide-sequence variation in human lipoprotein lipase. *Am. J. Hum. Genet.* **63,** 595–612.

129. Nickerson, D. A., et al., DNA sequence diversity in a 9.7-kb region of the human lipoprotein lipase gene. *Nat. Genet.* **19**, 233–240.

130. Sachidanandam, R., et al., A map of human genome sequence variation containing 1.42 million single nucleotide polymorphisms. *Nature* **409**, 928–933.

131. Xiao, W. and Oefner, J., Denaturing high-performance liquid chromatography, a review. *Hum. Mutat.* **17**, 439–474.

132. Johnson, G. C., et al., Haplotype tagging for the identification of common disease genes. *Nat. Genet.* **29**, 233–237.

133. Shi, M. M., Bleavins, M. R., and de la Iglesia, F. A. (1999) Technologies for detecting genetic polymorphisms in pharmacogenomics. *Mol. Diagn.* **4**, 343–351.

134. Tillib, S. V. and Mirzabekov, A. D. (2001) Advances in the analysis of DNA sequence variations using oligonucleotide microchip technology. *Curr. Opin. Biotechnol.* **12**, 53–58.

135. Patil, N., et al., Blocks of limited haplotype diversity revealed by high-resolution scanning of human chromosome 21. *Science* **294**, 1719–1723.

136. Kwok, Y. (2001) GENOMICS, Genetic association by whole-genome analysis? *Science* **294**, 1669–1670.

137. Helmuth, L. (2001) Genome research, map of the human genome 3.0. *Science* **293**, 583–585.

138. Lander, E. S. (1999) Array of hope. *Nat. Genet.* **21(suppl)**, 3–4.

139. Thomas, S. (1980) Hybridization of denatured RNA and small DNA fragments transferred to nitrocellulose. *Proc. Natl. Acad. Sci. USA* **77**, 5201–5205.

140. Roberts, L. (1991) Gambling on a shortcut to genome sequencing. *Science* **252**, 1618,1619.

141. Bonaldo, M. F., Lennon, G., and Soares, M. B. (1996) Normalization and subtraction, two approaches to facilitate gene discovery. *Genome Res.* **6**, 791–806.

142. Liang, P. and Pardee, A. B. (1992) Differential display of eukaryotic messenger RNA by means of the polymerase chain reaction. *Science* **257**, 967–971.

143. Liang, P., et al. (1995) Analysis of altered gene expression by differential display. *Methods Enzymol.* **254**, 304–321.

144. Liang, P. and A. B. Pardee (1998) Differential display. A general protocol. *Mol. Biotechnol.* **10**, 261–267.

145. Velculescu, V. E., et al. (1995) Serial analysis of gene expression. *Science* **270**, 484–487.

146. Velculescu, V. E., et al. (1997) Characterization of the yeast transcriptome. *Cell* **88**, 243–251.

147. Southern, E. M. (1975) Detection of specific sequences among DNA fragments separated by gel electrophoresis. *J. Mol. Biol.* **98**, 503–517.

148. Southern, E., Mir, K., and Shchepinov, M. (1999) Molecular interactions on microarrays. *Nat. Genet.* **21(suppl)**, 5–9.

149. Schena, M., et al. (1995) Quantitative monitoring of gene expression patterns with a complementary DNA microarray. *Science* **270**, 467–470.

150. Duggan, D. J., et al. (1999) Expression profiling using cDNA microarrays. *Nat. Genet.* **21(suppl)**, 10–14.

151. Cheung, V. G., et al. (1999) Making and reading microarrays. *Nat. Genet.* **21(suppl)**, 15–19.

152. Bowtell, D. D. (1999) Options available—from start to finish—for obtaining expression data by microarray. *Nat. Genet.* **21(suppl)**, 25–32.

153. Lipshutz, R. J., et al. (1999) High density synthetic oligonucleotide arrays. *Nat. Genet.* **21(suppl)**, 20–24.

154. Hughes, T. R., et al. (2001) Expression profiling using microarrays fabricated by an ink-jet oligonucleotide synthesizer. *Nat. Biotechnol.* **19**, 342–347.

155. Eisen, M. B., et al. (1998) Cluster analysis and display of genome-wide expression patterns. *Proc. Natl. Acad. Sci. USA* **95**, 14,863–14,868.

156. Baker, T. K., et al. (2001) Temporal gene expression analysis of monolayer cultured rat hepatocytes. *Chem. Res. Toxicol.* **14**, 1218–1231.

157. Gullans, S. R. (2000) Of microarrays and meandering data points. *Nat. Genet.* **26**, 4–5.

158. Tusher, V. G., Tibshirani, R., and Chu, G. (2001) Significance analysis of microarrays applied to the ionizing radiation response. (erratum appears in *Proc. Natl. Acad. Sci. USA* 2001 Aug 28;98,10515). *Proc. Natl. Acad. Sci. USA* **98**, 5116–5121.

159. Zhao, L. P., Prentice, R., and Breeden, L. (2001) Statistical modeling of large microarray data sets to identify stimulus–response profiles. *Proc. Natl. Acad. Sci. USA* **98**, 5631–5636.

160. Kerr, M. K. and Churchill, G. A. (2001) Statistical design and the analysis of gene expression microarray data. *Genet. Res.* **77**, 123–128.

161. Kerr, M. K. and Churchill, G. A. (2001) Bootstrapping cluster analysis, assessing the reliability of conclusions from microarray experiments. *Proc. Natl. Acad. Sci. USA* **98**, 8961–8965.

162. Siedow, J. N. (2001) Making sense of microarrays. *Genome Biol* **2**, 4003.

163. Brazma, A., et al. (2001) Minimum information about a microarray experiment (MIAME)–toward standards for microarray data. *Nat. Genet.* **29**, 365–371.

164. Decottignies, A. and Goffeau, A. (1997) Complete inventory of the yeast ABC proteins. *Nat. Genet.* **15**, 137–145.

165. Tatusov, R. L., Koonin, E. V., and Lipman, D. J. (1997) A genomic perspective on protein families. *Science* **278**, 631–637.

166. Lockhart, D. J. (1998) Mutant yeast on drugs. *Nat. Med.* **4**, 1235–1236.

167. Oliver, S. (1999) Redundancy reveals drugs in action. *Nat. Genet.* **21**, 245–246.

168. Marton, M. J., et al. (1998) Drug target validation and identification of secondary drug target effects using DNA microarrays. *Nat. Med.* **4**, 1293–1301.

169. Gray, N. S., et al. (1998) Exploiting chemical libraries, structure, and genomics in the search for kinase inhibitors. *Science* **281**, 533–538.

170. Corder, E. H., et al. (1993) Gene dose of apolipoprotein E type 4 allele and the risk of Alzheimer's disease in late onset families. *Science* **261**, 921–923.

171. Saunders, A. M., et al. (1993) Association of apolipoprotein E allele epsilon 4 with late–onset familial and sporadic Alzheimer's disease. *Neurology* **43**, 1467–1472.

172. Farlow, M. R., et al. (1996) Apolipoprotein E genotype and gender influence response to tacrine therapy. *Annu. NY Acad. Sci.* **802**, 101–110.

173. Altshuler, D., et al. (2000) The common PPARgamma Pro12Ala polymorphism is associated with decreased risk of type 2 diabetes. *Nat. Genet.* **26**, 76–80.

174. Drazen, J. M., et al. (1999) Pharmacogenetic association between ALOX5 promoter genotype and the response to anti–asthma treatment. *Nat. Genet.* **22**, 168–170.

175. Liggett, S. B. (2001) Pharmacogenetic applications of the human genome project. *Nat. Med.* **7**, 281–283.

176. Debouck, C. and Goodfellow, N. (1999) DNA microarrays in drug discovery and development. *Nat. Genet.* **21(suppl)**, 48–50.

177. Glynne, R., et al. (2000) B-lymphocyte quiescence, tolerance and activation as viewed by global gene expression profiling on microarrays. *Immunol. Rev.* **176**, 216–246.

178. Glynne, R., et al. (2000) How self–tolerance and the immunosuppressive drug FK506 prevent B-cell mitogenesis. *Nature* **403**, 672–676.

179. Ross, D. T., et al. (2000) Systematic variation in gene expression patterns in human cancer cell lines. *Nat. Genet.* **24**, 227–235.

180. Scherf, U., et al. (2000) A gene expression database for the molecular pharmacology of cancer. *Nat. Genet.* **24**, 236–244.

181. Pinkel, D. (2000) Cancer cells, chemotherapy and gene clusters. *Nat. Genet.* **24**, 208–209.

182. Wang, W., et al. (2001) Pharmacogenomic dissection of resistance to thymidylate synthase inhibitors. *Cancer Res.* **61**, 5505–5510.

183. Lam, L. T., et al. (2001) Genomic–scale measurement of mRNA turnover and the mechanisms of action of the anti–cancer drug flavopiridol. *Genome Biol.* **2**, 0041.

184. Der, S. D., et al. (1998) Identification of genes differentially regulated by interferon alpha, beta, or gamma using oligonucleotide arrays. *Proc. Natl. Acad. Sci. USA* **95,** 15,623–15,628.

185. Garber, K. (2000) Protein C may be sepsis solution. *Nat. Biotechnol.* **18,** 917–918.

186. Grinnell, B. W. and Joyce, D. (2001) Recombinant human activated protein C, a system modulator of vascular function for treatment of severe sepsis. *Crit. Care Med.* **29(suppl),** S53–S60; discussion S60–S61.

187. Joyce, D. E., et al. (2001) Gene expression profile of antithrombotic protein c defines new mechanisms modulating inflammation and apoptosis. *J. Biol. Chem.* **276,** 11,199–11,203.

188. DeFrancesco, L. (2001) First sepsis drug nears market. *Nat. Med.* **7,** 516–517.

189. Golub, T. R., et al. (1999) Molecular classification of cancer, class discovery and class prediction by gene expression monitoring. *Science* **286,** 531–537.

190. Sorlie, T., et al. (2001) Gene expression patterns of breast carcinomas distinguish tumor subclasses with clinical implications. *Proc. Natl. Acad. Sci. USA* **98,** 10,869–10,874.

191. Hammond, T. G., et al. (2000) Mechanical culture conditions effect gene expression, gravity-induced changes on the space shuttle. *Physiol. Genomics* **3,** 163–173.

192. Chen, B. P., et al. (2001) DNA microarray analysis of gene expression in endothelial cells in response to 24-h shear stress. *Physiol. Genomics* **7,** 55–63.

193. DeRisi, J. L., Iyer, V. R., and Brown, O. (1997) Exploring the metabolic and genetic control of gene expression on a genomic scale. *Science* **278,** 680–686.

14

Proteomics

Gerald W. Becker, Michael D. Knierman, Pavel Shiyanov,
and John E. Hale

1. Introduction

With the recent completion and publication of the first-pass sequence of the human genome *(1,2)*, scientific attention has been refocused on the roles of these genes, particularly the roles and functions of the products of these genes, the proteins. The study of the gene products or proteins of a cell or tissue or organism is now being referred to as the study of proteomics.

The term "proteomics" first appeared in print in 1995 in a paper published by Wilkins and his colleagues *(3)*, and was used to refer to all expressed proteins that arise from a genome. Initially, proteomics was associated solely with two-dimensional SDS-poly-acrylamide gel electrophoresis (2d-PAGE) and with methods to identify the proteins resolved by this technique.

In the years that followed, the definition of proteomics has been broadened substantially. A good current definition of proteomics is "the identification, characterization, and quantification of all proteins involved in a particular pathway, organelle, cell, tissue, organ, or organism that can be studied in concert to provide accurate and comprehensive data about the function of proteins within these systems." It is convenient to think of proteomics as the protein analog of genomics.

Proteomics is far more complex than genomics. Although still somewhat controversial, recent estimates place the number of genes in the human genome at approx 30,000–40,000 *(2)*. This surprisingly low number of predicted genes—the simple one-celled yeast, *S. cerevisiae*, has over 6000 genes *(4)*—suggests that much of the biological diversity seen in human beings must reside in the proteome. Indeed, estimates of the number of proteins range from tens of thousands to greater than one million chemically distinct proteins in the human proteome. The old tenet of "one gene, one protein" is no longer valid, as it is well-established now that a single gene can yield a multitude of proteins through the process of generating splice variants and by a variety of post-translational modifications of the polypeptide backbone. Examples of these modifications include glycosylation, phosphorylation, sulfation, oxidation, deamidation, N-terminal acylation, and proteolytic modification.

Proteomics can be applied to many aspects of drug discovery, including (a) novel protein discovery, (b) biomarker discovery and validation, (c) mechanisms of drug

From: *Handbook of Industrial Cell Culture: Mammalian, Microbial, and Plant Cells*
Edited by: V. A. Vinci and S. R. Parekh © Humana Press Inc., Totowa, NJ

action, (d) elucidation of biochemical pathways, (e) target identification and validation, (f) protein identification and validation, (g) lead optimization, (h) metabolism, (i) protein engineering, (j) differential protein expression, and (k) characterization of proteins and post-translational modifications of proteins.

An understanding of proteomics and the tools of proteomic analysis is of critical importance to those who are involved in industrial cell culture. The latter two aspects of drug discovery—differential protein expression and characterization of proteins and their post-translational modifications are of particular interest.

It is well-established that changing cell-culture conditions can have a significant impact on protein expression. Proteomics (particularly, expressional proteomics) can be used to evaluate the impact of such changes on protein expression by revealing proteins that are newly expressed or upregulated, as well as proteins whose expression is attenuated. Furthermore, changes in protein expression can result in differences in the starting material from which the product is produced, thereby impacting the downstream purification of that product. In many cases, the final purified product must be analyzed for the presence of detectable levels of any new contaminants resulting from alterations in protein expression.

Post-translational modifications of proteins, including glycosylation (both N-linked and O-linked) and phosphorylation can have a significant influence on protein activity and stability *(5,6)*, but other modifications such as sulfation, N-terminal acylation, and proteolytic cleavage are also important. Often, the choice of a host cell or culture conditions can influence the type and extent of these post-translational modifications. For example, most bacterial systems lack the enzymes to carry out modifications like glycosylation, whereas mammalian systems and baculovirus-infected insect cells are capable of adding sugars to proteins. However, large differences are possible in the extent or type of glycoslyation between insect cells and mammalian cells, or even among different mammalian systems.

Other examples include the phosphorylation of proteins by insect cells *(7)*. These cells are generally capable of phosphorylation, but the extent of this modification may vary greatly, giving rise to a multitude of phosphorylated species. If these cells are being used to produce a recombinant enzyme for high-throughput screening, for example, the phosphorylation pattern can have a significant effect on enzyme activity, and the pattern must be understood before designing strategies for optimizing enzyme activity in this cell system.

Clearly, proteomics can play an important role in the practice of industrial cell culture, in the choice of the best culture system, the control of the culture process, and the characterization of the products of the culture system.

Proteomics has been the subject of a large number of review articles and commentaries over the past few years *(8–12)*. This chapter is intended to provide an introduction to the topic, rather than a comprehensive review, including a discussion of the various types of proteomic experiments followed by a description of the tools available to perform these experiments.

2. Types of Proteomic Experiments

It is convenient to subdivide proteomics into three categories—functional proteomics, expressional proteomics, and structural proteomics—each with different research goals and different technologies to achieve those goals.

2.1. Functional Proteomics

Functional proteomics is the study of the relationship of proteins within functional complexes in cells. Most proteins do not function as individual proteins in a physiological situation but as members of multi-protein complexes. Fig. 1 shows the flow of genetic information from DNA to RNA and to protein. Proteins then assemble into functional complexes, forming the basis of biochemical pathways. Such complexes include well-studied examples such as ribosomal complexes *(13)* as well as signaling complexes *(14,15)*, multi-enzyme complexes *(16,17)*, and others *(18,19)*. The identification of the individual members of these protein complexes can lead to a better understanding of how individual proteins interact, and can provide clues to the function of the individual proteins.

A flow diagram for a functional proteomics experiment is shown in Fig. 2. Following a biological experiment, protein complexes of interest are isolated by affinity capture, using either a specific antibody or other non-antibody protein with a specificity for a functional group that has been introduced into a protein or a specificity for a class of proteins. A summary of affinity-capture techniques with examples is given in Table 1. Although affinity methods are generally very specific for individual proteins, or groups of proteins these methods are often plagued by high levels of nonspecific binding. This—coupled with the fact that protein functional complexes are often very complex—necessitates further separation using either an electrophoretic or chromatographic method. These methods include affinity, electrophoretic, and chromatographic methods, and are discussed in more detail in **Subheading 3.1.**

Most functional proteomics experiments use one-dimensional SDS polyacrylamide gel electrophoresis (1d-PAGE) to accomplish this separation. An example of a 1d-PAGE separation of proteins isolated by affinity-capture methods is presented in Fig. 3. In this experiment, proteins were immunoprecipitated with an antibody to a specific cell-surface receptor, and these proteins were applied to the gel and separated as shown. Following electrophoresis, the proteins were stained with Coomassie blue and the individual protein bands were excised and digested with trypsin, creating a set of tryptic peptides. These peptides were analyzed through mass spectrometry, leading to identification of the proteins and in some cases a quantification of the amount of protein present. The mass spectrometry methods commonly used in proteomics work are described in **Subheading 3.2.**, and quantification in **Subheading 3.4.**

The process of preparing a protein sample for mass spectrometry involves several steps. A general flow diagram for accomplishing this is shown in Fig. 4. The electrophoretic gel in this diagram can be 1d-PAGE, as in a functional proteomics experiment, or it can be 2d-PAGE (*see* **Subheading 3.11.**). The separated proteins are visualized using Coomassie blue, silver stain, or fluorescent dyes. The stained protein bands or spots are excised from the gel and prepared for trypsin digestion. The proteins are reduced using dithiothreitol (DTT) or β-mercaptoethanol, and alkylated using either iodoacetic acid, iodoacetamide, or another alkylating agent to modify the cysteine residues within the proteins. The proteins are then digested within the gel slice, using a protease to create a set of peptides that can be analyzed by mass spectrometry. The enzyme trypsin is commonly used for this purpose, but other proteases such as Lys-C or Arg-C may also be used. In-gel protein digestion can be very effective, but care must be taken to assure that the proteolysis is carried out under optimal conditions

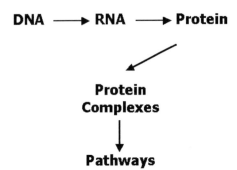

Fig. 1. The flow of genetic information from DNA to biochemical pathways. Genetic information residing in DNA is transcribed into RNA and ultimately translated into protein. Proteins assemble into active functional complexes and these complexes make up biochemical pathways. Information that can provide insight into the proteins that make up these complexes and the arrangement of the proteins in the complex can often place a novel protein in a known pathway and provide an insight into the physiological function of that protein.

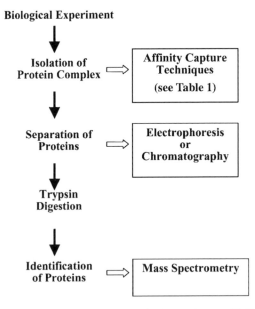

Fig. 2. Flow diagram for a functional proteomics experiment. Following a biological experiment, protein complexes of interest are isolated using one of the affinity-capture techniques described in Table 1. The protein components of the complex are separated using either an electrophoretic or chromatographic method. The resolved proteins are digested with the enzyme trypsin to create a set of tryptic peptides that are analyzed using mass spectrometry, leading to an identification of the proteins and in some cases a quantification of the amount of protein present.

Table 1
Affinity Capture Approaches

Recognition protein	Example
Antibody to a specific protein	Anti-insulin, anti-hGH
Antibody to a post-translational modification of a protein	Anti-phosphotyrosine
Antibody to an epitope tag that has been engineered into a protein	Anti-FLAG, His-Tag
Antibody to a specific protein that is part of a fusion protein	Anti GST (glutathione-S-transferase) Anti MBP (maltose-binding protein)
Non-antibody proteins that interact specifically and with high affinity with functional groups that can be attached to a protein	Avidin interacting with biotin
Non-antibody proteins that interact specifically and with high affinity with functional groups that occur naturally within proteins	Lectin interacting with specific carbohydrates

Avidin and the lectins are usually used attached to beads in order to facilitate collection of the affinity-captured proteins. Likewise, antibodies may be attached to beads. In some cases, a secondary antibody attached to a bead is used to capture the first antibody, which is soluble. Beads are typically made of agarose or sepharose, and can be collected by centrifugation. Magnetic beads, which can be collected through the use of a magnet, are gaining in popularity.

Abbreviations used: hGH: human growth hormone, FLAG refers to a specific artificial epitope with the amino acid sequence: Asp-Tyr-Lys-Asp-Asp-Asp-Asp-Lys containing an enterokinase cleavage site for removal of the epitope from the protein, His-Tag: a sequence of six histidine residues, which serve as a recognition epitope as well as a purification handle.

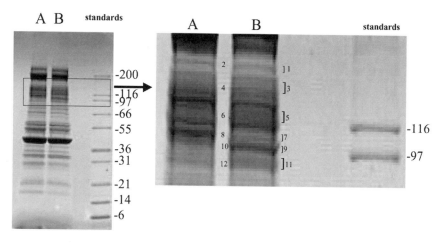

Fig. 3. 1-dimensional-gel of affinity-captured proteins stained with Coomassie blue. This gel shows the proteins that are immunoprecipitated from a cell lysate using an antibody to a specific receptor. Lane **A** is from the parental cell line, and Lane **B** is from the parental line transfected with a gene encoding the receptor. The proteins from such a gel are analyzed by excising the individual gel bands, digesting the proteins in the gel slice using the enzyme trypsin, extracting the tryptic peptides, and analysis by mass spectrometry.

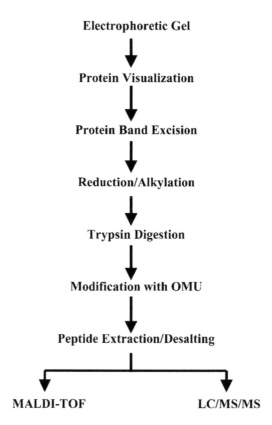

Fig. 4. Flow diagram for a general proteomics experiment starting with a protein separated by electrophoresis. This scheme applies to both 1d-PAGE and 2d-PAGE. Proteins are visualized, excised from the gel, and modified by reduction and alkylation. Proteolysis of the proteins is accomplished through the action of trypsin or a similar protease. At this point, the lysine residues may be converted to homoarginine by modification with O-methylisourea (OMU). This modification greatly enhances the ionization of lysine-containing peptides by MALDI *(21)*. All of these steps are carried out on the protein while it is still in the gel slice. Finally, the modified peptides are extracted from the gel, desalted, and analyzed by mass spectrometry.

to achieve a complete digestion *(20)*. A step that has recently been introduced into the processing of samples for mass spectrometric analysis is the modification of lysine residues with the reagent O-methylisourea (OMU), resulting in the conversion of lysine to homoarginine *(21)*. This conversion has the effect of increasing the ionization of lysine-containing peptides in matrix-assisted laser desorption/ionizaiton (MALDI) resulting in a richer mass spectrum and improving the identification of proteins through this technique. A final step prior to performing MALDI-TOF analysis of a peptide mixture is desalting. A convenient method for doing this is to make use of a small reversed-phase column packed in a pipet tip. Such columns are now commercially available, and are marketed as "ZipTips" by the Millipore Corporation. These ZipTips are available as 0.6- or 0.2-μL bed volumes packed in a 10-μL pipet tip. The reversed-phase resins are either C4 or C8.

Finally, the modified peptides are analyzed by mass spectrometry. A convenient strategy for protein identification is to first collect data from a MALDI-TOF experiment which generates the masses of peptides in the mixture. This type of experiment is commonly known as "peptide mass fingerprinting," and provides data that can be used to search a sequence database for matches. Proteins can often be identified with a high level of confidence using only this procedure. More often, identification is made, but corroborating data is needed to make the identification with confidence. The second level of mass spectrometric analysis is an LC/MS/MS experiment, which provides peptide-sequence information. These so-called "tandem mass spectra" can be used to search a sequence database for matches. Agreement between the two mass spectrometric methods provides a very high level of confidence in the protein identification. Mass spectrometry is discussed in more detail in **Subheading 3.2.1.**

2.2. Expressional Proteomics

Expressional proteomics (sometimes known as comparative proteomics) is the study of gene expression at the protein level, and is used to compare the expression levels of individual proteins in one sample with those in another sample. This experimental approach allows one to compare (a) diseased vs normal tissue, (b) treated vs control cells, (c) the impact of altering cell-culture conditions, or any other situation in which a comparison between two samples is desired.

There are three necessary components of an expressional proteomics experiment, separation, identification, and quantification.

The method that has traditionally been used to separate proteins in an expressional proteomics experiment is the technique of two-dimensional SDS polyacrylamide gel electrophoresis (2d-PAGE). This technique may offer the greatest resolving power of any separation method available to the protein chemist, and is described in **Subheading 3.11.** Briefly, the samples to be compared are applied to 2d gels, and electrophoresis is performed in each dimension. The methods for visualizing the proteins on the 2d gels are the same as for 1d gels: staining with Coomassie blue, silver staining, or staining with fluorescent dyes. The 2d gel provides a two-dimensional array of the proteins in the experimental sample with the proteins separated on the basis of charge and mol wt. An image of the gels can be obtained using densitometric scanning of the gels (described in **Subheading 3.4.1.**), providing a pattern of proteins that is characteristic of the experimental sample. By comparing patterns from different samples, differences in protein expression between the two samples can be detected. Identification of the proteins on the 2d gels is the same as that described here for 1d gels. The flow diagram in Fig. 4 applies to 2d-PAGE as well as to 1d-PAGE.

In order to be able to compare expression levels of the proteins separated by 2d-PAGE, a relative quantification of the proteins in the samples being compared must be made. Quantification is discussed in **Subheading 3.4.** For 2d-PAGE, the densitometric scans perviously discussed are used, and a "spot volume" (the integrated intensity across the protein spot) is calculated for each protein spot. This spot volume is the basis for quantitative comparisons between experimental samples, and is used to assess whether the proteins are up- or downregulated, or remain unchanged between samples.

Many practitioners of expressional proteomics are turning to chromatographic methods to replace the 2d-PAGE approach. These chromatographic methods are often multidimensional, meaning that multiple orthogonal column chromatographic methods are employed. Chromatography is discussed in **Subheading 3.1.2.** From a proteomics perspective, chromatography offers several advantages to the 2d-PAGE approach. Chromatography has a higher throughput than 2d-PAGE. Although it typically requires a period of several days to a week to perform a 2d-PAGE experiment, a chromatographic separation can be carried out in a matter of hours. If multidimensional chromatography is used, 2–3 d may be necessary to complete the study. Automation of the chromatographic process is relatively straightforward, with autosamplers and multiple switching valves, and a process can be set up to minimize human intervention. Electrophoresis, however, is labor-intensive, and is considerably more difficult to automate because it requires substantial time of a laboratory technician. One of the most vexing problems of 2d-PAGE is the limited dynamic range of the technique with at best, a 3-log range. Often, proteins of extremely high concentration in a sample must be removed before performing electrophoresis *(22)*. Although this is an effective strategy, it complicates the process considerably and slows the analysis even more.

Despite the drawbacks of 2d-PAGE, it is still a viable experimental technique that is widely used in expressional proteomics. An example of an expressional proteomics experiment using 2d-PAGE is shown in Fig. 5. The human endothelial-cell line, ECV304, was transfected with a gene of interest. The experiment was intended to pinpoint proteins that were co-regulated with this gene. In other words, which proteins were up- or downregulated in response to expression of the transfected gene? A series of gels was run on lysates from cells transfected with the gene and compared with a series of gels from control or mock-transfected cells. The results of this experiment are summarized in Fig. 5, which shows a master gel or composite gel containing all of the spots from all of the gels in both the experimental and control groups. The gel images were compared using the Kepler software for 2d-PAGE analysis. One protein, Spot 294, was found to be upregulated. The other numbered proteins were downregulated.

2.3. Structural Proteomics

Structural proteomics is the determination of the three-dimensional structures of the proteins in a proteome, using the techniques of X-ray crystallography and NMR and high-throughput protein production. The goals of structural proteomics include definition of all the key "functional" sites or domains of all proteins, support of lead optimization activities for the pharmaceutical industry, and increasing the number of viable drug targets.

The requirements for the proteins used in structural studies are rigorous. Crystallography necessitates highly purified proteins, which can form crystals suitable for diffraction studies. NMR has similar requirements for protein purity, but incorporation of stable isotopes (typically C^{13} and N^{15}) into the protein is often necessary for the required NMR experiments.

Currently, most proteins for structural proteomic studies are produced through the application of recombinant DNA technology, in which the proteins are produced by expression of the relevant genes in an appropriate cell-culture system. Conditions must be optimized to minimize differences in post-translational modifications—particularly glycosylation—arising from the cell-culture process. High-level expression is generally required to facilitate downstream purification. The purification process must

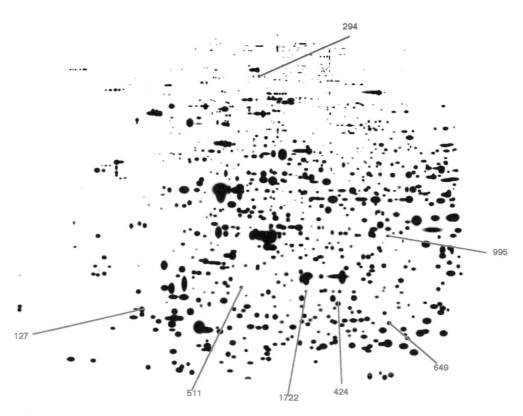

Fig. 5. Two-dimensional SDS polyacrylamide electrophoresis. This image represents a composite of several individual gels showing all of the protein spots visible in all of the gels. The gel images were processed and evaluated using the Kepler software (*see* Table 3). One protein, Spot 294, was found to be upregulated. The other numbered proteins were downregulated.

accomplish enrichment of the protein of interest in the original cell-culture medium to a point where the protein is highly homogenous. The stable isotope enrichment mentioned here is accomplished through metabolic labeling by providing a carbon and nitrogen source containing the stable isotopes to the recombinant organism in culture.

3. The Tools and Techniques of Proteomics

3.1. Separation Technologies

Separation technologies for protein analysis fall into two broad categories—electrophoretic and chromatographic. In the past, proteomics experimentalists usually applied the highest resolving technique as the first step in order to obtain highly purified protein for identification. New advances in mass spectrometric technologies have made the identification of proteins in mixtures possible, thus giving the proteomicist greater choice in separation technologies.

3.1.1. Two-Dimensional Gel Electrophoresis

Two-dimensional gel electrophoresis (2d-PAGE) is the separation technique offering the highest resolving power. This technique is much the same now as when it was introduced in the mid-1970s *(23,24)*. Proteins are separated in the first dimension according to charge on a thin isoelectric focusing tube gel or strip. This tube gel is then transferred to the top of a polyacrylamide gel, and sealed to the surface, and proteins are separated according to size by SDS-PAGE. Proteins are typically visualized by Coomassie or silver staining. The largest drawbacks to this technology are the limited sample size that can be applied to the IEF gel and problems with reproducibility introduced from stretching of the IEF gel upon transfer to the PAGE gel. Immobilized pH gradients introduced in the past few years have helped to resolve both of these issues *(25–27)*. These IEF strips, which are prepared by co-polymerization of ampholytes within a polyacrylamide matrix on a GelBond[tm] film, are commercially available in a number of pH ranges. Larger quantities of sample may be applied to these strips, and they are not elastic like their tube-gel predecessors. An additional advance in 2d-PAGE technology has occurred with the advent of more sensitive stains such as Sypro red *(28)*. The availability of high-throughput, high-sensitivity protein identification techniques, coupled with a number of database-searching algorithms, has positioned 2d-PAGE technologies to become an indispensable tool for proteomics. Despite all of its advantages, 2d-PAGE suffers from some additional drawbacks. One will necessarily visualize the more abundant proteins in a sample. Unmasking lower-abundance proteins in a mixture requires prefractionation *(22,29)* prior to iso-electric focusing (IEF), and because many prefractionation steps (such as chromatography) require conditions that interfere with IEF, additional sample preparation is needed. Integral membrane proteins and hydrophobic proteins do not separate well in the IEF step, and are not well-recovered in 2d-PAGE experiments. Even with all of the recent improvements, 2d-PAGE is a labor-intensive and time-consuming endeavor. Many companies that specialize in 2d-PAGE have begun to automate the process, which yields higher reproducibility and throughput.

In many instances, 1-dimensional gels (1d-PAGE) may be used in combination with other separation techniques in the visualization and identification of proteins. Immunoaffinity and other affinity techniques and chromatographic separations simplify protein mixtures enough that PAGE analysis resolve proteins well enough for identification of individual components. This approach has been applied to the identification of proteins in signaling pathways *(14,15,30,31)* in studying protein-protein interactions, and is a powerful technique in the identification of multiple components of a mixture. In addition, membrane proteins behave better in 1d-PAGE gels because there is no IEF step, and the proteins are kept in the presence of SDS.

3.1.2. Chromatography

Another powerful technology that has been applied to the proteomic characterization of proteins is chromatography. This can take as many forms, as protein fractionation technologies exist and a number of groups have utilized chromatographic prefractionation strategies with 2d-PAGE gels in an effort to increase the detection and identification of low-abundance proteins *(22,29)*. Many researchers have begun to couple chromatography directly to mass spectrometers in an effort to increase the sensitivity of protein identification. Perhaps the most fully exploited technology has been LC-mass spectrometry (LC-MS). It has become quite common to employ a reversed-phase separation for proteins or peptides using resins such as C-18 or C-4, and to couple these columns to an electrospray mass spectrometer for determination of the masses of eluting proteins or peptides. This chromatography fits particularly well with mass spectrometry, since the solvent systems employ volatile components that are compatible with the ionization process (32). This technology has been used extensively to characterize proteins by peptide mapping (32). Newer applications of LC-MS have been targeted at determining the identities of proteins in mixtures (33). Recently cation exchange and reversed-phase separations have been combined in a technique known as Mudpit (multi-dimensional protein identification technology). In this technique, cation-exchange resin is packed before reversed-phase resin in a capillary, and mixtures of peptides from protein digests are introduced at low pH. Peptides are sequentially released from the cation-exchange resin by steps in the pH or ionic strength, and these peptides are captured by the reversed-phase resin. The peptides are then fractionated with an acetonitrile gradient (34). Combination of this chromatography with an ion-trap mass spectrometer allows for automated data acquisition using programs such as Finnigan's "triple play," which detects peaks, determines their charge states, and collects MS/MS spectra. Another new technique developed for mass spectrometric applications is known as monolith columns. This technique involves packing a capillary with an activated resin that may be polymerized by heating the capillary (35). Once they have been polymerized, different functional groups may be coupled to the resin. This technology allows for the construction of multilayered chromatographic media, and may be customized to any particular application, as long as volatile solvent systems are used.

3.1.3. Affinity Techniques

The first step in any functional proteomics experiment is the isolation of the protein complex being studied. This is usually accomplished by affinity capture, taking advantage of the exquisite specificity of certain classes of proteins. The most obvious application of this technology is the isolation of protein by antibody-affinity purification. This technique may be tailored to specific cases, such as the isolation of an individual protein to which an antibody exists. The technique may be applied to less specific applications by using a class-specific antibody, such as an anti-phosphotyrosine antibody, to isolate all proteins containing a phosphorylated tyrosine residue. By adjustment of washing conditions, proteins associated with the affinity-isolated protein may also be recovered. This approach has been used to identify protein components of a receptor-signaling complex (15). Other affinity techniques have been applied to proteomics experiments. Overexpression of a protein of interest containing a purification handle such as the His tag (36) or Flag tag (37) or as a GST-fusion protein allows one to affinity isolate the protein and utilize it to monitor protein-protein interactions through pull-down approaches (36). Proteins may be chemically labeled with specific agents such as biotin, which have high affinity for some other agent (avidin for biotin), and utilized to incubate with protein mixtures to pull down complexes. A summary of various affinity strategies is provided in Table 1.

3.2. Analytical Technologies

3.2.1. Mass Spectrometry

There are currently two popular ways to identify the protein complement of a cell after separation. The most widely used option today is to digest the separated proteins with an enzyme and analyze the resulting peptides by mass spectrometry. This approach is known as peptide fingerprinting, and has proven to be robust and rapid. Another approach that is gaining popularity is to analyze the peptides from a protein digest on a capillary reversed-phase column connected to a mass spectrometer that is capable of generating mass and sequence information on the peptides as they elute from the reversed-phase column. This technique is generally referred to as LC/MS/MS, and can identify many components of a mixture.

3.2.1.1. PROTEIN IDENTIFICATION BY PEPTIDE FINGERPRINTING

After a separation, the protein is digested with an enzyme, (trypsin is the most popular enzyme, cleaving the protein backbone at the C-terminal side of lysine or arginine). The peptides are then analyzed by mass spectrometry. Usually the mass spectrometer used is a MALDI-TOF instrument because it is very sensitive, can acquire a mass spectrum very quickly, and generates a single peak for each peptide present in the digest. The current generation of MALDI-TOF instruments used for protein identification are equipped with delayed ion extraction and a reflectron analyzer. These additions to the instrument make it possible to measure the masses of the peptides with great accuracy *(38)*.

The peptide masses recorded in a MALDI-TOF spectrum represent a fingerprint for a particular protein. Many software packages exist—Profound *(39)*, Mascot *(40)*, and Protein Prospector *(41)*—which take the recorded masses and search the peptide-mass fingerprint across a database of protein sequences. The software uses the specificity of the digesting enzyme to create peptide-mass fingerprints of all the proteins in the database, and then determines the best match to the MALDI-TOF spectrum.

The peptide-fingerprint approach works best if the peptide masses are accurate and there is sufficient peptide coverage of the protein to make a confident assignment of identity. One of the limitations of this technique is that the protein separation must have a high degree of resolution—resulting in only one or two proteins in the sample to be digested—to allow confident assignment of identity. Another difficulty is that MALDI has significant discrimination between peptides with lysine and those containing arginine *(42)*. The arginine-containing peptides dominate the MALDI spectrum, and the lysine-containing peptides are weak or absent, thus limiting peptide coverage of the protein. A significant improvement in peptide coverage can be made to the protein fingerprint visible in the MALDI spectrum by conversion of the lysine-sidechain to homoarginine with OMU *(21)*. Modified peptides have an even greater chance of appearance in the MALDI spectrum, thus leading to greater peptide coverage of the protein of interest and increased confidence in the resulting identification. A typical MALDI spectrum of a protein digested from a gel slice is shown in Fig. 6.

3.2.1.2. PROTEIN IDENTIFICATION BY LC/MS/MS

If a more complex sample is to be analyzed, LC/MS/MS may be used to identify the constituent proteins without further fractionation *(43)*. The sample is digested into peptides (trypsin is still preferred) that are then loaded onto a reversed-phase column and eluted with an increasing organic gradient. For proteomic analysis, this is typically done using columns with diameters of 300 μm or smaller to improve the sensitivity of the analysis. The column effluent is directly introduced into an electrospray source of an ion trap, Q-TOF or other electrospray-type mass spectrometer. These mass spec-

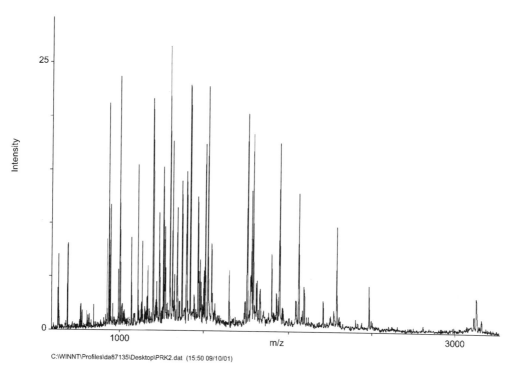

C:\WINNT\Profiles\da87135\Desktop\PRK2.dat (15:50 09/10/01)

Fig. 6. MALDI-TOF spectrum. This represents a typical MALDI-Tof spectrum recorded on a tryptic digest of the protein S6 kinase. The instrument used for this experiment was a Voyager DE Pro manufactured by Applied Biosystems. Sample preparation was as diagrammed in Fig. 4, including the modification with O-methylisourea (OMU).

trometers are capable of rapidly switching from MS to MS/MS mode, and are equipped with software capable of selecting ions to fragment in real time. Once a peptide ion is selected, the mass spectrometer performs a fragmentation of the peptide by collision-induced fragmentation or by modulation of the ion-trap voltage. Peptides typically fragment at the peptide bond and generate ions known as b or y ions (depending on whether charge is retained on the N- or C-terminus). Thus, a series of these daughter ions will yield sequence information about a peptide that can be used to search a protein database using software such as Sequest *(43,44)*, Mascot *(40)*, or Sonar *(45)* to identify the parent. Each peptide is an independent verification of the protein identity. All of the peptides are then correlated to identify the proteins present in the sample. This is an extremely powerful technique for identification of a complex mixture of proteins. An example of an LC/MS/MS experiment is provided in Fig. 7. The upper panel shows the total ion chromatogram of the peptide mixture, and the bottom panel shows a selected MS/MS spectrum from that chromatogram.

The technique can be extended by including another mode of separation online with the reversed-phase separation such as ion-exchange chromatography to greatly increase the ability to identify more complex mixtures of proteins *(46)*. Quantification can also be achieved along with identification with the use of labeling reagents such as the isotope-coded affinity tag (ICAT) reagent *(47)* (*see* **Subheading 3.4.2.**).

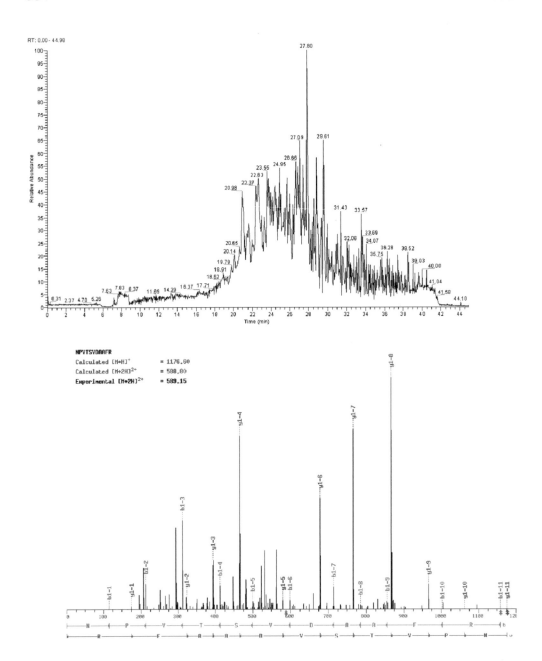

Fig. 7. LC/MS/MS chromatogram and spectrum. An LC/MS/MS experiment was performed on a tryptic digest of an unknown protein. Chromatography was on a capillary column (75 mm × 50 mm, LC Packings, Aquasil C18) developed with a gradient of 0.1% formic acid/acetonitrile. The upper panel shows the total ion chromatogram, and the bottom panel shows a selected MS/MS spectrum from the chromatogram. The spectrum was interpreted using the Sequest program to yield the peptide sequence Asn-Pro-Val-Thr-Ser-Val-Asp-Ala-Ala-Phe-Arg.

Newer mass spectrometers currently in use have the capacity to measure peptide masses to 1 ppm accuracy. These instruments, called Fourier transform ion cyclotron resonance (FTICR) mass spectrometers, may be used to identify proteins in extremely complex mixtures using accurate mass tags (the peptide mass measured to extremely high accuracy). Once a peptide has been identified by MS/MS, the accurate mass tag of that peptide may be used as a diagnostic ion for the presence of its parent protein *(48)*. This technique may ultimately lead to a wider dynamic range of peptide detection, which is one of the limiting factors in the sensitivity of proteomics experimentation.

3.2.2. Edman Sequencing

Edman sequencing refers to the sequential chemical modification and cleavage of individual amino acids from the amino terminus of a protein or peptide. First developed by Edman in the 1950s *(49)*, this technique has been modified and improved, and today the chemical reactions are highly efficient and have been automated *(50)*. Commercially available instruments are available to perform the chemistry and to identify the amino acids. The steps of the Edman reaction involve reaction of phenylisothiocyanate (PITC) with the free N-terminus of the protein under basic conditions to form a phenylthiocarbamyl derivative (PTC-protein). Cleavage of the modified N-terminal amino acid is affected, with trichloroacetic acid, releasing this amino acid as the anilinothiazolinone (ATZ-amino acid) and freeing the next amino acid for a subsequent reaction. The ATZ-amino acid is extracted and converted to the phenylthiohydantoin derivative (PTH-amino acid) by treatment with strong acid. This derivative is identified by reversed-phase HPLC and comparison of its retention time with those of PTH-amino acid standards. This process is repeated many times to sequence the protein. Under optimal conditions, it is possible to sequence as many as 75 amino acids.

In a proteomics experiment, Edman sequencing is typically applied to proteins that have been separated by SDS-PAGE. After electrophoresis, the proteins are transferred from the gel to a membrane of polyvinylidene difluoride (PVDF) *(51)*. After staining the membrane with a protein stain such as Coomassie blue, the protein spots are cut out of the membrane and placed in the reaction chamber of the sequencing instrument. In general, picomole quantities of protein or peptide are needed to obtain the sequence of a sufficient number of amino-acid residues to unambiguously identify the protein.

This technique has played a significant role in the proteomic identification of proteins separated by gels, but it is a time-consuming technique and with recent improvements in mass spectrometry, Edman sequencing is playing a diminished role today. Modern mass spectrometry methods offer significant advantages in sensitivity and throughput.

3.3. Protein–Protein Interactions

The study of protein-protein interactions can provide information about the function of a protein. In this era immediately following the sequencing of the human genome, there are a large number of proteins whose sequence has been predicted, but for which no physiological function has been described. Knowledge of the binding partners for these proteins provides both the basis for hypothesis generation concerning the function and the foundation for further experimentation. Three approaches are used to study protein-protein interactions: surface plasmon resonance, yeast 2-hybrid, and immunoaffinity approaches.

3.3.1. Surface Plasmon Resonance

Surface plasmon resonance (SPR) is an optical phenomenon that can be employed to study protein binding in real time *(52)*. The technique has been developed and commercialized by Biacore AB, a company that offers instruments and reagents to carry out proteomic experiments using this technique. The basis of SPR is illustrated in Fig. 8, in which a protein molecule has been immobilized on the surface of the sensor chip and exposed to test solutions passed through the microfluidic channels. Binding of analytes to the immobilized protein can be detected by monitoring changes in the refractive index and when these changes are monitored as a function of time, a sensorgram can be recorded (Fig. 9). The sensorgram contains information on both the binding or association phase and the release or dissociation phase of the interaction. Using curve-fitting algorithms, the two phases of the interaction can be processed to yield kinetic-rate constants for both association and dissociation. The ratio of these two kinetic constants yields the equilibrium or dissociation constant.

There are numerous applications of the SPR technology to drug discovery and development including proteomics (target identification and ligand fishing), target and assay validation for high-throughput screening, hit-to-lead characterization, lead optimization, and characterization of antibodies and other proteins being developed for therapeutic applications. This discussion will focus on the proteomic application of ligand fishing.

Ligand fishing is the term that has been applied to the process of screening for ligands that bind with specificity and high affinity to a target protein that has been immobilized on the sensor chip. Solutions containing purified proteins that are suspected of being ligands or cell extracts can be passed over the target protein quickly. If a binding response is detected, the specificity of this response can be assessed by co-injection of the suspected ligand with the target protein itself.

Applications of the ligand fishing approach may be found in the literature. For example, Fitz and his colleagues *(53)* used ligand fishing to identify a ligand for the murine Flt4 (or vascular endothelial growth-factor receptor-3) protein. A fusion of the extracellular domain of Flt4 and the constant domain of human IgG (Flt4-Fc) was immobilized to a sensor chip and conditioned media from over 100 different cell lines were screened for binding. Only 3 of the 100 media samples showed evidence of binding to Flt4-Fc, but not to a control protein (human IgG). Specificity of the binding interaction was demonstrated by co-injection of soluble Flt4-Fc along with the media sample, with abolishment of the binding response. Following purification of the newly discovered ligand by affinity chromatography, degenerate PCR primers were designed based on N-terminal Edman sequencing, and these primers were used for molecular cloning of the murine cDNA. The ligand was found to have a hydrophobic signal peptide, which, when removed, generated an N-terminal sequence identical to that determined by Edman sequencing. Expression of this cDNA produced a protein that had the property of activating Flt4 as measured by induction of tyrosine phosphorylation and induced mitogenesis in vitro of lymphatic endothelial cells. This work forms the foundation for further investigation into the function of Flt4 and its ligand in angiogenesis.

3.3.2. Yeast 2-Hybrid

The yeast 2-hybrid system is a yeast genetics/molecular biological approach to determining protein-binding partners for selected "bait" proteins. Yeast 2-hybrid systems are simple, sensitive, and amenable to high-throughput methods. Initially designed to find binding partners for single proteins, the system has evolved to the point that it

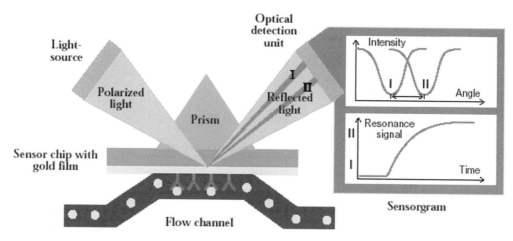

Fig. 8. Surface plasmon resonance. Protein molecules are immobilized on the sensor chip, which has a carboxymethylated dextran-gold surface. The chip is configured to expose the immobilized proteins to a microfluidic flow channel through which test solutions may be passed. A binding event results in an increase in mass in the aqueous layer next to the surface of the chip. This mass increase can be detected by measuring changes in refractive index of polarized light impinged upon the surface of the chip. The resulting sensorgram (*see* Fig. 9) provides quantitative information on specificity of binding, the concentration of active molecules in the sample, kinetics, and affinity.

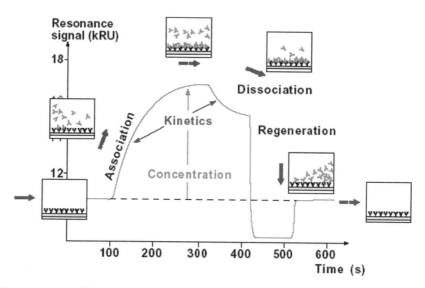

Fig. 9. Sensorgram. The sensorgram is a recording of changes in resonance signal as a function of time. A binding event results in an increase in resonance signal, whereas a dissociation results in a decrease of this signal. Binding of molecules to the protein immobilized on the surface can therefore be followed in real time as seen during the association phase shown above. The shape of this curve can be analyzed to yield the rate constant for the association phase. Likewise, the rate constant for the dissociation phase, shown above, can also be extracted. The surface of the sensor chip can be regenerated by stripping all bound analytes for subsequent reuse.

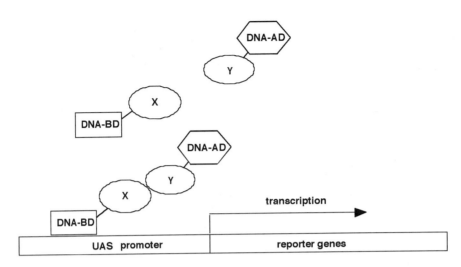

Fig. 10. Diagram illustrating the yeast 2-hybrid system. The bait protein (X) is fused to the DNA-binding domain of the yeast transcriptional activator protein GAL4. Potential binding partners (Y) derived from a cDNA library are fused to the Gal4 activation domain. If X and Y interact, the GAL4 protein is able to bind to the UAS promoter (upstream activation sequence), thereby activating transcription of the reporter genes.

can be applied to the determination of partners of a multitude of proteins. Indeed, the entire yeast genome, encoding over 6000 open reading frames (ORF), has been screened by a 2-hybrid approach resulting in the detection of 957 putative interactions involving 1004 yeast proteins *(54)*.

The yeast 2-hybrid method, first described by Fields and Song *(55)*, and illustrated in Fig. 10, takes advantage of the properties of the yeast GAL4 protein, a transcriptional activator of the genes encoding the enzymes required for galactose utilization in the yeast *Saccharomyces cerevisiae*. The GAL4 protein contains two domains—a DNA-binding domain and an activation domain—both of which are necessary to activate transcription. The binding domain is fused to a bait protein designated as "X" in the diagram. The activation domain is fused to proteins derived from a cDNA library containing coding sequences of potential binding partners of the bait protein. The library protein is designated as "Y" in the diagram. If X and Y interact to form a complex, the two domains of GAL4 are brought into proximity, reconstituting full activity of the protein and resulting in transcription of the reporter gene.

3.4. Quantification

An essential component of an expressional proteomics experiment is the measurement of the amounts of individual proteins in samples arising under different conditions. This quantification of the protein is seldom done in an absolute sense; rather it is the relative amount of protein in two samples that is measured. Two methods, depending on the separation technology employed, have been applied. Densitometry is the method used when the proteins are resolved by 2d-PAGE. The chromatographic methods for protein separation allow another approach for quantification that is based on stable isotope labeling.

Table 2
Densitometers

Type	Instrument	Manufacturer
Absorbance	GS-710 Imaging Densitometer	Bio-Rad
Fluorescence	Storm imager	Amersham Bioscience
Fluorescence	Typhoon imager	Amersham Bioscience
Both	Fluor-S MAX MultiImager System	Bio-Rad

Instruments are available for performing densitometric scans of gels using absorbance, fluorescence, or both. This table provides some examples of instruments of both types and list the manufacturers of these instruments.

Table 3
Software for Image Analysis

Software	Manufacturer
Kepler	
	Large Scale Proteomics
PDQuest	Bio-Rad
Melanie	GeneBio
ImageMaster	Amersham Biosciences

3.4.1. Densitometry

A protein separation by 2d-PAGE (discussed in greater detail in **Subheading 3.1.1.**) provides an array of the proteins in a mixture resolved on the basis of charge and mol wt. Proteins in this array are visualized by staining with Coomassie blue, silver stain, or a fluorescent dye such as Sypro red *(28)*. Quantitative information is obtained by scanning the protein array with a densitometer that measures either absorbance or fluorescence of the stained protein spot. Several instruments that are commercially available on the market measure absorbance, fluorescence, or both, and examples of these instruments are provided in Table 2.

The output of the densitometric scan is a digitized data file that can be processed with a variety of software packages designed specifically for this application (Table 3).

3.4.2. Stable Isotope Labeling

A recent development in the quantification of proteins in complex mixtures is the use of stable isotope labeling coupled with tandem mass spectrometry *(47)*. Termed "isotope-coded tandem affinity tags" (ICATs), these reagents are alkylating reagents designed to specifically label cysteine residues, contain the affinity tag, biotin, to allow capture of the labeled peptides using avidin affinity, and contain eight deuteriums in the so-called heavy form of the reagent. The light form does not contain deuterium. The structure of the ICAT reagent is shown in Fig. 11. A typical experiment using these reagents involves two samples from an expressional proteomics experiment, in which the determination of the relative amounts of proteins in the two samples is desired. One sample is labeled with the light reagent, and the other with the heavy reagent. The

Isotope-Coded Affinity Tags

heavy reagent: D8-ICAT Reagent (X=deuterium)
light reagent: D0-ICAT Reagent (X=hydrogen)

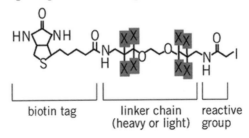

biotin tag   linker chain   reactive
                  (heavy or light)   group

Fig. 11. Structure of the ICAT reagent.

samples are mixed and proteolyzed with trypsin. The peptides containing either the heavy or light isotope tag are affinity-isolated using avidin-affinity methods. Analysis of the peptides by LC-MS provides quantitative information in the form of pairs of peptides, differing in mass by eight daltons. Measurement of the ratios of heavy to light provides the relative amounts of each protein in the two original samples, and identification of the peptides is provided by data from an MS/MS experiment.

A related approach, called "global internal standard technology" (GIST), has been advanced by Regnier and his colleagues *(56)*. This method also uses deuterated and non-deuterated reagents to modify the proteins or peptides. However, GIST makes use of a variety of reagents, including acylating reagents and esterifying reagents, to make the method more generally applicable than the ICAT approach. Whereas the ICAT reagent reacts with the sulfhydryl group of cysteine residues, the GIST reagents can react with the N-terminus and the ε-amino group of lysine (acylating reagents) or with the carboxyl groups of aspartic and glutamic acid or the C-terminus (esterification reagents). Acylating groups that have been tested include N-acetoxysuccinimide, succinic anhydride, and the N-acyloxysuccinimide derivatives of propionic and pentanoic acid. Esterification of carboxyl functional groups can be accomplished with methanol, which is available in deuterated form.

4. Future Directions

As attention focuses on proteomics as a key set of technologies to further understand the functions of proteins, biochemical pathways, and the networks that make up systems biology, a great deal of effort is being expended to improve these technologies. Improvements in sample throughput and analytical sensitivity are the primary areas impacted by these efforts. This section reviews some of these improvements, especially developments in protein-chip technology and in mass spectrometry.

4.1. Immobilized Protein Chips

Immobilized protein chips represent a new avenue in proteomics research. Immobilized protein chips can be divided into two major categories: protein chips designed for protein-protein interaction and those designed for enzymatic studies.

4.1.1. Chips for Protein-Protein Interaction Studies

The ProteinChip™ System developed by Ciphergen, Inc. is intended to facilitate the discovery of novel biomarkers and to monitor changes in known biomarkers. The technology utilizes a variety of surface activations to immobilize proteins. Solutions containing interacting proteins can then be incubated with the surfaces. Captured protein masses are determined by subsequent TOF MS. The use of chips with immobilized polyclonal antibodies for secreted amyloid peptides has been applied to the profiling of peptide variants in cell culture or biological fluids *(57)*. The assay allows detection of nanomolar amounts of amyloid β peptides *(57,58)*. Protein-chip arrays have been shown to be extremely effective in the identification of new biomarkers, and six potential biomarkers for prostate cancer have been identified in a short time *(59)*.

Protein chips have been constructed using printing technologies *(60)*, making possible very large arrays (as many as 10,000 proteins) in a small area. The limiting factor in this case becomes the availability of purified protein reagents. It would be useful to have protein chips with immobilized families of receptors or ligands that would allow quick identification of interacting partners. This would also enable the profiling of secreted and membrane-associated expressed ligands/receptors in cell culture under various growth conditions. Antibodies to different families/groups of proteins immobilized on chips are also of great interest. Arenkov et al. have applied this approach to rapidly detect specific proteins in protein mixtures *(61)*.

4.1.2. Enzyme Chips

A novel application of protein-chip technology is the immobilization of enzymes on a chip with subsequent activity assays performed on the array of enzymes in a high-throughput manner. Zhu and co-authors *(62)* have developed such a chip using yeast protein kinases. Of 122 known yeast protein kinases, 119 were expressed in yeast as glutathione-S-transferase (GST)-fusion proteins. The expressed proteins were purified and covalently attached to wells of the silicone elastomer, poly (dimethylsiloxane) using the crosslinker, 3-glycidoxypropyltrimethoxysilane. The immobilized proteins were tested for activity using standard in vitro kinase assays employing ^{33}Pγ-ATP and 17 different substrates, including the kinases themselves (autophosphorylation) with GST as the control. Signals were measured and quantified on a high-resolution phosphoimager.

Such an approach using chips containing immobilized kinases or potential substrates can facilitate new drug target discovery through the rapid identification of novel kinases or kinase substrates. Furthermore, these chips can be applied in monitoring known signal-transduction pathways, thereby speeding up the drug validation process. Other families of proteins with specific enzymatic activities such as extracellular matrix proteases or caspases, or their substrates can also be arrayed on chips.

A major issue with this new technology is the maintenance of enzyme activity through the immobilization process and after the protein is arrayed on the chip. Clearly, the work of Zhu et al. (62) has demonstrated that this is possible with at least one class of enzymes. It remains to be seen whether other classes of enzymes will behave as well using this technology.

A second issue concerns the development of suitable activity assays. These assays require high sensitivity, must be relatively rapid, and must be amenable to the high-throughput chip approach. Assays involving radioactive substrates fulfill these requirements, but are generally less attractive than other nonradioactive approaches. Mass spectrometry may ultimately be the answer to the chip-based assay requirements, but substantial development work must to be done before this conclusion can be reached.

The ability to carry out high-throughput functional assays makes the protein-chip approach attractive as a means for rapidly monitoring different signaling and metabolic pathways, for identification and validation of novel biomarkers and drug targets. This technology has the potential to transform the approach to the functional analysis of entire proteomes.

4.2. Advances in Mass Spectrometry

Advances in mass spectrometry are focused primarily on increasing throughput, enhancing sensitivity, improving accuracy, and making the overall process of acquiring mass spectrometric data easier and more robust. Several recent developments have occurred in this area, and hold promise for improving this type of analysis.

4.2.1. Electrospray Chips

An interesting approach to improving throughput is to create a so-called electrospray chip, in which one *(63)* or many *(64)* electrospray nozzles are fabricated on a silica chip. Chips exist with as many as 96 nozzles arrayed within a square inch of surface area. These devices have been shown to work in the infusion mode where the chip is interfaced with an infusion pump and the spray from the nozzles is directed into a mass spectrometer. The true potential of this approach, however, will come from the interface of the chip with a miniaturized chromatographic system where multiple peptide samples can be separated and analyzed in an automated fashion in a relatively short time period *(65)*.

4.2.2. TOF-TOF

Time-of-flight (TOF) mass analyzers are known for their resolution and high mass accuracy. The combination of two TOF analyzers with a collision cell between them provides the ability to analyze product ions produced by collision-induced dissociation with unprecedented resolution and accuracy *(66)*. Coupled to MALDI ionization, these so-called TOF-TOF instruments are just now becoming available. With the ability to perform high-mass accuracy MS/MS analyses in a high-throughput fashion, these instruments promise to greatly facilitate and enhance the identification of proteins in a proteomics experiment.

4.2.3. AP-MALDI

Traditionally, the MALDI ionization source has been operated in a high vacuum with a TOF analyzer. A recent advance has been the development of the AP-MALDI source, which operates at atmospheric pressure *(67)* and is used with an ion-trap analyzer *(68)*. This new source allows the user to generate the same ions that are seen on the traditional MALDI-TOF and perform MS/MS analyses in the ion trap.

4.2.4. Advanced MS Technology

Increasing the speed of the protein identification and quantification is a goal of many current research projects. Some interesting work is currently underway to identify proteins by mass spectrometry without first digesting the proteins to peptides. FTICR MS (*see* **Subheading 3.2.1.2.**) has been used to identify proteins by gas-phase fragmentation of intact proteins and searching the fragment masses against a protein database. The high accuracy and resolution of the FTICR MS is essential for this analysis to work *(69,70)*. If cheaper and less complex instruments become available that rival the accuracy and resolution of the FTICR MS, this method may become a practical way to achieve proteome identification and quantification.

5. Applications

5.1. Biomarkers

One area that shows great promise for proteomics is biomarker validation. Applications of biomarkers are numerous in the process of drug discovery. Protein markers of drug effectiveness are needed at multiple stages of pharmaceutical development, both preclinically and clinically. This application of proteomics typically would represent an example of expressional proteomics in which a drug treatment is examined in terms of its effects on the protein expression or metabolism of an organism or tissue. The primary challenge for this application of proteomics is the development of high-throughput, quantitative techniques, both for identification and quantification of proteins responding to drug treatment.

5.2. Expression Evaluation

Proteomics technologies may be applied to evaluate the expression and processing of recombinant proteins in heterologous systems. Parameters that may affect the activity of the protein can be followed. A simple example of this would be to monitor the phosphorylation state of a protein. Phosphorylated forms of proteins will separate on a 2-D gel, and some quantitative information about phosphorylation levels may thus be obtained. Beyond this, mass spectral analysis may be used to identify the sites of phosphorylation through careful analysis of the tryptic digest of the proteins separated in the gel. Proteomic technologies may be utilized to identify trace contaminants and related substances in bioproducts. Other post-translational modifications may also be evaluated (glycosylation state, proteolytic activation/inactivation state, and many others).

5.3. ADME/Toxicology

Another area of the pharmaceutical industry that may utilize proteomic technology is toxicology. The need for surrogate markers of drug efficacy and for toxic side effects may be addressed by expressional proteomics approaches. Two-dimensional gels were used for this purpose even before mass spectrometric techniques for protein identification were developed *(71)*. As with biomarker validation, a key consideration in this application of proteomics is the ability to quantify proteins and evaluate relative increases or decreases in protein expression in response to drug administration. A related area of pharmaceutical development is drug disposition and metabolism (ADME). In particular, in the study of protein therapeutics, functional proteomics approaches may lead to a greater understanding of protein stability and provide opportunities for protein engineering to optimize pharmacokinetic and pharmacodynamic properties.

5.4. Quality Control

As more protein therapeutics are developed—including monoclonal antibodies (MAbs)—regulatory agencies will require even more sophisticated analytical packages to be prepared to monitor the quality and reproducibility of production processes. Proteomic technologies will play an increasingly important role in this analysis. Similar to expression evaluation, this analysis will also control for protein parameters affected at all stages of production (including purification, formulation, and stability testing). Thus, proteomics will offer important tools to analytical chemists throughout the development organization.

6. Conclusion

The field of proteomics is undergoing an explosive expansion now that attention has turned from the sequencing of the genome to determining the function of all of the newly discovered proteins and their impact on biological systems. This chapter summarizes current technologies as of January 2002. As technologies are continually improved and expanded, proteomics technologies are sure to increase in sensitivity and throughput. Completely new technologies will be developed to meet these challenges, and orthogonal technologies will be applied to the study of protein chemistry. What the field of molecular biology was to the decade of the nineties and to the sequencing of the human genome, analytical and protein chemistry will be to the next decade and the study of proteomics. The most successful research efforts will be those that fully utilize the capabilities of proteomics and integrate these technologies into the earliest stages of discovery and development. These will be the technologies that ultimately lead to the fullest exploitation of the promise of genomics.

References

1. Venter, J. C., Adams, M. D., Myers, E. W. et al (2001) The sequence of the human genome. *Science* **291,** 1304–1351.
2. International Human Genome Sequencing Consortium (2001) Initial sequencing and analysis of the human genome. *Nature* **409,** 860–921.
3. Wasinger, V. C., Cordwell, S. J., Cerpa-Poljak, A., Yan, J. X., Gooley, A. A., Wilkins, M. R., et al. (1995) Progress with gene-product mapping of the Mollicutes: *Mycoplasma genitalium. Electrophoresis* **16,** 1090–1094.
4. Goffeau, A., Barrell, B. G., Bussey, H., Davis, R. W., Dujon, B., Feldmann, H., et al. (1996) Life with 6000 genes. *Science* **274,** 546–567.
5. Yarema, K. J. and Bertozzi, C. R. (2001) Characterizing glycosylation pathways. *Genome Biology* **2,** reviews 0004.1-0004.10.
6. Hubbard, S. R. and Till, J. H. (2000) Protein tyrosine kinase structure and function. *Annu. Rev. Biochem.* **69,** 373–398.
7. O'Reilly, D. R., Miller, L. K., and Luckow, V. A. (1992) in, *Baculovirus Expression Vectors: A Laboratory Manual,* W.H. Freeman and Company, New York, p. 216.
8. Pandey, A. and Mann, M. (2000) Proteomics to study genes and genomes. *Nature* **405,** 837–846.
9. Fields, S. (2001) Proteomics: proteomics in genomeland. *Science* **291,** 1221–1224.
10. Blackstock, W. (2000) in *Proteomics: A Trends Guide,* (Blackstock, W. and Mann, M., eds.), Elsevier Science, London, pp. 12–17.
11. Griffin, T. J. and Aebersold, R. (2001) Advances in proteome analysis by mass spectrometry. *J. Biol. Chem.* **276,** 45497–45500.
12. Mann, M., Hendrickson, R. C., and Pandey, A. (2001) Analysis of proteins and proteomes by mass spectrometry. *Annu. Rev. Biochem.* **70,** 437–473.
13. Green, R. and Noller, H. F. (1997) Ribosomes and translation. *Annu. Rev. Biochem.* **66,** 679–716.

14. Pandey, A., Fernandez, M. M., Steen, H., Blagoev, B., Nielsen, M. M., Roche, S., et al. (2000) Identification of a novel immunoreceptor tyrosine-based activation motif-containing molecule, STAM2, by mass spectrometry and its involvement in growth factor and cytokine receptor signaling pathways. *J. Biol. Chem.* **275,** 38633–38639.

15. Pandey, A., Podtelejnikov, A. V., Blagoev, B., Bustelo, X. R., Mann, M., and Lodish, H. F. (2000) Analysis of receptor signaling pathways by mass spectrometry: identification of vav-2 as a substrate of the epidermal and platelet-derived growth factor receptors. *Proc. Natl. Acad. Sci. USA* **97,** 179–184.

16. Neubauer, G., King, A., Rappsilber, J., Calvio, C., Watson, M., Ajuh, P., et al. (1998) Mass spectrometry and EST-database searching allows characterization of the multi-protein spliceosome complex. *Nat. Genet.* **20,** 5–6.

17. Voges, D., Zwickl, P., and Baumeister, W. (1999) The 26S proteasome: a molecular machine designed for controlled proteolysis. *Annu. Rev. Biochem.* **68,** 1015–1068.

18. Miyamoto, S., Teramoto, H. Coso, O. A., Gutkind, J. S., Burbelo, P. D., Akiyama, S. K., et al. (1995) Integrin function: Molecular hierarchies of cytoskeletal and signaling molecules. *J. Cell Biol.* **131,** 791–805.

19. Barth, A. I. M., Näthke, I. S., and Nelson, W. J. (1997) Cadherins, catenins and APC protein: interplay between cytoskeletal complexes and signaling pathways. *Curr. Op. Cell Biol.* **9,** 683–690.

20. Wilm, M., Shevchenko, A., Houthaeve, T., Breit, S., Schweigerer, L., Fotsis, T., et al (1996) Femtomole sequencing of proteins from polyacrylamide gels by nano-electrospray mass spectrometry. *Nature* **379,** 466–469.

21. Hale, J. E., Butler, J. P., Knierman, M. D., and Becker, G. W. (2000) Increased sensitivity of tryptic peptide detection by MALDI-TOF mass spectrometry is achieved by conversion of lysine to homoarginine. *Anal. Biochem.* **287,** 110–117.

22. Krishnan, S., Hale, J. E., and Becker, G. W. (2001) Proteomics: chromatographic fractionation prior to 2-dimensional polyacrylamide gel electrophoresis for enrichment of low abundance proteins facilitating identification by mass spectrometric methods, in: *Enzyme Technology for Pharmaceutical and Biotechnological Applications,* (Kirst, H. A., Yeh, W. K., and Zmijewski, M. J., eds.), Marcel Dekker, Inc., New York, NY, pp. 575–596.

23. Klose, J. (1975) Protein mapping by combined iso electric focusing and electrophoresis of mouse tissues a novel approach to testing for induced point mutations in mammals. *Humangenetik* **26,** 231–243.

24. O'Farrell, P. H. (1975) High resolution two-dimensional electrophoresis of proteins. *J. Biol. Chem.* **250,** 4007–4021.

25. Righetti, P. G. (1990) Immobilized pH gradients: theory and methodology, in: *Laboratory Techniques in Biochemistry and Molecular Biology,* vol. 20. (Burdon, R. H. and van Knippenberg, P. H., eds.), Elsevier Biomedical Press, Amsterdam, The Netherlands.

26. Righetti, P. G. and Bossi, A. (1997) Isoelectric focusing in immobilized pH gradients: an update. *J. Chromatogr. B* **699,** 77–89.

27. Gorg, A., Obermaier, C., Boguth, G., Csordas, A., Diaz, J.-J., and Madjar, J.-J. (1997) Very alkaline immobilized pH gradients for two-dimensional electrophoresis of ribosomal and nuclear proteins. *Electrophoresis* **18,** 328–337.

28. Valdes, I., Pitarch, A., Gil, C., Bermudez, A., Llorente, M., Nombela, C., et al. (2000) Novel procedure for the identification of proteins by mass fingerprinting combining two-dimensional electrophoresis with fluorescent SYPRO red staining. *J. Mass Spectrom.* **35,** 672–682.

29. Righetti, P. G., Castagna, A., and Herbert, B. (2001) Prefractionation techniques in proteome analysis. *Anal. Chem.* **73,** 320A–326A.

30. Rachez, C., Lemon, B. D., Suldan, Z., Bromleigh, V., Gamble, M., Naar, A.M., et al. (1999) Ligand-dependent transcription activation by nuclear receptors requires the DRIP complex. *Nature* **398,** 824–828.

31. Husi, H., Ward, M. A., Choudhary, J. S., Blackstock, W. P., and Grant, S. G. (2000) Proteomic analysis of NMDA receptor-adhesion protein signaling complexes. *Nature Neuroscience* **3,** 661–669.

32. Carr, S. A., Hemling, M. E., Bean, M. F., and Roberts, G. D. (1991) Integration of mass spectrometry in analytical biotechnology. *Anal. Chem.* **63,** 2802–2824.

33. McCormack, A. L., Schieltz, D. M., Goode, B., Yang, S., Barnes, G., Drubin, D., et al. (1997) Direct analysis and identification of proteins in mixtures by LC/MS/MS and database searching at the low-femtomole level. *Anal. Chem.* **69,** 767–776.

34. Washburn, M. P., Wolters, D., and Yates, J. R. (2001) Large-scale analysis of the yeast proteome by multidimensional protein identification technology. *Nat. Biotechnol.* **19,** 242–247.

35. Moore, R. E., Licklider, L., Schumann, D., and Lee, T. D. (1998) A microscale electrospray interface incorporating a monolithic, poly(styrene-divinylbenzene) support for on-line liquid chromatography/tandem mass spectrometry analysis of peptides and proteins. *Anal. Chem.* **70,** 4879–4884.

36. Poon, R. Y. and Hunt, T. (1994) Reversible immunoprecipitation using histidine- or glutathione S-transferase-tagged staphylococcal protein A. *Anal. Biochem.* **218,** 26–33.

37. Hopp, T. P., Prickett, K. S., Price, U., Libby, R. T., March, C. J., Cerretti, P., et al. (1988). A short polypeptide marker sequence useful for recombinant protein identification and purification. *Biotechnol. (NY)* **6,** 1205–1210.

38. Jensen, O. N., Podtelejnikov, A., and Mann, M. (1996) Delayed extraction improves specificity in database searches by matrix-assisted laser desorption/ionization peptide maps. *Rapid Commun. Mass Spectrom.* **10,** 1371–1378.

39. Zhang, W. and Chait, B. T. (2000) ProFound: an expert system for protein identification using mass spectrometric peptide mapping information. *Anal. Chem.* **72,** 2482–2489.

40. Perkins, D. N., Pappin, D. J., Creasy, D. M., and Cottrell, J. S. (1999) Probability-based protein identification by searching sequence databases using mass spectrometry data. *Electrophoresis* **20,** 3551–3567.

41. Clauser, K. R., Baker, P. R., and Burlingame, A. L. (1999) Role of accurate mass measurement (+/− 10 ppm) in protein identification strategies employing MS or MS/MS and database searching. *Anal. Chem.* **71,** 2871– 2882.

42. Krause, E., Wenschuh, H., Jungblut, P.R. (1999) The dominance of arginine-containing peptides in MALDI-derived tryptic mass fingerprints of proteins. *Anal. Chem.* **71,** 4160–4165.

43. Ducret, A., Van Oostveen, I., Eng, J. K., Yates, J. R. 3rd, and Aebersold, R. (1998) High throughput protein characterization by automated reverse-phase chromatography/electrospray tandem mass spectrometry. *Protein Sci.* **7,** 706–719.

44. Link, A. J., Eng, J., Schieltz, D. M., Carmack, E., Mize, G. J., Morris, D. R., et al. (1999) Direct analysis of protein complexes using mass spectrometry. *Nat. Biotechnol.* **17,** 676–682.

45. Beavis, R. C. and Fenyö, D. (2000) Database searching with mass spectrometric information, in: *Proteomics: A Trends Guide* (Mann, M. and Blackstock W., eds), Elsevier Science, New York, pp. 23–27.

46. Washburn, M. P., Wolters, D., and Yates, J. R. 3rd. (2001) Large-scale analysis of the yeast proteome by multidimensional protein identification technology. *Nat. Biotechnol.* **19,** 242–247.

47. Gygi, S. P., Rist, B., Gerber, S. A., Turecek, F., Gelb, M. H., and Aebersold, R. (1999) Quantitative analysis of complex protein mixtures using isotope-coded affinity tags. *Nat. Biotechnol.* **17,** 994–999.

48. Shen, Y., Zhao, R., Belov, M. E., Conrads, T. P., Anderson, G. A., Tang, K., et al. (2001) Packed capillary reversed-phase liquid chromatography with high-performance electrospray ionization Fourier transform ion cyclotron resonance mass spectrometry for proteomics. *Anal. Chem.* **73,** 1766–1775.

49. Edman, P. (1950) Method for the determination of the amino acid sequence in peptides. *Acta Chem. Scand.* **4,** 283–293.

50. Hewick, R. M., Hunkapiller, M.W., Hood, L.E., and Dreyer, W.J. (1981) A gas-liquid solid phase peptide and protein sequenator. *J. Biol. Chem.* **256 ,** 7990–7997.

51. Matsudaira, P. (1987) Sequence from picomole quantities of proteins electroblotted onto polyvinylidene difluoride membranes. *J. Biol. Chem.* **262,** 10,035–10,038.

52. Rich, R. L. and Myszka, D. G. (2000) Advances in surface plasmon resonance biosensor analysis. *Curr. Opin. Biotechnol.* **11,** 54–61.

53. Fitz, L. J., Morris, J. C., Towler, P., Long, A., Burgess, P., Greco, R., et al. (1997) Characterization of murine Flt4 ligand/VEGF-C. *Oncogene* **15,** 613–618.

54. Uetz, P., Giot, L., Cagney, G., Mansfield, T. A., Judson, R. S., Knight, J. R., et al. (2000) A comprehensive analysis of protein-protein interactions in *Saccharomyces cerevisiae. Nature* **403,** 623–627.

55. Fields, S. and Song, O. (1989) A novel genetic system to detect protein-protein interactions. *Nature* **340,** 245–246.

56. Regnier, F. E., Riggs, L., Zhang, R., Xiong, L., Liu, P., Chakraborty, A., et al (2002) Comparative proteomics based on stable isotope labeling and affinity selection. *J. Mass Spectrom.* **37,** 133–145.

57. Davies, H., Lomas, L., and Austen, B. (1999) Profiling of amyloid beta peptide variants using SELDI Protein Chip arrays. *Biotechniques* **27,** 1258–1261.

58. Frears, E. R., Stephens, D. J., Walters, C. E., Davies, H., and Austen, B. M. (1999) The role of cholesterol in the biosynthesis of beta-amyloid. *Neuroreport* **10,** 1699–1705.

59. Senior, K. (1999). Fingerprinting disease with protein chip arrays. *Mol. Med. Today* **5,** 326–327.

60. MacBeath, G. and Schreiber, S. L. (2000) Printing proteins as microarrays for high-throughput function determination. *Science* **289,** 1760–1763.

61. Arenkov, P., Kukhtin, A., Gemmell, A., Voloshchuk, S., Chupeeva, V., and Mirzabekov, A. (2000) Protein microchips: use for immunoassay and enzymatic reactions. *Anal. Biochem.* **278,** 123–131.

62. Zhu, H., Klemic, J. F., Chang, S., Bertone, P., Casamayor, A., Klemic, K. G., et al. (2000) Analysis of yeast protein kinases using protein chips. *Nat. Genet.* **26,** 283–289.

63. Lazar, I. M., Ramsey, R. S., Jacobson, S. C., Foote, R. S., and Ramsey, J. M. (2000) Novel microfabricated device for electrokinetically induced pressure flow and electrospray ionization mass spectrometry. *J. Chromatogr. A* **892,** 195–201.

64. Wachs, T. and Henion, J. (2001) Electrospray device for coupling microscale separations and other miniaturized devices with electrospray mass spectrometry. *Anal. Chem.* **73,** 632–638.

65. Li, J., Wang, C., Kelly, J. F., Harrison, D. J., and Thibault, P. (2000) Rapid and sensitive separation of trace level protein digests using microfabricated devices coupled to a quadrupole-time-of-flight mass spectrometer. *Electrophoresis* **21,** 198–210.

66. Medzihradszky, K. F., Campbell, J. M., Baldwin, M. A., Falick, A. M., Juhasz, P., Vestal, M. L., et al. (2000) The characteristics of peptide collision-induced dissociation using a high-performance MALDI-TOF/TOF tandem mass spectrometer. *Anal. Chem.* **72,** 552–558.

67. Laiko, V. V., Baldwin, M. A., and Burlingame, A. L. (2000) Atmospheric pressure matrix-assisted laser desorption/ionization mass spectrometry. *Anal. Chem.* **72,** 652–657.

68. Laiko, V. V., Moyer, S. C., and Cotter, R. J. (2000) Atmospheric pressure MALDI/ion trap mass spectrometry. *Anal. Chem.* **72,** 5239–5243.

69. Meng, F., Cargile, B. J., Miller, L. M., Forbes, A. J., Johnson, J. R., and Kelleher, N. L. (2001) Informatics and multiplexing of intact protein identification in bacteria and the Archaea. *Nat. Biotechnol.* **19,** 952–957.

70. Horn, D. M., Zubarev, R. A., and McLafferty, F. W. (2000) Automated de novo sequencing of proteins by tandem high-resolution mass spectrometry. *Proc. Natl. Acad. Sci. USA* **97,** 10,313–10,317.
71. Richardson, F. C., Strom, S. C., Copple, D. M., Bendele, R. A., Probst, G. S., and Anderson, L. (1993) Comparisons of protein changes in human and rodent hepatocytes induced by the rat-specific carcinogen, methapyrilene. *Electrophoresis* **14,** 157–161.

15

Metabolic Flux Analysis, Modeling, and Engineering Solutions

Walter M. van Gulik, Wouter A. van Winden, and Joseph J. Heijnen

1. Introduction

Microorganisms have been used for many decades to produce valuable chemicals for the food, pharmaceutical, and bulk industries. These include amino acids, vitamins, antibiotics, alcohols, and organic acids. Random classical mutation techniques have led to improvements in production properties.

Recently also cultures of higher eukaryotic organisms have been employed for product formation, for example, the production of monoclonal antibodies (MAbs) from animal cells and the production of alkaloids from plant cells.

The unraveling of complete genome sequences holds the potential to reveal the blueprint of industrially important microorganisms and cell cultures. The assignment of function to open reading frames (ORFs) is progressing, and functional assignment is well underway for some genomes. These experimental techniques are likely to further increase in speed and potential in the coming years. With the aid of available recombinant-DNA techniques, this raises the possibility of precise modifications in microbial metabolism, for improvement of microbial product formation, as well as for the production of entirely novel compounds.

In order to deal with the massive amounts of data generated, mathematical analysis and computer simulation are required. Because of the complexity of cellular metabolism, the application of mathematical models is a prerequisite for the accurate prediction of which modifications to the microbial metabolism are required to reach a desired goal. The foundation of such models is formed by the stoichiometry of all relevant biochemical reactions of the system under study. From steady-state balancing of input/output and intermediate compounds, a set of linear equations is obtained. The resulting system of linear equations can either be over-determined, determined, or underdetermined—i.e., the degree of freedom of the system is either smaller, equal to, or greater then the number of measured rates. Over-determined and determined systems can easily be solved. That is, all unknown rates can be calculated from the known rates. In case of an over-determined system, the redundancy of the data allows for statistical data analysis and minimization of error. In an underdetermined system, the number of possible solutions is infinite, and either additional measurements, such as C13-NMR flux measure-

From: *Handbook of Industrial Cell Culture: Mammalian, Microbial, and Plant Cells*
Edited by: V. A. Vinci and S. R. Parekh © Humana Press Inc., Totowa, NJ

ments or additional constraints combined with linear optimization techniques are required to obtain a solution.

Presently, stoichiometric modeling ("metabolic flux analysis") is well-developed *(1–4)*. In the first part of this chapter, the principles of metabolic flux analysis are explained. A number of examples of successful applications, as well as the limits of stoichiometric modeling of microbial metabolism, are provided. Finally, the potential and limits of C13-NMR flux analysis are examined.

It is clear that stoichiometric models alone are not sufficient to predict the targets for recombinant DNA-based improvement of microbial metabolism. To accomplish this goal, kinetic information is also required. Kinetic modeling of microbial metabolism is still in need of development, and here a variety of approaches exist. Savageau *(5)* developed his Biological Systems Theory (BST). Kacser and Burns *(6)* and Heinrich and Rapoport *(7)* developed the Metabolic Control Analysis (MCA), which was modified for large perturbations *(8)*. Kinetic models have also been developed using the complex nonlinear enzyme-kinetic-rate expressions determined from in vitro experiments *(7,9)*. Recently, the tendency modeling approach was proposed using a combination of mass action and power-law kinetics combined with time-scale analysis to minimize the number of parameters *(10)*. In the second part of this chapter, a general mathematical description for (non-) steady-state metabolic networks is presented, which unifies, generalizes, and extends the concepts of the metabolic control analysis and of kinetic modeling. This method will be illustrated with some examples.

2. Stoichiometric Modeling

2.1. Principles of Metabolic Flux Balancing (MFB)

2.1.1. Metabolic Balance Equations

The network of biochemical reactions representing microbial metabolism can be written in a concise way by defining a reaction stoichiometry matrix, S, and a vector, c, that contains all the biochemical compounds considered:

$$S \cdot c = 0 \qquad \text{[Eq. 1]}$$

For each compound, i, involved in a metabolic system with volume, V, e.g., a bioreactor with growing and/or producing cells, a mass balance can be defined:

$$\frac{dVCi}{dt} = V\left(r_{Ai} + \Phi_i\right) \qquad \text{[Eq. 2]}$$

where r_{Ai} denotes the net rate of conversion of compound, i, in the metabolic system and ϕ_i denotes the net rate of transport of compound, i, over the boundaries of the system, i.e., the boundaries of the bioreactor with volume V.

Assuming (pseudo) steady-state behavior, Eq. 2 simplifies to:

$$r_{Ai} + \Phi_i = 0 \text{ or } \Phi_i = -r_{Ai} \qquad \text{[Eq. 3]}$$

Only a small number of compounds is exchanged between the metabolic system and the environment, including nutrients, O_2, CO_2, H_2O, biomass, and products. For compounds that are exchanged with the environment, Eq. 3 holds. The net conversion rate of these compounds is determined by the rates of the biochemical reactions in which compound, i is consumed or produced. For all compounds, which remain intracellular, $\phi_i = 0$, and thus Eq. 3 simplifies to:

$$r_{Ai} = 0 \qquad \text{[Eq. 4]}$$

Defining a rate vector, r, containing both the reaction rates (r_i) and the net conversion rates (r_{Ai}),

$$r = [r_1...r_i...r_n, r_{A1}...r_{Ai}...r_{Am}] \qquad \text{[Eq. 5]}$$

and an ($m \times (m + n)$) matrix A composed of an ($n \times m$) reaction stoichiometry matrix S and an ($m \times m$) identity matrix I_m:

$$A = [S^T - I_m] \qquad \text{[Eq. 6]}$$

The balance equations for the m compounds of a system consisting of n biochemical reactions are denoted by:

$$A \cdot r = 0 \qquad \text{[Eq. 7]}$$

Thus, a homogeneous system of m balance equations is obtained to describe the metabolic network. The number of independent equations equals Rank A. Dependent equations are typically balance equations for cofactors, such as adenosine diphosphate (ADP), nicotinamide adenine dinucleotide (NAD), flavin adenine dinucleotide (FAD), nicotinamide adenine dinucleotide phosphate (NADP), coenzyme A (COA).

2.1.2. Solving Determined Metabolic Networks

From compounds that are not exchanged with the environment, the net conversion rate is known (note that in these cases $r_{Ai} = 0$). When the metabolic network is made up of n biochemical reactions and p compounds are exchanged with the environment, the total number of unknown rates equals ($n + p$). Thus, the system can be solved when ($n + p$ - Rank A) rates are defined.

Matrix A and vector r can be partitioned, respectively, into A' and A^0 and r' and r^0 where A^0 is associated with the net exchange rates which are equal to zero, r^0, and A' is associated with all other non-zero conversion and exchange rates, r':

$$A' \cdot r' + A^0 \cdot r^0 = 0 \qquad \text{[Eq. 8]}$$

Because $r^0 = 0$ this is equivalent to:

$$A' \cdot r' = 0 \qquad \text{[Eq. 9]}$$

This operation results in a considerable reduction of the size of matrix A. The solution of the metabolic network represented by the matrix A' is obtained by determining the reduced row echelon form. The solution consists of a system of ($n + p$) linear equations describing the unknown rates as a function of ($n + p$ - Rank A') rates to be defined.

2.1.3. Solving Underdetermined Metabolic Networks

Underdetermined metabolic networks can be solved by linear optimization. With this technique, the numerical values of all unknown rates are calculated for a particular constraint (e.g., maximization of the biomass yield) for a given set of measured rates.

Therefore, matrix A' is partitioned into A_m and A_c, where A_m is associated with the measured fluxes represented by the vector r_m and A_c is associated with the fluxes to be calculated, represented by the vector r_c. This results in:

$$A_c \cdot r_c + A_m \cdot r_m = 0 \qquad \text{[Eq. 10]}$$

which can be written as:

$$A_c \cdot r_c = b \qquad \text{[Eq. 11]}$$

where

$$b = -A_m \cdot r_m$$

By formulating the solution as a linear programming problem, the objective function

$$Z = c^T \cdot r_c \qquad \text{[Eq. 12]}$$

that is, a linear combination of the fluxes r_c is minimized under the constraint that

$$A_c \cdot r_c = b \qquad \text{[Eq. 13]}$$

The vector c contains the weight factors. When, for example, the objective is maximization of the biomass yield, the weight, c_i, for the biomass production rate is set to -1, and all other weight factors are set to zero.

2.1.4. Solving Over-Determined Metabolic Networks

In case of over-determined metabolic networks, more rates have been measured than is strictly necessary to calculate the unknown rates. The redundant information is used to improve the quality of the flux estimates. The weighed least-squares solution to Eq. (9) is

$$r_c = \left(A^T \cdot \psi^{-1} \cdot A\right)^{-1} \cdot A^T \cdot \psi^{-1} \cdot r_m \qquad \text{[Eq. 14]}$$

Where ψ denotes the covariance matrix associated with the measured rates, represented by vector r_m.

2.2. Applications of Metabolic Flux Balancing, Case 1: Energetics of Growth and Penicillin Production in a High-Producing Strain of *Penicillium chrysogenum*

2.2.1. Estimation of Maximum Yields

Reliable estimation of maximum theoretical yields of product on substrate is of great value in industrial strain-improvement programs. If properly done, such estimations may provide a good indication for the available room for further improvement. A prerequisite is that sufficient information is available on the metabolism of the particular microorganism, not only with respect to product biosynthesis but also with respect to the central metabolic pathways. An essential aspect for the proper estimation of biomass and/or product yields is sufficient information on the stoichiometry of adenosine triphosphate (ATP)-producing and consuming reactions and thermodynamic/biochemical irreversibility constraints. From this information, a stoichiometric model can be constructed that allows a straightforward calculation of the maximum yields of biomass and product on substrate and the maintenance coefficient *(2,11,12)*—the parameters of the well-known linear equation for substrate consumption:

$$qs = \frac{\mu}{Y_{SX}^{max}} + \frac{q_P}{Y_{SP}^{max}} + m_S \qquad \text{[Eq. 15]}$$

A general problem in calculating these maximum yields is that the ATP-stoichiometry of some metabolic processes is either unknown or unclear. Examples of such processes are intracellular transport *(13)*, turnover of macromolecules and organelles, and enzyme regulation by phosphorylation/dephosphorylation *(14,15)*. To drive these processes, unknown amounts of energy, ATP, are required. A practical solution to this problem is the assumption that an unknown amount of additional energy is required for growth-related and non-growth-related maintenance, leading to two unknown ATP stoichiometry parameters (K_X, m_{ATP}) on the consumption side. On the ATP production side we have a third parameter, namely the ratio between oxygen consumption in oxidative phosphorylation and ATP formation—i.e., the P/O-ratio. This approach has been applied in the metabolic modeling of the growth of yeast cultures on a large number of different carbon sources *(2)* and growth on glucose/ethanol mixtures *(12)*. It has been shown that experimental data for growth on at least two different substrates at different growth rates allows the estimation of the unknown ATP-stoichiometry parameters P/O, K_X and m_{ATP} *(2,12)*. In this case, this approach has been extended to microbial growth with product formation that is the production of penicillin-G in *Penicillium chrysogenum*.

Although the energy (ATP) needed for the biosynthesis of a molecule of penicillin-G can be calculated directly from the well-known biosynthesis pathway, it is evident that additional energy is required e.g. for intracellular transport and for product excretion. Recently, it has been found that the energy demand for the formation of the tripeptide L-α-aminoadipyl-L-cysteinyl-D-valine (ACV), which under optimal conditions needs the hydrolysis of 3 mol ATP to AMP per mol of tripeptide formed, may require more than 20 mol of ATP under unfavorable conditions *(16)*. It was reported that the additional energy consumed was caused by the hydrolytic loss of activated intermediates.

To account for additional energy consumption (mol ATP/mol penicillin) associated with penicillin production, a parameter K_P is introduced. The value of this parameter represents an additional amount of ATP needed for penicillin-G biosynthesis, which is in addition to the amount dictated by the known ATP-stoichiometry of the biosynthesis pathway specified in the metabolic model. The result is that the metabolic network model, described in detail in vanGulik et al. *(17)*, contains four unknown ATP-stoichiometry parameters (P/O, K_X, K_P, m_{ATP}), which have to be estimated from experimental data.

2.2.2. Model for the ATP Stoichiometry of the Metabolic Network

With the four unknown ATP-stoichiometry parameters, the ATP-balance can be expressed as:

$$P/O \cdot q_{2e} - \Sigma q_i^{ATP} - K_X \cdot \mu - K_P \cdot q_{Pen} - m_{ATP} = 0 \qquad \text{[Eq. 16]}$$

where the first term represents the production of ATP in oxidative phosphorylation, Σq_i^{ATP} represents the net rate of ATP consumption in the part of the metabolic network model of which the ATP stoichiometry is known (i.e., the result of all stoichiometrically fixed ATP usage, as well as production in substrate level phosphorylation), q_{Pen} is the specific growth rate and μ is the specific β-lactam production rate.

It is important to realize that the *P/O*-ratio as defined in Eq. 16 can not be considered a fixed parameter because this ratio is determined by the division of the electron flux over the different proton-translocating complexes (1, 3, and 4) of the respiratory chain, which have different H⁺/2*e* stoichiometries. The growth conditions—including the carbon substrate used, the growth rate, the rate of product formation—determine the relative amounts of mitochondrial (NADH and FADH) and cytosolic (NADH) electron fluxes. Because the electrons derived from mitochondrial NADH pass the complete respiratory chain, whereas the electrons derived from FADH and cytosolic NADH generally pass only complexes 3 and 4, the division of the electron flux over the different complexes is a function of the growth conditions. Therefore, the operational *P/O*-ratio will also be a function of the growth conditions. If the metabolic model is sufficiently detailed—e.g., intracellular compartmentation is included—the origin of the electrons generated in microbial catabolism is known, as well as the relative contributions of complexes 1, 2, and 4 of the respiratory chain to oxidative phosphorylation. To include this, the first term of Eq. 16 must be replaced by:

$$P/O \cdot q_{2e} = \delta \cdot \left(q_{2e}^{NADH:mit} + \alpha \cdot q_{2e}^{NADH:cyt} + \beta \cdot q_{2e}^{FADH}\right) \qquad \text{[Eq. 17]}$$

Where α and β represent the relative contributions to proton translocation of electrons delivered by cytosolic NADH and FADH, respectively. The values of these parameters depend on the configuration of the electron transport chain.

A combination of Eq. 16 and Eq. 17 with $\alpha = \beta = 0.6$ *(18)* yields the ATP-balance as it has been used for the estimation of the four ATP stoichiometry parameters (δ, K_X, K_P, and m_{ATP}):

$$\delta \cdot \left(q_{2e}^{NADH:mit} + 0.6 \cdot q_{2e}^{NADH:cyt} + 0.6 \cdot q_{2e}^{FADH}\right) - \Sigma q_i^{ATP} - K_X \cdot \mu - K_P \cdot q_{Pen} - m_{ATP} = 0 \qquad \text{[Eq. 18]}$$

2.2.3. Estimation of ATP Stoichiometry Parameters

From steady-state chemostat experiments performed on various carbon substrates (i.e., glucose, ethanol, and acetate) at different dilution rates linear equations of the form of Eq. 18 were derived. The (reconciled) specific fluxes, i.e. $q_{2e}^{NADH:mit}$, $q_{2e}^{NADH:cyt}$, q_{2e}^{FADH}, Σq_i^{ATP}, μ, and q_{Pen}, have been obtained for each experiment from metabolic flux balancing using the measured uptake rates of substrate, oxygen, ammonia and the side-chain precursor phenylacetic acid (PAA) and the production rates of biomass, penicillin, carbon dioxide, and by-products. From a chi-square test, it appeared that in all cases the fit of the experimental data to the stoichiometric model was satisfactory (confidence level for errors in data and/or model less than 75%). For further details, refer to vanGulik et al. *(17)* and vanGulik et al. *(18)*.

With the assumption that the ATP stoichiometry parameters can be considered as constants and thus independent of the growth rate and the substrate used, their values can be estimated by linear regression. This was done using experimental data from 20 steady-state glucose, ethanol, and acetate limited chemostat cultures performed at different dilution rates. The results of the estimations are shown in Table 1. Compared to literature data, no exceptional values have been found for the P/O-ratio, growth-dependent and non-growth-dependent maintenance energy requirements. Clearly, a very surprising result was the estimated large additional amount of ATP needed for product formation expressed as the high value of KP of 73 mol ATP per mol of penicillin-G produced. This finding must have important consequences for the maximum theoretical yield of penicillin on C-source.

2.2.4. Maximum Biomass and Product Yield and Maintenance on Substrate

From the metabolic model, expressions can be derived for the parameters of the linear equation of substrate consumption (Eq. 15) Y_{SX}^{max}, Y_{SP}^{max}, and m_s and as a function of

Table 1
Estimated Values of the ATP-Stoichiometry Parameters
with Their 95% Confidence Intervals

Parameter	Value (measurement unit)	
δ (*P/O* ratio)	1.84 ± 0.08	mol ATP/mol O
K_x	0.38 ± 0.11	mol ATP/Cmol biomass
K_P	73 ± 20	mol ATP/mol penicillin
m_{ATP}	0.033 ± 0.012	mol ATP/Cmol biomass/h

the ATP-stoichiometry parameters δ, K_X, K_P, and m_{ATP}. This is accomplished by solving the set of linear equations representing the mass balances for all compounds present in the metabolic network model for the specific growth rate, μ, and the specific rate of penicillin production, q_{Pen}, as input variables and the ATP-stoichiometry parameters δ, K_X, K_P and m_S as parameters. The result is a set of linear equations that expresses all intracellular fluxes and net conversion rates as a function of μ and q_{Pen}. As an example, the linear equation expressing the specific substrate uptake rate, q_S, as a function of μ and q_{Pen} for ethanol as the carbon source is derived as:

$$q_S = \frac{0.732\delta + 0.357K_X + 0.82}{\delta - 0.179} \cdot \mu + \frac{7.71\delta + 0.357K_P + 9.29}{\delta - 0.179} \cdot q_{Pen} + \left(\frac{5 - 2\delta}{13\delta - 2.32} + 0.154 \right) \cdot m_{ATP} \quad \text{[Eq. 19]}$$

Clearly, the first term on the right-hand side of Eq. 19 can be recognized as substrate uptake for growth, $\frac{1}{Y_{SX}^{max}} \cdot \mu$, the second term represents substrate uptake for penicillin production, $\frac{1}{Y_{SP}^{max}} \cdot q_{Pen}$ and the last term represents substrate uptake for maintenance. Similar equations were derived for glucose and acetate as carbon sources. For the obtained expressions for Y_{SX}^{max}, Y_{SP}^{max}, and m_s for the three carbon sources we refer to vanGulik et al. *(18)*. An analogous approach was used to derive expressions for the parameters of the linear equation for oxygen consumption Y_{OX}^{max}, Y_{OP}^{max} and m_o.

By substituting the estimated values of these parameters (Table 2) in the obtained expressions the stoichiometric coefficients Y_{SX}^{max}, Y_{SP}^{max}, and m_s of the linear equation for substrate consumption (Eq 15) for glucose, ethanol, and acetate can be calculated. The results are shown in Table 2.

No exceptional values were found for Y_{SX}^{max} and m_s. However, with respect to the maximum product yield, Y_{SP}^{max} comparison with previously calculated values (0.43–0.60 mol penicillin per mol of glucose, (*see* refs. in vanGulik et al. *[18]*) shows that the present calculated maximum yield of penicillin on glucose is more than a factor of 2 lower. This is a direct consequence of the high extra energy costs (KP) estimated in this study.

2.2.5. Validation of Estimated Additional Energy Needs for Penicillin Production

If the estimated extra energy needs for penicillin production of 73 mol ATP per mol of penicillin indeed represents a realistic figure, changes in the specific penicillin production rate should have a significant effect on the operational biomass yield, and thus on the mycelium concentration in carbon-limited chemostat culture. Moreover, from the relationship between these two rates the validity of the estimated value of K_P can be

Table 2
**Calculated Yield and Maintenance Parameters of Penicillin and Biomass
on Carbon Source with Their 95% Confidence Intervals**

C-source	Y_{SX}^{max} (Cmol/Cmol)	Y_{SP}^{max} (mol/Cmol)	m_S (Cmol/Cmol/h)
Glucose	0.663 ± 0.013	0.029 ± 0.004	0.0088 ± 0.0032
Ethanol	0.721 ± 0.015	0.034 ± 0.005	0.0071 ± 0.0026
Acetate	0.425 ± 0.010	0.020 ± 0.003	0.0117 ± 0.0042

verified. This was done in two ways: by performing glucose-limited steady-state chemostat experiments without the side-chain precursor PAA, resulting in the absence of Pen-G production and only formation of small amounts of the by-products 6-aminopenicillanic acid (6-APA), 8-hydroxypenicillanic acid (8-HPA), isopenicillin-N (IPN) and 6-oxopiperide-2-carboxylic acid (OPC), and by studying the relation between the pseudo steady-state biomass and Pen-G concentration in a degenerating culture, where the penicillin-G production rate gradually declines to low values.

From the results of these experiments it was concluded that the estimated high extra amount of ATP for product formation was indeed realistic *(18)*.

2.3. Case 2: Flux Calculations and Estimation of Maximum Yields for Mixed Substrate Growth

2.3.1. Metabolic Model for Growth of S. cerevisiae on Glucose/Ethanol Mixtures

By extension of a metabolic model for growth on glucose with the biochemical pathways necessary for growth on ethanol, that is gluconeogenesis and glyoxylate shunt *(2)*, a model for growth on glucose/ethanol mixtures was obtained. The resulting metabolic network model appeared to be underdetermined—and, in addition to the feed rates of glucose and ethanol, five additional rates had to be defined in order to calculate all unknown rates.

It is likely that from the complete set of reactions of such an underdetermined model, several subsets of reactions may be chosen, each yielding a valid determined stoichiometric model. The mathematical strategies to find these sub-models are null space and convex cone analysis *(19)*. This is essentially the same as finding all elementary flux modes *(1)*, being all non-decomposable flux distributions admissible in steady state. Through the calculation of all possible elementary flux modes and subsequently choosing the mode giving the highest biomass and or product yield, the optimum metabolic flux pattern can be found *(20)*.

An alternative method to reach the same goal is constrained linear optimization. This was also applied to estimate the metabolic flux pattern as a function of the ethanol/glucose ratio in the feed. The constraint that was chosen for the optimization was maximum biomass yield on the mixed carbon substrate. In this way, for each ethanol/glucose ratio the subset of fluxes > 0 is computed, resulting in the highest possible biomass yield. The fluxes through the metabolic network of *S. cerevisiae* were estimated for growth on glucose and ethanol alone and for growth on a range of glucose/ethanol mixtures using the previously estimated ATP-stoichiometry parameters K_X and δ (= the P/O-ratio) *(2)*. Changes in the metabolic flux pattern—switching on and switching off of metabolic pathways, were found to occur at ethanol fractions of the feed of 0.09, 0.48, 0.58, and 0.73 C-mol/C-mol (Fig. 1).

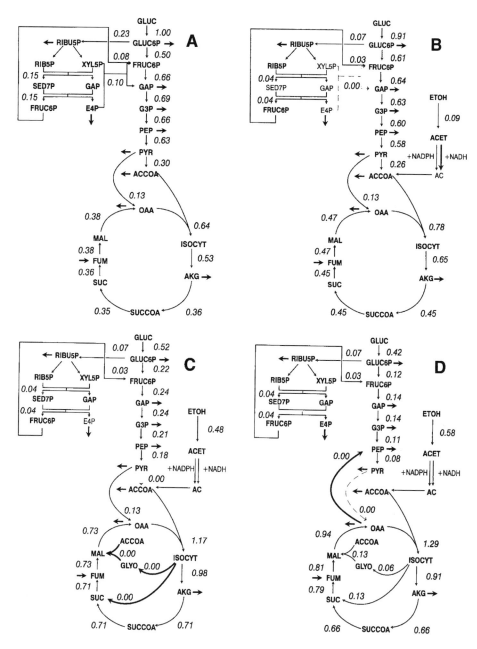

Fig. 1. Estimated optimal metabolic flux patterns for aerobic growth of *Saccharomyces cerevisiae* on a mixture of glucose and ethanol. All fluxes are given in Cmol carbon transferred, and are presented as fractions of the consumption rate of mixed carbon substrate (Cmol/h). (**A**) growth on 100% glucose. (**B**) cessation of NADPH production in the pentose phosphate pathway at 0.09 Cmol/Cmol ethanol in the feed. (**C**) cessation of the flux through pyruvate decarboxylase and start of glyoxylate cycle at 0.48 Cmol/Cmol ethanol. (**D**) cessation of the flux through pyruvate carboxylase and instead reversal of the carbon flux via PEP-carboxykinase at an ethanol fraction of 0.58 Cmol/Cmol. (**E**) reversal of several reversible steps in glycolysis, cessation of the carbon flux through phosphofructokinase and instead reversal of the carbon flux via fructose-1,6-biphosphatase at an ethanol fraction of 0.73 Cmol/Cmol. (**F**) growth on 100% ethanol (*see* next page).

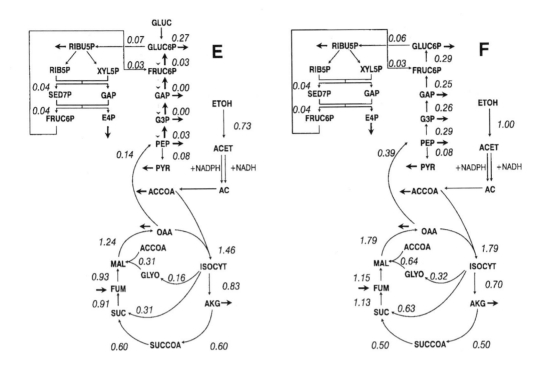

Fig. 1. continued.

2.3.2. Metabolic Flux Analysis

Fig. 1A shows the calculated metabolic fluxes through the glycolysis, the pentose phosphate pathway, and the citric acid cycle for growth on glucose. The first change, which was predicted by the metabolic network when the ethanol content of the feed was increased, was the cessation of transketolase converting xylulose 5-phosphate + erythrose 4-phosphate into glyceraldehyde 3-phosphate + fructose 6-phosphate (Fig. 1B). This occurred at an ethanol fraction in the feed of 0.09 (Cmol/Cmol). The second change was predicted to occur at an ethanol content of the feed of 0.48 Cmol/Cmol. At this point the flux through pyruvate dehydrogenase falls to zero, indicating that all acetyl CoA is now completely synthesized from ethanol (Fig. 1C).

When the ethanol content of the feed is increased further, the filling-up of the citric acid cycle can no longer be provided for by pyruvate carboxylase alone, and the metabolic network predicts that the glyoxylate shunt (i.e., isocytrate lyase and malate synthase) becomes operative. A further increase of the ethanol content of the feed resulted in a predicted cessation of the flux through pyruvate carboxylase at an ethanol fraction of 0.58 Cmol/Cmol. Instead, the carbon flux was channeled through PEP-carboxykinase to convert oxaloacetate to PEP (Fig. 1D). Thereafter, increase of the ethanol fraction resulted in a reversal of several reversible steps in glycolysis until at an ethanol content of 0.73 Cmol/Cmol the calculated flux through phosphofructokinase fell to zero and was replaced by the reversed reaction through fructose-1,6-bisphosphatase (Fig. 1D). With these last changes, the metabolic network for growth of S. cerevisiae on ethanol was finally obtained. Fig. 1F shows the metabolic flux pattern for growth on 100% ethanol.

Comparison of the calculated changes in the metabolic flux pattern for growth of *S. cerevisiae* on ethanol/glucose mixtures with measurements of the activities of key enzymes may give insight to their regulation. De Jong-Gubbels et al. *(21)* found that the activities of isocytrate lyase, malate synthase, PEP-carboxykinase, and fructose-1,6-biphosphatase in cell-free extracts were negligible in glucose-limited chemostat cultures. Malate synthase and fructose-1,6-bisphosphatase activities were detected at ethanol contents of the feed of 0.4 Cmol/Cmol and above and 0.7 Cmol/Cmol and above, respectively. This corresponds well with the predictions obtained from the metabolic network. However, the activities of isocytrate lyase and PEP-carboxykinase were already detectable at low ethanol fractions of the feed. It was found that pyruvate kinase activity decreased at increasing ethanol content. This is indeed predicted by the metabolic network (Fig. 1A to 1F), although it is also predicted to reach a constant level above an ethanol content of 0.58 Cmol/Cmol. Unfortunately, the measurements on pyruvate kinase activity contained too much scatter to draw further conclusions.

Metabolic network calculations predicted a constant flux through pyruvate carboxylase up to an ethanol content of 0.58 Cmol/Cmol. At this point, the flux decreases to zero when PEP must be synthesized partly from ethanol. However, the Jong-Gubbels et al. *(21)* did not observe changes in the activity of pyruvate carboxylase upon transition from 100% glucose to 100% ethanol.

The flux through phosphofructokinase was predicted to decrease at increasing ethanol fractions and to fall to zero at an ethanol fraction of 0.73 (Fig. 1E). Also in this case, the measured enzyme activity was not influenced by the ethanol fraction in the feed. However, it was concluded by the authors that the actual fluxes through the enzymes may well be modulated by intracellular concentrations of substrates and effectors, and not at the level of enzyme synthesis alone.

2.3.3. Yield of Biomass on Mixed Substrate and Oxygen

For each of the five metabolic regimes predicted from the linear optimization studies (Fig. 1A–E), a determined metabolic network was constructed. This resulted in five metabolic networks, each covering a part of the entire glucose/ethanol range. All networks had a degree of freedom of 2—each needed two input variables, namely the consumption rates of glucose and ethanol.

Note that the solutions of these five determined metabolic networks consist of linear equations expressing all unknown rates as a function of the consumption rates of glucose and ethanol. For example, the biomass production rate can be expressed as:

$$r_{Ax} = \alpha_e \cdot r_{A,etoh} + \alpha_g \cdot r_{A.gluc} \qquad \text{[Eq. 20]}$$

where α_e and α_e are the coefficients of the linear equation and $r_{A,etoh}$ and $r_{A,gluc}$ are the net conversion rates of ethanol and glucose, respectively. Dividing Eq. 20 by the consumption rate of the mixed substrate results in an equation for the yield of biomass on the mixed substrate as a function of the ethanol fraction in the feed, *f*.

$$Y_{SX} = \alpha_e \cdot f + \alpha_g \cdot (1 - f) \qquad \text{[Eq. 21]}$$

The linear equations for Y_{SX} as a function of the ethanol fraction in the feed, which were obtained from the five different networks are:

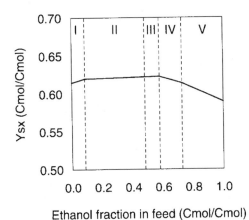

Fig. 2. Metabolic network prediction of the yield of biomass on mixtures of glucose and ethanol as a function of the ethanol fraction in the feed.

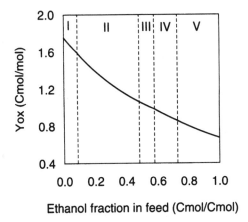

Fig. 3. Metabolic network prediction of the yield of biomass on oxygen for growth on mixtures of glucose and ethanol.

$$Y_{SX} = j \cdot 0.672 + (1 - f) \cdot 0.615 \quad 0.00 \le f \le 0.09$$

$$Y_{SX} = j \cdot 0.625 + (1 - f) \cdot 0.619 \quad 0.09 < f \le 0.48$$

$$Y_{SX} = j \cdot 0.625 + (1 - f) \cdot 0.615 \quad 0.48 < f \le 0.58$$

$$Y_{SX} = j \cdot 0.598 + (1 - f) \cdot 0.657 \quad 0.58 < f \le 0.73 \qquad \text{[Eq. 22]}$$

$$Y_{SX} = j \cdot 0.598 + (1 - f) \cdot 0.679 \quad 0.73 < f \le 1.00$$

The result is shown in Fig. 2. It can be seen from this figure that the yield of biomass on the mixed substrate increases until where the pentose phosphate pathway is no longer providing NADPH for biosynthesis (at 0.09 Cmol/Cmol ethanol in the feed). Thereafter, the yield continues to increase only slightly at an increasing ethanol fraction in the feed. Note that the point at which the glyoxylate cycle is incorporated ($f = 0.48$) has no influence on the relation between the ethanol fraction in the feed, f, and the biomass yield. However, after incorporation of PEP-carboxykinase ($f = 0.58$) Y_{SX} starts to decline, as a result of the expenditure of 1 ATP per PEP synthesized from oxaloacetate. The incorporation of fructose-1,6-bisphosphatase converting fructose-1,6-bisphosphate into fructose-6-phosphate by hydrolysis of the phosphate ester at C-1 results in a more pronounced decline of Y_{SX} for $f > 0.73$.

The yield of biomass on oxygen, Y_{OX}, can be calculated in a similar way from the metabolic network solutions for the rates of biomass production and oxygen consumption. Alternatively, it can also be derived from the five linear equations describing Y_{SX} as a function of the ethanol fraction of the feed and a generalized degree of reduction balance (obtained equations not shown). A plot of the calculated yield of biomass on oxygen, Y_{OX}, as a function of the ethanol fraction in the feed is shown in Fig. 3. As expected, the calculated Y_{OX} decreases dramatically at increasing ethanol fraction in the feed.

2.4. Case 3: Theoretical Yield Limits to Overproduction of Amino Acids in Saccharomyces cerevisiae

When non-growth-associated maintenance energy needs are negligible, the well-known linear equation for substrate consumption for growth and product formation can be written as:

$$q_s = \frac{\mu}{Y_{SX}^{max}} + \frac{q_P}{Y_{SP}^{max}} \qquad \text{[Eq. 23]}$$

The operational yield of product on substrate is then given by:

$$Y_{SP} = \frac{q_P}{q_S} = \frac{q_P}{\dfrac{\mu}{Y_{SX}^{max}} + \dfrac{q_P}{Y_{SP}^{max}}} \qquad \text{[Eq. 24]}$$

By applying a metabolic network model for growth of *S. cerevisiae (2)*, possible stoichiometric limits to amino acid production were studied. Using the estimated values for the ATP-stoichiometry parameters and glucose as the substrate, the metabolic network model provides values for the maximum theoretical yield for each of the 20 amino acids that can be produced.

From Eq. 24 it follows that at zero growth rate, μ, the maximum theoretical value of the operational product yield, Y_{SP}, is equal to the parameter Y_{SP}^{max}. For each amino acid, the value of Y_{SP}^{max} can be calculated from the metabolic network. However, it was found that calculation of the fluxes through the metabolic network for the production of each of the 20 amino acids at zero growth rate ($r_{Ax} = 0$) resulted, in some cases, in biochemical inconsistencies (e.g., backward operation of the citric acid cycle). It appeared that these biochemical inconsistencies occurred only for amino acids when the production was accompanied by a net production of ATP. These biochemical inconsistencies could be avoided by dissipating the excess ATP produced. In these cases, biomass production may be a sink for excess ATP produced. Another possibility the cells might have is hydrolysis of ATP in futile cycles. However, the importance of futile cycles in yeast should not be overestimated *(22)*.

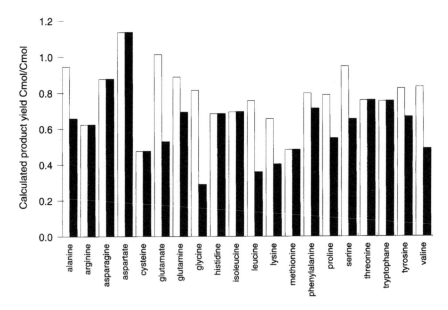

Fig. 4. Metabolic network estimation of maximum theoretical yields for amino acid production in *Saccharomyces cerevisae*. White bars: maximum theoretical yield of product on carbon source under the assumption of zero biomass growth. Black bars: limits to the theoretical product yields resulting from biochemical constraints.

For this example, it was assumed that excess ATP could only be consumed through biomass production. For each amino acid produced, the minimum biomass production rate was calculated for which no biochemical inconsistencies occurred. From Eq. 24, it can be inferred that when biomass growth is required for production of these amino acids, and thus part of the carbon substrate is necessarily consumed for biomass formation, this will result in a limit to the maximum theoretical yield, $Y_{SP}^{\lim}$, where:

$$Y_{SP} \leq Y_{SP}^{\lim} < Y_{SP}^{\max} \qquad \text{[Eq. 25]}$$

In such cases, ATP-dissipation by other means, such as increased maintenance energy requirements, would increase . These limits have been calculated for all 20 amino acids. The results are shown in Fig. 4. It can be seen from this figure that for some amino acids, $Y_{SP}^{\lim}$ may reach values of only 50% or less than $Y_{SP}^{\max}$.

2.5. Case 4: Limit Functions for Lysine Production in Corynebacterium glutamicum

As a case study for the applicability of metabolic network analysis, it has been investigated whether a similar metabolic limit may exist for lysine production in *Corynebacterium glutamicum*. Experimental data on the growth and Lysine production of a genetically modified strain of *C. glutamicum* in threonine-limited chemostat culture were obtained from Kiss and Stephanopoulos *(23)*. To describe the metabolism of the *C. glutamicum* strain, the yeast metabolic network *(2)* was modified with respect to amino acid biosynthesis (lysine biosynthesis is different, and this strain is auxotrophic for threonine, methionine, and leucine) and glucose transport.

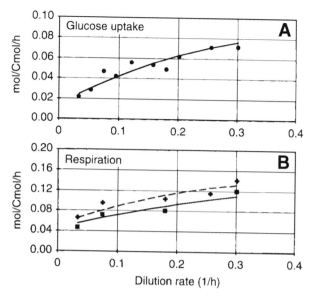

Fig. 5. Fit of metabolic network to describe specific glucose consumption and specific respiration as a function of dilution rate in chemostat cultures of *Corynebacterium glutamicum*. Experimental data have been obtained from Kiss and Stephanopoulos *(23)*. (**A**) filled circles: specific glucose consumption, solid line: metabolic network description. (**B**) filled squares: specific oxygen consumption, crosses: specific carbon dioxide production, solid line: metabolic network description of specific oxygen consumption, dashed line: metabolic network description of specific carbon dioxide production.

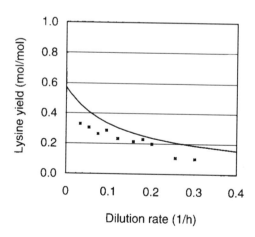

Fig. 6. Yield of lysine on glucose as a function of the dilution rate in chemostat cultures of *C. glutamicum*. Filled squares: experimental data[10]. Solid line: limit function based on biochemical constraint.

The general metabolic network solution for the specific rates of glucose consumption, q_S, oxygen consumption, q_O, and carbon dioxide production, q_c, as a function of the growth rate, μ, specific product formation, q_P, maintenance energy requirements, and effective *P/O*-ratio, δ, is shown in Eqs. 26 to 28.

$$-q_s = \frac{(0.147 + 0.153\,\delta')\,\mu + (0.838 + 1.128\,\delta')\,q_P + 0.967\,(K\mu + m_{ATP})\,X_P}{(0.387 + 0.967\,\delta')} \quad \text{[Eq. 26]}$$

$$-q_o = \frac{0.515\,\mu + 2.320\,q_P + 0.483\,(K\mu + m_{ATP})\,X_P}{(0.387 + 0.967\,\delta')} \quad \text{[Eq. 27]}$$

$$q_c = \frac{(0.513 + 0.040\,\delta')\,\mu + (2.707 + 0.967\,\delta')\,q_P + 0.483\,(K\mu + m_{ATP})\,X_P}{(0.387 + 0.967\,\delta')} \quad \text{[Eq. 28]}$$

From a fit of the metabolic network solution, Eqs. 26 to 28, to the growth data (glucose consumption, oxygen consumption, and carbon dioxide production as a function of the growth rate), the effective *P/O*-ratio and the maintenance energy requirements were estimated. It was found that the assumption of growth-associated maintenance resulted in unrealistic values for the effective *P/O*-ratio ($\delta' > 3$). The assumption that maintenance energy requirements were non-growth-associated resulted in more realistic estimates: $\delta' = 2.13$ and $m_{ATP} = 0.43$ mol/Cmol protein/h. This is markedly different from the parameters obtained for the yeast network. However, it should be considered that the organism used and the growth conditions are different. The yeast is grown C-limited, and *C. glutamicum* is grown N-limited.

Fig. 5 shows that the obtained fit of the calibrated metabolic network to the growth data is satisfactory. Using the fitted parameter values, the influence of the lysine production rate on metabolic flux distribution was investigated. It was found that the first reaction that reversed at increasing lysine production rates was the conversion of succinyl CoA to succinate by succinyl CoA synthetase. This was not surprising, because the biosynthesis of lysine in C. glutamicum results in the conversion of an equimolar amount of succinyl CoA into succinate. Thus, lysine biosynthesis serves as a bypass of succinyl CoA synthethase and will eventually result in the exhaustion of succinyl CoA. However, from a thermodynamic point of view, succinyl CoA synthethase is readily reversible. The rates of lysine production and substrate consumption at the point when the flux through succinyl CoA synthethase starts to reverse can be calculated from the solution of the metabolic network (derivation not shown). Both are a function of the growth rate, μ, the effective *P/O*-ratio, δ', and the growth-independent maintenance on ATP, m_{ATP}. The ratio q_P/q_s, that is, the yield of product (lysine) on substrate at this point is equal to:

$$Y_{SP}^{lim} = \frac{(0.153 - 0.058\,\delta')\,\mu + 0.166\,m_{ATP}}{(0.207 + 0.116\,\delta')\,\mu + 0.290\,m_{ATP}} \quad \text{[Eq. 29]}$$

From the result shown in Fig. 6, it can be seen that the experimental Y_{SP} values are just below the limit predicted by the assumed metabolic constraint. It should be emphasized that because succinyl CoA synthethase is thermodynamically reversible, the example given here is rather hypothetical. However, it is not known whether succinyl CoA synthethase is easily reversible under physiological conditions in *C. glutamicum*. It should be noted that when such a metabolic limit exists, the increase of enzyme levels in the product pathway has no effect.

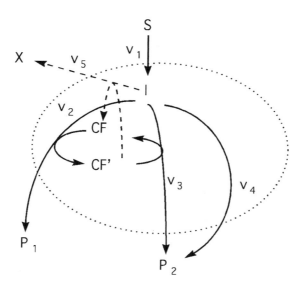

Fig. 7. A metabolic network in which substrate S is converted to intermediate I that is converted to biomass X and products P_1 and P_2. CF is a cofactor; the prime indicates its activated state.

2.6. 13C-NMR-Based Metabolic Flux Analysis

2.6.1. The Inobservability Problem

As outlined in the introduction to this chapter, the calculation of fluxes in underdetermined metabolic networks requires additional measurements, or can be performed by constrained linear optimization. However, in some cases the structure of the metabolic network prohibits determination of all intracellular fluxes, regardless of the measurements that have been performed. An example of such a feature is the occurrence of parallel pathways in the cell—compound A is converted into compound B via two different routes, and the net result is exactly the same. In such cases, mass balances only yield the sum of the fluxes through the parallel pathway branches but never the separate fluxes. A real-life example of a parallel pathway is found in lysine biosynthesis, (*see* Sonntag et al.) *(24)*.

Another phenomenon that is not observable from traditional metabolic flux balancing is the occurrence of bidirectional fluxes *(25)*. Mass balancing only yields net fluxes between the metabolite pools, but it does not give information on the forward and backward fluxes. However, from a cell-regulatory point of view, it is interesting to know whether reactions in the cell are uni- or bidirectional. Reversibility assumptions can be made on the basis of thermodynamic calculations. However, thermodynamic data that are determined in vitro are often found to be invalid in vivo, because of the complex properties of the viscous, protein-rich cytosol of the cell.

2.6.2. Solutions to the Inobservability Problem

In order to determine fluxes that cannot be found from mass balances alone, additional information is needed about the reactions in the cell. Enzyme assays are one of the additional sources of information. By determining which enzymes are expressed in the metabolic system under investigation, it can be inferred which pathways are active. However, enzyme assays do not give information about the magnitude of the fluxes through these pathways unless the exact enzyme kinetics and intracellular metabolite concentrations are known. At present, the validity of in vitro enzyme kinetics under in vivo circumstances is under debate, and methods for the accurate measurements of intracellular metabolites are under development. Therefore, determining all intracellular fluxes from enzyme and metabolite measurements is not yet feasible.

Alternatively, the problem of parallel pathways can be solved if different cofactors are involved in the two branches. (Note that some researchers state that pathways employing different cofactors may not be called "parallel.") A hypothetical example of how a cofactor balance can make the fluxes through two parallel pathways observable is shown in Fig. 7. In this figure, the fluxes through the parallel reactions v_3 and v_4 leading from intermediate I to product P_2 are different with respect to their cofactors. If the activated cofactor is assumed to be ATP, one can imagine that reaction v_4 is efficient and does not involve the hydrolysis of ATP, whereas reaction v_3 does. If the cofactor balance is taken into account, it can be seen that the rate at which product P_1 is produced, determines the rate at which ATP is generated. This rate fixes the fluxes requiring ATP (i.e., v_3 and v_5). Now all intracellular fluxes can be determined, since the set of flux constraints is determined. Equation 30 shows that the rank of the combined stoichiometric and measurement matrices equals the number of fluxes.

$$
\begin{array}{l}
\text{Mass balance } I \\
\text{Mass balance } CF \\
\text{Measured consumption } S \\
\text{Measured production } X \\
\text{Measured production } P_1 \\
\text{Measured production } P_2
\end{array}
\begin{pmatrix}
-1 & 1 & 1 & 1 & 1 \\
0 & -1 & 1 & 0 & 1 \\
-1 & 0 & 0 & 0 & 0 \\
0 & 0 & 0 & 0 & 1 \\
0 & 1 & 0 & 0 & 0 \\
0 & 0 & 1 & 1 & 0
\end{pmatrix}
\begin{pmatrix}
v1 \\
v2 \\
v3 \\
v4 \\
v5
\end{pmatrix}
\begin{pmatrix}
0 \\
0 \\
\varphi_s \\
\varphi_x \\
\varphi_{P_1} \\
\varphi_{P_2}
\end{pmatrix}
$$

$$
rank\left(\frac{S}{\underline{\underline{M}}}\right) = 5 \qquad\qquad \text{[Eq. 30]}
$$

An example of this situation is encountered in the primary carbon metabolism. The glycolytic and pentose phosphate pathways may be considered as parallel pathways converting glucose 6-phosphate to pyruvate. Whereas the pentose phosphate branch of this parallel pathway yields ATP, NADH, and NADPH, the glycolytic branch yields more ATP and NADH, but no NADPH. In contrast to NADH, NADPH is known to be involved in the biosynthesis of biomass and (in the case of *Penicillium chrysogenum*) penicillin. Therefore, equating the consumption of NADPH in biosythetic reactions to its production in the oxidative steps of the pentose phosphate is one way to determine how much glucose is consumed via the pentose phosphate pathway. Mass balances for the cofactors NADH and NADPH are often used in metabolic flux balancing; however, they have some disadvantages. Clearly, cofactor balances do not enable determination of the separate forward and backward fluxes in bidirectional reaction steps. Furthermore, in some cases assumptions must be made with respect to NAD(P)H specificities of various enzymes and the presence or absence of transhydrogenases that interconvert NADH and NADPH *(2,25–29)*. Finally, cofactors are involved in many reactions in

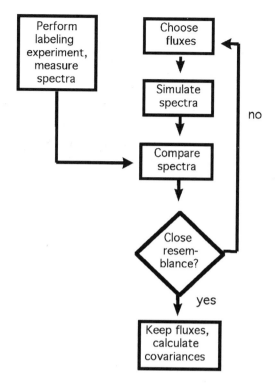

Fig. 8. The overall algorithm used for determining a set of fluxes from measured NMR spectra.

the cell so that it is difficult to take them all into account. As a result, the inclusion of cofactor mass balances may lead to erroneous flux estimates.

2.6.3. Application of 13C-NMR for Metabolic Flux Analysis

^{13}C-NMR-based metabolic flux analysis is a method that has the potential to resolve fluxes flowing through a complex metabolic network. This method exploits the fact that the pathways in the cell break and synthesize carbon-carbon bonds in a well-known way. Experimentally the method comes down to feeding a ^{13}C-labeled compound to a cell culture and subsequently measuring the amounts of labelled carbon atoms at various carbon positions of the metabolic intermediates. These measurements either rely on the different magnetic momentum of [12]C- and ^{13}C-atoms (Nuclear Magnetic Resonance spectroscopy, NMR) or on the different atomic masses of the two isotopes (Mass Spectrometry, MS).

The intracellular metabolic fluxes can then be calculated from the ^{13}C-labeling data that are obtained from NMR or MS *(28,30)*. The general approach of ^{13}C NMR based metabolic flux analysis is outlined in Fig. 8. As the figure shows, fluxes cannot be directly determined from the measured labeling data. Because of non-linearities in the model that relates the ^{13}C-distributions of metabolites to the steady-state metabolic fluxes, the fluxes have to be found using an iterative optimization.

2.6.4. Results of 13C-Labeling Studies

Returning to the case of the parallel glycolytic and pentose phosphate pathway, it can be shown that both branches lead to a different rearrangement of the carbon atoms that enter the cell in the form of glucose. As a case study, the fluxes through the glycolysis and the pentose phosphate pathway in *P. chrysogenum* have been estimated from NMR measurements of the ^{13}C-label distributions of intermediates of both pathways. The question was to determine whether these labeling studies result in different flux estimations than those that were previously determined using NADPH-balances *(17)*. This may shed new light on the question of whether the availability of the cofactor NADPH is a potential bottleneck in the production of penicillin.

From preliminary experiments, the steady-state fluxes in cultures of *P. chrysogenum* growing on defined medium containing either NO^{3-} or NH^{4+} as a nitrogen source and containing 10% uniformly labeled [$^{13}C_6$]-glucose have been estimated. The analyses indicate that fluxes into the pentose phosphate pathway may be higher than previously determined. However, thus far the procedure shown in Fig. 8 has not yet yielded satisfactory fits between measured and simulated ^{13}C-labeling data. Therefore, the primary focus of our research has been on some important methodological aspects of the ^{13}C-labeling technique.

Problems we have addressed include the sensitivity of the outcomes of ^{13}C-labeling studies to modeling errors *(31)*. Two potential errors in metabolic models are the oversimplification of the non-oxidative part of the pentose phosphate pathway and the ignorance of the effects of so-called "channeling"—i.e, the direct enzyme to enzyme transfer of intermediates. Furthermore we have studied which fluxes in a metabolic network are theoretically identifiable from a given set of ^{13}C-labeling data *(32)*. Finally, spectal analysis software has been developed in order to improve the quality of the measurement data that are extracted from the NMR spectra *(33)*. Apart from accurate labeling data, this spectral analysis software yields covariances of the measurement data that are indispensable for the determination of the reliability of the fluxes that are found using the ^{13}C-labeling technique. These new developments will result in more reliable ^{13}C based flux estimates of primary carbon metabolism.

2.7. Stoichiometric Modeling: Conclusions and Outlook

At present, the theory and practice of metabolic flux analysis has been well-developed. In the most recent papers on metabolic flux analysis, no new techniques are presented, but developed techniques are used for new applications. Pramanik et al. *(34)* applied metabolic flux analysis combined with linear optimization to model the metabolism of enhanced biological phosphorus removal (EBPR) from wastewater. The modeling provided better insight into the metabolism and an improved understanding of the EBPR process. Moreover, it allowed description of biopolymer synthesis and degradation in the context of the entire metabolism of the cell.

A very detailed stoichiometric model of *E. coli*, based on biochemical literature, genomic information, and metabolic databases, was constructed by Edwards and Palsson *(35)*. Using this model, metabolic flux balancing was applied for the *in silico* analysis of gene deletions. It was concluded that this model could be used to interpret mutant behavior and that the modeling results lead to a further understanding of the complex genotype-phenotype relation. A similar study was conducted for Haemophilus influenzae *(36)*. Here, the stoichiometric model was completely derived from the annotated genome sequence of the organism. This study showed that the synthesis of in silico metabolic genotypes from annotated genome sequences is possible. The results

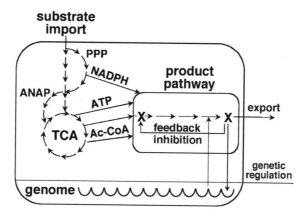

Fig. 9. Organization of metabolism in microorganisms.

indicated that to accomplish growth on mixed substrates, six different optimal metabolic phenotypes could be defined, that arose from the same metabolic genotype, each with different constraining features.

It can be inferred from numerous applications described in the literature that stoichiometric modeling is a powerful technique to calculate otherwise unmeasurable intracellular metabolic fluxes, to estimate ATP-stoichiometry parameters and maximum theoretical yields, to obtain a deeper insight in metabolism, to derive optimal metabolic flux patterns for maximum biomass or product yield, and to investigate genotype-phenotype relationships. However, metabolic engineering of industrial strains for improved product formation or the production of novel compounds also requires insight into the kinetic properties of the relevant pathways. The state of the art in the kinetic modeling of microbial metabolism is discussed in **Subheading 3.**

3. Kinetic Modelling

3.1. The Need for In Vivo Kinetic Modeling in Rational Metabolic Engineering

With the development of recombinant-DNA tools and the explosion of genetic information from many microbial genome-sequencing projects, it has become feasible to implement precise changes in the metabolism of microorganisms. These changes aim to introduce new product pathways or improve existing product formation. Using such metabolic engineering, it is theoretically possible to use microorganisms (e.g., *E. coli* or *S. cerevisiae*) for the production of a multitude of bulk and fine chemicals.

During the past decade, these opportunities have been explored with moderate success *(37)*. Most of the efforts have been directed toward increasing enzyme activities in product pathways and increasing the availability of molecules entering the product pathway by increasing substrate import, thus expecting increased product formation, which was often not realized. Recently, it has been recognized that this might be caused by negligence of the effect of this "push" approach on metabolite levels, in particular, the importance of product export may have been underestimated. Knowing that end-product feedback inhibition is a general property of many product pathways, it is essential to keep the intracellular product concentration low. This calls for a "pull" approach, as opposed to the generally used "push" approach *(38)*.

An intuitive approach to increase product formation would then be the following:

1. Increase the export of the product (*see* Fig. 9) by increasing the export capacity (pull out the product). The resulting lower intracellular product concentration will often naturally lead to higher product formation rates because of the cellular control mechanisms (end-product feedback control of enzyme activity and genetic control of enzyme level).

2. Diminish the end-product inhibition on the first enzyme of the product pathway (using protein engineering). It should be noted that in the absence of sufficient product export capacity, diminishing feedback inhibition may be counteracted by the increased intracellular end-product concentration. The result may be that end-product inhibition still plays a role, which may lead to lower enzyme levels in the product pathway caused by genetic control mechanisms (preservation of homeostasis). Having first increased the product export capacity allows the full product-flux benefit of decreased feedback inhibition. It can be expected that the levels of enzymes of the product pathway have also increased following increased product export capacity and decreased end-product inhibition (homeostatic response of genetic control mechanisms to decreased metabolite levels).

3. It can be expected that increase of all enzyme levels in the product pathway will further increase productivity.
   Following this approach, the inevitable point will be reached where changes in the product formation pathway (including export) will have only a marginal effect on productivity.

4. Further increase in product formation must then be realized by changes in primary metabolism. Obvious candidates here are the anaplerotic reactions feeding the TCA-cycle for those products of which the precursors are supplied by the TCA-cycle.

In general, however, changing primary metabolism is much less straightforward than this approach for a linear product pathway. This is because of the high pathway entanglement, (complex interactions, and presence of branches, cycles, and conserved moieties).

Given this situation, the following general questions are encountered.

- How to estimate the maximum theoretical product yield on substrate? As such, this maximum theoretical yield provides an indication of the absolute limit to a further improvement of the overall product yield. This can be estimated from a reliable stoichiometric model (*see* **Subheadings 2.2.–2.5.**).
- How to determine, for a given organism, whether the product pathway or primary metabolism is rate limiting? A relatively simple method to obtain an answer to this question, applied to penicillin-G production in an industrial strain of *Penicillium chrysogenum*, will be presented in **Subheading 3.2.**
- How to determine, in a simple way without knowledge of the enzyme kinetics of the various steps, the distribution of the flux control in a product biosynthesis pathway? This is illustrated for the biosynthesis pathway of Penicillin-V in *P. chrysogenum* in **Subheading 3.3.**
- How to construct reliable, albeit not overly complicated, kinetic models of primary metabolism based on in-vivo kinetics. This has not been achieved so far. In **Subheading 3.4.**, we present our ideas to tackle this problem and provide an overview of the mathematical and experimental tools developed so far.

3.2. Is Primary Metabolism a Bottleneck for the Biosynthesis of Penicillin-G ?

3.2.1. Identification of Principal Nodes for Penicillin Production

A priori analysis of a metabolic network model for growth and product formation of *P. chrysogenum (17)* was carried out in order to identify the subset of metabolic nodes

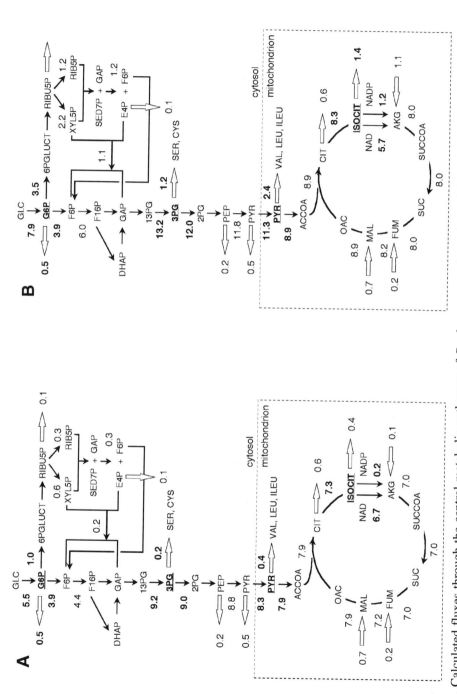

Fig. 10. Calculated fluxes through the central metabolic pathways of *P. chrysogenum* from the stoichiometric model. The model contains two degrees of freedom, namely the specific growth rate and the specific penicillin production rate. Fluxes were calculated for a growth rate of 0.01 h^{-1}: **(A)** without production of penicillin; **(B)** with production of penicillin at a hypothetical rate of 1 mmol/Cmol biomass/h. Branch points showing significant changes in flux partitioning are underlined.

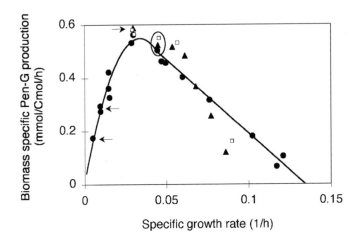

Fig. 11. Measured specific production of penicillin-G in glucose (●), ethanol (▲) and acetate (□) limited cultures of *Penicillium chrysogenum* DS12975. Points indicated with arrows represent no stable steady states, but observed maximum specific penicillin production rates during the experiments. Solid line: qPen - μ relationship for the glucose-limited chemostats.

for which the flux partitioning strongly depends on the rate of product formation. These nodes are referred to as principal nodes *(4)*.

The analysis was carried out by calculating the flux distribution through the complete metabolic network as a function of the production rates of biomass and penicillin-G. For this *a priori* analysis, literature values for the ATP stoichiometry parameters (P/O-ratio and maintenance energy requirements) have been used *(39)*. Fluxes through the metabolic network were calculated for a growth rate of 0.01 h^{-1} without production of penicillin (Fig. 10 A) and with production of penicillin (Fig. 10 B). For the second case, the specific penicillin production rate was set to a hypothetical value of 1 mmol/Cmol/h, which is twice the maximum value observed by Jørgensen et al. *(39)* during fed-batch cultivation of *P. chrysogenum*. Although this seems to be a high value, it can be calculated that the yield of penicillin on glucose under these conditions is still only 30% of the published theoretical maximum of 0.43 mol/mol glucose *(39)*. It can be inferred from a comparison of the calculated flux patterns for both conditions that penicillin production leads to an increase of the (biomass-specific) glucose uptake rate, and accordingly, to increased specific fluxes through the central metabolic pathways. It can be seen from Fig. 10 that all additional glucose consumed is channeled through the pentose phosphate pathway, thus increasing the flux through this pathway significantly.

Also the flux through the glycolysis is found to increase as a result of penicillin production, especially through the upper part, above 3-phosphoglycerate (3PG). Because of the withdrawal through cysteine of 3PG for penicillin synthesis, the increase in the glycolytic flux below 3PG is smaller. Because part of the pyruvate flux is channeled (via valine) toward penicillin biosynthesis, the smallest flux increase is found to occur in the TCA-cycle.

As a result of these changing flux patterns through central metabolism, four branch points show significant changes in flux partitioning. These are underlined in Fig. 10 and are located at 1) glucose-6-phosphate, because penicillin production requires an increased flux through the pentose phosphate pathway to meet the increased demands for cytosolic NADPH for cysteine biosynthesis, 2) 3-phosphoglycerate, which is the carbon precursor of cysteine, 3) mitochondrial pyruvate, which is the carbon precursor of valine, and 4) mitochondrial isocitrate, because mitochondrial NADP-dependent isocitrate dehydrogenase is the source of mitochondrial NADPH for valine biosynthesis. Significant penicillin production requires relatively large changes in flux partitioning around these four principal nodes. The rigidity of these nodes determines whether they are potential bottlenecks for a further increase of penicillin production. Metabolic flux calculations for glucose, xylose, ethanol and acetate as carbon sources and ammonium and nitrate as nitrogen sources revealed, that growth and product formation on these substrates leads not only to dramatic changes in fluxes through the central metabolic pathways, but also in large changes in flux partitioning around the four principal nodes for penicillin production (results not shown). Chemostat cultivation on these substrates was used as an experimental tool to manipulate the fluxes through the central metabolic pathways of *P. chrysogenum*, and thus to investigate the rigidity of these principal nodes for increased penicillin production.

3.2.2. Calculation of the Flux Distribution Around Principal Nodes from Experimental Data

In Fig. 11, the observed relation between the biomass-specific production of penicillin-G and the specific growth rate is plotted for steady-state, carbon-limited chemostat growth on glucose, ethanol, and acetate.

In glucose-limited chemostats, penicillin production was observed to increase strongly at an increasing dilution rate (which is equal to the growth rate in steady-state chemostat culture) to a maximum of 0.56 mmol Pen-G / Cmol biomass/h at a growth rate of 0.03 h^{-1}. When the dilution rate was further increased, specific Pen-G production declined in a linear fashion. The same general trend was observed for both the ethanol- and acetate-limited chemostats.

The calculated fluxes around the four principal nodes for steady-state glucose, ethanol, and acetate limited chemostats performed at the same dilution rate of 0.045 h^{-1} are shown in Figs. 12A to D. At this dilution rate specific Pen-G production was almost identical for the three substrates (*see* Fig. 11).

Figure 12A shows the different flux patterns around the glucose-6-phosphate branch point for growth on glucose, ethanol, and acetate. If glucose is the carbon source, the glucose-6-phosphate produced by hexokinase is partitioned between glycolysis, the pentose phosphate pathway, and anabolism. For C-2 carbon sources such as ethanol and acetate, glucose-6-phosphate is generated by gluconeogenesis. The flux is now partitioned between two main pathways—the pentose phosphate-pathway and anabolism. For growth on ethanol, it is assumed that cytosolic NADPH is generated mainly through NADP-dependent acetaldehyde dehydrogenase and that the pentose phosphate-pathway has only an anabolic function. This leads to a significantly lower flux of glucose-6-phosphate entering the pentose phosphate-pathway than for growth on acetate and glucose.

The fluxes around the 3-phosphoglycerate node are shown in Fig. 12B. Because growth on the C-2 carbon sources ethanol and acetate requires gluconeogenesis, this leads to a reversal of the flux. Nevertheless, the flux of 3-phosphoglycerate via cys-

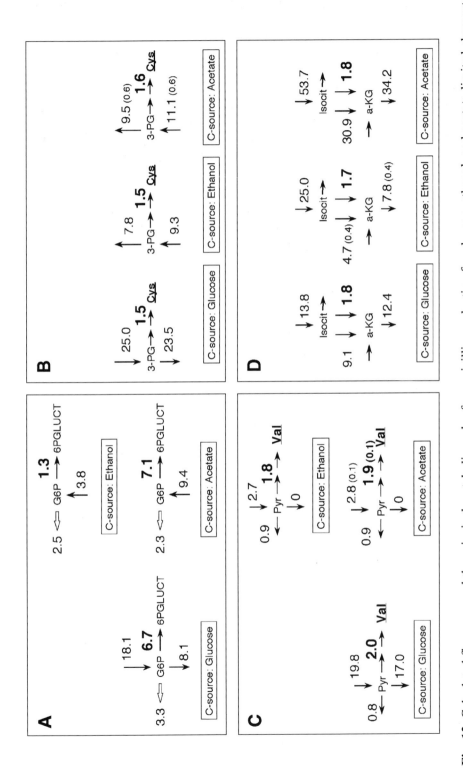

Fig. 12. Calculated fluxes around the principal metabolic nodes for penicillin production for glucose, ethanol, and acetate-limited chemostat growth at a dilution rate of 0.045 h^{-1} (**A**) glucose-6-phosphate node; (**B**) 3-phosphoglycerate node; (**C**) mitochondrial pyruvate; node (**D**) mitochondrial isocitrate node. The standard deviations of the calculated fluxes generally varied between 0.7% and 2.5%. For fluxes with standard deviations higher than 2.5%, the corresponding standard deviations are indicated between brackets.

teine to penicillin is approximately the same in all cases. The flux partitioning around the mitochondrial pyruvate node is shown in Fig. 12C. For growth on C-2 carbon sources, carbon enters the TCA cycle as acetyl CoA, and therefore the fluxes around mitochondrial pyruvate are largely reduced when the cells are grown on ethanol or acetate instead of glucose as the sole carbon source. This has no apparent influence on the flux of pyruvate via valine to penicillin.

In Fig. 12D the fluxes around isocitrate dehydrogenase are shown. Also, in this case, large differences occur, here the flux for growth on acetate is the largest because of the relatively large energetic cost of biomass and product formation from acetate.

From these observations, it can be concluded that despite the fact that growth on the carbon substrates glucose, ethanol, and acetate introduced large changes in flux partitioning around principal nodes for penicillin production in primary metabolism, the observed changes in the specific penicillin production rate were marginal. This indicates that the four principal nodes are highly flexible and are unlikely to form potential bottlenecks for (a further increase of) penicillin production.

3.3. Metabolic Flux-Control Analysis of the Penicillin-V Biosynthesis Pathway

3.3.1. Introduction

The discipline of metabolic engineering aims to adapt the metabolism of a microorganism to improve the productivity and/or yield of certain cellular products. In order to tell the geneticist which enzyme activities in the cell should be up- or downregulated, a metabolic engineer must analyze which enzymatic steps are limiting the overall performance of the cell. This process is called "Metabolic Control Analysis." This analysis has been described extensively elsewhere *(1,40)*. In this section, some of the fundamental elements of metabolic control analysis are presented.

The central concepts in the control analysis are normalized sensitivities of metabolic variables to other variables or parameters in the system. For example, the normalized sensitivity of the metabolic flux through a pathway (J) to the activities of the enzymes that constitute the pathway (e_i) are termed flux-control coefficients and are defined as:

$$C_{e_i}^J = \frac{e_i \cdot \partial J}{J \cdot \partial e_i}$$

[Eq. 31]

These coefficients show how strongly the increase or decrease of a given enzyme activity affects the flux through the pathway. The flux-control coefficients of all the enzymes in a linear metabolic pathway are related by the flux-control summation theorem *(41)*:

$$\left(C_{e_{1,0}}^J \cdots C_{e_{n,0}}^J \right) \cdot \begin{pmatrix} 1 \\ \vdots \\ 1 \end{pmatrix} = 1$$

[Eq. 32]

This theorem states that the flux-control coefficients in a pathway sum up to one, which means that if one of the pathway enzymes has a flux-control close to one, the others have negligible control. Increasing the activity of this enzyme is a good first step in increasing the pathway flux. However, this will lead to a decrease of the flux-control coefficient of the concerning enzyme that evidently entails an increase of the flux control of all other enzymes. By consequence, another enzyme may become flux-limiting

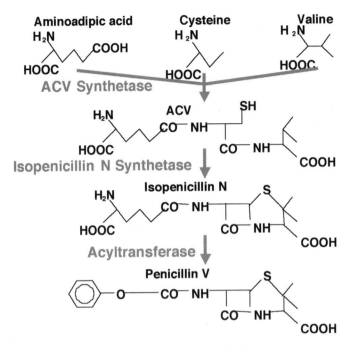

Fig. 13. The penicillin-V biosynthesis pathway.

and will therefore form the next target for genetic modification. It has become clear that flux increase often requires the simultaneous increase in several (if not all) pathway enzymes.

A second important group of entities in control analysis are the elasticity coefficients,—the normalized sensitivities of the reaction rates (v_j) to metabolite levels in the cell (x_i). Elasticity coefficients are defined as:

$$\varepsilon^{v_j}_{x_i} = \frac{x_i \cdot \partial v_j}{v_j \cdot \partial x_i} \qquad \text{[Eq. 33]}$$

An elasticity may represent the influence of the concentrations of substrates or products of a reaction on the rate of the concerning reaction. Alternatively, they may represent the allosteric influence (inhibition/activation) that a nonreacting metabolite has on the reaction rate. Elasticity coefficients and flux-control coefficients are related by the flux-control connectivity theorem *(6)*:

$$\left(C^J_{e_{1,0}} \cdots C^J_{e_{n,0}} \right) \cdot \begin{pmatrix} \varepsilon^{v_1}_{x_{1,0}} & \cdots & \varepsilon^{v_1}_{x_{m,0}} \\ \vdots & \ddots & \vdots \\ \varepsilon^{v_n}_{x_{1,0}} & \cdots & \varepsilon^{v_n}_{x_{m,0}} \end{pmatrix} = 0 \qquad \text{[Eq. 34]}$$

Flux-control coefficients and elasticity coefficients can be determined by combining Eqs. 32 and 34 with a mathematical description of the enzyme kinetics and measurements of metabolic rates, enzyme activites, and metabolite levels. There are alternative ways to do this, as shown in **Subheading 3.3.2.**

3.3.2. The Penicillin Biosynthesis Pathway

The application of metabolic control analysis to a product pathway is illustrated by a case study involving the biosynthetic pathway of the antibiotic penicillin, which is produced as a secondary metabolite by *P. chrysogenum*. In this discussion of the biosynthesis of the penicillin molecule only the formation of this antibiotic from its amino acid precursors is discussed—the biosynthesis of the precursors themselves is not included.

The biosynthetic pathway of pencillin V consists of three enzymatic steps (*see* Fig. 13). In the first step the amino acids valine, cysteine and α-aminoadipic acid are condensed by the enzyme ACV synthetase (ACVS) to form the tripeptide ACV. The following step is the closure of the α-lactam ring structure and the formation of isopenicillin N (IPN) by the enzyme IPN synthase (IPNS). Finally, the α-aminoadipic acid side-chain of the IPN molecule is exchanged for a phenyl acetic-acid side chain by the enzyme acyl transferase (AT).

In order to find out whether one or more of the enzymatic steps in this linear biosynthesis pathway are limiting the penicillin production rate, we should to determine the flux control of the enzymes in this route. First it is shown how the flux control is traditionally determined. Subsequently, a new method is proposed that requires considerably less assumptions regarding the exact enzyme kinetics.

3.3.3. Traditional Method

The traditional way of determining flux-control coefficients was applied to the penicillin biosynthesis pathway by Nielsen et al. *(42)*. They formulated kinetic equations for the enzymes ACVS and IPNS based on the extensive research on the kinetic mechanisms of these two enzymes. The exact kinetics of AT—the enzyme that catalyzes the last step in the pathway—is less well-known. However, based on the fact that IPN was not observed to accumulate in cells, Nielsen et al. concluded that this enzyme could not be rate-limiting, so its flux-control coefficient was assumed to be equal to zero.

Algebraic derivatives of the kinetic expressions of AVCS and IPNS to the metabolites that appear in the expressions yielded the elasticity coefficients as functions of metabolite levels and kinetic parameters. Using these equations, the values of the elasticity coefficients were calculated from the metabolite levels that were measured during the course of a fed-batch cultivation of *P. chysogenum* and the literature values for the kinetic parameters. When the elasticities are known, Eqs. 32 and 34 yield enough constraints to calculate all flux-control coefficients.

The determination of control coefficients as the derivatives of the detailed kinetic expressions as described here has some important disadvantages. First, the kinetic expressions of a pathway often contain many kinetic parameters, such as maximal rates, affinity constants, and inhibition constants. If these are to be estimated from in vivo data, a rich and high-quality set of measurement data of all the involved metabolites (including cofactors) should be available. As this is seldom the case, parameter values are often taken from the literature *(42)*.

Literature data are often based on in vitro studies in which purified enzymes are used. However, care should be taken when using the outcomes of these studies for predicting in vivo kinetics. In vitro experiments are generally not performed under physiological conditions with respect to the pH, concentrations of ionic solutes, the presence of (unidentified) regulatory metabolites, substrate and product concentrations, and the presence of (high concentrations of) proteins that may alter the activity of the studied enzyme. For example, Teusink et al. *(9)* showed that many of the in vitro determined kinetic parameters of the glycolytic enzymes had to be adapted in order to make their mechanistic model correctly predict the metabolite concentrations measured in vivo.

Finally, kinetic expressions are never complete. When the detailed kinetics of the enzymes are introduced, assumptions are made about which metabolites exert an effect on each reaction and which do not. The omission of an important regulatory compound in a kinetic equation leads to a misestimation of the reaction rate.

3.3.4. Alternative Method

For the reasons outlined here, we propose the use of approximate kinetic expressions for describing in vivo kinetics. It will be shown that the use of a general linear-logarithmic approximate rate equation for the kinetics of all enzymes in a metabolic pathway has the advantage that the flux-control coefficients can be estimated without making a single assumption about the exact kinetics and regulatory structure of the pathway. Moreover, the only measurement data needed to determine the control coefficients in this way are the pathway flux and the enzyme activities. No metabolite levels need to be measured. Based on the control coefficients that are calculated for a reference steady state, the pathway flux and control coefficients in other steady states can be predicted from measured enzyme activities alone. This shows that the linlog-approximation has predictive as well as descriptive power.

The only assumption that has been made is that the reaction rate (v) is proportional to the enzyme activity (e). It is related to the levels of metabolites x (represented by a vector containing the metabolite concentrations) in an arbitrary way:

$$v\,(t) \;=\; e\,(t) \cdot f\,[x(t)] \qquad\qquad \text{[Eq. 35]}$$

In traditional metabolic control theory, Eq. 35 is completely linearized around a reference state (indicated by subscript "0" and characterized by levels of enzymes e_0, fluxes v_0 and metabolites X_0). This linearization results in:

$$v\,(t) \;=\; e\,(t) \cdot f\,(x_0) + e_0 \left[\frac{\partial f(x)}{\partial x} \right]_0^T \cdot [x(t) - x_0] \qquad\qquad \text{[Eq. 36]}$$

Note that because x is a vector of metabolite levels, the last term in Eq. 36 represents the sum of the derivatives towards the various metabolite concentrations multiplied with the finite differences of those concentrations from the reference state. When rate Eq. 36 is divided by the rate in the reference state (v_0), the equation reads:

$$\frac{v\,(t)}{v_0} \;=\; \frac{e\,(t)}{e_0} + \left[\frac{x \cdot \partial f(x)}{f(x) \cdot \partial x} \right]_0^T \cdot \left\langle \frac{|x(t) - x_0|}{x_0} \right\rangle \qquad\qquad \text{[Eq. 37]}$$

Again, the last term is in fact a sum of terms for the various elements of vector x. The products and quotients within the bracketed terms should be read as element-by-element operations. Making the generally accepted assumption that the enzyme amount (e) is independent of the metabolite levels (x), it is clear that the normalized derivative in Eq. 37 is the normalized derivative of the reaction rate in Eq. 35. This derivative is exactly the definition of an elasticity coefficient in the reference state $(\varepsilon_{x,0}^v$, *see* Eq. 33)! Therefore, we can rewrite the equation as follows:

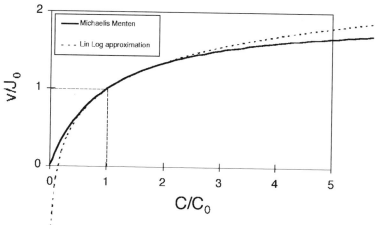

Fig. 14. Lin-log approximation of hyperbolic Michaels Menten kinetics.

$$\frac{v(t)}{v_0} = \frac{e_0}{e(t)} \cdot \underline{\varepsilon}_{x,0}^{v} \cdot \left\langle \frac{[x(t) - x_0]}{x_0} \right\rangle \qquad \text{[Eq. 38]}$$

Eq. 38 is only applicable to very small changes around the reference. A practical application to real experimental situations, where fluxes and enzyme levels change manifold, is therefore not possible. To solve this problem, Small and Kacser *(43,44)* proposed a theoretical extension of traditional metabolic-control analysis toward large changes of enzymes and fluxes by introducing deviation indices. The repelling mathematics, however, are highly intricate and difficult to apply.

Here we will show that an alternative extension of metabolic control analysis to accommodate large changes is easily attained, leading to simple equations. The key is the realization that only the nonlinear $f[x(t)]$—not the complete Eq. 35—needs to be linearized. It is always better to approximate a nonlinear function by another nonlinear function. A proper choice is to use a linear combination of logarothmic terms (lin-log format). For example, if reaction rate v is some nonlinear function of the metabolite concentrations X_1 and X_2—that is $v = f(X_1, X_2)$—this can be approximated as $v = a + b \cdot ln(X_1) + c \cdot ln(X_2)$.

Fig. 14 shows for Michaelis Menten kinetics the quality of the lin-log approximation. This format is also inspired by the concept that the enzyme rate is related to the thermodynamic driving force of the reaction. The linearization procedure leads then to:

$$\frac{v(t)}{v_0} = \left[\frac{e(t)}{e_0} \right] \cdot \left\langle 1 + \underline{\varepsilon}_{x,0}^{v} \cdot ln \left[\frac{x(t)}{x_0} \right] \right\rangle \qquad \text{[Eq. 39]}$$

where $\dfrac{e(t)}{e_0}$ is a square diagonal matrix.

This equation can be rewritten as:

$$\frac{v(t)}{v_0} \frac{e_0}{e(t)} = 1 + \underline{\varepsilon}_{x,0}^{v} \cdot ln \left| \frac{x(t)}{x_0} \right| \qquad \text{[Eq. 40]}$$

Multiplying the left and right hand sides with the flux control matrix C^J and using the summation and connectivity relations yields the general flux equation:

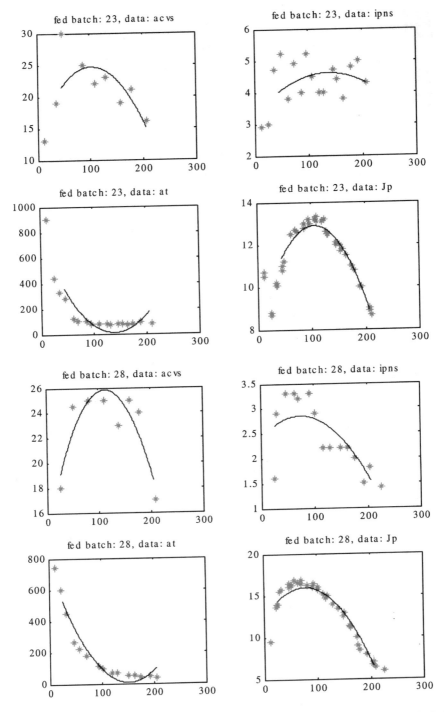

Fig. 15. The raw data of Nielsen (1995) (*) and polynomial approximations (line). Upper four: fed-batch 23, lower four: fed-batch 28. Shown are the enzyme activities of ACVS, IPNS, and AT, and the penicillin production flux. The horizontal axes show the fed-batch duration in hours, the vertical axes show enzyme activities and flux in mmoles/g DW/h (except for INPS where the unit is mmol/g DW/h).

$$C^J \left[\frac{v(t)}{v_0} \frac{e_0}{e(t)} \right] = 1 \qquad \text{[Eq. 41]}$$

The term between brackets is an n-dimensional vector with elements $\dfrac{v_i(t)}{v_{i,0}} \dfrac{e_{i,0}}{e_i(t)}$.

For a linear penicillin pathway Eq. 39 can be rewritten as:

$$C^J_{e1,0} \cdot \frac{v_1(t)}{v_{1,0}} \cdot \frac{e_{1,0}}{e_1(t)} + \cdots + C^J_{en,0} \cdot \frac{v_n(t)}{v_{n,0}} \cdot \frac{e_{n,0}}{e_n(t)} = 1 \qquad \text{[Eq. 42]}$$

If the microorganism of which the flux-control coefficients are to be determined is grown in a fed-batch fermentation, one may assume that at any time during the fed batch the cells, are in a pseudo steady state. Because of the depletion of some of the nutrients and the accumulation of the products in the vessel, this steady state will change during the course of the fed batch. From Eq. 42, we can derive a relationship that predicts the flux through the n-step linear pathway for any steady state that is attained during the fed batch. Because of the steady-state condition all reaction rates (v) equal the pathway flux (J), so we can rewrite Eq. 42 as:

$$J(t) = J_0 / \left[C^J_{e1,0} \cdot \frac{e_{1,0}}{e_1(t)} + \cdots + C^J_{e(n-1),0} \cdot \frac{e_{n-1,0}}{e_{n-1}(t)} + \left(1 - C^J_{e1,0} - \cdots - C^J_{e(n-1),0} \right) \cdot \frac{e_{n,0}}{e_n(t)} \right] \text{[Eq. 43]}$$

The last term between brackets in this equation is again based on the flux-control summation theorem (Eq. 32). The expression requires as inputs the enzyme activities at the specific time t plus the flux-control coefficients and enzyme activities in the reference steady state.

From Eq. 43, we can also obtain relationships that give the control coefficients for any time during the fed batch by determining the normalized derivatives of Eq. 43 toward the various enzyme activities:

$$C^J_{e1}(t) = \frac{e_{1,0}}{e_1(t)} \cdot C^J_{e1,0} / \left[C^J_{e1,0} \cdot \left(\frac{e_{1,0}}{e_1(t)} - \frac{e_{n,0}}{e_n(t)} \right) + \cdots + C^J_{e(n-1),0} \cdot \left(\frac{e_{n-1,0}}{e_{n-1}(t)} - \frac{e_{n,0}}{e_n(t)} \right) + \frac{e_{n,0}}{e_n(t)} \right] \text{[Eq. 44]}$$

In **Subheading 3.3.5.**, this theory is applied to the flux control in the penicillin biosynthesis pathway.

3.3.5. Application: Control Analysis of the Penicillin-V Biosynthesis Pathway

The data used for the metabolic control analysis based on the linlog approximation discussed here were taken from two fed-batch cultivations (numbers 23 and 28) of *P. chysogenum* published by Nielsen *(45)*. For both fed batches, the feed medium consisted of glucose, ammonia, and phenoxyacetic acid. The difference between the two fed-batch cultures was that in case of fed batch 28 the three precursor molecules of penicllin (α-amino adipic acid, valine and cysteine) were added to the medium.

In order to interpolate between the data points, second-order polynomials were used. The raw data and the approximations are shown in Fig. 15.

The flux-control coefficients were determined for each possible reference state by nonlinear minimization of the sum of squared deviations between the fluxes measured during the fed batch and the fluxes calculated by filling in the measured enzyme activi-

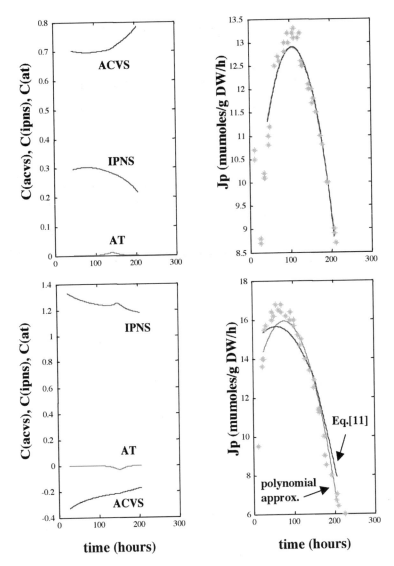

Fig. 16. The calculated flux-control coefficients (left) and the penicillin production flux (right): raw data of Nielsen (*), polynomial approximation and calculated according to Eq. 11 (lines).

ties in Eq. 43. The reference state was chosen as the time t, at which the minimized sum of squares between the measured and calculated flux was minimal.

We determined the following flux-control coefficients. For fed-batch 23: $C^J_{ACVS,0}=$ 0.70, $C^J_{IPNS,0}= 0.30$ and $C^J_{AT,0}= 0.00$ (reference: t = 42 h). For fed-batch 28: $C^J_{ACVS,0}=$ –0.28, $C^J_{IPNS,0}= 1.28$ and $C^J_{AT,0}= 0.00$ (reference: t = 31 h). Based on the flux-control coefficients in the reference state and the measured enzyme activities (Fig. 15), the flux-control coefficients and penicillin production flux in all other steady states can be calculated using Eqs. 43 and 44. The outcomes are shown in Fig. 16.

In Fig. 15, it can be seen that the polynomials used to interpolate between the data points are relatively rough approximations. Therefore, the outcome of the calculations in Fig. 16 should be interpreted with caution. The outcomes in the left-hand side graphs of Fig. 16 confirm Nielsen's assumption that the enzyme AT has no control over the flux through the pathway.

In his study, Nielsen et al. *(42)* found a shift of flux control from AVCS in the beginning of the fed batch to IPNS at the end of the fed batch in both fed batches 23 and 28. We find a very different flux-control profile: in fed-batch 23, ACVS has 70–80% of the control during the entire fermentation, whereas the same enzyme has –30 to –18% (negative) control in fed-batch 28. In the latter experiment, IPNS has full control of the penicillin production flux. The fact that we found that ACVS has no control in the fed-batch, in which the substrates of this enzyme were present in the feed (fed-batch 28), whereas it has control in the experiment without precursor feeding (fed-batch 23) seems reasonable. Another support for the validity of the proposed method is that the graphs on the right-hand side of Fig. 16 show that the penicillin production flux calculated using Eq. 43 fits well with the polynomial approximations of Nielsens data. This case clearly shows that detailed kinetic-rate equations are not needed to calculate flux-control coefficients.

3.4. New Experimental and Theoretical Tools to Direct Metabolic Engineering of Microorganisms and Cell Cultures

At present, an increasing gap is developing between our experimental capabilities in metabolic engineering of cellular metabolism and our quantitative theoretical understanding of the kinetic interaction between primary metabolism and product pathways. Such a theoretical understanding is absolutely essential for a rational reprogramming of metabolism.

3.4.1. The Problem of Identifying Rate Limiting Steps in Complex Metabolic Networks

In case the rate of product formation is limited by primary metabolism, the question arises: which enzyme(s) are to be targets for recDNA modification? This is a very difficult question for a number of reasons. An intuitive guess is hardly possible because of the complex network structure (branches, cycles, bypasses, conserved moieties) and the complex interactions (allosteric and/or kinetic effects, cofactor-based interactions). Also, it is very likely that more than one enzyme must be changed simultaneously *(46)* to achieve a substantial increase in the rate of product formation. Finally, it has been recognized that an increase in metabolic fluxes through changing enzyme levels should not result in excessive increases in intracellular metabolite and protein levels. Intracellular concentrations of metabolites that are too high generally cause undesirable effects as kinetic interference and/or degradation into toxic by-products. Also, strong changes in metabolite concentrations trigger genetic-control mechanisms to bring them back to acceptable levels (principle of homeostasis). This will result in undesirable changes of enzyme levels.

Excessive increases in enzyme levels can lead to such high intracellular protein concentrations that unwanted effects such as protein interactions and protein crystallization occur, all leading to unpredictable changes in enzyme conformation and enzyme kinetic properties.

Consequently, in kinetic modeling of microbial metabolism it should be realized that the intracellular metabolite levels and the total protein concentration should remain between physiological feasible levels. This is achieved by defining constraints *(52)*, such as:

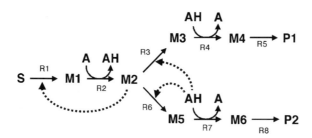

- **Reversible Hill kinetics (7 parameters/eq) (R1,3,6)**
- **Ordered bi bi kinetics (10 parameters/eq) (R2,4,7)**
- **Uni uni kinetics (4 parameters/eq) (R5,8)**

Fig. 17. Artificial metabolic network, as proposed by Mendes (1998).

- Only moderate changes in metabolite levels are allowed (e.g., +/– 50%).
- The total protein content—the sum of all enzymes—should not change too much. Thus, if the level of an enzyme goes up, than the levels of other enzymes should go down.

The problem is therefore to find the set of enzymes to be changed under these constraints to provide the highest increase in rate of product formation.

This problem can only be solved by using kinetic models of primary metabolism, and then using computer-based constrained optimizations to determine the best enzyme targets for recDNA changes.

This immediately raises the question of how to arrive at these kinetic models of metabolism. Traditionally, these models have been based on kinetic properties determined from experiments with purified enzymes (in vitro kinetics). It is now becoming clear that the in vitro conditions differ significantly from the intracellular conditions. Under in vitro conditions, the enzyme concentration is generally much lower, the substrate concentration is higher, and other proteins and metabolites, which may have kinetic effects, are absent. Wright et al. *(47)* and Teusink *(48)* have found that applying in vitro determined kinetic properties in mathematical modeling does indeed lead to large differences between measured and simulated concentration profiles and fluxes.

These findings indicate that the kinetic properties of metabolic pathways can only be determined in a reliable way from in vivo experiments. These are typically perturbation experiments *(49)*, in which a culture growing under steady-state conditions is perturbed through the addition of a pulse of a substrate or an inhibitor. During a short time period, the intra- and extracellular concentrations are measured. A time window of about 200 s is considered to be short enough to avoid changes in enzyme levels. From the measured concentration patterns in time, the in vivo kinetic properties can be extracted. However, the range of possible concentration changes under in vivo conditions is limited because of the homeostatic mechanisms that operate in microorganisms. This will severely limit the number of identifiable kinetic parameters. Therefore, the kinetic format used must be sparse in parameters. In conclusion, in vivo kinetic models with a small number of parameters are needed in order to find (using mathematical techniques) enzyme(s) in primary metabolism that significantly determine

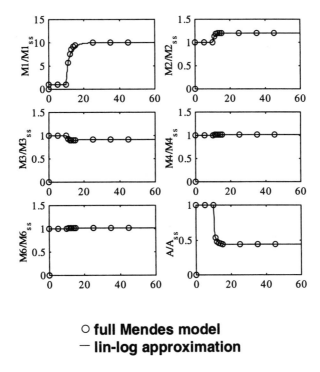

○ **full Mendes model**
─ **lin-log approximation**

Fig. 18. Comparison of simulated metabolite profiles, full kinetic model and Lin-log approximation.

product fluxes, while maintaining constraints on metabolite levels and total enzyme amounts.

3.4.2. Theoretical and Experimental Methods for in vivo Kinetics

The application of linear logarithmic-rate equations has been proposed recently *(50)* as a suitable approximate kinetic format. It is suitable because the number of parameters is minimal and the approximation quality is excellent. The format is a generalization of the concept of thermodynamic driving force, which creates speed *(51)*. Fig. 16 shows the lin-log approximation of hyperbolic Michaelis Menten kinetics. The rate approximation is surprisingly good over a large concentration range. Recently, Visser and Heijnen *(52)* have applied a lin-log kinetic model to a simple artificial metabolic network, as proposed by Mendes and Kell *(53)*, to show the performance of lin-log kinetics. This metabolic network (Fig. 17) contains realistic features as a branch point, conserved moieties and complex highly nonlinear (reversible) kinetics.

The original kinetic-rate equations were approximated by the lin-log format (in which the number of kinetic parameters dropped from 59 to 33). Subsequently, the original full kinetic model of Mendes and Kell and the lin-log model were used to simulate the dynamics of the metabolic system as response to a substrate pulse (Fig. 18). The lin-log kinetic format clearly shows a good performance. Furthermore, the lin-log kinetic format has another very useful property. Because the influence of metabolite concentrations (X_i) on the rate (v) is given as a linear sum of logarithmic concentration terms ($v = a + b \ln X_1 + c \ln X_2 + ...$), full analytical solutions *(50)* can be obtained of:

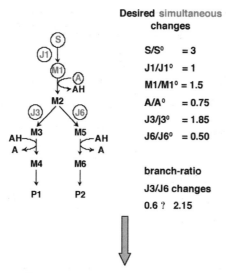

Desired simultaneous
changes

$S/S^0 = 3$

$J1/J1^0 = 1$

$M1/M1^0 = 1.5$

$A/A^0 = 0.75$

$J3/j3^0 = 1.85$

$J6/J6^0 = 0.50$

branch-ratio

J3/J6 changes

0.6 ? 2.15

Design-equation immediately gives
the desired simultaneous enzyme changes

Fig. 19. Redesigning a metabolic network: desired simultaneous changes in fluxes/metabolites.

- All steady-state network reaction rates as function of all enzyme levels;
- All steady-state metabolite levels as function of all enzyme levels;
- The changes in enzyme levels that are needed to obtain a desired flux redistribution in the network (Metabolic design equation)

This design equation was tested *(52)* using the network of Mendes and Kell. Desired large changes in fluxes and metabolite levels were specified (Fig. 19). The design equation directly provides the enzyme levels to be modified (Fig. 20, upper panel). These enzyme levels were implemented in the original full kinetic model, and the resulting metabolite levels and fluxes (Fig. 20) obtained are compared with the design specification. The agreement was found to be very good. Clearly, the approximate lin-log kinetic format and the derived design equation have proven to be very useful.

Finally, the lin-log kinetic format has recently *(54)* been used in a simulation study to optimize the glucose uptake flux in *E. coli* glycolysis under constraints for metabolite levels and total enzyme amount. A complex kinetic model for the glycolysis was approximated by using lin-log rate equations. Both kinetic models were used to calculate the optimal enzyme levels for maximal glucose uptake flux using the same constraint. It was found that both models gave virtually the same optimal solution.

So, in conclusion, it appears from the performed simulation studies that the lin-log kinetic format is very promising as a theoretical basis for in vivo kinetics. In the near future, this concept will be applied to construct in vivo kinetic models of metabolism. It should be clear that the concept is not limited to microbial cells, but can also be applied for plant and animal cells.

The experimental data used to construct these kinetic models, are obtained from pulse-response experiments, performed within a short time interval (typically, 100–200 s). Although the experiments are performed very rapidly, the subsequent sample handling and the analysis of a large number of intracellular metabolites has proven to be very

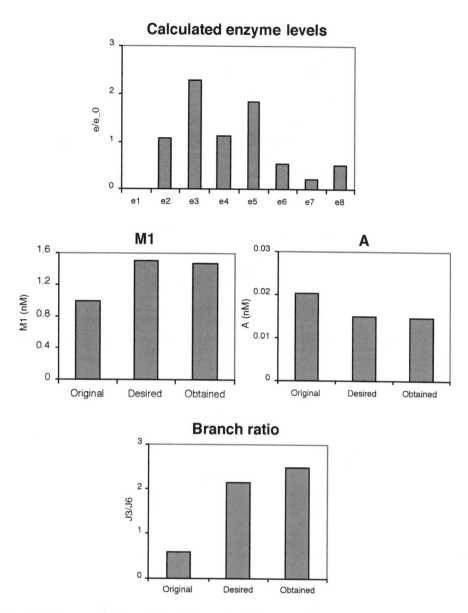

Fig. 20. Enzyme levels calculated from the design equation (upper) and comparison between desired and obtained result.

laborious *(49)*. To facilitate this type of research, high-throughput analysis methods are essential. Recently, a new method for the measurement of intracellular metabolites using LC-MSMS has been developed *(55)*. The value of this approach is the very small sample volume (10 μL), in which a large number of glycolytic intermediates can simultaneously be analyzed. Furthermore, Visser et al. *(56)* developed a new device, the BioSCOPE, (Fig. 21) which can perform pulse-response experiments, including sampling, outside the chemostat. This is a valuable alternative method for the method of Rizzi et al. *(49)*, where the perturbing agent is added to the fermentor. The BioSCOPE is actually a stopped-flow

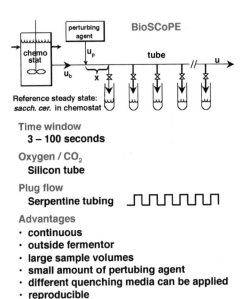

Time window
3 – 100 seconds

Oxygen / CO_2
Silicon tube

Plug flow
Serpentine tubing

Advantages
- **continuous**
- **outside fermentor**
- **large sample volumes**
- **small amount of pertubing agent**
- **different quenching media can be applied**
- **reproducible**
- **many perturbations on 1 day are possible**

Fig. 21. The Bioscope sampling device for rapid sampling of biomass during a pulse experiment.

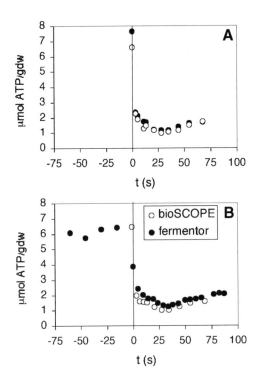

Fig. 22. Measurement of ATP intracellular concentration during a glucose pulse (1 g/L) to a steady-state chemostat culture of *Saccharomyces cerevisiae* (glucose, aerobic, D = 0.05 h^{-1}). (**A**) Pulse experiment performed in the BioSCOPE, in duplicate. (**B**) Comparison between pulse experiment performed in the BioSCOPE and directly in the fermentor, according to Reuss/ Rizzi method.

device, and consists of a plug-flow reactor with sampling ports along the length of the reactor. The sampling time corresponds to the port location. The BioSCOPE is fed continuously with steady-state broth taken from a chemostat. This broth stream is combined with a small stream of perturbation agent, and thus, the reaction proceeds along the length of the plug-flow reactor. Typically, 10–15 samples can be taken in a time frame of about 100 s. The BioSCOPE is a continuous method, and in principle, it allows the withdrawal of unlimited amounts of sample and needs only a limited amount of perturbing agent. Furthermore, the steady-state chemostat is not perturbed, and thus, several perturbation experiments can be performed in a limited amount of time using the same chemostat. For further details on the construction of the BioSCOPE, and operational parameters such as residence-time distribution and oxygen supply, *see* Visser et al. *(56)*. The reproducibility is shown in Fig. 22A, where the measured response of the ATP level during a glucose pulse is illustrated. Fig. 22B shows a comparison of the measured ATP response between a glucose pulse experiment performed in a steady-state chemostat and a similar experiment performed in the Bioscope. This graph clearly indicates that the agreement was very satisfactory.

In performing the pulse experiments for in vivo kinetics, the response of the intracellular metabolites as well as the response of the extracellular metabolites must be measured in order to calculate their rates of consumption (e.g., glucose, O_2 and production (e.g., ethanol, acetate, and glycerol, CO_2).

O_2 and CO_2 pose great problems because of their exchange to the aeration gas used for O_2/CO_2 transfer. Recently, a new method has been developed to calculate the in vivo oxygen uptake rate (q_{O_2}, mmol O_2/gDMh) and CO_2 production rate (q_{CO_2}, mmol/gDMh) from continuous offgas concentration measurement of O_2/CO_2 during pulse experiments *(57)*.

4. Conclusion

Successful metabolic engineering of microbial, plant, and animal-cell metabolism requires quantitative insight in its properties and its interaction with product pathways. Recently, significant progress has been made in both theoretical and experimental techniques to tackle the formidable problem of unraveling in vivo kinetic properties of metabolic pathways. However, the presented techniques have still not resolved the problem of how to predict the control of enzyme levels by genetic regulation. Here, such additional techniques as DNA-arrays (mRNA levels) and proteome measurement must be made operational in the near future. Also, the theoretical techniques (modeling of mRNA expression levels) to predict enzyme levels must be developed. Although this is also a formidable problem, the benefits of being able to design a cell factory for a wide range of products is appealing enough to enter this adventure.

References

1. Heinrich, R. and Schuster, S. (1996) The regulation of cellular systems. Chapman and Hall, New York, NY.
2. vanGulik W. M. and Heijnen J. J. (1995) A metabolic network stoichiometry analysis of microbial growth and product formation. *Biotechnol. Bioeng.* **48,** 681–698.
3. Varma, A. and Palsson, B. O. (1994) Stoichiometric flux balance models quantitatively predict growth and metabolic by-product secretion in wildtype *Escherichia coli* w3110. *Appl. Environ. Microbiol.* **60,** 3724–3731.

4. Stephanopoulos, G. and Vallino, J. J. (1991) Network rigidity and metabolic engineering in metabolite production. *Science* **252,** 1675–1681.

5. Savageau, M. A. (1976) Biochemical systems analysis: A study of function and design in molecular biology. Addison-Wesley, Reading, MA.

6. Kacser, H. and Burns, J. A. (1973) The control of flux. *Symp. Soc. Exp. Biol.* **27,** 65–104.

7. Heinrich, R. and Rapoport, T. A. (1974) A linear steady state treatment of enzymatic chains. *Eur. J. Biochem.* **42,** 89–95.

8. Small, J. R. and Kacser, H. (1993) Responses of metabolic systems to large changes in enzyme activities and effectors. 2. The linear treatment of branched pathways and metabolite concentrations. Assessment of the general non-linear case. *Eur. J. Biochem.* **213(1),** 625–640.

9. Teusink, B., Passarge, J., Reijenga, C. A., Esgalhado, E., Van der Weijden, C. C., Schepper, M., et al. (2000) Can yeast glycolysis be understood in terms of in vitro kinetics of the constituent enzymes? Testing biochemistry. *Eur. J. Biochem.* **267,** 5313–5329.

10. Visser, D., v. d. Heijden, R., Mauch, K., Reuss, M., and Heijnen, J. J. (2000) Tendency modeling: a new approach to obtain simplified kinetic models of metabolism applied to *Saccharomyces cerevisiae. Metab. Eng.* **2,** 252–275.

11. Roels, J. A. (1983) Energetics and kinetics in biotechnology. Elsevier Biomedical, Amsterdam.

12. Vanrolleghem, P. A., deJong-Gubbels, P., vanGulik, W. M., Pronk, J. T., vanDijken, J. P., and Heijnen J. J. (1996) Validation of a metabolic network for *Saccharomyces cerevisiae* using mixed substrate studies. *Biotechnol. Prog.* **12,** 434–448.

13. Andre, B. (1995) An overview of membrane transport proteins in *Saccharomyces cerevisiae. Yeast* **11,** 1575–1611.

14. Hinnebusch, A. G. and Liebman S. W. (1991) Protein synthesis and translational control in Saccharomyces cerevisiae, in *The Molecular Biology of the Yeast Saccharomyces* (Broach, J. R., Pringle, J. R., and Jones, E. W., eds.), Cold Spring Harbor Laboratory Press, New York, NY, pp. 627–735.

15. Rigoulet, M., Leverve, X., Fontaine, E., Ouhabi, R., and Guerin, B. (1998) Quantitative analysis of some mechanisms affecting the yield of oxidative phosphorylation: dependence upon both fluxes and forces. *Mol. Cell. Biochem.* **184,** 35–52.

16. Kallow, W., von Döhren, H., and Kleinkauf, H. (1998) Penicillin biosynthesis—energy requirement for tripeptide precursor formation by delta-(l-alpha-aminoadipyl)-l-cysteinyl-d-valine synthetase from Acremonium chrysogenum. *Biochemistry* **37,** 5947–5952.

17. vanGulik, W. M., deLaat, W. T. A. M., Vinke, J. L., and Heijnen, J. J. (2000) Application of metabolic flux analysis for the identification of metabolic bottlenecks in the biosynthesis of penicillin-G. *Biotechnol. Bioeng.* **68,** 602–618.

18. vanGulik, W. M., Antoniewicz, M. R., deLaat, W. T. A. M., Vinke, J. L., and Heijnen, J. J. (2001) Energetics of growth and penicillin production in a high producing strain of *Penicillium chrysogenum. Biotechnol. Bioeng.* **72,** 185–193.

19. Mauch, K., Buziol, S., Schmid, J., and Reuss, M. (2001) Computer aided design of metabolic networks, in AIChE Symposium Series. Chemical Process Control-6 Conference, Tucson, AZ.

20. Schuster, S., Dandekar T., and Fell, D. A. (1999) Detection of elementary flux modes in biochemical networks: a promising tool for pathway analysis and metabolic engineering. *Trends Biotechnol.* **17,** 53–60.

21. De Jong-Gubbels, P., Vanrolleghem, P. A., Heijnen, J. J., van Dijken, J. P., and Pronk, J. T. (1995) Metabolic fluxes in chemostat cultures of *Saccharomyces cerevisiae* grown on mixtures of glucose and ethanol. *Yeast* **11,** 407–418.

22. Gancedo, J. M. (1986) Carbohydrate metabolism in yeast, in *Carbohydrate Metabolism in Cultured Cells* (Morgan, J. M., ed.), Plenum Press, New York, NY, pp. 245–286.

23. Kiss, R. D. and Stephanopoulos, G. (1992) Metabolic characterization of a L-lysine producing strain by continuous culture. *Biotechnol. Bioeng.* **39,** 565–574.

24. Sonntag, K., Eggeling, L., De Graaf, A. A., and Sahm, H. (1993) Flux partitioning in the split pathway of lysine synthesis in *Corynebacterium glutamicum*. *Eur. J. Biochem.* **213**, 1325–1331.

25. Wiechert, W. and de Graaf, A. A. (1997) Bidirectional reaction steps in metabolic networks: I. modeling and simulation of carbon isotope labeling experiments. *Biotechnol. Bioeng.* **55**, 101–117.

26. Marx, A., de Graaf, A. A., Wiechert, W., Eggeling, L., and Sahm, H. (1996) Determination of the fluxes in the central metabolism of *Corynebacterium glutamicum* by nuclear magnetic resonance spectroscopy combined with metabolic balancing. *Biotechnol. Bioeng.* **49**, 111–129.

27. Marx, A., Eikmans, B. J., Sahm, H., de Graaf, A. A., and Eggeling, L. (1999) Response of the central metabolism in *Corynebacterium glutamicum* to the use of an NADH-dependent glutamate dehydrogenase. *Metab. Eng.* **1**, 1, 35–48.

28. Schmidt, K, Carlsen, M., Nielsen, J., and Villadsen, J. (1997) Modelling isotopomer distributions in biochemical networks using isotopomer mapping matrices. *Biotechnol. Bioeng.* **55**, 831–840.

29. Schmidt, K., Marx, A., de Graaf, A. A., Wiechert, W., Sahm, H.,Nielsen, J., et al. (1998) ^{13}C tracer experiments and metabolite balancing for metabolic flux analysis: comparing two approaches. *Biotechnol. Bioeng.* **58**, 254–257.

30. Szyperski, T. (1998) ^{13}C-NMR, MS and metabolic flux balancing in biotechnology research. *Quart. Rev. Biophys.* **31**, 41–106.

31. van Winden, W. A., Verheijen, P. J. T., and Heijnen, J. J. (2001a) Possible pitfalls of flux calculations based on ^{13}C-labeling. *Metab. Eng.* **3**, 151–162.

32. van Winden, W. A., Heijnen, J. J., Verheijen, P. J. T., and Grievink, J. (2001) A priori analysis of metabolic flux identifiability from 13C-labeling data. *Biotechnol. Bioeng.* **74**, 505–516.

33. van Winden, W. A., Schipper, D., Verheijen, P. J. T., and Heijnen, J. J. (2001) Innovations in the generation and analysis of 2D [^{13}C,^{1}H] COSY spectra for flux analysis purposes. *Metab. Eng.* **3**, 322–343.

34. Pramanik, J., Trelstad, P. L., Schuler, A. J., Jenkins, D., and Keasling, J. D. (1999) Development and validation of a flux-based stoichiometric model for enhanced biological phosphorus removal metabolism. *Wat. Res.* **33**, 462–476.

35. Edwards, J. S. and Palsson, B. O. (2000) The *Escherichia coli* MG1655 in silico metabolic genotype: its definition, characteristics, and capabilities. *Proc. Natl. Acad. Sci. USA* **97**, 5528–5533.

36. Edwards, J. S. and Palsson, B. O. (1999) Systems properties of the *Heamophilus influenzae Rd* metabolic genotype. *J. Biol. Chem.* **274**, 17,410–17,416.

37. Bailey, J. E. (1998) Mathematical modelling and analysis in biochemical engineering: past accomplishments and future opportunities. *Biotechnol. Prog.* **14**, 8–20.

38. Cornish-Bowden, A. and Hofmeyr, J. H. S. (1994) Determination of control coefficients in intact metabolic systems. *Biochem. J.* **298**, 367–375.

39. Jørgensen, H. S., Nielsen, J., Villadsen, J., and Møllgaard, H. (1995) Metabolic flux distributions in *Penicillium chrysogenum* during fed-batch cultivations. *Biotechnol. Bioeng.* **46**, 117–131.

40. Stephanopoulos, G. N., Aristidou, A. A., and Nielsen, J. (1998) Metabolic control analysis, in *Metabolic Engineering, Principles and Methodologies*, Academic Press, San Diego, CA.

41. Heinrich, R., Rapoport, S. M., and Rapoport, T. A. (1977) Metabolic regulation and mathematical models. *Prog. Biophys. Mol. Biol.* **32**, 1–82.

42. Nielsen, J. and Jørgensen, H. S. (1995) Metabolic control analysis of the penicillin biosynthetic pathway in a high-yielding strain of *Penicillium chrysogenum*. *Biotechnol. Prog.* **11**, 299–305.

43. Small, J. R. and Kacser, H. (1993) Responses of metabolic systems to large changes in enzyme activities and effectors. 1. The linear treatment of unbranched chains. *Eur. J. Biochem.* **213**, 613–624.

44. Small, J. R. and Kacser, H. (1993) Responses of metabolic systems to large changes in enzyme activities and effectors. 2. The linear treatment of branched pathways and metabolite concentrations. Assessment of the general non-linear case. *Eur. J. Biochem.* **213**, 625–640.

45. Nielsen, J. (1995) "Physiological Engineering Aspects of *Penicillium chrysogenum*", DSc. thesis Technical University of Denmark, Lyngby, Denmark.

46. Fell, D. A. and Thomas, S. (1995) Physiological control of metabolic flux: the requirement for multisite modulation. *Biochem. J.* **311,** 35–39.

47. Wright, B. E. and Kelly, P. J. (1981) Kinetic models of metabolism in intact cells, tissues and organisms. *Curr. Top. Cell. Regul.* **19,** 103–158.

48. Teusink, B. (1999) Exposing a complex metabolic system: glycolysis in *Saccharomyces cerevisiae*. PhD thesis, Universiteit van Amsterdam.

49. Rizzi, M., Baltes, M., Theobald, U., and Reuss, M. (1997) In vivo analysis of metabolic dynamics in *Saccharomyces cerevisiae*, II. Mathematical model. *Biotechnol. Bioeng.* **55,** 592–608.

50. Heijnen, J. J. (2000) Unified kinetic and MCA based models in metabolic engineering. Paper presented at Metabolic Engineering III, Colorado Springs, CO.

51. Westerhoff, H. V. and Van Dam, K. (1987) Thermodynamics and Control of Biological Free-Energy Transduction. Elsevier, Amsterdam.

52. Visser, D. and Heijnen, J. J. Dynamic simulation and metabolic re-design of a branched pathway using LinLog kinetics. *Metab. Engin.,* submitted.

53. Mendes, P. and Kell, D. B. (1998) Non-linear optimization of biochemical pathways: applications to metabolite engineering and parameter estimation. *Bioinformatics* **14,** 869–883.

54. Visser, D., Schmid, J. W., Mauch, K., Reuss, M., and Heijnen, J. J. Optimization of *Escherichia coli*'s primary metabolism using lin-log kinetics. *Metab. Engin.,* submitted.

55. VanDam, J. C., Eman, M. R., Frank, J., Lange, H. C., vanDedem, G. W. K., and Heijnen, J. J. (2002) Analysis of glycolytic intermediates in *Saccharomyces cerevisiae* using anion exchange chromatography and electrospray ionization with tandem mass spectromic detection. *Anal. Biochim. Acta.* **460,** 209–218.

56. Visser, D., van Zuylen, G. A., van Dam, J. C., Oudshoorn, A., Eman, M. R., Ras, C., et al. (2002) Rapid sampling for analysis of in-vivo kinetics using the BioSCoPE: a system for continuous pulse experiments. *Biotechnol. Bioeng.* **79,** 674–681.

57. Wu, L., Van Gulik, W. M., and Heijnen, J. J. Dynamic measurements of off-gas signals in a perturbed chemostats in a 300 seconds time window for in-vivo kinetics analysis of *Saccharomyces cerevisiae*. *Biotechnol. Bioeng.,* in press.

16

Advances in Analytical Chemistry for Biotechnology

Mass Spectrometry of Peptides, Proteins, and Glycoproteins

Jeffrey S. Patrick

1. Introduction

1.1. Background

The analysis and characterization of biomacromolecules presents numerous challenges to the analytical chemist. For proteins, this stems from the general complexity of the polypeptide backbone, and is confounded by the complex processes that may occur in biological systems to further modify the peptide core. The number of inherent possibilities (20 common amino acids), and the fact that bioactivity is inherently linked to protein primary, secondary, tertiary, and, often, quaternary structure creates a daunting challenge for the bioanalytical and biophysical chemist. A simple decapeptide composed of a random combination of the 20 common amino acids has more than of 100 billion possible combinations. If the composition is limited to five amino acids but the order is random, this number becomes 24 possible peptides. This problem becomes even more challenging when the myriad of post-translational modifications is added to the possible peptides. This level of complexity often requires more than one dimension of analysis.

Until the past 20 yr or so, proteins to be characterized were isolated in small amounts in a slow and agonizing process. The advent of techniques for high-level and controlled expression of native and mutated proteins outside of their normal system has created both opportunities and challenges for the analytical biochemist. The ability to rapidly express and isolate multiple, closely related proteins in a comparatively short period of time creates the need for detailed characterization in relatively brief periods of time. This requires more sensitive, higher-throughput methods of biomolecular analysis and characterization. Significant information about which proteins may potentially be expressed has been gleaned from genetic analyses such as PCR, including the characterization of entire genomes from organisms. However, the information obtained from these analyses does not address the cellular machinery responsible for translating and processing the gene into a functional protein, but does provide clues with which to search for selected proteins. This information has spawned the rapid identification of proteins as expressed and processed by organisms in a collection of experiments known as proteomics. The confirmation of the presence of a protein under a particular set of

From: *Handbook of Industrial Cell Culture: Mammalian, Microbial, and Plant Cells*
Edited by: V. A. Vinci and S. R. Parekh © Humana Press Inc., Totowa, NJ

physiological conditions only begs for more detailed characterization to further define the function of the protein in the system of interest through experimentation rather than comparison with database information. This is particularly important for proteins that may be candidates for drug therapies, or may be drug candidates themselves.

Together, these developments in molecular biology have created increasing demands on physical and analytical techniques to more rapidly provide more information on less material. Indeed, analytical chemists and physical biochemists are becoming the information scientists of the biotechnology world. These demands have been met through a range of technical advances, including the enhancement of capillary electrophoretic techniques *(1)*, improvements in capillary HPLC systems and columns *(2)*, miniaturization of entire analytical experiments to effectively operate in arrays the size of a computer processor *(3)*, the use of powerful computational methods to rapidly acquire and decipher complex multi-dimensional NMR data *(4)*, and the simplification and extension of mass spectrometric experiments to accommodate the challenges of biomacromolecular characterization *(5)*. Of these, the technology which has seen the most rapid advances and provides the most information in a short period of time is mass spectrometry. This technology provides an inherently high level of sensitivity and selectivity. These advances, reflected in the amount of work done on peptides and proteins by MS, are evident from the increased percentage of abstracts at the Annual Conference of the American Society for Mass Spectrometry—this number has more than quadrupled between 1998 and 2001 *(6)*. The experiments available to the scientist investigating proteins and peptides of biotechnological origin have revolutionized the capacity to characterize these molecules and reduce the time from months and years to weeks or less. As the number of therapeutic proteins of recombinant origin increases, the application of mass spectrometric tools to the determination of the physical attributes and critical properties of proteins and the safety and efficacy is of growing importance.

1.2. Scope

This chapter provides an overview of the fundamentals of the characterization and analysis of peptides, proteins, and glycoproteins using mass spectrometry. Topics covered include ionization methods and mass analyzers common to biomolecular analyses, nomenclature used in describing the molecules and their ions, and a broad range of common experiments (CE/MS, HPLC/MS, and MS/MS) used in the mass spectrometric analysis and characterization of peptides, proteins, oligosaccharides, and glycoproteins. The focus is on recent advances (1996–2001), which have contributed to the extended application of mass spectrometry to the characterization and analysis of glycoproteins, oligosaccharides, proteins, and peptides. The goal is to provide researchers in molecular biology and biochemistry who are interested in applying mass spectrometry with key background and a general understanding of the tools and techniques available and some of the opportunities and caveats of these experiments. In addition, the chapter provides a perspective on the future possibilities of biomolecular mass spectrometry.

2. Fundamental Aspects of Mass Spectrometry

The mass spectrometry (MS) experiment has five steps: sample introduction, ionization, mass analysis (i.e., ion separation), ion detection, and data recording/manipulation. A depiction of a typical MS experiment is shown in Fig. 1. These steps are the

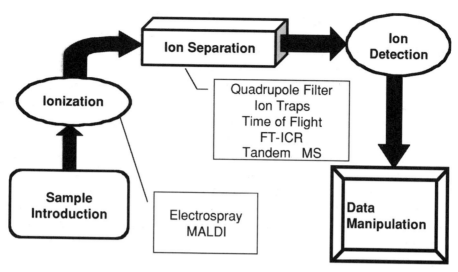

Fig. 1. Schematic diagram of the necessary components of a mass spectrometric experiment.

same regardless of the analyte, its size, or its matrix. Introduction and ionization can be coincidental (i.e., ESI), or they may be separate processes that together influence the quality of the mass spectral information obtained. Mass analysis can be achieved using any of several different instruments, each with its own assets and liabilities, and the mass analyzer often defines the mode of ion detection. The final component of the experiment is data acquisition and manipulation, which is becoming increasingly important but less difficult as software for the interpretation of spectra and databases to identify biomolecules becomes more readily available *(7)*. Although each element of the analysis is important, only ionization and mass analysis are discussed here. Below, the principles and strengths/weaknesses of the two ionization methods, ESI and MALDI, and the four mass analyzers (quadrupole, ion traps, TOF, FT-ICRs) most commonly used in biomacromolecular analysis are discussed. More detailed examinations of their operating principles and utility are referenced.

2.1. Principles of Ionization and Common Techniques for the Ionization of Biomolecules

Sample introduction and ionization are the first and most significant components to a mass spectrometry experiment. This determines whether or not ions will be observed and whether they are representative of the actual sample. If ionization fails, is unpredictable, or is non-representative, the experiment will provide far less information than the mass analyzer may possibly provide. However, it is important to remember that different mass analyzers are more or less compatible with different means of sample introduction and ionization. In general, biomolecules (here assume a mol wt in excess of 1000) are introduced by one of two mechanisms. The first is as a free-flowing solution such as the effluent from high-performance liquid chromatography (HPLC) or CE. Sample introduced by this means is most often ionized using the spray techniques generally known as Atmospheric Pressure Ionization (API), which includes atmospheric pressure chemical ionization (APCI), electrospray ionization (ESI), and ionspray (or pneumatically assisted ESI). Each of these is amenable to varying introduction conditions and a different class of compounds and range of mol wts, and each has its advantages and disadvantages in biological mass spectrometry. The second method of sample introduction is as a mixture of the sample and a non-volatile matrix. Samples introduced by this mechanism are ionized by one of the desorption ionization (DI) methods *(8)*, which include plasma desorption ionization (PD), fast-atom bombardment (FAB), secondary-ion mass spectrometry (SIMS) and matrix-assisted laser desorption/ionization (MALDI) as the most common methods for biomolecules. MALDI has achieved the highest level of utilization in modern analyses. Among the matrices used in these experiments are glycerol, nitrocellulose, sorbitol, ammonium chloride, organic polymers, and small organic molecules that absorb UV/Vis/IR radiation, among others. The sample/matrix mixture is applied to a probe tip or plate that is then introduced into an evacuated ionization source.

The details of ionization by API and DI methods have been the subject of exhaustive reviews and books *(8,9)*.

2.1.1. Atmospheric Pressure Ionization (API)

This is a group of related experiments in which the sample is introduced as a fluid or solution to an electric field at approximately atmospheric pressure. This includes ESI and APCI, although ESI is the preferred method for biomacromolecules. Depending on the design of the source, liquid flow rates of between a few nL per min up to a few mL per min can be used with commercially available ESI and APCI sources. The effluent or solvent in the API experiment can vary widely in composition, and this directly influences the ability to distribute charge onto the analyte of interest.

In the ESI experiment, the effluent is passed through a "needle" of capillary internal diameter. A typical ESI source is shown in Fig. 2. The needle is typically charged to between 500 and 5000 volts relative to the mass analyzer dependent upon the source design, analyte, and matrix, and can have flows of inert gas (typically nitrogen) used to assist the evaporation of solvent. The applied voltage creates potential fields that generate charged droplets. The charged droplets then lose solvent through evaporation until the repulsive forces between like charges on the droplet surface destabilize the droplets and eject analyte ions through a so-called Coulombic explosion (Fig. 3). As the droplets condense and explode, charge is transferred from the solvent molecule to the analyte based on the capability of an analyte to carry charge in the condensed or solution phase. The considerations include size of the biomolecule, basicity (pK) of the analyte, and the pH of the solvent used in the ESI experiment. The particulars for optimizing the generation and transfer of ions generated by ESI into a mass spectrometer are exemplified in a study by Tempst and colleagues *(10)*. This translates into analyte ions with multiple charges (up to many 10s of charges), which may result from proton transfers ($[M+nH^+]^{n+}$ or $[M-n\ H^+]^{n-}$), cationization ($[M+nX^{m+}]^{n*m+}$ where X is Na, K, Ca, or NH_4^+ among others), or combinations thereof. This process creates a distribution of ions with z (charge) values across a narrow range (e.g., 12–20), which creates a so-called charge envelope (*see* Fig. 4) so that for any one analyte there are several pseudo-molecular ions produced, each with a different charge-state. Fig. 4A and 4B

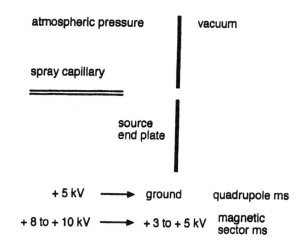

Fig. 2. Diagram of a typical (simple) electrospray ion source indicating some standard conditions *(25)*. Figure reproduced with permission of John Wiley and Sons, Inc.

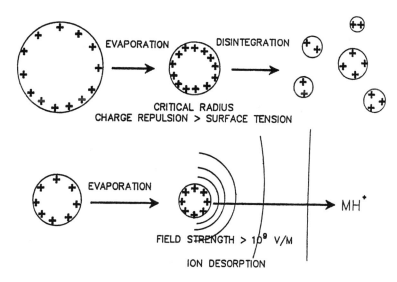

Fig. 3. Schematic description showing the processes of ion generation during electrospray ionization according to the Charged Residue Model (CRM; top) and Ion Desorption Model (IDM; bottom) for ion formation *(11)*. Figure reproduced with permission of the American Chemical Society.

depict the electrospray mass spectra obtained from solutions of single proteins that are presented in the simplified (deconvoluted) form in the figure insets. Fig. 5A provides a more extreme mass spectrum with the deconvoluted spectrum (Fig. 5B) showing not less than 13 different protein isoforms. In general, as a rough estimate, one might expect one charge per 500–1000 daltons of size.

The advantage of this process is that the resulting multiple charges create ions with mass-to-charge ratios (m/z), which are proportionately lower than the molecule itself

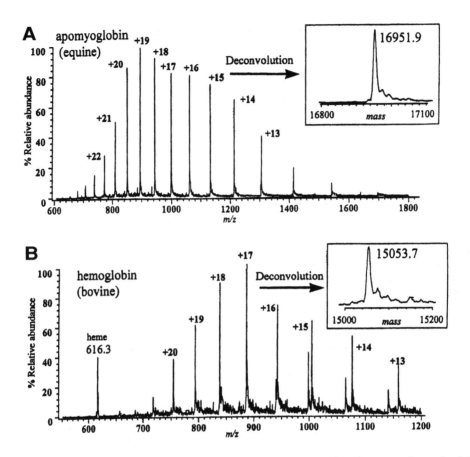

Fig. 4. The electrospray ionization mass spectra and deconvoluted spectra (insets) of (**A**) equine apomyoglobin and (**B**) bovine hemoglobin obtained using a Finnigan LCQ™ ion trap mass spectrometer *(23)*. The insets show the average mol wts obtained from application of the deconvolution algorithm. Figure reproduced with permission of John Wiley and Sons, Inc.

and make it possible to mass-analyze ions from proteins and biomolecules in excess of 100,000 Daltons using a mass spectrometer with an m/z range of 4000 or less. The disadvantage of ESI is that deciphering which analyte produced the detected ions is not trivial in mixtures (*see* Fig. 5A). This is overcome by a process in which the resulting mass spectrum and distribution of ions from a given analyte, referred to as a "charge envelope," is then treated with a mathematical algorithm that deconvolutes the spectrum to produce a single m/z or mol wt for each component in the mixture. For protonation, this process is represented mathematically as:

$$m/z(1) = [M(analyte) + n_1 * H^+]/ n_1 \text{ and } m/z(2) = [M(analyte) + n_2 * H^+]/ n_2$$

Where M(analyte) is the mol wt of the analyte, n_x is the number of protons or charges, and H^+ is the mol wt of a proton. If one now solves both equations for M(analyte) and assumes that neighboring ions differ by a single charge so that $n_2 = n_1 + 1$, one is left with the following relationship:

$$M(analyte) = n_1 * m/z(1) + n_1 * H^+ = n_2 * m/z(2) + n_2 * H^+ = (n_1 + 1) * m/z(2) + (n_1 + 1) * H^+, \text{ etc.}$$

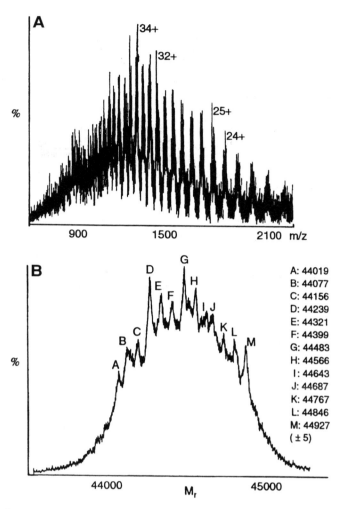

A: 44019
B: 44077
C: 44156
D: 44239
E: 44321
F: 44399
G: 44483
H: 44566
I: 44643
J: 44687
K: 44767
L: 44846
M: 44927
(± 5)

Fig. 5. The electrospray mass spectrum (**A**) and deconvoluted mass spectrum (**B**) of bovine ovalbumin are shown to demonstrate the need for deconvolution of electrospray mass spectra *(12)*. Figure reproduced with permission of John Wiley and Sons, Inc.

Here, m/z (1) and m/z (2) are determined experimentally from the mass spectrum and n_1 is determined algebraically, leading to the discovery of M(analyte). This process, known as deconvolution, is now routinely available as part of the software packages of most instrument vendors, but it is important for the analyst to be familiar with this calculation to confirm determinations, because deconvolutions can be highly dependent upon the parameters used. The practical results of deconvolution can be seen by comparing the raw mass spectrum in Fig. 5A and the deconvoluted mass spectrum in Fig. 5B, in which the MW values extrapolated by deconvolution are listed. Clearly, identification of any single species in the raw mass spectrum with any confidence would be difficult at best. The MWs are provided with an uncertainty that comes from the statistics applied to the multiple determinations mentioned previously in the deconvolution process. Additional rigor has been added to this process in what is referred

Table 1
Common Solvent Mixtures for ESI-MS of Biomolecules

0.1% trifluoroacetic acid in 5–95% acetonitrile, methanol, or propanol
10–100 mM ammonium hydroxide in 5–95% acetonitrile, etc. (negative ion analysis)
1–5% acteic acid in 5–95% acetonitrile, methanol, or propanol
0.1–1% formic acid in 5–95% acetonitrile, methanol, or propanol
5–200 mM ammonium acetate (or formate) in 5–95% acetonitrile, methanol, or propanol
5–50 mM ammonium acetate (or formate) ca. pH 6 containing 0.1–5 mM sodium, calcium or
   potassium acetate in 5–95% acetonitrile, methanol, or propanol
5–200 mM ammonium acetate (or formate) in 5-95% acetonitrile, methanol, or propanol

to as the maximum entropy approach to deconvolution *(13,14)*. With this addition, deconvoluted spectra are highly quantifiable.

A volatile matrix and controlled conditions are critical to the success of electrospray experiments. Typically the solvent entering the ionization source is between 25% and 90% (v/v) organic (usually methanol, propanol and acetonitrile) containing a relatively low concentration (<100mM) of a proton-donating component such as an organic acid. Although typically avoided, nonvolatile salt additives such as sodium, potassium, or calcium acetate can be added when protonation of the analyte is unfavorable. Some typical effluent compositions are shown in Table 1. Since the extent of ion formation in ESI largely reflects the solution phase processes, a range of pHs are utilized to accommodate both positive and negative ion formation.

Consistency of conditions for the experiment of interest is critically important to the analysis of biomolecules. For example, if the pH of the infused sample is changed across a reasonable range, the charge distribution changes appreciably. This is shown in Fig. 6A–E for cytochrome c. Although multiple factors are involved in the observed effect, the correlation between solution pH and ion distribution is clear. For proteins and peptides, the charge is largely associated with the basic N-terminus as well as arginine and lysine residues in positive-ion mode, and with the acidic C-terminus as well as glutamate and aspartate residues in negative-ion mode. For oligosaccharides, the charges are formed on the amines, amides, hydroxyls, and carboxylic acid functional groups. However, because of the relatively low difference in gas-phase basicity of oligosaccharides compared to the solvents of interest (e.g., water and methanol clusters), among other reasons, oligosaccharides provide lower ionization yields than proteins or peptides in either positive or negative ion modes. In any case, the total charges can exceed the total number of basic (acidic) residues because of other sources of charge formation. Low levels of inorganic salts are useful in some ESI experiments. However, if the amount of salts or other nonvolatile buffers is not limited and controlled, the distribution and relative signal will fluctuate throughout an experiment and Na, K, and H-dependent ionization will compete in an uncontrolled fashion. Too much salt (greater than a few mM) will cause mechanical problems (plugging of the tip), and can cause suppression of ionization.

2.1.2. Desorption Ionization (DI)

Desoprtion ionization is a family of techniques *(8)* in which the analyte or sample is applied to a surface as a mixture with a nonvolatile matrix followed by insertion of the

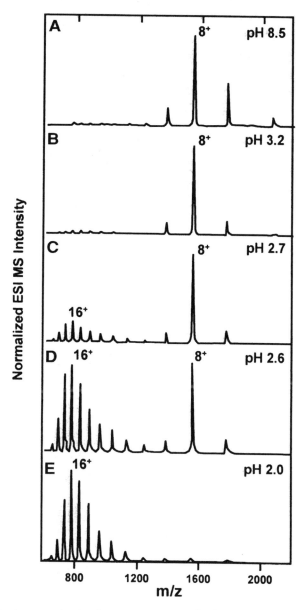

Fig. 6. The electrospray ionization mass spectra obtained for cytochrome c are shown as obtained from solutions of the protein at varying pH values *(63)*. The pHs of the protein solutions (obtained using ammonium hydroxide and acetic acid) are indicated in the individual figures. The figures show the sensitivity of the ESI process to solution conditions. Figure reproduced with permission of Wiley Interscience, Inc.

surface into a source with moderate vacuum (a few mTorr) and bombardment of the sample with an energetic beam. The beam is either a keV beam of atoms (FAB) or ions (SIMS), a laser (MALDI) or laser desorption/ionization (LDI), or a beam of subatomic particles (PD). To provide a signal of longer duration and stable intensity, the beam is delivered at a relatively low flux or in a pulsed mode (e.g., MALDI and PD). Imping-

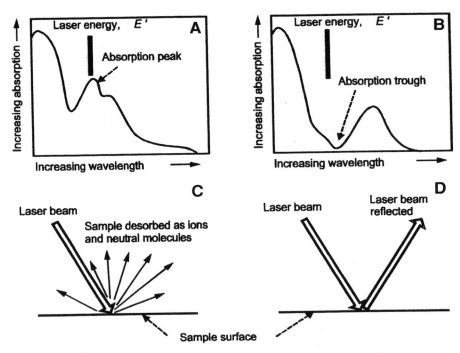

Fig. 7. Fundamental processes in the MALDI. The fundamentals of the MALDI process are depicted for **(A)** and **(C)** a matrix that absorbs strongly at the wavelength (energy E') and results in ion production. The processes are similarly depicted, **(B)** and **(D)**, for a matrix which absorbs poorly at the wavelength (energy E') and results in no ion production *(21)*. Figure reproduced with permission of Micromass, Ltd.

ing upon the sample spot, the beam acts in two ways: 1) it imparts energy into the matrix to create a chain of proton (ion) transfer reactions in a gas-phase region known as a selvegde *(8)*; 2) the beam provides the substantial energy required to launch the large, nonvolatile analyte into the gas-phase with a minimum of undesired thermal energy deposited into the analyte.

The matrix and energy beam are what defines the DI experiment, and MALDI is the most common of these for biomolecules. In MALDI, the sample is deposited into a matrix that absorbs light at a wavelength emitted by a moderate-powered laser. The ionization process is characterized in Fig. 7 *(15)* for MALDI, but is typical for any DI method. In the case of MALDI, the absorbance of the laser electromagnetic energy (compare Fig. 7A,B) by the matrix is critical to obtaining effective ion production (Fig. 7C,D). A nitrogen laser that emits at 337 nm is most commonly used on commercial MALDI instruments. The most commonly used matrices are provided in Table 2 along with the analytes for which they are typically used. It is important to note that new matrices are constantly being applied to these analyses. In addition, as new lasers become more readily available, the number of potential matrices will grow exponentially, creating the potential for "custom" matrices. The selection of the matrix that provides the best signal-to-noise ratio, is most compatible with the components of the sample, minimizes adduct formation, and provides the optimal signal

Table 2
Common Matrices for MALDI

Matrix	Application
2,5-dihydroxybenzoic acid	Peptides, lipids, oligosaccharides
3,5-dimethoxy-4-hydroxycinnamic acid (sinapinnic acid)	Larger peptides, proteins, glycoproteins
α-cyano-4-hydroxycinnaimic acid (CHCA)	Peptides, smaller proteins, oligonucleotides
Ferrulic acid	Larger proteins (>100 kD), glycoproteins, oligonucleotides
Trihydroxyacetophenone (THAP)	Oligosaccharides, glcyoproteins, other

and resolutions may require some experimentation. It is recommended that in preliminary experiments more than on matrix be investigated. The reasons for this are apparent in Fig. 8 *(16)*, which shows the analysis of a tryptic digest of recombinant human erythropoietin in dihydroxybenzoic acid (DHB) (Fig. 8A), 2-aminobenzoic acid (Fig. 8B), and a-cyano-4-hydroxycinnamic acid (Fig. 8C). The ions apparent, resolution, background, and overall quality of the spectra vary considerably in the various matrices. This is further reinforced in Fig. 9 *(17)*, where spectra for b-galactosidase are provided after preparation in different solutions and using different MALDI matrices. The complete absence of ions represents an obvious extreme in results, but one which the experimenter may encounter and should be aware of. This can be addressed by preparing and analyzing the sample using different solution systems and MALDI matrices in preliminary experiments.

In order to produce ions effectively, in addition to absorbing the radiation produced by the laser, the matrix *must* form crystals. It is the disruption of the crystals by the absorbed light that initiates the ionization process and induces proton-transfer reactions, which result in the protonation/deprotonation of the analyte molecules. The implications of matrix crystallization are shown in Fig. 10A–F, in which the different mechanisms of matrix crystalization are matched with the resulting MALDI spectra *(26)*. Factors that influence matrix crystallization include the purity of the chemical used as the matrix, the buffer or salt content of the sample to be analyzed, analyte concentration, and matrix and sample deposition method. Although crystallization is important, the process is relatively robust, and therefore MALDI—unlike ESI—is relatively robust to sample contaminants, including low levels of buffers, detergents, chaotropes, and other reagents commonly used in biochemical manipulations. For typical samples, the tolerable amounts of some common additives found in biological samples are shown in Table 3. The experimenter should consider that these amounts are typical, and the actual amount permissible in any experiment will depend upon the factors listed above, the propensity of the analyte(s) to form ions, and other factors, including the adsorption of peptide (and protein) analytes to the desorption surface *(18)*.

MALDI ionization produces predominantly singly charged pseudomolecular ions with lesser amounts of multiply charged analyte ions and analyte-cluster ions $M_xH_y^{y+}$ (where M is the analyte and x and y are the numbers of analyte and protons in the pseudomolecular

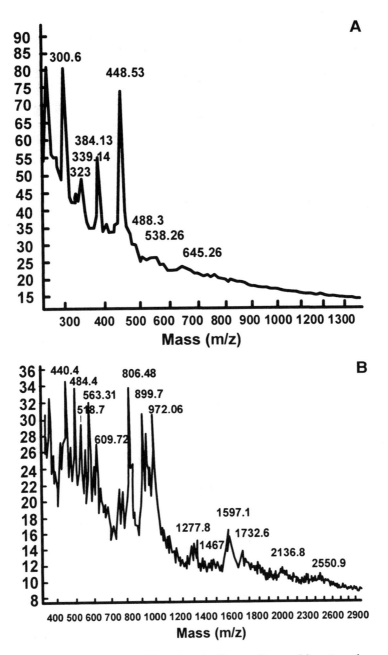

Fig. 8. MALDI-TOF mass spectra for a tryptic digest of recombinant erythropoietin using different matrices *(16)*. Spectra were obtained from: **(A)** DHB (dihydroxybenzoic acid); **(B)** 2-aminobenzoic acid; **(C)** α-cyano-4-hydroxycinnamic acid matrices. Spectra indicate changes in signal-to-noise and ionization selectivity possible with different matrices. Figure reproduced with permission of WJG Press.

ion complex, respectively). Conditions that influence the relative amounts of the multiply charged species and the extent of cluster ion formation as well as the quality (signal-to-noise) of a spectrum include analyte concentration, laser flux, MALDI matrix, sample

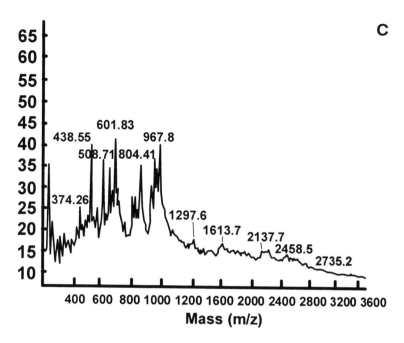

Fig. 8. (continued)

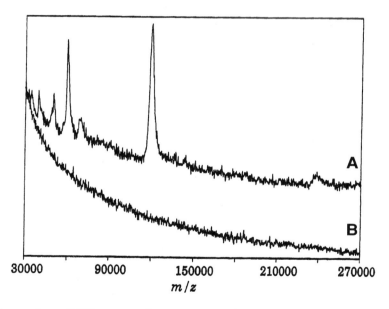

Fig. 9. Comparison of different matrix systems with β-galactosidase as analyte. (**A**) Ferrulic acid with 2-propanol and water. (**B**) Sinapinic acid with acetonitrile *(17)*. Figure reproduced with permission of Humana Press.

matrix, physical properties of the molecule of interest (hydrophobicity or pK), mol wt of the target molecule, and its state in solution (monomer or dimer).

The advantage of producing singly charged ions (facile mol-wt determination) is offset by the need for a mass analyzer with the ability to separate and detect ions hav-

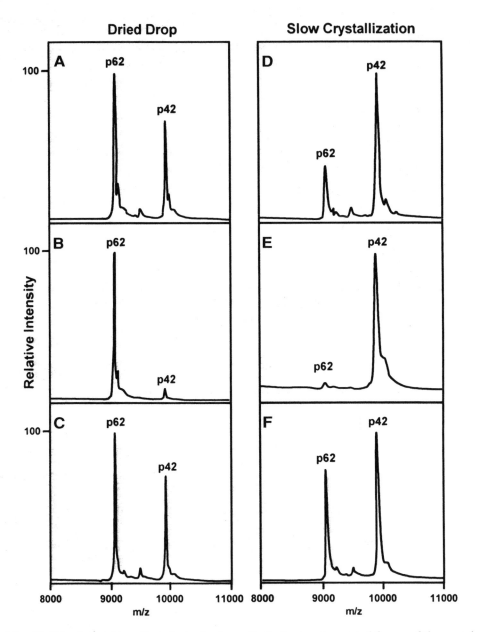

Fig. 10. Discrimination effects in ionization effects for two truncated forms of the proteins p42 and p62 using different matrix/solvent combinations. Spectra were obtained using solutions of 4HCCA matrix dissolved in: (a,d) formic acid/water/2-propanol (1:3:2 v/v/v); (b,e) 0.1% trifluoroacetic acid.acetonitrile (2/1 v/v); (c,f) 0.1% trifluoroacetic acid/acetonitrile (2/1 v/v) with 8 m*M* N-octylglucoside *(26)*. Figure reproduced with permission of the American Chemical Society.

ing very high m/z ratios (e.g., >100,000). Furthermore, since MALDI is a pulsed ionization technique, mass analyzers amenable to pulsing are most often used. This is also in contrast to ESI, which is amenable to any instrument with minor vacuum or ion optical modifications.

Table 3
Acceptable Additives for MALDI Analysis

Additive	Acceptable concentration
Urea	1 M
NaCl (and similar monovalent salts)	50 mM
Guanidine (HCl)	1 M
Sodium dodecylsulfate	0.01% (w/v)
Phosphate buffers	10 mM
TRIS and similar amine-based buffers	50 mM
Glycerol	1% (w/v)
n-Octyl-β-glucopyranoside	1% (w/v)

Note that concentrations below this are preferred to provide optimal signal and mass resolution.

2.2. Fundamental Aspects of Modern Instrumentation

The fundamental design of mass spectrometers has not changed appreciably during the past few decades. Details of each instrument have been modified to optimize sensitivity, resolution, and reliability. Most significant, however, is the user-friendliness of the operation, which has largely been the result of computer and software advances. Each instrument has unique physical principles of operation, and thus, advantages and disadvantages for selected applications and fundamental limitations. The instruments most commonly used in modern biochemical mass spectrometry include the quadrupole mass filters (including the triple quadrupole), the time-of-flight mass analyzer (linear and reflectron), and the quadrupole-ion trap. Instruments such the tandem TOF (TOF-TOF) *(19)* and the hybrid quadrupole time-of-flight (Q[q]TOF) *(20)* are now more frequently used, but their operation and performance are extensions of their simpler predecessors. Another instrument that is not used at a high rate among analytical biochemists, but merits mentioning for its unique capabilities is the Fourier Transform Ion Cyclotron Resonance (FT-ICR) mass spectrometer (i.e., FTMS) *(21)*. Each of the instruments mentioned is listed in Table 4, in which the capabilities of the systems are provided. The principles behind the operation of the quadrupole-ion traps and mass filters, and TOF mass analyzers, the most commonly utilized mass analyzers for biomolecules, are discussed briefly to provide a fundamental level of understanding of each system.

2.2.1. Time of Flight

Time-of-flight (TOF) mass analysis is founded on the principle that for a given applied electric potential the kinetic energy of any ion, regardless of mass, will be constant, and in the absence of other accelerating forces the velocity will be a function of the m/z of the ions of interest. As a result, the time required for an ion to travel a fixed distance in the absence of electric potentials and under relatively high vacuum (i.e., approaching zero collisions) is defined by the following equation:

$$T = L \cdot (m/(2KE)z)^{1/2}$$

Where T is time, L is the distance the ions travel from the point of acceleration to detection, m is the mass of the ion, z is the charge on the ion, and KE is the kinetic

Table 4
Considerations of Mass Analyzers

Instrument	Mass range	Mass resolution	MS/MS	Ionization	Misc.
Quadrupole	2000–4000	Unit resolution	No	ESI/APCI; poor match to MALDI	Rugged and cost-effective; simple to use $75–150,000
Triple quadrupole	2000–4000	Unit resolution	CID; neutral loss/precursor/product ion scans	ESI/APCI; poor match to MALDI	Rugged and cost effective; simple to use $175–300,000
Ion trap	2000–4000 mass range extension available	Unit resolution …may be higher in special cases ("zoom scan")	Product ion scans; CID; amenable to laser-induced dissociation; MS^n available	ESI/APCI; compatible with MALDI but limited by mass range	Rugged and cost effective; simple to use; v. small footprint $90–250,000
Time-of-flight	Unlimited mass range (in practice around 500,000)	v. good for reflectron (>5,000 if m/z < 10,000) poor if linear	No true linked scan capabilities Post-source-decay rTOF	Compatibility depends upon design OaTOF is compatible with ESI or MALDI Linear - MALDI	Simplest operation; High resolution/accurate mass available $75–300,000

Table 4 (continued)
Considerations of Mass Analyzers

Instrument	Mass range	Mass resolution	MS/MS	Ionization	Misc.
Q-TOF	2000–4000 (limited by quad)	Unit mass resolution in quadrupole; v. good resolution and mass accuracy with rTOF	CID; Neutral loss/ precursor/product ion scans	Compatible with ESI/APCI and MALDI	Flexibility of QqQ with the high resolution/accurate mass; Q limits utilization of rTOF $250–500,000
FT-ICR	Unlimited mass range	v. good mass accuracy and resolution (>100,000 capable with << 10 ppm mass error)	Product ion scans CID; also amenable to laser-induced dissociation and electron-capture dissociation; MS^n	Ultra-high vacuum required; special considerations in interfacing API methods	Large footprint and very costly; magnet upkeep >$500,000
TOF-TOF	See TOF	High mass accuracy and resolution in both stages of MS/MS Combine with PSD for effectively MS^3	Low- and high-energy CID available Linked scanning available through software	ESI or MALDI interfaces available	Best operated in high vacuum; off-angle Introduction required for API methods est. $600,000

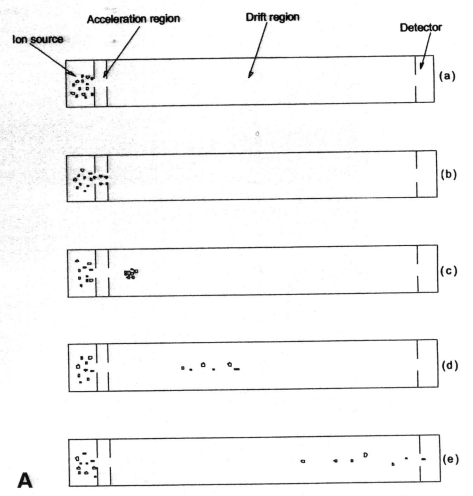

Fig. 11. Depiction of the separation of ions in a TOF mass spectrometer. Figure A depicts the separation of ions in a linear TOF. The ions (same mass) separate through the flight tube (drift region) because of different initial accelerating forces. This results in a dispersed ion bundle and poor mass spectral resolution. Fig. 11B depicts the separation of ions in a reflectron TOF (rTOF). The ions are separated in the linear flight tube (drift region) as described above. In the reflectron the dispersed kinetic energy distribution is corrected in the electrostatic mirror (reflectron). The result is a discrete ion bundle traveling in the second drift region after the reflectron and enhanced mass spectral resolution *(21)*. Figure reproduced with permission of Micromass, Ltd.

energy imposed by electrical potentials to accelerate the ions. For a fixed L, the flight time (T) is proportional to $(m/z)^{1/2}$. The implications of this relationship are that calibration of time and mass is not linear, and the accurate calibration comes from the application of a multipoint calibration that covers the range of interest. Accurate mass analysis requires the use of calibrants of m/z known with an accuracy of 0.001 m/z unit or less. Generic diagrams of TOF mass analyzers are shown in Fig. 11A,B.

The TOF analyzer can be operated in either of two modes. These are known as linear mode as shown in Fig. 11A (low resolution; higher mass range; high sensitivity) and

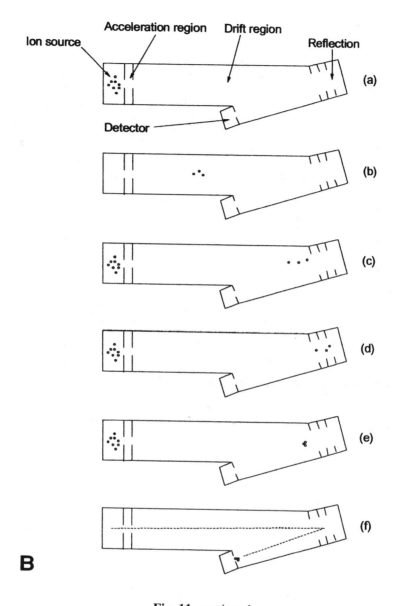

Fig. 11. continued.

reflectron mode as shown in Fig. 11B (high resolution; lower mass range; reduced sensitivity). In the linear mode, the flight path of the analytes is shorter, and terminates at the end of the flight tube at a microchannel plate detector. Analytes move from the source to the detector in a linear path. In this mode, represented in Fig. 11A, resolution is compromised from optimal by peak broadening because of positional and kinetic-energy differences caused in the ionization source across the ion packet. For example, if two ions start from two points in the ion source of different distances from the detector when the pulse is applied to accelerate the ions down the flight tube, they will arrive at the detector at different times, yet they will still be the same m/z. This problem has

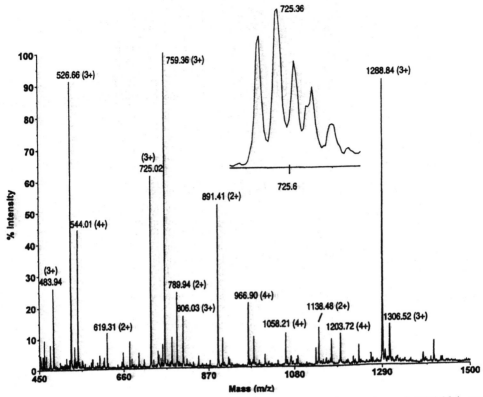

Fig. 12. Raw mass spectrum from a peptide digest. The ion observed at m/z 725.02 is suggested to be triply charged. The expanded region in the inset shows the ion of interest is in fact triply charged based on the spacing of the isotopes in the spectrum (ca. 0.3 daltons) *(38)*. Figure reproduced with permission of John Wiley and Sons, Inc.

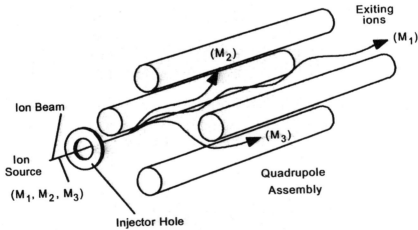

Fig. 13. Depiction of a quadrupole-mass filter and the path of ions through a quadrupole mass filter. The three ions of different mass-to-charge ratios (M_1, M_2, M_3) travel between the four rods. M_2 and M_3 do not have stable trajectories under the scan conditions, while M_1 is stable under these conditions and passes completely through the quadrupoles. The RF and DC potentials applied to the quadrupoles defines which m/z will be stable and pass through the quadrupoles *(21)*. Figure reproduced with permission of Micromass, Ltd.

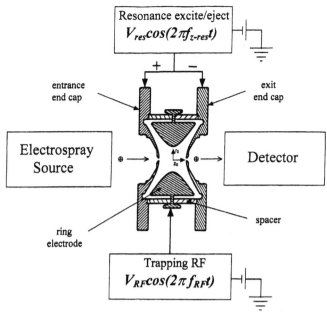

Fig. 14. A three-dimensional quadrupole-ion trap consisting of two hyperbolic end caps and a central hyperbolic electrode positioned between an ESI source and a detector. Equations to define trapping (ring electrode) and excitation/ejection potentials (end cap electrodes) are shown *(23)*. Figure reproduced with permission of John Wiley and Sons, Inc.

been thoroughly examined by delayed extraction *(22)*, in which ions are collisionally cooled and compressed in space before acceleration down the flight tube. The mass range in the linear mode is also theoretically infinite, and ions exceeding 1,000,000 Daltons have been successfully analyzed.

In the reflectron mode, represented in Fig. 11B, the ions are reflected using an electrostatic mirror or reflectron. This enhances performance in two ways—the effective flight length (L in equation) is roughly doubled; and the mirror acts to focus ion packets with the kinetic energy distribution of ions of a given m/z entering the mirror being broader than the energy distribution of the same ions leaving the mirror (depicted in Fig. 11B). The result is that the width of the peak for a given m/z is reduced and the ability to distinguish ions of similar m/z (resolution) is improved. The compromise of the improved resolution in the reflectron mode is a reduction in the available mass range because of difficulties in turning ions of high kinetic energy (m/z) in a reasonably sized reflectron. The high resolution obtained in the reflectron TOF analyzers (rTOF) along with the high temporal resolution of modern digitizers (better than nanosecond resolution) permit rTOF mass analyzers to be utilized for accurate mass experiments. The use of calibrants that closely bracket the analyte of interest and are of accurately known m/z, can regularly determine the m/z of an unknown to within <10 ppm accuracy (0.001 m/z for an ion at 1000 m/z). As demonstrated later in this chapter, this capability is highly beneficial in many experiments for the characterization of biomolecules. The most trivial of these, but of great significance in the analysis of peptide maps, is the ability to determine the charge state of an ion by simply observing

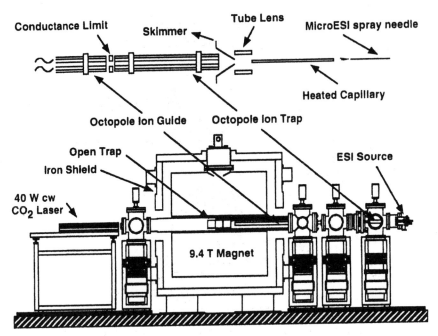

Fig. 15. Schematic diagram of the electrospray 9.4 Tesla FT-ICR mass spectrometer at the National Magnetic Field Laboratory including a laser for photodissociation. This represents all of the critical components of an FT-ICR (FTMS) instrument to be applied to biomolecule analysis by mass spectrometry. The details of the electrospray ionization source and inlet are shown in the top portion of the figure *(28)*. Figure reproduced with permission of Humana Press.

the isotope distribution. This is demonstrated in Fig. 12, where the isotope distribution of a typical peptide is shown. The spacing of the isotopes at approx 0.3 m/z units, along with a reasonable isotope envelope, indicates that this ion is most likely carrying three charges ($z = 3$).

2.2.2. Quadrupole Mass Filters and Ion Traps

Mass analyzers in the quadrupole family operate through the combined use of DC (direct current) and RF (radio-frequency) electric fields to guide ions in two-dimensional (mass filter) *(21)* (Fig. 13) or trap ions in three-dimensional (ion trap) *(23)* (Fig. 14) space. In either case the RF (U) and DC (V) potentials are scanned as a function of time, resulting a time-correlated mass separation. The scan coordination of the DC and RF voltages is coordinated and described by the Mathieu equation. The equation and its details are discussed elsewhere (*see* refs. *23,24*). The magnitude for the two potentials and their relationship to one another are also dependent upon the frequency of the RF potential, as described by the equation.

Mass filters are four cylindrical or hyperbolic rods that are spaced uniformly with respect to one another with RF and DC potentials applied to all four rods but phases synchronized on opposing rods (Fig. 13). In the case of the mass filter, ions pass through the quadrupole rods at only the appropriate (stable) combination of RF and DC potentials. Ions that are not solutions to the Mathieu equation under the conditions of the scan at a particular time collide with the rods and are not detected. This mode of operation is referred to as mass-selected stability.

The ion trap (Fig. 14) is composed of two end caps and a single ring electrode. Ions are typically generated external to the trap, injected into the trap at low energy, and trapped for mass analysis. The trap contains He gas at 1 mTorr pressure to facilitate the trapping of the ions through low-energy collisions. The ring electrode typically has DC potential applied to it, and the end caps have an RF potential applied to them. Similar to the mass filter, the ion trap permits a slightly different mode of mass analysis, in which ions are retained in the trap and then selectively destabilized (mass-selected instability) to permit mass analysis. The ion trap can also be operated in the mass-selected stability mode, but is more efficiently operated in the mass-selected instability mode in which all ions (or a desired range) are trapped and then selectively destabilized, or ejected from the trap and detected. In both cases the combination of DC and RF conditions are applied to provide a stable path for an ion of specific m/z. In addition to selective ejection, ions can be selectively retained in the ion trap *(23,24)*. This permits further analysis of the ions and, usually, collision-induced dissociation of the trapped ions by the application of a supplemental RF potential. The resulting fragment ions can then be retained in the ion trap and mass-selectively ejected *(23,24)*. This ability to selectively retain and eject ions makes ion traps an extremely powerful mass spectrometer, which is capable of performing what is referred to as MS^n *(23,24)*.

The quadrupole family represents the most common group of mass analyzers because of their flexibility, ruggedness, size, and their relatively early commercial availability. The mass filters provide at least unit resolution across the entire mass range (typically 2000–4000 m/z).

2.2.3. Fourier Transform Ion Cyclotron Resonance

Fourier Transform Ion Cyclotron Resonance (FT-ICR, e.g., FTMS) is the highest-resolution means of mass analysis, with resolutions in excess of 100,000 fwhm observed regularly. It is also the highest-sensitivity means of mass analysis, and zeptomole amounts of peptides have been mass-analyzed *(27)*. It is also the most space consuming and costly of the methods. A typical FT-ICR unit is depicted in Fig. 15. The most important component of an FT-ICR is the magnet that permits the high resolution according to Eq. 4. The magnet is the reason for both the costly nature of the instrument ($300,000 up to in excess of $1,500,000) as well as the space requirements (isolation of the magnetic filed for optimum performance; liquid nitrogen tanks for magnet cooling). FT-ICR is an ion-trapping technique in which ionization for biomolecules occurs external to the mass-analysis cell, and the ions formed are introduced into the mass analyzer as a guided ion beam of low kinetic energy.

The mass analysis itself is based on the cyclic motion—ion cyclotron motion—that an ion follows in the gas phase in the presence of an electromagnetic field. This motion is caused by the application of electrostatic pulse that sends the ions out into the cell towards the electrodes. The current produced by the circulating ions is recorded as a function of time after the pulse is applied, and a Fourier Transform is applied to provide the transition from frequency to mass. The cyclic motion then restores the ion to the center of the cell with a decay in signal that is characteristic of the m/z of each ion. The motion creates a current, which is detected through the cell, and the magnitude of the current is proportional to the abundance of the ion of interest. The signal in the frequency (of cyclotron decay) domain is treated by Fourier Transform to create a time and m/z signal. The relationship between cyclotron frequency and m/z is shown in the

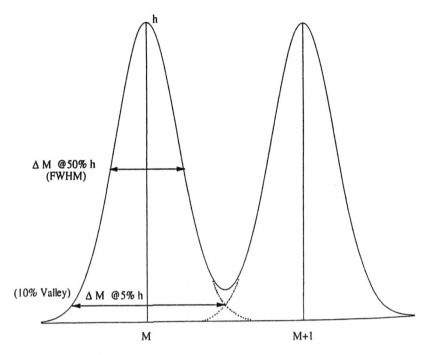

Fig. 16. Diagram of the measurements used to determine mass spectral resolution. Most commonly used is M @ 50% h (FWHM = full-width, half maximum). Note that the absolute measurement is dependent only upon the properties of the ion of interest *(34)*. Figure reproduced with permission of Humana Press.

following equation, where ω is the cyclotron frequency, v is the velocity, r is the radius, z is the charge on the ion, m is the mass, and B is the magnet-field strength, therefore the frequency is inversely proportional to m/z.

$$\omega = (v/r) = (z/m)*B$$

Each orbit that an ion makes provides an opportunity to measure its frequency, and thus m/z. Unlike the quadrupole-ion trap, the ion does not need to be ejected to be detected (nondestructive), which permits nearly infinite measurements on its cyclotron frequency. This permits extensive statistical and mathematical treatment of the signal, raw or transformed, which is the source of the extremely high resolution and mass accuracy observed in FT-ICR analyses. Central to the FT-ICR experiment is the use of high-powered computing systems to both control the many experiments (such as pulsing or scanning) which occur on sub-microsecond time-scales, and to permit the acquisition of the data generated in this way. Similarly to the quadrupole-ion trap, ions can be selectively retained by FT-ICRs for further manipulation and analysis, including dissociation via collisional activation, photodissociation, and electron-capture dissociation *(29,30)*.

2.3. Basic Definitions and Terminology in Mass Spectrometry

For clarity and consistency, a brief review of some of the key nomenclature of mass spectrometry is provided, which includes a readily accessible summary of some of the most common terms for the reader who may not be well-versed in mass spectrometry. For more rigorous treatment and discussion of this background information, the interested reader is referred to texts edited by Caprioli *(31)*, Roboz *(32)*, McLafferty *(33)* as well as other works directed toward the application of mass spectrometry to biomolecules *(34–35)*.

The key definitions common in biomolecular mass spectrometry and used throughout this chapter are provided in this section in bullet format. These, supplemented by the descriptions of the modes of operation of the various types of mass analyzers (**Subheading 2.2.**) should provide the reader with all the information needed to understand the experiments in this chapter and determine whether further investigation is warranted.

2.3.1. Mass-to-Charge Ratio (m/z)

This is the mass (Daltons) of the molecule or complex that has been ionized divided by the total number of charges on that species. This measure is particularly important for molecules generated by soft ionization methods, which may produce ions having more than one charge per molecule expressed as a.m.u. per Coulomb; although this is different than the mol wt, with the correct information it can lead to the determination of the mol wt.

2.3.2. Resolution (R)

This defines the capability of a mass analyzer to distinguish ions of varying m/z. Unlike R in separation sciences, this is measured on a single analyte (ion of a given m/z), and is usually referred to as the resolution at full-width at half-height. A typical measurement is shown in Fig. 16, and is described by the equation: R (full-width at half-height) = $M/w_{1/2}$, where M is the m/z of the ion of interest and $w_{1/2}$ is the width of the ion at one-half of its amplitude (both in the same units). For example, an ion of m/z = 1500 with a $w_{1/2}$ of 0.2 would have a resolution of 7500. (This resolving power would be achieved using rTOF or FT-ICR mass analyzers.) The term "unit resolution" is often applied, and represents an ion with a width of 1 Dalton (m/z). The resolution observed is dependent upon the mass analyzer used, and ranges from approx <500 for linear TOF mass spectrometers at higher m/z to more than 10,000 for FT-ICR instruments.

2.3.3. Mass Accuracy

This term defines the relative error in the target mass, and can be expressed as an uncertainty (e.g., similar to a %rsd). It is typically expressed in parts per million (ppm). Although mass accuracy is different from resolution, resolution can impact the mass accuracy because a system with poor resolution may not be capable of resolving closely related ions (e.g., isotopes), and this results in an apparent mass error. This is estimated using peptide or protein calibrants, typically other than those used in the mass calibration. For example, a peptide has a target mol wt ($M+H^+$) of 1257.2569, and in an MALDI-rTOF experiment the measured mass is 1257.2795. The difference is 0.0226 m/z units, which when divided by the target mass provides a relative difference of $(0.02226/1257.2569)*10^6 = 17.97$ ppm. This calculation is usually applied to data from rTOF, FT-ICR and sector experiments.

2.3.4. Pseudo-Molecular Ion

This term refers to the adduct between an analyte molecule and a charged species such as a proton (H^+), sodium ion or similar or in the negative-ion mode $M-H^+$ or

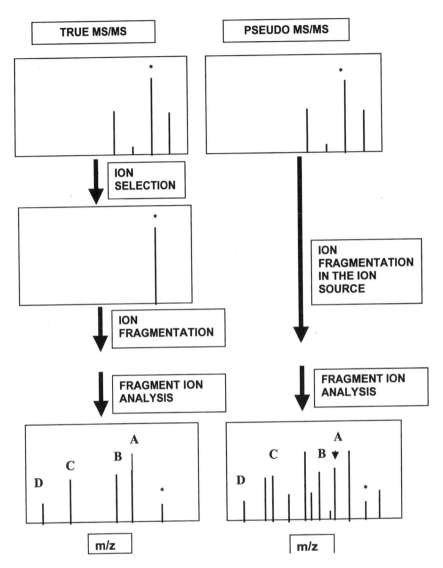

Fig. 17. Comparison of true and pseudo MS/MS (source CID or PSD) experiments. The ion of interest (*) in the mass spectrum (top) yields fragment ions (bottom). In the true MS/MS experiment (left) the ion (*) is selected and fragmented to produce a "clean" product ion spectrum (ions **A–D**). In the pseudo MS/MS experiment, the ion (*) is not selected and product ions are observed for both the ion of interest (ions **A–D**) and other ions (unlabeled).

M+Cl⁻. The term is derived from chemical ionization, and is applied to biomolecules through the various techniques known as API.

3. Tandem Mass Spectrometric Analysis of Peptides, Proteins, and Oligosaccharides

3.1. The MS/MS Experiment and Instrumentation

The analysis of the structural elements of biopolymers is achieved through a series of related experiments in which the biopolymer is ionized and energy is deposited to induce the fragmentation of the ions in a fashion that reflects their smaller monomer components. This is known as a tandem mass spectrometric, or MS/MS, experiment, and can take many forms using a variety of instruments. The principal means of achieving this are through the true MS/MS experiment in which an ion or ions of interest are mass-selected in a first mass analysis, and are then activated, typically by collision with an inert gas (nitrogen, argon, or helium). The collisions provide internal energy to the ions, which is released through several mechanisms including bond dissociation. The resulting fragment ions are mass-analyzed in a separate mass-analysis experiment. Alternatively, a pseudo-MS/MS experiment can be utilized in which a mixture of ions are activated and fragmented during ionization and the fragments mass-analyzed along with residual unfragmented ions in a single-dimensional mass analyzer. The true MS/ MS experiment is most familiar, and has been used traditionally in the analysis of small organic and biomolecular analytes. However, the pseudo MS/MS experiments are gaining in popularity. In **Subheadings 3.2.** and **3.3.**, the elements of true and pseudo-tandem MS experiments are discussed on a theoretical basis, and applications of the experiments to the analysis and characterization of proteins, peptides, and glycoforms are provided.

3.2. Tandem Mass Spectrometry Using Collisional Activation

Traditionally, the MS/MS experiment has been performed using collisional activation or collision-induced dissociation (CAD; CID) on an array of instruments including sectors, triple (and higher-order) quadrupoles, and ion-trapping instruments (quadrupole ion trap and FT-ICR). In this experiment, the ion(s) of interest are mass-selected from the other ions in the sample and are energetically collided with an inert gas such as He, Ar, or nitrogen to introduce energy into the ion and induce fragmentation through covalent-bond dissociation. This experiment is depicted in a general manner in Fig. 17, where it is compared to the pseudo MS/MS experiment.

In collisional activation, the mass-selected ion is accelerated at a neutral gas with from a few to several thousand eV of kinetic energy. The position in this range of energies depends upon the mass analyzer being used, and is approximately grouped as low-energy CID at <100 eV and high-energy CID at >1000 eV. These inelastic collisions occur in a pressurized cell in which ion focusing but no mass selection occurs (during activation). The nature of the gas target and its pressure (as well as collision-cell geometry) significantly influences the resultant ion fragmentation. After activation, the fragment ions are extracted from the collision cell, and are mass analyzed to produce the MS/MS spectrum. Other modes of activation, which are also utilized to lesser extents on select instruments, offer their own unique advantages: electron capture in FT-ICRs *(29,30)*, and photodissociation primarily on trapping instruments *(30)*, among others *(24,36)*. Depending upon the design of the instrument, more than two passes of mass analysis may be performed or special experiments in which only selected ions are monitored and associated with ions in the original sample (parent or neutral-loss scans).

These individual experiments, which are outlined here, offer their own unique advantages including selectivity and fragmentation efficiency. Recently, new instruments have been added to the repertoire of tandem-mass spectrometers and include the Q-TOF *(20,37)*, and TOF-TOF *(19)*. These instruments provide unique advantages including high resolution, high-throughput MS/MS, and flexibility with respect to ionization. In all cases, the activation achieves nominally the same results—fragmentation of the target ion into smaller ions (product ions), which can be used to elucidate the structure of the original ion. Each instrument offers different advantages and different opportunities for the MS/MS experiment.

3.3. Pseudo MS/MS: Source CID and Post-Source Decay with One-Dimensional Mass Analysis

The analysis of peptides, proteins, and glycoforms by MS/MS has been outlined above. The simplest but least selective approach to this analysis is in-source fragmentation or source CID *(38)*. In this experiment, increasing the skimmer and/or tip potentials accelerates ions through the relatively high pressure of the source region, and thereby elevates the internal energy of the ions and induces fragmentation while the analyte ions are in transit to the mass analyzer *(see* Fig. 2). This experiment is typically done with ESI or related ionization modes, and a single quadrupole-mass analyzer or orthogonal acceleration time-of-flight (oaTOF) mass analyzer, although it is also possible to introduce additional internal energy using MALDI or other ionization methods. (The MALDI experiment, which has special considerations, is discussed in this subheading.) The pitfall to this analysis is that for a direct infusion analysis or in the case of unresolved or poorly resolved analytes in HPLC, the precursor ion of any of the fragment ion is not readily identifiable without some effort. If the analyte is relatively pure or the separation provides adequate resolution, this experiment can yield considerable information about the structure (e.g., sequence or composition) of the biopolymer.

In the extremes of application, this experiment is used to screen for functional groups in the analyte pool of interest, with a focus on relatively labile adducts by monitoring selected fragment ions in the mass spectrum. It has proven particularly useful in identifying post-translational modifications such as glycosylation and phosphorylation in peptide-mapping experiments. For example, Thibault and colleagues have used the CZE-nanoelectrospray-MS experiment with source CID to characterize protein glycoforms *(37)*. In this application, CZE under acidic conditions is used to separate proteolytic fragments of several glycoproteins containing N- and O-linked glycosylation sites. Using elevated orifice potentials, selected fragment ions generated in the source are monitored as an indicator of peptides containing possible glycosylation sites. The results obtained for this experiment on a lectin from *L. tetragonolobus* are shown in Fig. 18A,D. In this experiment, the total ion response between 500 and 1700 m/z is compared with the fragment ions at m/z 163, 204, and 366 which are indicative of hexoses (163), N-acetyl-hexosamine (204) and N-acetyl-lactosamine (366). If the analytes eluting in the CZE effluent are ionized under different cone voltages, information can be provided on the mol wts of the analytes as well as some degree of fragmentation. Spectra obtained from a single CZE peak obtained in this manner are shown in Fig. 19A and 19B. These were obtained from two separate CZE analyses at 50V and 120V orifice potentials. The substantial fragmentation provides information on the glycans present (in this case, core mannose). As mentioned previously, this is most readily applied to relatively labile compounds such as carbohydrates, which have much lower bond energies that link them to proteins compared to the peptide bonds themselves. This difference in bond energies actually permits some level of selectivity in the source

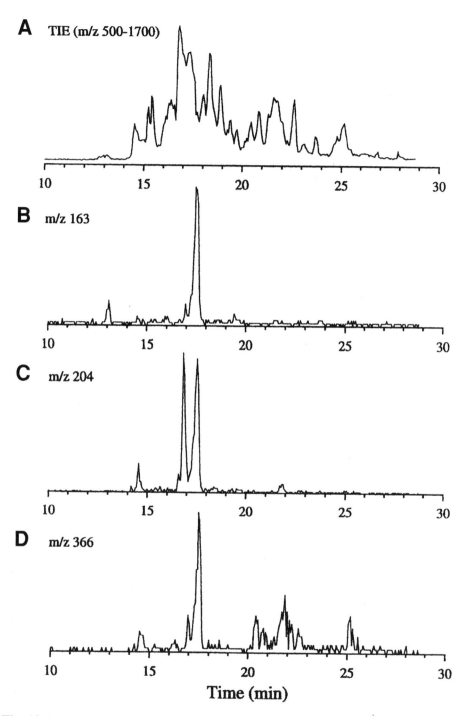

Fig. 18. Nanoelectrospray CZE-MS analysis of a tryptic digest of L. tetragonolobus showing: (**A**) Total Ion Electropherogram; (**B**) selected ion electropherogram for m/z 163 (hexose/mannose); (**C**) selected ion electropherogram for m/z 207 (N-acetylgalactosamine); (**D**) selected ion electropherogram for m/z 366 (N-acteyllactosamine) *(39)*. Figure reproduced with permission of Elsevier Science.

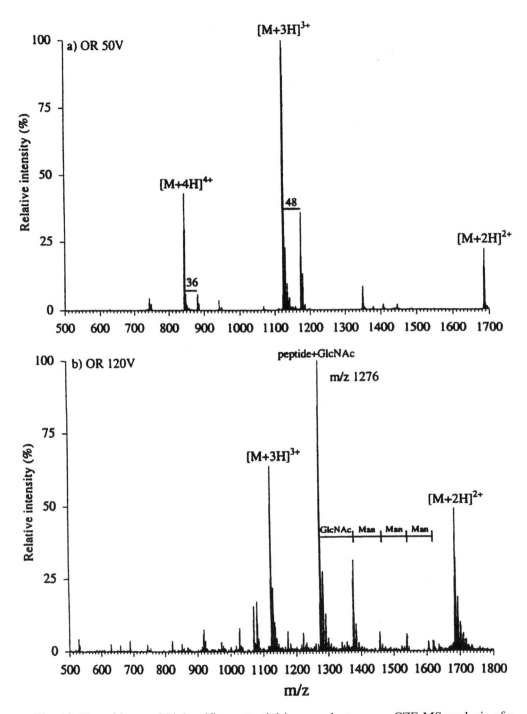

Fig. 19. Use of low and high orifice potential in nanoelectrospray CZE-MS analysis of a tryptic digest of L. tetragonolobus showing a) mass spectrum of the peak migrating at 17 min (Fig. 18) obtained at low (50 V) orifice voltage; b) mass spectrum of the peak migrating at 17 min (Fig. 18) obtained at high (120 V) orifice voltage. GlcNAc, N-acetylglucosamine; Man, mannose *(39)*. Figure reproduced with permission of Elsevier Science.

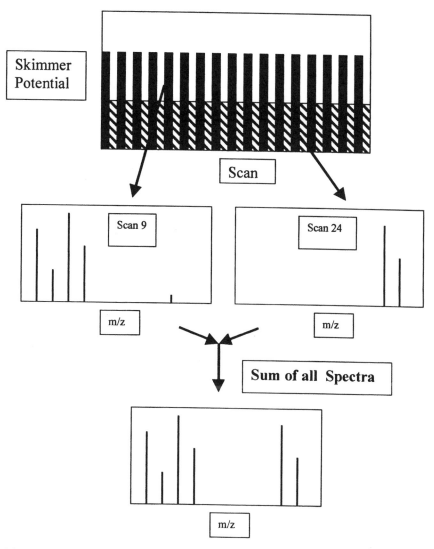

Fig. 20. Depiction of the source CID experiment in the "toggle" mode. Dark and cross-hatched bars (top panel) represent high and low orifice potentials. Typical scans are shown under the high (right) and low (left) orifice potentials. The spectra obtained for all scans at both high and low potentials are summed (bottom panel) to provide both mW and fragment ion information.

CID experiment, because fragmentation of the peptide bond is more difficult to achieve, and the fragmentation will follow the most thermodynamically favored pathway (in this case, the glycosidic bond). Compared to collisional activation accompanied by true tandem-mass analysis, this experiment is cost-effective yet limited in the information it can provide.

Although performed in two separate experiments above (Figs. 18 and 19), many instruments offer the opportunity to change the orifice voltage within a single analysis

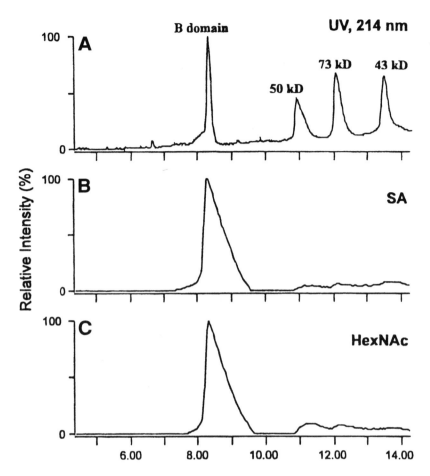

Fig. 21. Chromatograms from HPLC-UV-MS analysis of the B domain of recombinant human Factor VIII glycoprotein. (**A**) UV chromatogram; (**B**) selected ion chromatogram of fragment ions characteristic of sialic acid (m/z = 274, 292); (**C**) selected ino chromatogram of fragment ions characteristic of N-acetylhexosamine (m/z = 168, 186, 204, 366) *(40)*. Figure reproduced with permission of the American Chemical Society.

by rapidly changing between high and low voltages based on time, the number of scans/spectra acquired, or based on data-dependent feedback. In this case, the skimmer potential is alternated between high and low voltages in an automated fashion, in a so-called toggle experiment. This experiment permits information on both the intact species (low potential) and fragments (high potential) to be obtained on the pseudomolecular ions present in the ion source (*see* Fig. 18). (The limitation in energy deposition is the result of several factors including source ion optics, compromise in sensitivity, and typical source designs.) The rate at which the potentials can be varied depends upon the instrument being used. In some cases, relatively intelligent experiments can be performed. For example, in an LC/MS source CID toggle experiment, specific fragment ions can be monitored as the source is toggled. When the ion of interest is detected above a threshold level, the potential can be held at the high level until the species

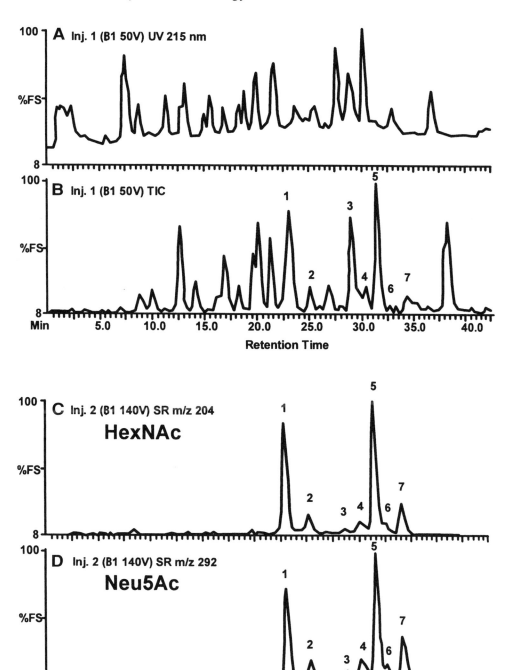

Fig. 22. LC/ESI/MS and SIM (selected ion monitoring) analysis of bovine fetuin tryptic digest. Panel (**A**) absorbance at 215 nm; (**B**) TIC chromatogram; (**C–D**) selected ion chromatograms of carbohydrate ions at m/z 204 and 292, respectively *(41)*. Figure reproduced with permission of the American Chemical Society.

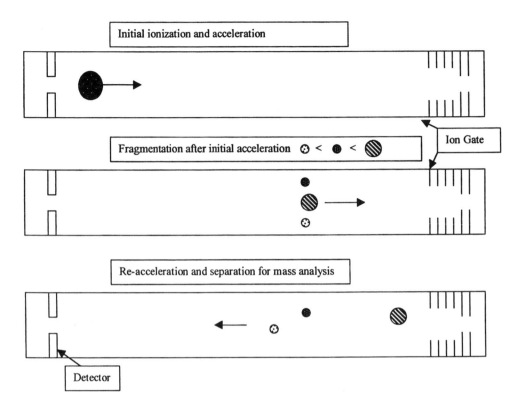

Fig. 23. Depiction of the post-source-decay (PSD) experiment. Ion in the source is accelerated with excess internal energy (top panel). The ion fragments with all fragments moving In
unison and maintaining same velocity as parent ion (middle panel). The fragment ions are then
accelerated through the reflectron with each fragment ion, acquiring a velocity indicative of its
m/z and thus being resolved in time in the final drift region (bottom panel).

generating the fragment of interest has completely eluted. This permits additional spectra to be available for signal averaging or summing. In this case, summed spectra will
have both molecular-ion and fragment-ion information, or the spectra can be selectively
accumulated to emphasize either the fragment or molecular ions. This experiment would
be similar to a precursor-ion scan typically performed on triple-quadrupole, Q-TOF, or
sector instruments. The toggle experiment is diagrammed conceptually in Fig. 20. The
advantage of the toggle experiment is that it requires only a single mass analyzer.

Mazsaroff and colleagues *(40)* have used the source CID experiment to quantitatively compare sialic acid and N-acetyl-hexosamine on recombinant human factor VIII
(rhfVIII), a large recombinant glycoprotein. A typical set of chromatograms is shown
in Fig. 21, in which the response at 214 nm, the response caused by fragments characteristic of sialic acid (m/z = 274, 292), and the response caused by the fragments characteristic of N-acetyl-lactosamine (m/z = 168, 186, 204, 366) after appropriate
background corrections are provided. Clearly, the selected ion traces indicate that only
the B-domain contains components that produce the fragment ions indicative of
glycosylation. The resulting fragment ion ratio (FIR) provided a measure of the extent
of sialylation of, in this case, the B domain of rhfVIII.

Similarly, this approach has been utilized in characterizing the tryptic peptides produced from the archetypal glycoprotein, bovine fetuin. Burlingame et al. *(41)* have used LC/MS with source CID and selected ion monitoring to identify glycopeptides. The various chromatograms are shown in Fig. 22, and provide the clear identity of the potential peptides and glycopeptides of interest. Clearly, this experiment provides a powerful tool to identify peptides of a known or desired modification, but may require additional characterization by true MS/MS, chemical, or biochemical methods to confirm the exact nature of the identity. This experiment has seen applied to a variety of problems *(34,35,42)*.

In a fashion similar to the source CID discussed here for ESI, additional internal energy can also be imparted to analyte ions generated through desorption ionization methods. The most common of these experiments employs MALDI. In this case, the laser flux is increased and/or the matrix is changed, which causes a higher internal energy deposition during ionization, creating a higher probability for a transient or metastable fragmentation. A timing gate at the reflectron is used to select an ion, typically with low resolving power. The mass-selected, energized ions of the designated m/z are then accelerated down the flight tube. As they proceed, the ions fragment but travel as a group as if they were all of the original m/z. The ions then enter the reflectron, where new accelerating forces are applied, and the fragment ions then travel toward the detector at velocities characteristic of their unique masses. This process is depicted in Fig. 23. This experiment is known as post-source decay (PSD) *(43)*, and is done using a reflectron TOF mass analyzer. The experiment offers the ability for a moderate resolution mass isolation of the excited ions prior to fragmentation using a timed ion gate. Typically, the mass selection range is up to ten Daltons, with the compromise of reduced signal-to-noise at the narrow ranges. An example of the resolution obtained during ion selection by the timing gate is shown in Fig. 24 *(43)*, where, in separate experiments, two ions of interest are selected from a mixture of peptides. In this case, the ions differ from their neighbor by an amount between 9 and 18 m/z. The isolated ion spectra indicate some undesirable ions (none more than 5%) which would ultimately contribute to the product ion spectra obtained. Mass analysis of the fragment ions occurs in segments, and between these the potential on the reflectron is changed to accommodate the kinetic-energy differences in the fragment ions and focus them for proper mass analysis in the second TOF region. The mass spectra acquired at each reflectron potential are then added together to provide a complete spectrum of the fragment ions. A typical MALDI-PSD spectrum is shown in Fig. 25D *(19)* for the M+H+ ion of the peptide KLDVLQ (m/z = 715.4), where it is contrasted with more traditional CID-MS/MS spectra. The MALDI-PSD experiment is less ambiguous than the API source CID experiment, because in this case there is some element of mass selection. By contrast, Fig. 25 indicates that the PSD spectrum has poorer resolution (parent ion) and signal-to-noise, and less informative fragmentation information than the CID spectra (compare Fig. 25D with 25 A–C). Application of the PSD experiment is demonstrated by the MS characterization of peptides and oligosaccharides.

PSD has a similar potential to obtain structural information comparable to that from a true MS/MS experiment. This is demonstrated by the differentiation of simple oligosaccharides similar to those from glycoproteins as examined using MALDI-PSD-

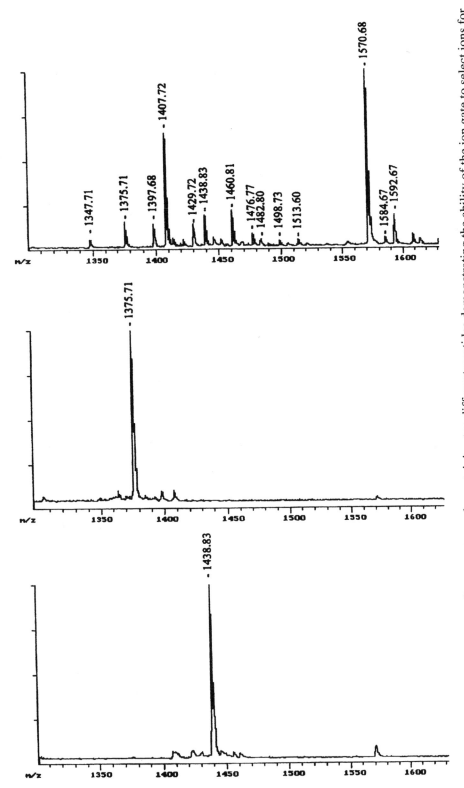

Fig. 24. MALDI mass spectrum of a contest sample containing ten different peptides demonstrating the ability of the ion gate to select ions for PSD. Upper spectrum: ion gate off. Middle spectrum: ion gate on and set to transmit the peptide at m/z = 1375.8. Lower spectrum: ion gate on and set to transmit the peptide at m/z = 1438.7. Spectra acquired on ALADIM I MALDI-TOF instrument (43). Figure reproduced with permission of John Wiley and Sons, Inc.

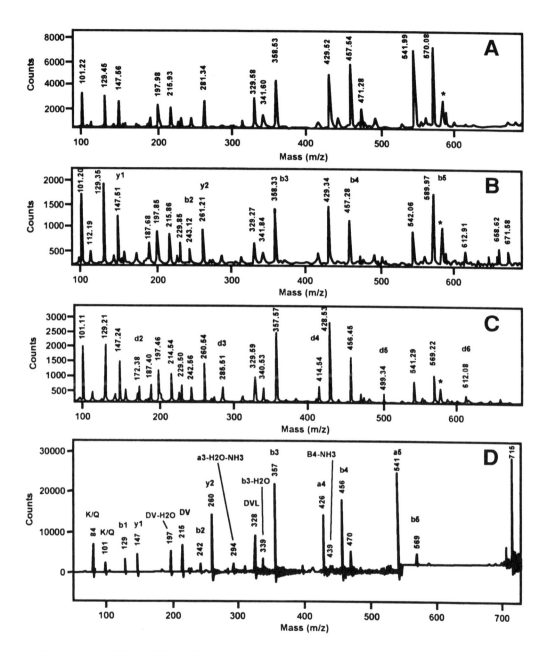

Fig. 25. MALDI-TOF/TOF CID spectra of synthetic peptide KLDVLQ (MH⁺ at m/z 715.4). The reflectron voltage was 26.5 kV for each acquisition. (**A**) Spectrum acquired using CHCA as the matrix and He as the collision gas; (**B**) Spectrum acquired using DHB and He; (**C**) Spectrum acquired using DHB/Ar; (**D**) MALDI-PSD spectrum of the same peptide acquired on a Voyager DE STR (MALDI-TOF) mass spectrometer with CHCA as matrix, recorded in eight steps, lowering the voltage by 25% for each subsequent frame *(19)*. Figure reproduced with permission of the American Chemical Society.

TOF by Yamagaki and Nakanishi *(44)*. The model trisaccharides differ only in their branching structure, and are known as Lewis type a (Le^a) and Lewis type x (Le^x). This study examined the PSD fragmentation of the analyte using both fragment-ion analysis and relative fragment abundance to demonstrate the ability to distinguish these ions, which are positional isomers. Both fragment-ion analysis and relative fragment-ion abundance demonstrated the ability to distinguish these linkage isomers. Similarly, Endo and colleagues *(45)* have examined the MALDI-PSD mass spectra obtained from a range of oligosaccharides that have been derivatized using 2-aminobenzamide, a common reagent to facilitate the HPLC analysis of glycans through fluorescence detection. Significantly, the authors did not modify the hydroxyl groups on the glycan, as in other instances *(46)*. Using 2–3 pmol of analyte spotted on a plate and a DHB matrix, investigators obtained significant sequencing and, more significantly, linkage information. Notably, in contrast to amino acids, monosaccharides can be linked in several different ways through different hydroxyls with different linkage stereochemistries. This was observed for the PSD spectra of two tetrasaccharide 2-aminobenzamide derivatives, which provided significant differences in the abundance of fragment ions. Although this permits the determination that two glycans are different in linkage—or more appropriately, two glycan-derived ions are composed of different linkage compositions—it does not permit the absolute assignment of linkages without the availability of glycan standards analyzed under the same conditions.

Significant application of the PSD experiment to peptides also exist. An exemplary study by Rouse and colleagues *(47)* compares the information obtained from high-energy CID MS/MS, low-energy CID MS/MS, and PSD (metastable). Typical results from this set of experiments are shown in Fig. 26. In this experiment, the fragmentation of a model peptide, Des-Arg-1-bradykinin (PPGFSPFR), observed in the three MS/MS modes is compared. Clearly, the most information (greatest breadth of fragmentation), is obtained from the high-energy (7 keV) CID experiment, in which a higher proportion of ions derived from the C-terminus (arginine) and ions derived from cleavage of the peptide backbone (d_n, w_n, v_n ions) are observed compared to the low-energy CID and PSD experiments. The observation of the C-terminal ions in this case is energetically more favorable because of the arginine at the C-terminal residue. However, although it does not provide detailed information about the peptide backbone or amino acid side-chains, the PSD experiment strikes a balance between N- and C-terminal ions and internal fragmentation. It also provides sufficient information for complete sequence coverage of this octapeptide.

3.4. Information Obtained from the MS/MS of Biomolecules

Fragmentation of biopolymers provides substantial information on the composition of the biopolymer and the order in which the monomer units are connected. This is true for both oligosaccharides and proteins. Each of these biopolymers has its own rules for fragmentation and nomenclature. These are discussed in this section for the principal fragmentations. The rules are also diagrammed as they relate to ion fragmentation. As

---

Fig. 26. *(opposite page)* Mass spectral comparison of Des-Arg(1) bradykinin (avg. mW 904.0) PPGFSPFR. (**A**) high-energy CID (7 keV); (**B**) low-energy CID (45 eV); (**C**) MALDI-PSD in a reflector TOF-MS. Glycerol and metastable peaks are designated with an asterisk (*) in the LSIMS and CID spectra, respectively *(47)*. Figure reproduced with permission of Elsevier Science.

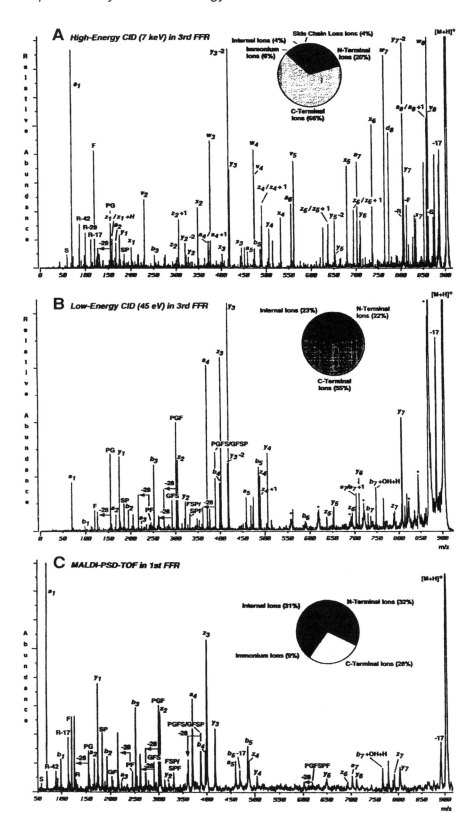

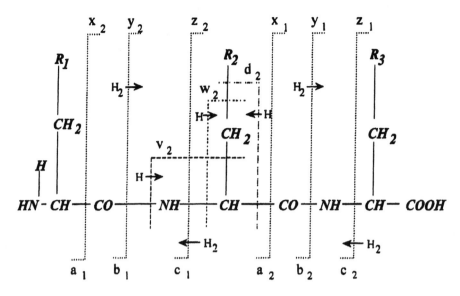

Fig. 27. Nomenclature for peptide fragment ions according to Biemann *(43)*. Figure reproduced with permission of John Wiley and Sons, Inc.

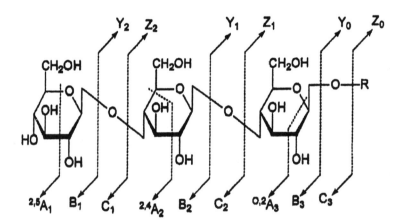

Fig. 28. Nomenclature for fragment ions from carbohydrate (after Domon and Costello *[51]*). *(86)*. Figure reproduced with permission of Wiley-VCH.

a general rule, fragmentation of multiply charged ions is often accompanied by a change (typically reduction) of the charge. In high-resolution instruments such as the oaTOF, FTMS, and Q-TOF, the charge state of the ion is readily identified.

The rules for the nomenclature of peptide-fragment ions have been clearly established *(48,49)*, and are regularly updated to accommodate new fragmentations. The primary fragmentations are depicted in Fig. 27 *(43)*. The observation and relative abundance of these ions depends mostly upon the composition of the peptide, the mechanism and energetics of ionization and fragmentation, and the charge-state of the peptide. Typically, b and y'-fragment ions are observed in greatest abundance. However, chemical methods to direct the fragmentation have been developed *(50)*. In these instances, a fixed charge is placed at either the N- or C-terminal amino acid of a peptide. This method provides a mechanism by which the fragment ions observed in the MS/MS or pseudo-MS/MS spectrum are largely limited to one or a few series of ions.

The nomenclature rules for the fragmentation of oligosaccharides have been established by Domon and Costello *(51)*. These rules are complicated by the fact that, unlike peptides, oligosaccharides can be connected in stereochemically different ways, and a single monosaccharide may have more than two additional monomer units attached to it. In addition, the sugar rings have a tendency to fragment internally to create a series of A-ions. The stereochemistry is not addressed in the nomenclature rules, because under normal conditions of fragmentation the stereoisomers will produce the same ions, but perhaps at different relative levels. The nomenclature of oligosaccharide fragment ions is depicted in Fig. 28. As with peptides, the fragmentation observed is dependent upon the oligosaccharide composition. However, for carbohydrate monomers, there is a relative lack of variability in the basicity or acidity of the monomer side chains.

4. Applications of Mass Spectrometry to Biotechnology

The tools and instruments discussed in **Subheading 2.2.** provide the potential to identify and characterize complex biologicals in a rapid fashion and augment—yet do not totally replace—many traditional methods, including SDS-PAGE for mW estimation and Edman degradation for amino-terminal sequencing. Their refinement and utilization in the study of problems of biotechnological origin continues to advance the quantity and quality of information available to the cell and molecular biologist, as well as the biochemical engineer.

4.1. Molecular Weight Determination

The determination of the mol wt of a peptide or protein is one of the first steps in establishing its identity. As demonstrated here, a tremendous amount of information can be obtained from this apparently simple determination. Prior to 1990, this was achieved primarily through the use of techniques such as sodium dodecyl sulfate-polyacrylamide gel electrophoresis (SDS-PAGE) and size-exclusion chromatography, which are typically limited to errors greater than 20% for a protein of mW >20,000 *(52)*. Currently, mW determinations are achieved in a simple MS experiment that typically applies MALDI, although electrospray or other techniques may also be used. Since the initial application of MALDI-TOF to the analysis of proteins in complex mixtures, analytes ranging from a few Daltons to millions of Daltons (intact viruses) have been analyzed. The utility of these experiments and scope of application can be seen in the table of mW determinations provided in a recent review of the current literature in *Analytical Chemistry* by Burlingame, Boyd and Gaskell *(5)*.

The advantage to MALDI as an ionization method is the relative simplicity in its interpretation of the resultant spectra and the robustness of the ionization method toward additives such as buffers, detergents, and chaotropic agents. Table 3 provides the tolerable concentrations of many of the common reagents found in biological samples from which a useful MALDI spectrum may be attained. The success of the MALDI experiment is largely dependent upon both the sample matrix and matching the sample with an appropriate matrix. If this is not done properly, inefficient ionization and artifactual

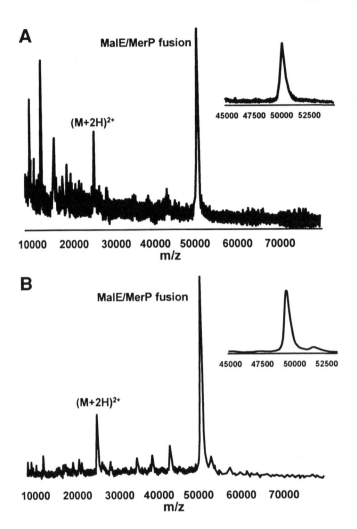

Fig. 29. (A) Mass spectrum of MalE/MerP fusion protein obtained from nonsonicated bacterial cells using linear mode detection. **(B)** Mass spectrum of the same protein obtained from sonicated cells *(53)*. Figure reproduced with permission of the American Chemical Society.

adducts are likely to occur. The primary considerations in this approach are: 1) the compatibility of the MALDI matrix with the analyte matrix (poor crystal formation will inhibit ionization); 2) the propensity of the analyte(s) of interest to adduct formation, either as simple sodiation or more elaborate complexes (polarity or hydrophobicity); 3) the capabilities of the instrument used (laser power, ion optics, or detector); and 4) the mW of the analyte. However, in most instances, minimal sample cleanup will result in meaningful spectra. An example of the relative robustness of MALDI(-TOF) as a method is demonstrated in Fig. 29, where a recombinant protein (MalE/MerP) is mass-analyzed from the cellular supernatant of a bacterial culture *(53)*. The only treatment to the sample was as follows: after application, the spot was rinsed with purified water to remove excess salts. Examples of complex sample melieu analyzed directly by MALDI include serum, urine, cell-culture media, and intact microbial cultures.

Table 5
Websites for Information on MALDI Sample Preparation

Organization	Website
Association of Biological Research Facilities (ABRF)	www.abrf.org
Vanderbilt University	ms.mc.vanderbilt.edu/tutorial.htm
Penn State University	www.hmc.psu.edu/core/Maldi/malditofprotocols.htm
Applied Biosystems	www.appliedbiosystems.com
MicroMass	www.micromass.co.uk/

In addition to the direct analysis of samples by MALDI, there are other options for sample cleanup prior to analysis. Among these, the most commonly used are cleanup using pipet tips loaded or coated with adsorptive material (e.g., ZipTips™), desalting using a sizing column or membrane (gel column or sizing membrane), and cleanup on the plate (*see* ref. *53*). Details of these methods are not included here. A caveat to any cleanup attempt is that, depending upon the nature of the analyte, there may be irreversible adsorption to the sorptive surface. A second caveat to this approach is that during the loading the analyst is fundamentally concentrating the protein *and* any other chemical that may be present in the sample matrix. This can produce artifacts including covalent and noncovalent modification of the protein, resulting in either higher or lower apparent mol ws. Typical conditions for sample cleanup prior to MALDI analysis are available from numerous websites, including those listed in Table 5. Remember that analyte/sample matrix may have unique concerns and may require minor adjustment of the typical conditions to meet the desired result.

A particularly interesting area of development in mol-wt determination from complex or dirty samples is the modification of the surface of the MALDI target to permit either selective retention of the analyte of interest *(54,55)* or to enhance the detection of the analytes of interest *(56–58)*. Further advances in enhanced detection and the use of an affinity surface have the potential to provide selective, simple, and rapid methods by which to analyze many classes of analytes, using relatively simple and well-characterized chemistries. In the simplest case, the surface has been modified by the addition of adsorptive beads directly onto the sample wells at the probe surface *(59)*. In this experiment, the sample is applied (along with trypsin) onto absorptive beads into the sample wells. The beads permit the sample to be concentrated by repeated application of small volumes and salts, or removal of low-mol-wt interferences (e.g., urea or sodium chloride). The sample is digested by the trypsin that was applied with the sample. Comparison of this approach, in which bovine serum albumin (BSA) was digested on the plate with the direct analysis of a solution digest of BSA in various matrices, is depicted in the mass spectra in Fig. 30. These spectra show that the on-plate approach provides significantly more BSA peptides (i.e., sequence coverage) at equivalent or superior signal-to-noise ratios, even from samples with high concentrations of matrix components.

In the electrospray experiment for mW determination, the sample must be mostly free of salt and other nonvolatiles, with particular attention to phosphate and sulfate prior to introduction into the ion source. Once this is achieved, the analysis can be

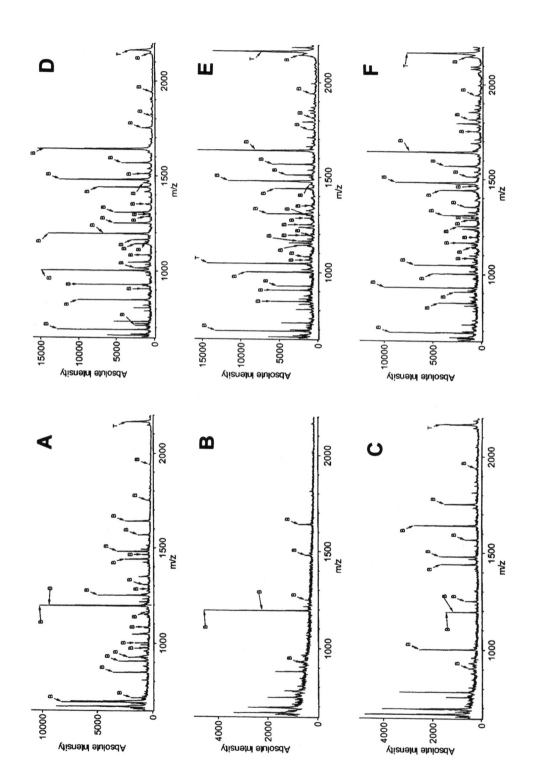

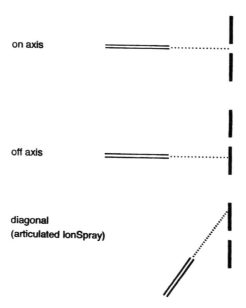

Fig. 31. Arrangements for the position of the ES capillary in the ion source. (**A**) on axis; (**B**) off axis; (**C**) diagonal (articulated) *(55)*. Figure reproduced with permission of John Wiley and Sons, Inc.

made, with analytical conditions and resulting mass spectra dependent upon a variety of factors including source design, mass analyzer, and data manipulations. Advances in electrospray-source design have permitted the analysis of analyte directly from solutions containing relatively high salt concentrations including sodium chloride, phosphate, and ion-pairing reagents. Examples of these sources *(55)* include the off-axis (Fig. 31B) and the angular entry (Fig. 31C), which are both available commercially and compared with the traditional on axis source (Fig. 31A). Each of these uses slightly different approaches to facilitate the use of salt-containing solutions while keeping the salt out of the mass analyzer and eliminating build-up from the source tip. In general, most orthogonal source designs will tolerate moderate amounts of salts for brief periods.

Key among the data manipulations for mol-wt determination by ESI-MS is deconvolution. As spectra become exceedingly complex—a process that is observed for proteins that are glycosylated—the deconvolution of the multiple multiply charged components, which vary by relatively small amounts, becomes less trivial. This is exemplified by the spectra in Fig. 5A and Fig. 5B in, which the electrospray mass spectrum of ovalbumin, a moderately complex glycoprotein, is shown along with the deconvoluted mass spectrum. The math behind deconvolution has been discussed in **Subheading 2.1.1.** The development of deconvolutions based on maximum entropy calculations *(13)* has permitted more robust deconvolutions from which quantitative information can be obtained.

Fig. 30. *(opposite page)* MALDI spectrum obtained from the free solution digestion of 500 nM BSA in (**A**) pure water; (**B**) 2 *M* NaCl; (**C**) 2 *M* Urea; MALDI spectra obtained from microbead preconcentration and digestion of 100 nM bovine serum albumin in (**D**) pure water; (**E**) 2 *M* NaCl; (**F**) 2 *M* Urea. B represents a peptide from BSA *(59)*. Figure reproduced with permission of American Chemical Society.

ESI experiments offer certain advantages over MALDI. One advantage is that since the observed charge-state distribution may vary for different protein isoforms, ESI-MS provides the ability to distinguish and resolve closely related species—through both mass and onization behavior—in a single sample, yet MALDI-TOF may not provide this capability because of resolution limitations. A second advantage is that the single ESI experiment actually offers multiple estimates of the mol wt through the appearance of several ions of varying m/z ratio (*see* ESI section above). Deconvolution of the spectra provides an opportunity to average these mW estimates ($M+n_1H^+$; $M+n_2H^+$, $M+n_3H^+$,....) and arrive at a reliable estimate of the mW of the protein on interest. However, the experimenter is cautioned to avoid reliance on a single MS tool or set of conditions for the determination of mol mW. Dependending upon source conditions and sample matrix (composition and pH), labile adducts that are inherent to the protein (such as sialic acid, pyridoxal, phosphate, and gamma carboxylation) may be eliminated during the course of ionization, or covalent and noncovalent adducts may be formed artifactually with the protein. Each of these may result in an inaccurate estimate of the protein mW or the modifications inherent to the protein's activity. From this perspective, it is critically important for the analyst to understand the nuances of the ionization method as well as the instrument being used for the determination.

In addition to the ionization method, the mass analyzer can make a significant difference in the ability to identify peptides and proteins using mass. The more ready availability and application of high-resolution instruments such as the FT-ICR, Q-TOF, and oaTOF have provided the tools to help the process considerably by reducing the uncertainty of an identification and reducing the dependence on software for the MW determination.

One problem that is historically answered using either N-terminal sequencing or high-energy MS/MS, or these tools combined with enzymatic and chemical means, is the differentiation of glutamine (Q) and lysine (K) in a peptide. These two amino acid residues contribute 128 Daltons to the nominal mass of a peptide, but have an actual mass difference of 0.0364 Daltons. An example of the differentiation of these two residues is shown in Fig. 32 *(60)* in which the ESI mass spectra of MICHAELKARAS and MICHAELQARAS are provided as obtained in a nanospray-oa-TOF experiment. The accurate mass determinations on the triply charged ions deconvolute to provide $M+H^+$ values of 1328.6368 (Q) and 1328.6757 (K). The difference between these two values is 0.0389 Daltons as compared to the theoretical difference of 0.0364. The provides a mass measurement error of 0.0025 out of 1328.636 Daltons or 1.8 ppm. The mass difference at this level is absolutely significant, but very much dependent upon appropriate internal mass calibration with calibrants on the high and low sides of the analyte. (The closer the m/z of the calibrant to the analyte, the better—TOF calibration is not linear across broad ranges, but approximates linear in small mass increments.) In the case of a mixture (co-eluting peaks in HPLC), the observed resolution would not be adequate to define this substitution. The spectrum would show only uncharacteristically poor resolution of the peptide isotopes. If better definition is required, then either the generation of ions at lower-charge states (singly charged, for example) may be attempted or ultra-high-resolution instruments may be required (e.g., FT-ICR).

A second advantage of the higher resolution, accurate-mass analyzers is the ready availability of information on the charge-state of the peptide/protein ion and the ability

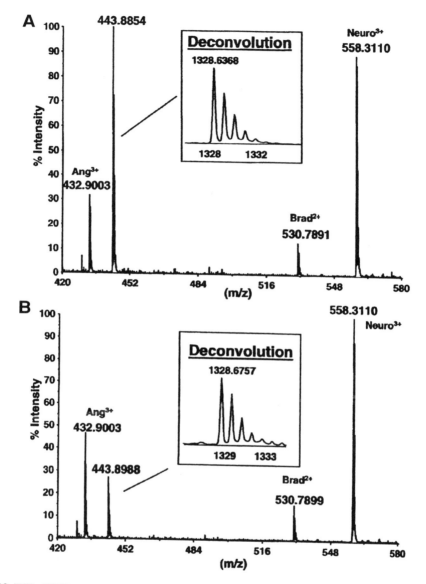

Fig. 32. ESI-oTOF mass spectra of (**A**) peptide MICHAELQARAS and (**B**) MICHAELKARAS. The signals of the triply protonated molecules are shown together with the peaks from the calibration solution. Triply charged molecules of angiotensin I and neurotensin are used for internal calibration. The deconvoluted peaks show the masses for the uncharged molecules *(60)*. Figure reproduced with permission of John Wiley and Sons, Ltd.

to differentiate with confidence amidated/deamidated and reduced/nonreduced forms of moderately sized peptides/proteins. An example of this is shown in Fig. 33 *(38)*, in which the isotope pattern in Fig. 33A indicates a doubly charged species (isotope separation of approx 0.5 Daltons and that of Fig. 33B clearly indicates a different charge-state with isotope separation of approx 0.3 Daltons, indicating a triply charged species.

Tertiary and quaternary structure are important aspects of protein functionality. This includes the three-dimensionality of the folded form, the assembly of multiple protein

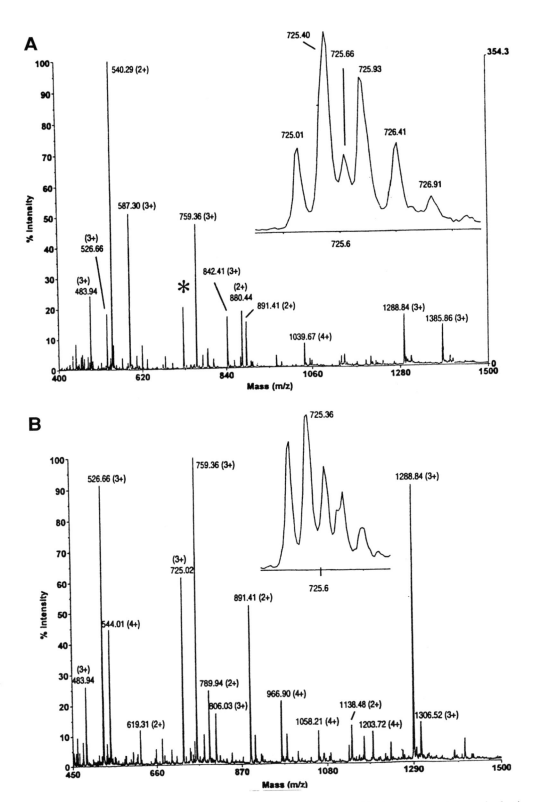

Fig. 33. Raw and deconvoluted mass spectra from peptide digests. The ions observed at nominal m/z 725 are compared and suggested as triply charged. The expanded regions in the insets show the ions of interest are not both triply charged based on the spacing of the isotopes in the spectra: (**A**) 0.5 m/z units; (**B**) ca. 0.3 m/z units *(38)*. Figure reproduced with permission of John Wiley and Sons, Inc.

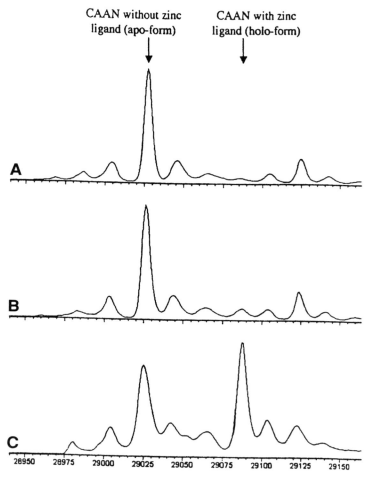

Fig. 34. Zinc-binding to carbonic anhydrase (CAAN). (**A**) Deconvoluted MS spectrum obtained after CAAN dissolved and washed in 0.1% trifluoroacetic acid; (**B**) Deconvoluted MS spectrum obtained after CAAN dissolved and washed in 100 m*M* calcium acetate; (**C**) Deconvoluted MS spectrum obtained after CAAN dissolved and washed 100 mM zinc acetate. Protein concentrations were normalized to 1 pmol/mL and acidified with acetic acid prior to acquisition of mass spectra *(61)*. Figure reproduced with permission of Wiley-VCH.

subunits to make a functional protein complex, and the binding of prosthetic groups and metal ions critical to the function of the protein of interest. This is significant for both proteins isolated from natural sources and those from recombinant sources. In the case of proteins produced as inclusion bodies from *E. coli*, proper protein folding is required when disulfide bonds are present in the functional protein. Traditional methods such as electrophoresis or chromatography often disrupt these complexes or provide insufficient information (e.g., the differences between properly and improperly folded forms of a 20-kD protein may not be substantial). The gentle nature of ESI makes it possible to often obtain detailed mol wt information on protein complexes and folding intermediates.

A simple example of this is the binding of metal ions to carbonic anhydrase *(61)*. When native enzyme is treated with aqueous acid to remove nascent Zn, the apo-

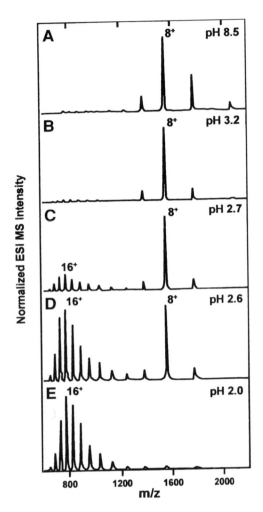

Fig. 35. (A) ESI mass spectra of cytochrome c recorded at different pH values: (top) 8.5; 3.2; 2.7; 2.6; (bottom) 2.0. The pH was adjusted by the addition of ammonium hydroxide and/ or acetic acid. **(B)** relative abundance of the cytochrome c charge states +7, +8 and +9 (top); relative charge states +10, +13, +15, and +16 (middle); dependence of average charge state on the pH (bottom). (63) Figure reproduced with permission of John Wiley and Sons, Inc.

enzyme provides a mW of 29,025. Equilibration of the apo-enzyme with calcium acetate provides no perceivable change in mW. However, the same experiment using zinc acetate produced a change in the apparent mol wt of the protein consistent with the complexation of Zn by the protein to produce the holo-enzyme. The results of this experiment are in shown Fig. 34. The selective binding of zinc is consistent with the known behaviors of carbonic anhydrase.

Similar experiments have also provided information of the quaternary structure of multi-subunit proteins. For example, Smith and colleagues *(62)* have examined the composition of alcohol dehydrogenase isozymes, using capillary isoelectric focusing with ESI-FTMS. In this case the investigators were able to utilize the gentle nature of ESI and the resolving power of FTMS to evaluate the subunit composition of the alcohol dehydrogenase isoenzyme complexes, which had monomeric mol wts between 39670.7 and 39729.9 Daltons and isoelectric points between 8.26 and 8.67. The mass spectrometry experiment coupled with capillary isoelectric focusing provided the ability to evaluate the distribution of the isoenzyme forms.

The capability of electrospray to monitor the folding of proteins is exemplified by the pH-dependent denaturation of cytochrome c *(63)*. This is shown in Fig. 35A and 35B. The electrospray spectra indicate a substantial change in the charge distribution of the ions as a function of time. As the pH is increased and the cytochrome c changes its three-dimensional structure, the number of protons retained under the conditions of the ESI experiment changes. The deconvoluted spectra (not shown) would provide the same effective mol wt for the cytochrome c in either experiment, so in this case it is the raw mass spectrum that provides the key information. The dynamics of denaturation or folding can be followed when selected ions indicative of a given folded form or charge-state are monitored as either a function of time or, in the present case, pH. In the case of cytochrome c (Fig. 35B), the changes in several ions are monitored as a function of pH, and these show a multiphasic folding phenomenon. This suggests that multiple, stable intermediates are involved in the denaturation, which is consistent with other physical data. The authors *(63)* examined the behavior of several other proteins as a function of pH. In a similar fashion, the folding of a denatured protein into its properly folded, biologically active form can also be monitored using ESI-MS. This method has been applied to the characterization of the folding pathway of recombinant human macrophage-colony-stimulating factor-β (rhM-CSF β) *(64)*, using the formation of adducts of melarsen oxide with folding intermediates to trap the intermediates through free disulfides and induce a small but measurable mass shift. This compound bridges neighboring cysteinyl groups, and thus traps the intermediate and provides a mass shift based on the number of melarsen oxide molecules bound to the rhM-CSF β. Fig. 36 shows the spectra obtained when rhM-CSF β is folded in the presence of melarsen oxide. The ESI-MS spectra are compared for unfolded protein (Fig. 36A), after folding the protein for 6 h and 42 h (Fig. 36B and C), and native, presumably properly folded rhM-CSF β (Fig. 36D). As the folding progresses, the charge envelope clearly progresses toward the native protein. Also apparent in Fig. 36 are the ions caused by dimer (D), which can be differentiated from the monomeric protein (M) by the binding of 50% of the melarsen oxide molecules per molecule of rhM-CSF β. This suggests, and is consistent with, the involvement of a dimeric form of the protein in the folding pathway.

A more thermochemical approach to protein unfolding was undertaken by Foti and colleagues *(65)*, in which they utilized ESI-MS and a spectrophotometric method to monitor the unfolding of wild-type azurin and wild-type amicyanin under different conditions over time. The ΔG° of unfolding the azurin and amicyanin obtained by ESI-MS, 38.3 and 30.5 kJ/mol and spectrophotometry, 33.9 and 28.3 kJ/mol, showed good relative agreement and further supports the idea that the ESI-MS provides some quantitative evaluation of protein folding when appropriate experimental precautions are taken. For a more extensive review of some of the work on the study of noncovalent interactions by electrospray mass spectrometry, *see* the review by Loo *(66)*.

A more complicated yet elegant experiment for the evaluation of protein structural differences involves the use of proton/deuterium exchange in solution followed by mass spectral analysis *(67)*. This methodology has a long history in nuclear magnetic reso-

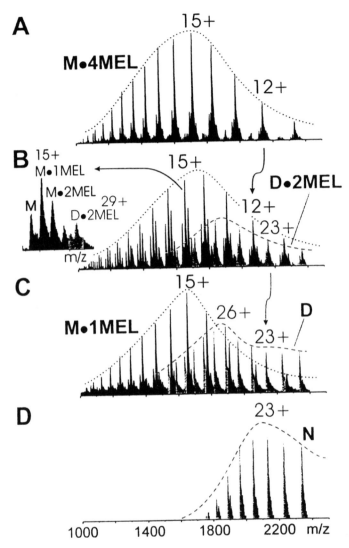

Fig. 36. ESI-MS monitoring of in vitro refolding of recombinant human macrophage-colony stimulating-factor β. ESI spectra of Mel-trapped intermediates at (**A**) 0, (**B**) 6 and (**C**) 42 hours of refolding, and (**D**) of mature (folded) homodimer *(64)*. Figure reproduced with permission of Wiley-Liss, Inc.

nance (NMR) studies of proteins, and has been followed by mass spectrometry *(68)*. The isotope-exchange method has been applied to the study of insulins of differing sequences in which the exchange is monitored over time using ESI-MS. The five different insulins examined—recombinant human, porcine, bovine, LysPro analog, and denatured bovine—showed different numbers of exchanged protons and slightly different exchange rates. The most intriguing finding is that recombinant human insulin and LysPro analog (in which the sequence of Lys and Pro have only been switched) show different numbers of exchangeable protons (Fig. 37). This is probably a result of different three-dimensional conformations for the two closely related polypeptides,

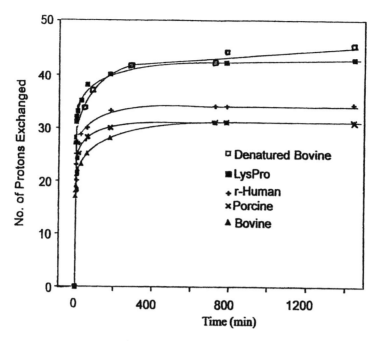

Fig. 37. Kinetics of H/D exchange for LysPro, r-human, bovine, denatured bovine and porcine insulins *(67)*. Figure reproduced with permission of the American Chemical Society.

Table 6
Observed Molecular Weights for "Known" Proteins *(16)*

Protein	Theoretical M (protein only)	Measured M	Relative error (%)	Carbohydrate (w/w %)
IL-2	15478	15610	2.8	n.a.
GM-CSF	14477	15451	5.56	n.a.
IFN-α2b	19378	19533	0.79	n.a.
TNF-α	17484	17517	0.18	n.a.
EPO	n.a.	28707	n.d.	35.6
IFN-α1	n.a.	20465	n.d.	5.2

which are isobaric and differ in only the order of the two amino acids. Future extended applications have not yet been determined, but may contribute to future methods for the release of pharmaceutically active peptide analogs that are either isobaric or exceedingly close (e.g., peptide analogs and homologs).

Many of these applications and experiments use the integrity and definitive nature of the mol-wt information to derive secondary information about a protein. In the course of analyzing recombinant proteins, the mol wt is often predicted from the molecular biology (cDNA or mRNA sequences). In these cases other information can be obtained from the determination of the mol wts. With the growing use of mammalian cells, yeast, and other systems capable of extensive post-translational modifications, the need to rapidly evaluate the presence or absence of these entities is highly beneficial in the development of culture conditions. For example, Table 6 shows how the mol-wt determinations can be used to simply confirm identity. Even in cases in which the sequence and mW are predictable, the cellular processing is often not. Most of the errors observed in this experiment are typical ($< 1\%$) for a determination in a linear TOF or simple ESI-MS experiment. Errors in excess of 1% are probable indicators of real modifications such as sulfation, phosphorylation, N-terminal modification, amino acid misincorporation, or genetic errors. The glycosylation of proteins can be critically important to the biological or pharmacological activity of a protein, and therefore it is important that the extent or existence of carbohydrate is established as quickly as possible. This can be done in the experimental determination of the mol wt of a protein when a target protein mW is available. The percent glycosylation is calculated as the 100*[measured mol wt—theoretical mol wt]/[theoretical mol wt]. Alternatively, the presence and extent of glycosylation of carbohydrate can also be supported based on other information. For example, the width of a MALDI peaks from a desalted mixture (best case) can be a good semiquantitative indicator of the presence of carbohydrate. The extra width of the peak is caused by the heterogeneity of the glycoforms resulting from the variable biochemistry that occurs in all mammalian cell systems. The width of the peak is correlated to glyco-heterogeneity, and the percentage of glycosylation can be estimated as shown for EPO and IFN-α1 in Table 6 *(16)*. Care should be take with this approach, because excessively broad peaks (compared to calibrants of comparable mol wt) is an indicator of post-translational modifications that may include glycosylation, lipidation, sulfation, other forms of post-translational modification, or ragged N- or C-terminal ends on the protein.

These tools for mol-wt determination have often been applied for the near complete characterization of proteins. When only MS is used in conjunction with HPLC or other separation tools, the disulfide bond patterns of human epidermal growth-factor-receptor *(69)* and the disulfide bonding and carbohydrate composition of human β-hexosaminidase B *(70)* and bovine lactoperoxidase *(71)* have been determined. In these instances, the investigators utilized mass spectrometry together with the selectivity of chemical treatments—reduction with dithiothreitol (DTT)—and enzymes including trypsin and peptidyl-N-glycanase F (PNGase F) to identify the location and identity of these specific structural entities. In all three of these examples, MALDI-MS was used as the tool of choice to generate spectra off-line from HPLC, although ESI-MS can also be used when concentrations and sample matrix are appropriate. There are an increasing number of other examples of this type of thorough characterization of proteins in which mass spectrometry is the principal tool.

4.2. Sequencing Using Molecular Weights

Although the mol wts of peptides and proteins can provide considerable information about the analytes being studied, the characterization of peptides, proteins, and oligosaccharides must ultimately include the analysis of the sequence of monomeric units in the case of amino acids and linkage of the monomers in the case of carbohydrates. The combined use of highly selective, or highly non-selective, enzymes with mass spectrometry can provide the information to confirm or determine the sequence of the peptide or oligosaccharide of interest. In addition, in the case of peptides or proteins, one can obtain information on the carboxyl terminal sequence of the peptide or protein.

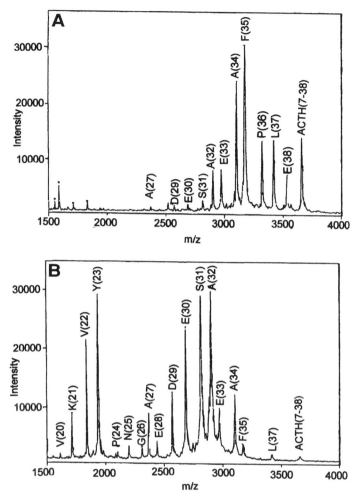

Fig. 38. MALDI spectra of on-plate concentration-dependent carboxypeptidase Y digestions of ACTH 7-38 fragment. (**A**). digest using 6.1×10^{-4} U/mL; (**B**) digest using 1.53×10^{-3} U/mL. * indicates doubly charged ions *(72)*. Figure reproduced with permission of the American Chemical Society.

Although instruments to perform this determination using solution-phase chemistry are commercially available, the amount of any one peptide required to obtain C-terminal sequence information is often too high to be useful, and the throughput is often prohibitive for simple screening of C-terminal information. The inherent selectivity and sensitivity of mass spectrometry makes this a very useful tool for these applications.

Known as C-terminal ladder sequencing *(72)*, the time- and concentration-dependent digestion of a peptide using carboxypeptidase Y (CPY) with analysis of the digest using MALDI provides a tool for the C-terminal sequencing of a peptide or protein. Martin and colleagues *(72)* pioneered this approach. By monitoring the sample as a function of time, one can monitor the sequence based on the mass-difference between the parent ion and the various other ions in the mass spectrum. An example of this approach is shown in Fig. 38, in which the model peptide ACTH 7-38 is digested with

CPY at two concentrations, and the resulting MALDI mass spectra are used to indicate which amino acid is removed during the digestion. This digestion can occur either in solution or on the plate where digestion time is limited by the rate of solvent evaporation. Clearly, in Fig. 38A and B, only a portion (C-terminal residues 20–38) of the peptide is covered in the reaction. To create a composite spectrum in which artifacts and calibration error can be eliminated, the time-points can be pooled after quenching to provide one spectrum containing all of the peptides and provide the so-called ladder. Alternatively, the concentration of the enzymatic digestion can be monitored as a function of time. One caveat to this approach is that this is MS and Ile and Leu cannot be differentiated, and Gln and Lys are only differentiable on a high-resolution instrument when relatively small peptides are used. Alternatively, in order to differentiate the release of Gln and Lys in the spectrum, the Lys residues can be converted to homoarginine *(73)*, which provides a unique mass difference for Lys relative to Gln. The Ile/Leu issue is not amenable to this type of approach.

The application of chemical of enzymatic methods may also be undertaken for N-terminal sequence analysis, but the need is not as great when the sensitivity and reliability of the Edman method are considered. James and colleagues *(74)* have developed chemical methods for obtaining N-terminal sequence information from MALDI spectra using thiol organic acids. Application to proteins from the *E. coli* ribosomes indicates that this is a first-pass method for the determination of the N-terminal residues of proteins in a rapid fashion. The chemistry involved parallels that of Edman chemistry.

Similar to peptides and proteins, oligosaccharides can be characterized through the combination of enzymes and selected mass spectrometric experiments. This is a common experiment in the current research to screen for the presence of N-linked oligosaccharides and to characterize the carbohydrate features if their presence is detected. The flexibility of MALDI often permits the enzymatic reactions to occur directly on the MALDI plate, if desired, so that the handling of the sample is minimized and the investigator may actually run several screening reactions at one time on precious little sample and permit the analysis of the deglycosylated protein as well as the oligosaccharide. One problem with this approach is the differences in ionization yields between the peptide/protein and the oligosaccharide caused by mol wt and basicity. On occassions this may require spotting the sample in two wells and using two different matrices after the digestion (e.g., perform two digestions and apply sinapinnic acid matrix for protein and dihydroxybenzoic acid for the oligosaccharides). In addition, it is strongly advised that information on relative amounts of different glycoforms be qualified when obtained by those methods. Some examples of this application follow.

Ribonuclease B (RNase B) is a common model glycoprotein. Novotny and colleagues *(46)* have explored the utilization of on-plate enzymatic digestion to characterize the oligosacchcarides released from model glycoproteins. As seen in Fig. 39A, it provides a significant level of heterogeneity in native MALDI spectrum, with a series of peaks separated by approx 162 Daltons (hexose monomer). Upon treatment of 5 mcg of protein with PNGase F on the plate for 1 h and analysis from DHB matrix (Fig. 39B) the heterogeneity of the molecular ion is diminished and (Fig. 39B inset) the appearance of ions at lower mol wts characteristic of two N-acetylglucosamines with from 5 up to 9 mannose (hexose) residues attached. This is a well-characterized, high-mannose structure known to be present in RNase B.

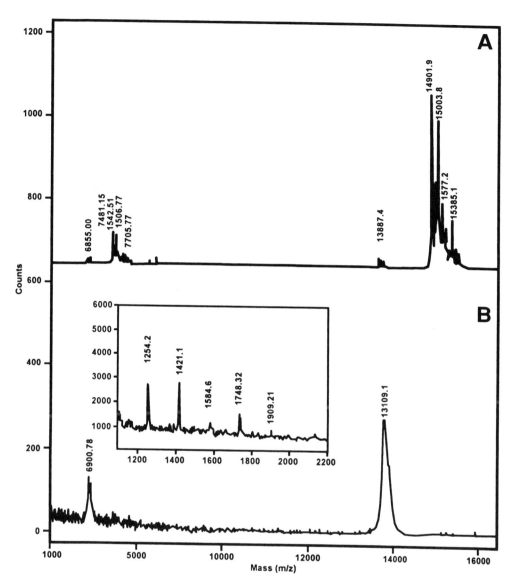

Fig. 39. MALDI mass spectrum of ribonuclease B before (**A**) and after (**B**) treatment with PNGase F. The inset represents the profile of N-glycans after digestion of 5 μg of ribunuclease B for 1 h on the plate *(46)*. Figure reproduced with permission of the American Chemical Society.

In a slightly different approach, the oligosaccharides can be released from bovine fetuin, a more complex glycoprotein, and reacted with various mixtures of glycosidase. The presence and characteristics of the oligosaccharides can be established by MS *(46)*. In Fig. 40 A, B, and C, the oligosaccharide mol wts are seen after release and desialylation (A), the additional removal of terminal galactose residues (B), and the removal of terminal galactose and N-acteyl-glucosamine residues (C). A more thorough study involving glycans, glycopeptides, and glycolipids was undertaken by Geyer, et al. *(75)*. The digestion of model N-linked glycans by a range of enzymes is shown in Fig. 41A–F. The treatments were by galactosidase (B), N-acetylglucosaminidase (C),

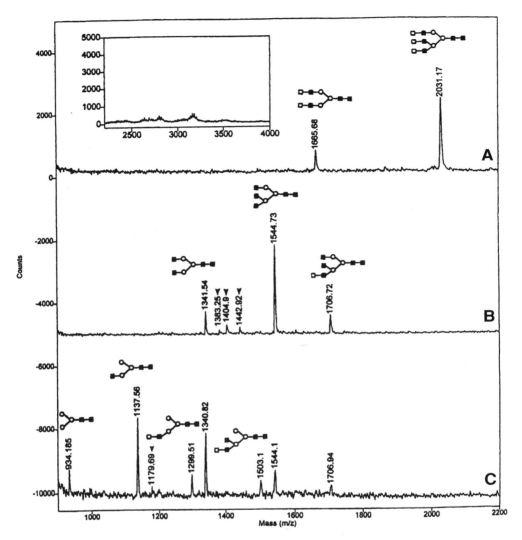

Fig. 40. MALDI mass spectra of the N-linked oligosaccharide profile recorded from 5 µg of bovine fetuin after treatment with PNGase F and neuraminidase (**A**); PNGase F, neuraminidase, and galactosidase (**B**); and PNGase F, neuraminidase, galactosidase, and N-acetylglucosaminidase (**C**). (All digests were at pH 7.5.) The inset represents the m/z range where the sialylated species appear as multiple sodium adducts. The peaks marked with an arrow are contaminants originating from the enzyme preparations *(46)*. Figure reproduced with permission of the American Chemical Society.

N-acetylhexosaminidase (D), fucosidase (E), and mannosidase (F), sequentially. Treatments with hexosamindase and glucosamindase differentiate between N-acetyl-glucosamine and N-acetyl-galactosamine residues.

The use of PNGase F in many applications precludes the necessity of information about any O-linked oligosaccharides that may be present. Treatment with PNGase F is the key to distinguishing N- and O-linked glycans. The inability of O-glycosidases to uniformly release all O-linked oligosaccharides makes a parallel universal application difficult. Kionka and colleagues *(76)* have utilized enzymatic treatments to evaluate

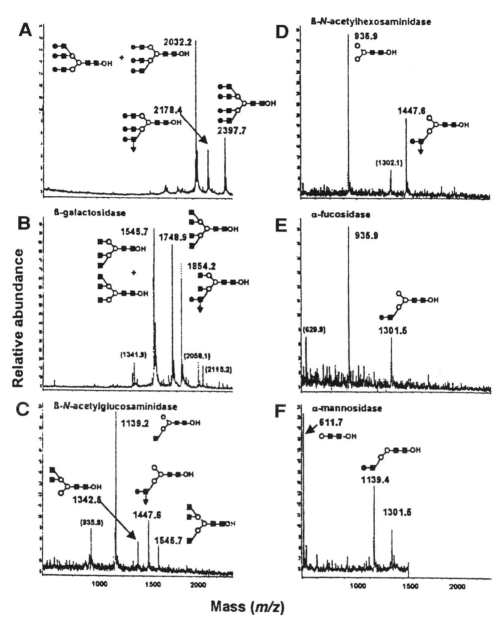

Fig. 41. Mass spectra of a mixture of oligosaccharide alditols (80 pmol) before and after sequential enzymic treatment: native glycans (**A**); incubation with β-galactosidase from *D. pneumoniae* (**B**); β-acetylglucosaminidase from *D. pneumoniae* (**C**); β-acetylhexosaminidase from jack beans (**D**); 0.08 munit α-fucosidase (5 h) (**E**); and α-mannosidase from jack beans (F) *(75)*. Figure reproduced with permission of the American Chemical Society.

the glycans on a recombinant F(ab')₂ fragment. In this case, the intact protein was treated with neuraminidase (sialidase) with and without O-glycosidase. The deconvoluted ESI-MS spectra are shown in Fig. 42. The losses of approx 307 and 365 Daltons from various peaks are indicative of sialic acid and HexNAcHex, in this case GalNAcGal based on the specificity of O-glycosidase. This demonstrates the ability to monitor the

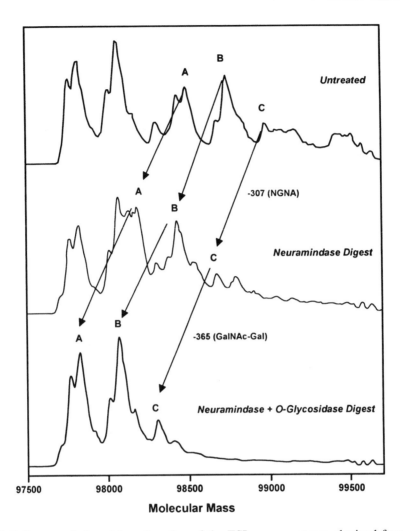

Fig. 42. Influence of glycosidase digestion of the ESI mass spectrum obtained from murine F(ab')$_2$ fragment. Sequential treatments with neuraminidase with or without O-glycosidase induce mass shifts of ca. 307 and 672 (365 + 307) m/z units *(76)*. Figure reproduced with permission of Elsevier Science.

intact protein and serves as an example of the use of ESI-MS in addition to the more often used MALDI-MS experiment. Huang and Riggin *(77)* and Mazsaroff et al. *(40)* have also used this type of approach in characterizing protein-derived carbohydrates by ESI-MS. These two approaches, however, provide for quantitative analysis of the contributions of various glycoforms to the carbohydrates on glycoproteins and rely on the release of the oligosaccharides prior to analysis.

4.3. Protein, Peptide, and Oligosaccharide Sequencing by MS/MS

Historically, protein identification and *de novo* peptide sequence determinations have been achieved using Edman chemistry on an automated platform. In this scenario, Edman chemistry is applied to intact protein as well as peptides isolated from proteolytic digests of the protein, and the information is assembled over a period of weeks

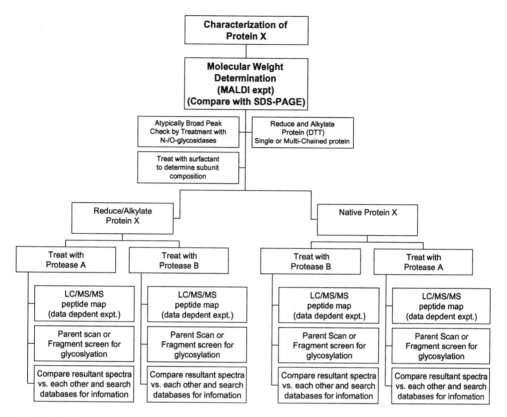

Fig. 43. Typical process for the characterization of a protein by proteolytic digestion and mass spectrometry.

to provide the identification of the protein of interest. Sequence information, combined with mW determination of the intact protein, permits the establishment of the complete primary structure of the protein under study. If the secondary and tertiary structure are also desired, additional experimentation is required. A typical roadmap for the utilization of mass spectrometry for the complete characterization of a protein is shown in Fig. 43. Mass spectrometric experiments now provide opportunities to obtain the same information in a fraction of the time, with limited sample handling and far smaller amounts of material relative to the historical methods of collecting fractions and analyzing the peaks by Edman degradation. Coverage of the peptide sequence for the types of experiments in Fig. 43 are typically greater than 50% for a single enzymatic digest and greater than 75% for two proteases. Factors that influence the sequence coverage by LC/MS include the extent of glycosylation and other post-translational modifications, number of subunits, N-terminal blockage (formyl or pyroglutamyl termini), amount of material, and purity, among others. Mass spectrometry is not a complete replacement for Edman chemistry, but provides the opportunity to substantially reduce the amount of time required to confirm or identify a protein or peptide. Each technique has its limitations, and can provide information that is complementary to the other.

This approach has been utilized in numerous reports that detail the characterization of new or modified proteins. In addition, three modern variations on this approach are being increasingly used as substitutes or complements to the LC/MS/MS experiment—the application of CE/MS and CE/MS/MS *(78–81)* and MALDI peptide mapping *(82–83)*.

The application of these MS and MS/MS tools also allows for the determination of post-translational modifications and their location in the protein or peptide primary structure. Precursor-ions scans and neutral loss scans (MS/MS) have been applied extensively to screening proeolytic digests of proteins for phosphorylation *(84,85)* and glycosylation *(41,86)*. A specific example of this is the selective determination of phosphopeptides, in which two dimensions of mass spectrometric analysis are utilized. The first dimension is a peptide map with UV, full-scan MS, and precursor ion scans, which are applied to the analysis of tryptic digest of the protein of interest. In the case of Sic1p in vitro phosphorylation, Carr and colleagues *(84)* characterized phosphorylation sites that utilize both HPLC-ESI-MS and MS/MS (both positive and negative ion modes) and off-line nanospray MS/MS. Negative-ion precursor-ion scans and positive-ion full-scan MS were used to identify which peptides in a tryptic digest were phosphorylated and to identify the mol wt of the peptide of interest, respectively. Fractions were collected and characterized off-line by nanospray MS/MS (positive ion) to provide the appropriate signal for the sequence determination in the MS/MS experiment. The experiment is shown in Fig. 44. The identification of a peptide at 19 min, which produces m/z 79 (PO_3^-), is clear (Fig. 44A). The positive ion MS experiment indicates that the ions under that peak are at m/z 637, 955 and 1049 (Fig. 44B). The MS/MS experiment then provides a higher level of selectivity and indicates that there are five species which produce m/z 79 during fragmentation (Fig. 44C). The off-line MS/MS experiments then provide details of the sequence of the two unknown species. The details permit the identification of the sequence within the Sic1p protein, and specifically, the site of phosphorylation.

In a similar fashion, precursor ion scans have been used extensively to screen peptide maps for glycosylated peptides *(41)*. This is typically achieved by monitoring m/z 162, 291, or 365, which are carbonium ions characteristic of hexoses (typically galactose), sialic acid, and N-acetyllactosamine, respectively. Detection of the peak of interest permits identification of the glycosylated peptide and more facile and effective MS characterization (product-ion MS/MS). An application of this type has been provided earlier in this chapter. Unlike previous examples *(38–40)*, which use source CID, the MS/MS experiment permits the identification of the actual ion that produces the fragment ion being monitored. Burlingame and colleagues *(41)* have used this to identify the N-linked glycopeptides in Factor VIII.

The TOF-TOF instrument offers unique capabilities compared to the other commercially available tandem mass spectrometers. The ability to explore both high- and low-energy collisions affords the experimenter access to a more exhaustive array of fragmentation, which permits the potential for more extensive characterization of the protein or peptide of interest. In addition, high mass-resolving power of the reflectron TOF permits accurate mass evaluation of the fragment ions, providing the potential to differentiate between Gln- and Lys-containing peptides, for example, without the aid of a sequencer. In operations using either MALDI or ESI for analyte ionization, the TOF-TOF has significant flexibility and power. The capabilities of the TOF-TOF in an MS/MS experiment are demonstrated and compared with MALDI-PSD analysis in Fig. 25 *(19)*. Aside from the fact that the spectra obtained on the TOF-TOF are obtained in a continuous experiment while the PSD spectrum is acquired in an automated step mode, the spectra from the TOF-TOF show significantly more depth in the fragmentation. The TOF-TOF provides the ability to rapidly analyze peptide maps or other peptide/protein mixtures using the relatively quick MALDI mass-mapping technique and acquire MS/MS spectra in a data-dependent manner on singly charged ions. Although still limited in its utilization in the general research world, the TOF-TOF will provide significant opportunities for the rapid characterization of peptides and proteins.

Perhaps the most revealing and useful application of MS and MS/MS in the analysis of proteins of recombinant origin is the identification of the unusual and the unpredictable. Post-translational modifications occupy much of this uncertainty, and have been discussed throughout this chapter. In addition to "normal" and "useful" modifications from the cellular machinery, many organisms used in the production of recombinant proteins have a propensity for genetic instability that may manifest itself through the production of a protein molecule with an incorrect amino acid sequence. The two paths toward this end are codon wobble and misloading of the transfer RNA during peptide synthesis. The codon wobble is caused by small differences in the codons for various amino acids. The other mechanism, amino acid misloading, is even less predictable, but usually results in a like-for-like substitutition (e.g., aliphatic for aliphatic amino acid). The implications of these undesired processes result in mass differences of between nominally 0 (Ile/Leu; Lys/Gln) and 186 Daltons in the peptide or protein. Issues related to the zero mass difference have been discussed elsewhere in the chapter.

An example of the successful identification of a genetic issue is provided by Yates and colleagues *(87)*. This approach utilized LC/MS/MS peptide mapping to identify a mutation resulting in change from E to K (a difference of 1 Dalton) at the 26 position in the beta subunit of hemoglobin E. Although MS/MS provided the route to identification, this particular mutation changes both the sequence of the protein and the net charge (polarity) of the peptide, and also provides a new cleavage site for trypsin—creating not one but two new peptides (before and after the site of mutation).

In addition to the improper translation of protein or the modification of its genetic coding material, properly generated protein can be modified through a variety of chemical processes (in vivo and in vitro). The most common of these are deamidation of Q and N, oxidation of H, W, M, and C, and transamidation. One form of transamidation is the internal rearrangement of asparagine residues to form an isoaspartyl residue (β peptide bond) resulting in a net change of 1 Dalton in the peptide or protein. However, based on mass alone, this cannot be distinguished from traditional N (or Q) deamidation. However, as shown in Fig. 45 *(88)* it is possible to distinguish α- and β-aspartyl residues in an MS/MS experiment. The loss of 46 m/z units clearly differentiates the spectra of these two peptides. The low relative yields of this fragment suggests that the experiment may not be universally applicable, and may encounter problems if the peptides in question co-elute or occur as some other form of mixture. However, it does clearly demonstrate the potential of the MS/MS experiment.

5. Conclusions and Future Directions

The tools, principles, and application of mass spectrometry to the problems and challenges of the analysis of proteins of recombinant origin have been discussed. This chapter demonstrates the what and the how of the mass spectrometric experiments common to protein, peptide, glycoprotein, and glycan characterization. Other topics that have

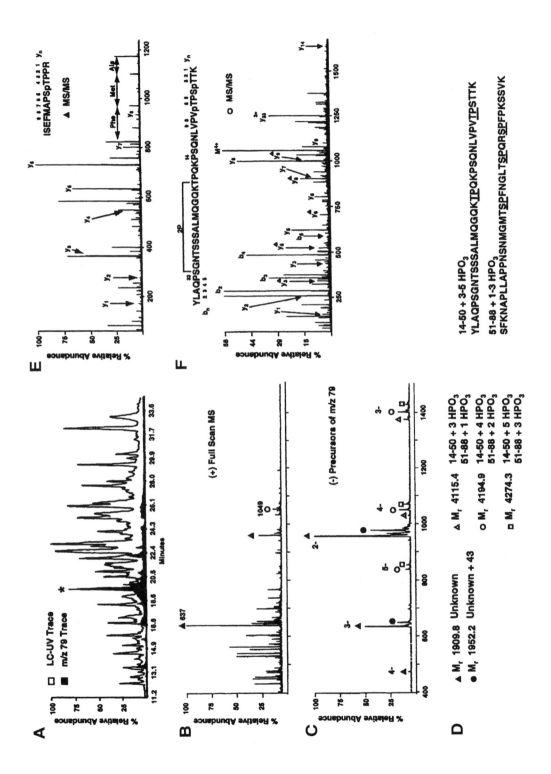

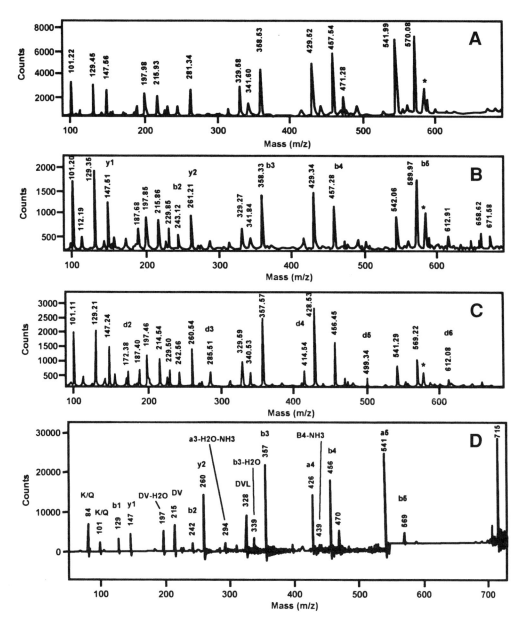

Fig. 45. Full ESI-MS/MS spectra of the peptides LVFFAE-(L-α-Asp)-VGSNK and LVFFAE-(L-β-Asp)-VGSNK are shown in (**A**) and (**C**), respectively. The expanded regions (m/z 500-625) between the y"₅ and y"₆ ions of the α- and β-Asp tryptic peptides are shown in (**B**) and (**D**), respectively *(88)*. Figure reproduced with permission of John Wiley and Sons, Inc.

Fig. 44. *(opposite page)* Multidimensional electrospray MS mapping of G1 Cdk-dependent in vitro phosphorylation of Sic1p. (**A**) LC/UV and selected ion spectrum of m/z 63 and 79⁻; (**B**) full scan mass spectrum of peak (*); (**C**) precursor scan for m/z 79⁻; (**D**) proposed identities of observed ions; (**E**) MS/MS product ion spectra of ISEFMAPSpTPPR; (**F**) MS/MS product ion spectra of YLAGPpSGNTSSSALMQGQKpTPQKPSQNLVPVpTPSpTTK *(84)*. Figure reproduced with permission of the American Chemical Society.

been mentioned, although not discussed in depth include the application of databases to the problems of biotechnology, cutting-edge instrumentation (presently being applied to problems in a academic setting but is not yet commercially available), and the application of high-throughput technologies to problems of biotechnological origin. In many ways, the first and last of these topics have been addressed and are not fully described in this chapter. Rather, the use of databases and high-throughput technologies developed to support the identification of proteins based on high-sensitivity detection of their pieces will lead to the development of tools to investigate glycomics (e.g., the identification of the glycoforms that constitute the population of a specific glycoprotein). The continued evolution of applications for instruments such as the TOF-TOF, and the continuing commercial development of instrumentation the ion trap—TOF hybrid (trap-TOF) *(89)* will provide the new tools to address the issues presented by problems such as glycomics. The key to future developments may be found in the sample pretreatment—for example, interfacing MS experiments with affinity-based sensors, as done by Nelson *(90)* using a nickel-affinity surface (Fig. 46). Also, analyses aimed at single-cell analysis to achieve information on cellular proteins, as done by Feng and Yeung *(91)*.

Clearly, mass spectrometry cannot address all of the problems inherent in the analysis and characterization of macromolecules of biotechnological origin. In this respect, the continuing development of hybrid tools including the refinement of the interface for capillary electrophoresis to—for example—ESI, which provides the opportunity for high-resolution separation and high-resolution mass analysis. In addition, miniaturization of the interfaces will afford the scientist the opportunity to utilize minimal amounts or samples. Also, the development of new forms of instrumentation will afford the opportunity to obtain information on biomolecules at several levels in a single experiment. The future for the utilization of mass spectrometry in the analysis and characterization is exceedingly bright, and will encompass more routine analyses of key quality attributes as well as more detailed structural information at earlier stages in the development of a recombinant protein.

References

1. Lagu, A. L. (1999) Applications of capillary in biotechnology. *Electrophoresis* **20**, 3145–3155; Lagu, A. L. and Patrick, J. S. (2001) Review: applications of capillary electrophoresis to the analysis of biotechnology-derived therapeutic proteins. *Electrophoresis* **22**, 4179–4196; Kennedy, R. T., German, I., Thompson, J. E., and Witkowski, S. R. (1999) Fast analytical-scale separations by capillary electrophoresis and liquid chromatography. *Chem. Rev.* **99(10)**, 3081–3131.
2. Novotny, M. (1997) Capillary biomolecular separations. *J. Chromatogr., B: Biomed. Appl.* **689(1)**, 55–70; Kennedy, R. T., German, I., Thompson, J. E., and Witkowski, S. R. (1999) Fast analytical-scale separations by capillary electrophoresis and liquid chromatography. *Chem. Rev.* **99(10)**, 3081–3131.
3. Manz, A. (1997) Ultimate speed and sample volumes in electrophoresis. *Biochem. Soc. Trans.* **25(1)**, 278–281; Sassi, A. P., Xue, Q., and Hooper, H. H. (2000) Making analysis in the life sciences faster through miniaturization. *Am. Lab.* **32(20)**, 36–41; Cowen, S. (2000) Lab on a chip. *Educ. Chem.* **37(4)**, 96–98.
4. Detken, A., Ernst, M., and Meier, B. H. (2001) Towards biomolecular structure determination by high-resolution solid-state NMR: assignment of solid peptides. *Chimia* **55(10)**, 844–851; Prestegard, J. H. and Kishore, A. I. (2001) Partial alignment of biomolecules: an aid to NMR characterization. *Curr. Opin. Chem. Biol.* **5(5)**, 584–590.

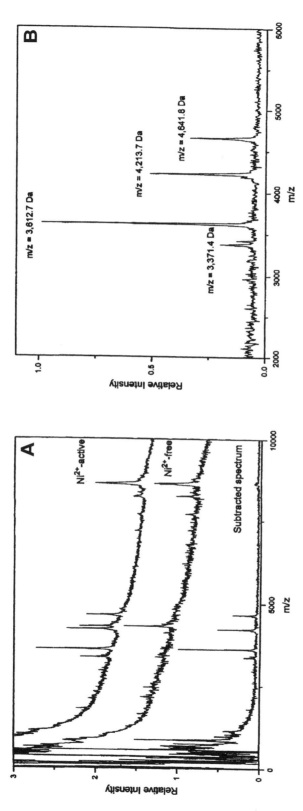

Fig. 46. (A) MALDI-TOF mass spectra of material eluted from the NTA sensor chip using a MALDI matrix solution. Several peptide species are observed in the spectrumresulting from Ni^{2+}-active SPR-BIA, whereas fewer are observed to result from Ni^{2+}-free SPR-BIA. The subtracted spectrum (bottom) shows four signals resulting from specific interaction with the Ni^{2+} chelate surface. (B) Expanded view of the subtracted spectrum shown in (A). Four peptides are retrieved from the unfractionated, trypsinized E. coli lysate as a result of the interaction with the Ni^{2+} chelate (90). Figure reproduced with permission of American Chemical Society.

5. Burlingame, A. L., Boyd, R. K., and Gaskell, S. J. (1998) Mass spectrometry. *Anal. Chem.* **70,** 647R–716R.

6. Filmore, D. (2001) "Mass (spec) Growth," *Today's Chemist at Work* **October 2001,** 45.

7. Tempst, P., Erdjument-Bromage, H., Posewitz, M. C., Geromanos, S., Freckleton, G., Grewal, A., et al. (2000) in *Mass Spectrometry in Biology and Medicine* (Burlingame, A. L., Carr, S. A., and Baldwin, M. A., eds.), Humana Press, Totowa, NJ, pp. 121–142.

8. Lyons, P. A., ed. (1985) *Desorption Mass Specctrometry*, ACS Symposium Series, Vol. 291, American Chemical Society, Washington, DC.

9. Bruins, A. P. (1991) Atmospheric pressure chemical ionization. *Mass Spectrom. Rev.* **10,** 53–77.

10. Geromanos, S., Freckleton, G., and Tempst, P. (2000) Tuning of an electrospray ionization source for maximum peptide-ion transmission into a mass spectrometer. *Anal. Chem.* **72,** 777–790.

11. Fenn, J. B., Rosell, J., Nohmi, T., Shen, S., and Banks, F. J., Jr. (1996), Electrospray ion formation: desorption versus desertion, in *Biological and Biotechnological Applications of Electrospray Ionization Mass Spectrometry* (Snyder, A. P., ed.), ACS Symposium Series Volume 619, American Chemical Society, Washington, DC.

12. Saba, J. A., Shen, X., Jamieson, J. C., and Perreault, H. (2001) Investigation of different combinations of derivatization, separation methods and electrospray ionization mass spectrometry for standard oligosaccharides and glycans from ovalbumin. *J. Mass Spectrom.* **36,** 563–574.

13. Green, B. N., Hutton, T., and Vinograd, S. N. (1996), Analysis of complex protein and glycoprotein mixtures by electrospray ionization mass spectrometry with maximum entropy processing. *Methods Mol. Biol.* **61,** 279–294.

14. Almudaris, A., Ashton, D. S., Beddell, C. R., Cooper, D. J., Craig, S. J., and Oliver, R. W. A. (1992) The assignment of charge states in complex electrospray spectra. *Eur. Mass Spectrom.* **2(1),** 57–67.

15. Bakhtiar, R. and Tse, F. L. S. (2000) Biological mass spectrometry: a primer. *Mutagenesis* **15(5),** 415–430.

16. Zhou, G.-H., Luo, G.-A., Sun, G.-Q., Cao, Y.-C., and Zhu, M.-S. (1999) Study of the quality of recombinant proteins using matrix-assisted laser desorption ionization time-of-flight mass spectrometry. *World J. Gastroenterology* **5(3),** 235–240.

17. Börnsen, K. O. (2000) Influence of salts, buffers, detergents, solvents and matrices on MALDI-MS protein analysis in complex mixtures, in *Mass Spectrometry of Proteins and Peptides, Methods in Molecular Biology, Vol. 146* (Chapman, J. R., ed.), Humana Press, Totowa, NJ.

18. Chen, C., Walker, A. K., Wu, Y., Timmons, R. B., and Kinsel, G. R. (1999) Influence of sample preparation methodology on the reduction of peptide matrix-assisted laser desorption/ionization ion signals by surface-peptide binding. *J. Mass Spectrom.* **34,** 1205–1207.

19. Medzihradszky, K. F., Campbell, J. M., Baldwin, M. A., Falick, A. M., Juhasz, P. M., Vestal, M., et al. (2000) The characteristics of peptide collision-induced dissociation using a high-performance MALDI-TOF-TOF tandem mass spectrometer. *Anal. Chem.* **72,** 552–558.

20. Morris, H. R., Paxton, T., Panico, M., McDowell, R., and Dell, A. (1997) A Novel geometry mass spectrometer, the Q-TOF, for low-femtomole/attomole-range biopolymer sequencing. *J. Protein Chem.* **16(5),** 469–479.

21. "Back to Basics", Micromass, Ltd., 1999, www. micromass. co. uk

22. Takach, E. J., Hines, W. M., Patterson, D. H., Juhasz, P., Falick, A. M., Vestal, M. L., et al. (1997) Accurate mass measurement using a MALDI-TOF with delayed extraction. *J. Protein Chem.* **16,** 363–369.

23. Bier, M. E. and Schwartz, J. C. (1997), Electrospray-ionization quadrupole ion-trap mass spectrometry, in *Electrospray Ionization Mass Spectrometry: Fundamental Instrumentation and Applications* (Cole, R. B., ed.), John Wiley and Sons, Inc., New York, NY, pp. 235–289.

24. March, R. E. and Todd, J. F. J., eds. (1995) *Practical Aspects of Ion Trap Mass Spectrometry: Volume III—Chemical, Environmental and Biomedical Applications.* CRC Press, Boca Raton.

25. Bruins, A. P. (1997), ESI source design and dynamic range considerations, in *Electrospray Ionization Mass Spectrometry: Fundamental Instrumentation and Applications* (Cole, R. B., ed.), John Wiley and Sons, Inc., New York, NY, pp. 107–136.

26. Cohen, S. L. and Chait, B. T. (1996), Influence of matrix solution conditions on the MALDI-MS analysis of peptides and proteins. *Anal. Chem.* **68**, 31–37.

27. Belov, M. E., Gorshkov, M. V., Udseth, H. R., Anderson, G. A., and Smith, R. D. (2000) Zeptomole-sensitivity electrospray ionization-Fourier transform ion cyclotron resonance mass spectrometry of proteins. *Anal. Chem.* **72**, 2271–2279.

28. Burlingame, A. L., Carr, S. A. Baldwin, M. A., eds. (2000) *Mass Spectrometry in Biology and Medicine*, Humana Press, Totowa, NJ.

29. Zubarev, R. A., Kruger, N. A., Fridriksson, E. K., Lewis, M. A., Horn, D. M., Carpenter, B. K., et al. (1999) Electron capture of gaseous multiply-charged proteins is favored at disulfide bonds and other sites of hydrogen atom affinity. *J. Am. Chem. Soc.* **121**, 2857–2862; Axelsson, J., Palmblad, M., Hakansson, K., and Hakansson, P. (1999) Electron capture dissociation of substance P using a commercially available Fourier transform ion cyclotron resonance mass spectrometer. *Rapid Commun. Mass Spectrom.* **13**, 474–477; Kelleher, N. L., Zubarev, R., Bush, K., Furie, B., Furie, B., McLafferty, F. W., et al. (1999) Localization of labile post-translational modifications by electron capture dissociation: the case of g-carboxyglutamic acid. *Anal. Chem.* **71**, 4250–4253.

30. Hakansson, K. Cooper, H. J., Emmett, M. R., Costello, C. E., Marshall, A. G., and Nilsson, C. L. (2001) Electron capture dissociation and infrared multiphoton dissociation MS/MS of an N-glycosylated tryptic peptide to yield complementary sequence information. *Anal. Chem.* **73**, 4530–4536.

31. Caprioli, R. M. III, Malorni, A., and Sindona, G. (1996) *Mass Spectrometry in Biomolecular Sciences*, Kluwer Academic Publishers, New York, NY.

32. J. A. McCloskey, ed. (1990), Mass spectrometry, in *Methods in Enzymology, Vol. 193*, Academic Press, San Diego, CA.

33. McLafferty, F. W. (1980) *Interpretation of Mass Spectrometry*, University Science Books, Mill Valley, CA.

34. Burlingame, A. L., Carr, S. A. and Baldwin, M. A., eds. (2000) *Mass Spectrometry in Biology and Medicine*, Humana Press, Totowa, NJ, Appendix.

35. Larsen, B. S. and McEwen, C. N., eds. (1998) *Mass Spectrometry of Biological Materials: Second Edition*, Marcel Dekker, Inc., New York, NY.

36. Gross, M. L. Tandem mass spectrometry: multisector magnetic instruments, in *Mass Spectrometry, Methods in Enzymology, Vol. 193* (McCloskey, J. A., ed.), Academic Press, San Diego, CA, pp. 131–153; Yost, R. A. and Boyd, R. K. Tandem mass spectrometry: quadrupole and hybrid instruments, in *Mass Spectrometry, Methods in Enzymology, Vol. 193* (McCloskey, J. A., ed.), Academic Press, San Diego, CA, pp. 154–200.

37. Chernushevich, I. V., Loboda, A. V., and Thomson, B. A. (2001) An introduction to quadrupole-time-of-flight mass spectrometry. *J. Mass Spectrom.* **36**, 849–865.

38. Medzihradszky, K. F., Besman, M. J., and Burlingame, A. L. (1998) Reverse-phase capillary high performance liquid chromatography/high performance electrospray ionization mass spectrometry: an essential tool for characterization of complex glycoprotein digests. *Rapid Commun. Mass Spectrom.* **12**, 472–478.

39. Bateman, K. P., White, R. L., Yaguchi, M., and Thibault, P. (1998) Characterization of protein glycoforms by capillary-zone electrophoresis-nanoelectrospray mass spectrometry. *J. Chromatogr. A* **794**, 327–344.

40. Mazsaroff, I., Yu, W., Kelley, B. D., and Vath, J. E. (1997) Quantitative comparison of global carbohydrate structures of glycoproteins using LC-MS and in-source fragmentation. *Anal. Chem.* **69**, 2517–2524.

41. Burlingame, A. L., Medzihradsky, K. F., Clauser, K. R., Hall, S. C., Maltby, D. A., and Walls, F. C. (1995) From protein primary sequence to the gamut of covalent post-translational modifications using mass spectrometry, in *Biochemical and Biotechnological Applications of Electrospray Ionization Mass Spectrometry* (Snyder, A. P., ed.), ACS Symposium Series Volume 619, American Chemical Society, Washington, DC, pp. 472–511.

42. Chapman, J. R., ed. (2000) Mass spectrometry of proteins and peptides, in *Methods in Molecular Biology, Vol. 146*, Humana Press, Totowa, NJ.

43. Spengler, B. (1997) Post-source decay analysis in matrix-assisted laser desorption/ionization mass spectrometry of biomolecules. *J. Mass Spectrom.* **32,** 1019–1036.

44. Yamagaki, T. and Nakanishi, H. (2001) Ion intensity analysis of post-source decay fragmentation in a curved filed reflectron matrix-assisted laser desorption/ionization time-of-flight mass spectrometry of carbohydrates: for structural characterization of glycosylation in proteome analysis. *Proteomics* **1,** 329–338.

45. Sato, Y., Suzuki, M., Nirasawa, T., Suzuki, A., and Endo, T. (2000) Microsequencing of glycans using 2-aminobenzamdie and MALDI-TOF mass spectrometry: occurrence of unique linkage-dependent fragmentation. *Anal. Chem.* **72,** 1207–1216.

46. Mechref, Y. and Novotny, M. V. (1998) Mass spectrometric mapping and sequencing of N-linked oligosaccharides derived from submicrogram amounts of glycoproteins. *Anal. Chem.* **70,** 455–463.

47. Rouse, J. C., Yu, W., and Martin, S. A. (1995) A comparison of the peptide fragmentation obtained from a reflector matrix-assisted laser desorption-ionization time-of-flight and a tandem four sector mass spectrometer. *J. Am. Soc. Mass Spectrom.* **6,** 822–835.

48. Roepstorff, P. (1984) Proposal for a common nomenclature for sequence ions in mass spectra of peptides, *Biol. Mass Spectrom.* **11,** 601; Johnson, R. S., Martin, S. A., and Biemann, K. (1988) Collision-induced fragmentation of (M+H)+ ions of peptides. Side chain specific sequence ions. *Inter. J. Mass Spectrom. Ion Proc.* **86,** 137–154.

49. Biemann, K. (1990) Nomenclature for peptide fragment ions (positive ions), in *Mass Spectrometry, Methods in Enzymology, Vol. 193* (McCloskey, J. A., ed.), Academic Press, San Diego, CA, pp. 886–887.

50. Roth, K. D. W., Huang, Z.-H., Sadagopan, N., and Watson, J. T. (1998) Charge derivatization of peptides for analysis by mass spectrometry. *Mass Spectrom. Reviews* **17,** 255–274.

51. Domon, B. and Costello, C. E. (1988) A systematic nomenclature for carbohydrate fragmentations in FAB-MS/MS spectra of glycoconjugates. *Glycoconjugate J.* **5(4),** 397–409.

52. Fenseleau, C. and Kelly, M. (1995) Complications in the determination of molecular weights of proteins and peptiudes using electrospray ionization mass spectrometry, in *Biochemical and Biotechnological Applications of Electrospray Ionization Mass Spectrometry* (Snyder, A. P., ed.), ACS Symposium Series Volume 619, American Chemical Society, Washington, DC, pp. 425–431.

53. Easterling, M. L., Colangelo, C. M., Scott, R. A., and Amster, I. J. (1998) Monitoring protein expression in whole bacterial cells with MALDI time-of-fight mass spectrometry. *Anal. Chem.* **70,** 2704–2709.

54. Bundy, J. L. and Fenseleau, C. (2001) Lectin and carbohydrate affinity capture surfaces for mass spectrometric analysi of microorganisms. *Anal. Biochem.* **73(4),** 751–757.

55. Bruins, A. P. (1997) ESI source design and dynamic range considerations, in *Electrospray Ionization Mass Spectrometry: Fundamental Instrumentation and Applications* (Cole, R. B., ed.), John Wiley and Sons, Inc., New York, NY, pp. 107–136.

56. Shen, Z., Thomas, J. J., Averbuj, C., Broo, K. M., Engelhard, M., Crowell, J. E., et al. (2001) Porous silicon as a versatile platform for laser desorption/ionization mass spectrometry. *Anal. Chem.* **73,** 612–619.

57. Jacobs, A. and Dahlman, O. (2001) Enhancement of the quality of MALDI mass spectra of highly acidic oligosaccharides by using Nafion-coated probe. *Anal. Chem.* **73,** 405–410.

58. Hng, K. C., Ding, H., and Gao, B. (1999) Use of poly(tetrafluoroethylene)s as a sample support for the MALDI-TOF analysis of DNA and Proteins. *Anal. Chem.* **71,** 518–521.

59. Doucette, A., Craft, D., and Li, L. (2000) Protein concentration and enzyme digestion on microbeads for MALDI-TOF peptide mass mapping of proteins from dilute solutions. *Anal. Chem.* **72,** 3355–3362.

60. Bahr, U. and Karas, M. (1999) Differentiation of "Isobaric" peptides and human milk oligosaccharides by exact mass measurements using electrospray ionization orthogonal time-of-flight analysis. *Rapid Commun. Mass Spectrom.* **13,** 1052–1058.

61. Kristensen, D. B., Imamura, K., Miyamoto, Y., and Yoshizato, K. (2000) Mass spectrometric approaches for the characterization of proteins on a hybrid quadrupole time-of-flight (Q-TOF) mass spectrometer. *Electrophoresis* **21,** 430–439.

62. Martinovic, S., Pasa-Tolic, L., Masselon, C., Jensen, P. K., Stone, C. L., and Smith, R. D. (2000) Characterization of human alcohol dehydrogenase isoenzymes by capillary isoelectric focusing-mass spectrometry. *Electrophoresis* **21,** 2368–2375.

63. Konermann, L. and Douglas, D. J. (1998) Equilibrium unfolding of proteins monitored by electrospray ionizatin mass spectrometry: distinguishing two-state from multi-state transitions. *Rapid Commun. Mass Spectrom.* **12,** 435–442.

64. Happersberger, H. P., Stapleton, J., Cowgill, C., and Glocker, M. O. (1998) Characterization of the folding pathway of recombinant human macrophage-colony stimulating factor-β (rhM-CSF β) by bis-cysteinyl modification and mass spectrometry. *Proteins: Structure, Function and Genetics* **Suppl. 2,** 50–62.

65. Cunsolo, V. Foti, S., La Rosa, C., Saletti, R., Canters, G. W., and Verbeet, M. P. H. (2001) Free energy for blue copper protein unfolding determined by electrospray ionization mass spectrometry. *Rapid Commun. Mass Spectrom.* **15,** 1817–1825.

66. Loo, J. A. (1997) Studying noncovalent protein complexes by electrospray ionization mass spectrometry. *Mass Spectrom. Rev.* **16,** 1–23.

67. Ramanathan, R., Gross, M. L., Zielinski, W. L., and Layloff, T. P. (1997) Monitoring recombinant protein drugs: A study of insulin by H/D exchange and electrospray ionization mass spectrometry. *Anal. Chem.* **69,** 5142–5145.

68. Miranker, A., Robinson, C. V., Radford, S. E., Aplin, R. T., and Dobson, C. M. (1993) Detection of transient protein folding populations by mass spectrometry. *Science* **262,** 896–900; Stevenson, C. L., Anderegg, R. J., and Borchardt, R. T. (1993) Probing the helical content of growth hormone-releasing factor analogs using electrospray ionization mass spectrometry. *J. Am. Soc. Mass Spectrom.* **4,** 646–651.

69. Abe, Y., Odaka, M., Inagaki, F., Lax, I., Schlessinger, J., and Kohda, D. (1998) Disulfide bond structure of human epidermal growth factor receptor. *J. Biological Chem.* **273(18),** 1150–1157.

70. Schuette, C. G., Weisgerber, J., and Sandhoff, K. (2001) Complete analysis of the glycosylation and disulfide bond pattern of human β-hexosaminidase by MALDI-MS. *Glycolbiology* **11(7),** 549–556.

71. Wolf, S. M., Ferrari, R. P., Traversa, S., and Biemann, K. (2000) Determination of the carbohydrate composition and the disulfide bond linkages of bovine lactoperoxidase by mass spectrometry. *J. Mass Spectrom.* **35,** 210–217.

72. Patterson, D. H., Tarr, G. E., Regnier, F. E., and Martin, S. A. (1995) C-terminal ladder sequencing via Matrix-assisted laser desorption mass spectrometry coupled with carboxypeptidase Y time-dependent and concentration dependent digestions. *Anal. Chem.* **67,** 3971–3978.

73. Bonetto, V., Bergman, A-C., Jörnvall, H., and Sillard, R. (1997) C-terminal sequence determination of modified peptides and proteins by MALDI MS. *J. Prot. Chem.* **16(5),** 371–374.

74. Hoving, S., Munchbach, M., Schmid, H., Signor, L., Lehmann, A., Staudenmann, W., et al. (2000) A method for the chemical generation of N-terminal peptide sequence tags for rapid protein identification. *Anal. Chem.* **72,** 1006–1014.

75. Geyer, H., Schmitt, S., Wuhrer, M., and Geyer, R. (1999) Structural analysis of glycoconjugates by on-target enzymatic digestion and MALDI-TOF-MS. *Anal. Chem.* **71,** 476–482.

76. Hagmann, M.-L., Kionka, C., Schreiner, M., and Schwer, C. (1998), Characterization of the $F(ab')_2$ fragment of murine monoclonal antibody using capillary isoelectric focusing and electrospray ionization mass spectrometry. *J. Chrom. A* **816,** 49–58.

77. Huang, L. and Riggin, R. M. (2000) Analysis of nonderivatized neutral and sialylated oligosaccharides by electrospray mass spectrometry. *Anal. Chem.* **72,** 3539–3546.

78. Boss, H. J., Watson, D. B., and Rush, R. S. (1998) Peptide capillary zone electrophoresis mass spectrometry of recombinant human erythropoietin: an evaluation of the analytical method. *Electrophoresis* **19,** 2654–2664.

79. Kristensen, D. B., Imamura, K., Miyamoto, Y., and Yoshizato, K. (2000) Mass spectrometric approaches for the characterization of proteins on a hybrid quadrupole time-of-flight (Q-TOF) mass spectrometer. *Electrophoresis* **21,** 430–439.

80. Liu, T., Shao, X-X., Zeng, R., and Xia, Q-C. (1999) Analysis of recombinant proteins by capillary zone electrophoresis coupled with electrospray ionization tandem mass spectrometry. *J. Chrom. A* **855,** 695–707.

81. Li, J., Kelly, J. F., Chernushevich, I., Harrison, D. J., and Thibault, P. (2000) Separation and identification of peptides from gel-isolated membrane proteins using a microfabricated device for combined capillary electrophoresis/nanoelectrospray mass spectrometry. *Anal. Chem.* **72,** 599–609.

82. Hanisch, F.-G., Jovanovic, M., and Peter-Katalinic, J. (2001) "Glycoprotein identification and localization of O-glycosylation sites by mass spectrometric analysis of deglycosylated/alkylaminylated peptide fragments. *Anal. Biochem.* **290,** 47–59.

83. Keough, T., Lacey, M. P., Fieno, A., Grant, R. A., Sun, Y., Bauer, M., et al. (2000) Tandem mass spectrometry methods for definitive protein identification in proteomics. *Electrophoresis* **21,** 2252–2265.

84. Annan, R. S., Huddleston, M. J., Verma, R., Deshaies, R. J., and Carr, S. A. (2001) A multidimensional electrospray MS-based approach to phosphopeptide mapping. *Anal. Chem.* **73,** 393–404.

85. Hoffmann, R., Metzger, S., Spengler, B., and Otvos, L., Jr. (1999) Sequencing of peptides phosphorylated on serines and threonines by post-source decay in matrix-assisted laser desorption/ionization time-of-flight mass spectrometry. *J. Mass Spectrom.* **34,** 1195–1204.

86. Harvey, D. J. (2001), Identification of protein-bound carbohydrates by mass spectrometry. *Proteomics* **1,** 311–328.

87. Gatlin, C. L., Eng, J. K., Cross, S. T., Detter, J. C., and Yates, J. R. III (2000) Automated identification of amino acid sequence variations in proteins by HPLC/microspray tandem mass spectrometry. *Anal. Chem.* **72,** 757 – 763.

88. Gonzalez, L. J., Shimizu, T., Satomi, Y., Betancourt, L., Besada, V., and Padron, G. (2000) Differentiating α- and β-aspartic acids by electrospray ionization and low-energy tandem mass spectrometry. *Rapid Commun. Mass Spectrom.* **14,** 2092–2102.

89. Collings, B. A., Campbell, J. M. Mao, D., and Douglas, D. J. (2001) A combined linear ion trap time-of-flight system with improved performance and MS^n capabilities. *Rapid Commun. Mass Spectrom.* **15,** 1777–1795; Hoaglund-Hyzer, C. S. and Clemmer, D. E. (2001) Ion trap/ion mobility/quadrupole/time-of-flight mass spectrometry for peptide mixture analysis. *Anal. Chem.* **73,** 177–184.

90. Nelson, R. W., Jarvik, J. W., Taillon, B. E., and Tubbs, K. A. (1999) BIA/MS of epitope tagged peptides directly from E. coli lysate: multiplex detection and protein identification at low-femtomole to subfemtomole levels. *Anal. Chem.* **71,** 2858–2865.

91. Fung, E. N. and Yeung, E. S. (1998) Direct analysis of single rat peritoneal mast cells with laser vaporization/ionization mass spectrometry. *Anal. Chem.* **70,** 3206–3212.

17

DNA Shuffling for Whole Cell Engineering

Steve del Cardayre and Keith Powell

1. Introduction

Pasteur patented the use of whole microbial cells in fermentation in the latter part of the nineteenth century. However, there are much earlier references to microbial fermentations, dating back to alcohol fermentation thousands of years ago. Although more recent times have seen the introduction of many novel processes. We have yet reached a full understanding of even the simplest fermentation. Today the commercial application of whole cells is broad and impacts industries as varied as pharmaceuticals, fuels, foods, and chemicals. Despite the complexity of whole cells, they remain the most commercially successful of the "biocatalysts." Still, as with the commercialization of all biological systems, the challenge is both identifying new and useful properties of commercial interest, and the performance of these biological systems under conditions that are commercially relevant. The most promising biocatalysts never find commercial success simply because they do not perform under conditions that would make them cost effective.

A great deal of money, resources, and effort is committed to the improvement of promising biocatalysts so that they might reach or maintain commercial success. Protein, enzyme, and metabolic engineering are all dedicated to this ultimate goal. Although great advances have been made in each of these disciplines, none have proven robust in their ability to deliver new commercial biocatalysts in a timely fashion. This is primarily because these rational approaches rely on models and assumptions that discount the complexity of biological systems. As an alternative to rational approaches, directed evolution was developed as a productive and robust means of purposely manipulating biological systems. Directed evolution is based upon a natural model of systems engineering, and asks the question "what works?" as opposed to "will this work?" The recursive process of diversification and selection mimics the natural evolutionary process and rapidly explores many potential solutions, as opposed to gambling on only a few.

In addition to plant and animal breeding, the earliest application of directed evolution lies in industrial classic strain improvement. The strategy of chemical or physical mutagenesis and screening has been and remains the primary method for improving commercial microorganisms. A similar strategy of sequential random mutagenesis and

From: *Handbook of Industrial Cell Culture: Mammalian, Microbial, and Plant Cells*
Edited by: V. A. Vinci and S. R. Parekh © Humana Press Inc., Totowa, NJ

screening has successfully been applied to the improvement of single genes, proteins, and enzymes. However, these approaches suffer because they are asexual processes. Mutations arise linearly one at a time, useful genetic diversity within a population is wasted since only one improved gene advances, and deleterious mutations that are acquired over time are difficult to lose. In 1994, Stemmer combined this approach with the theories of breeding and introduced DNA shuffling technology (22). DNA shuffling provides a means to breed a poplation of subgenomic DNA fragments in vitro (Fig. 1). The shuffling reaction produces a combinatorial library of new genes containing the possible permutations and combinations of the genetic diversity that exists within a starting population of genes (21). This innovation allowed directed evolution to become a sexual process as opposed to an asexual process, and to realize advantages of a sexual process immediately (Fig. 2). Multiple advantageous mutations are acquired simultaneously, evolution occurs at the population as opposed to the individual level, and deleterious mutations are easily segregated from beneficial mutations. Stemmer and his colleagues at Maxygen, the company that commercialized the shuffling technology, demonstrated that DNA shuffling in combination with high-throughput screening dramatically accelerates directed evolution (23). The technology has since revolutionized the field and the business of directed evolution. This chapter explores the application of DNA shuffling to the improvement of single genes, enzymes, and metabolic pathways, and describes how this technology has been extended to the improvement of whole cells for commercial use.

2. Functional Improvement of Genes and Enzymes through DNA Shuffling

2.1. DNA Shuffling and Directed Molecular Evolution

The application of DNA shuffling and other methods of directed evolution to the improvement of enzyme properties has been documented extensively in the primary literature, and numerous reviews of the success of these approaches are available (14,16,19,23). There are actually three fundamentally different methods of directed evolution, each differing by the method of diversification in the evolutionary process. Sequential random mutagenesis is analogous to classic strain improvement at the single DNA-fragment level. Diversity originates as random mutations. A mutant population is screened and the single best is taken forward for further diversification and screening (2,7,15). This process works, but is impeded by its asexual nature and the low intrinsic value of a random point mutation. Single gene shuffling also relies on random mutations, but after each round of screening, a population of improved variants is advanced to the next round of diversification as opposed to the single best. Shuffling catalyzes recombination within this population, producing a library of amplified diversity with new mutant combinations (Fig. 2). Recombination amplifies the diversity in a binomial fashion. If there are 10 point mutations in the population, shuffling theoretically produces the $2^{10} = 1024$ possible combinations of those mutations (21). Family shuffling differs from these two approaches because it takes advantage of the natural diversity that exists between homologous genes (5,17). Recombination within a family of homologous genes allows the exchange of large segments of divergent sequence, thus generating a gene library that spans a much larger portion

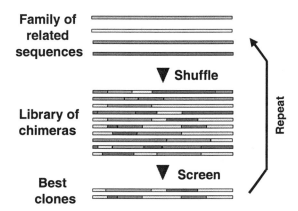

Fig. 1. DNA shuffling catalyzes recombination between a family of related gene sequences. These sequences can be DNA fragments that differ only by several random point mutations, as in single gene shuffling, or can be naturally occurring gene homologs, as in family shuffling. Shuffling generates a library of chimeric genes with genetic information from many of the parental sequences. Screening of this library identifies those genes that combine the best traits from numerous parents. These genes can then be further improved by additional rounds of shuffling and screening.

of sequence space than can be accessed simply by the random mutagenesis of a single gene. Since the sequence diversity that is generated derives from natural sequences that for millions of years have been selected for function and adaptability, the resulting family-shuffled library is both more active and more functionally diverse than one based on random mutations. Numerous permutations of the shuffling methods have been developed, but most aim to accomplish the same end result—recombination between structurally and functionally diverse sequences.

2.2. DNA Shuffling of Genes and Proteins

DNA shuffling has been applied to the improvement of numerous proteins and enzymes. The properties that have been improved can be generally organized into five groups: improved expression, environmental adaptation, improved selectivity, improved activity, and novel selectivity. These each have a profound implication for the enabling of enzyme as catalysts in organic synthesis, but may provide more application in the engineering of cellular systems.

2.3. Improved Expression

One of the earliest published reports of DNA shuffling was its application to the improvement of green fluorescent protein (GFP) activity in *Escherichia coli*. The gene was shuffled by single gene shuffling, and the library was screened visually by seeking *E. coli* colonies that fluoresced more intensely. Analysis of the improved variants demonstrated that the evolved GFPs were not more fluorescent, but rather were expressed more efficiently. Expression of the original gene resulted in the formation of inclusion bodies, and the new gene resulted in the formation of a higher proportion of properly folded and functional protein. Whether this was the result of folding kinetics or protein solubility is unknown, but the colonies expressing these variants appeared to fluoresce

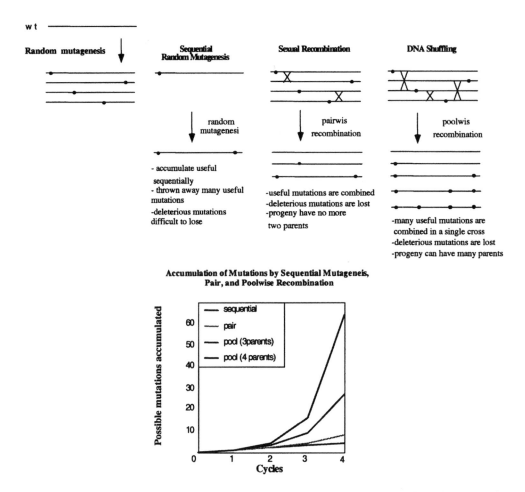

Fig. 2. The rate of evolution can be limited by the number of useful mutations that a member of a population can accumulate between selection events. In sequential random mutagenesis, useful mutations are accumulated at a rate of one per selection event. Many useful mutations are discarded in each cycle in favor of the best performer, and deleterious mutations that survive are difficult to lose and accumulate. In sexual evolution, pairwise recombination allows mutations from two different parents to segregate and recombine in different combinations. Useful mutations can accumulate, and deleterious mutations can be lost. Poolwsie recombination, such as that effected by DNA shuffling, has the same advantages as pairwise recombination, but allows mutations from many parents to recombine into a single progeny. Poolwise recombination provides a means to increase the number of useful mutations that can accumulate for each selection event. The graph shows a plot of the possible number of mutations an individual can accumulate by each process. Recombination is exponentially superior to sequential random mutagenesis, and this advantage increases exponentially with the number of parents that can recombine.

more intensely *(6)*. Often the most prevalent mutations found are "expression mutations" that affect improved transcription, translation, or folding and result in more active protein in the cell as opposed to a change in the stability, function, or kinetic parameters of the protein. Indeed, a mutation may often result in only a change in DNA sequence, with no change in amino acid sequence. Expression mutants are extremely useful in a practical sense, but can be problematic when seeking variants that affect other properties such as specificity or environmental tolerance, since they represent "false positives" that must be ruled out in subsequent hit characterization.

2.4. Environmental Adaptation

Perhaps the most frequent application of directed evolution described in the literature is the improvement of enzyme environmental tolerance, generally thermal stability. The application of such enzyme variants is broad, and ranges from use in organic synthesis to laundry detergent. Environmental tolerance here is meant to include variations in the physical and chemical properties of the medium in which an enzyme is expected to optimally perform, and includes temperature, pressure, pH, ionic strength, solvent, and solute concentrations. Perhaps the best example of the utility of DNA family shuffling in this regard was the family shuffling of 26 subtilisin variants by Ness et al. *(17)*. Subtilisin is one of the most thoroughly studied and engineered enzymes. The Maxygen group shuffled Savinase, a commercial subtilisin produced by Novo Nordisk for use in laundry detergent, with 25 homologs. The library was plated on casein plates, and active clones—those resulting in a halo as a result of secreted active enzyme—were picked for further analysis. Approximately 20% of the library was active, and sequence analysis of random clones (active and nonactive) revealed that 95% were complex chimeras, with >4 crossovers between parental sequences. Of the active clones, 654 were studied for "functional diversity." Each was tested for improvements in thermal stability, activity in organic solvent, and activity at high and at low pH, and 4–12% of the active shuffled variants (depending on the parameter screened) demonstrated improved activity over the best starting parental sequence used (Fig. 3). Thus 1–24 clones, from a thousand members of the crude library, had significant improvements in each of the properties investigated. Even if there had been no simple plate assay for active clones, a manual screening of 3000 members of the library would have identified several with improvements in each of the parameters. Further, improvements were up to two- to threefold relative to the best parent, and were not a result of expression mutations. This high level of functional diversity is significant. It means that more sophisticated, lower-throughput screens that reveal more useful information about each clone can be used to identify useful hits from a library. There is a frequent misconception in the literature that a genetic selection is required for the purpose of directed evolution. Screens are sufficient; they generally provide much more reliable results than a selection, and can be specific. Selections allow a cell to exercise the numerous mechanisms available to circumvent lethal events and ensure survival, and hence can be problematic.

An additional insight from this study came from the analysis of the library for clones with improvements in multiple properties. The population showing improvements at pH 10 was analyzed for improvements in thermal and solvent stability. A plot of the performance of this population is shown in Fig. 4. The diverse distribution of performance tells a great deal about the functional diversity of the library. First, the population with improved performance at high pH is diverse, it is not a redundant (clustered) library but a population of genetic variants that has converged on this phenotype. When exposed to different environmental challenges, these differences become apparent as each clone performs differently under the other environmental conditions. Second, there

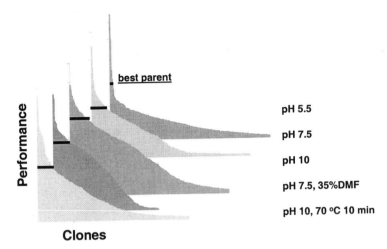

Fig. 3. Ness et al identified 654 shuffled subtilisin enzymes from a family-shuffled library and measured the activity of each under different environmental conditions: pH 5.5, pH 7.5., pH 10, pH 7.5, and 35% dimethylformamide (DMF), and pH 10 after incubation at 70°C for 10 min. For every condition, there were >20 library members with activity better than the best parent under the measured conditions.

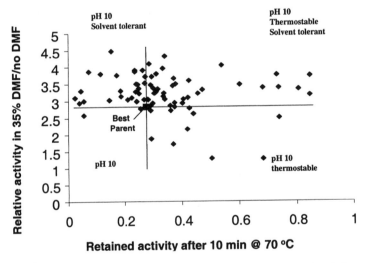

Fig. 4. Seventy-seven shuffled subtilisin enzymes that had improved activity at pH 10 were identified from a family-shuffled library. The activity of these enzymes were then measured in 35% dimethyl formamide (DMF), and after incubation at 70°C for 10 min. The functional diversity of the population demonstrates that numerous solutions to the pH 10 challenge were identified, and that within this small population, there are enzymes that are tolerant to all of the environmental challenges, pH, solvent, and heat.

are clones that perform with all combinations of the screened activity. Some have improved thermal stability, some improved solvent stability, some have neither, and some have both. The important point is that within this small population, all the complex phenotypes are present. It is simply a matter of screening for them.

2.5. Improved Selectivity

Enzymes, although exquisitely selective, often do not show exclusive selectivity, and will function on a variety of substrates with varying preference. Broad specificity can be of value when one hopes to use a single enzyme for a variety of applications. More often, however, a lack of selectivity can result in more than one reaction product. This is often the case when the enzyme is part of a metabolic pathway and the desired product is contaminated with shunt products. However, in the case of organic synthesis, regio- and enantio-selectivity are intrinsic to the value of enzymes as catalysts, and improving the selectivity of natural catalysts is one of the most promising opportunities in the field of directed evolution.

One of the early publications by Stemmer's group described the application of DNA shuffling for the improvement of the fucosidase activity of a galactosidase. The *E. coli* lacZ β-galactosidase was shuffled and the library was screened for activity on ortho- and para-nitrophenyl fucopyranoside esters *(25)*. Those with improved activity were taken forward for additional rounds of shuffling and screening. After 7 cycles, a new enzyme having 13 base changes and six amino acid changes was identified as having a 1000-fold improvement in (k_{cat}/K_M for the o-nitrophenyl fucoside)/ (k_{cat}/K_M for the o-nitrophenyl galactoside) and a 300-fold improvement for the (k_{cat}/K_M for the p-nitrophenyl fucoside)/ (k_{cat}/K_M for the p-nitrophenyl galactoside) relative to the *E. coli* enzyme. In a striking example of the utility of directed molecular evolution, May et al. overcame the rigid stereoselectivity of known hydantoinases, one of the key drawbacks to hydantoinase-mediated amino acid production *(13)*. Using directed evolution, this group inverted the stereospecificity of an *Arthrobacter hydantoinase* from the D to the L enantiomer of methionine hydantoin. Degussa now employs this evolved enzyme in a commercial process for the production of L-methionine. The important observation is that enzyme specificity—whether for distinct chemicals, distinct functional groups of a single chemical, or for a specific stereoisomer of a chiral compound—can be rapidly adapted through the process of directed molecular evolution.

2.6. Improved Activity

Although it is often possible to identify an enzyme that catalyzes a reaction of interest, the enzyme may not catalyze the reaction at a rate that is acceptable for commercial application. Thus, improving the catalytic efficiency of the enzyme is what is required. To use an enzyme in a commercial situation, one may require properties that are clearly divergent from those required to enable the organism to use that enzyme during growth. Thus, the enzyme can be evolved to adapt to this new function simply by screening for the new function. An example of the latter can be found again in the subtilisin evolution study by Ness et al. Although the group screened for improvements in activity under a variety of conditions, a two- to threefold improvement in activity at the native enzyme's pH optimum, pH 7.5, was identified within the family-shuffled library (Fig. 3). Similar improvements in activity were identified from a family-shuffled library of atrazine hydrolases discussed in **Subheading 2.6.** (Fig. 5) *(20)*.

2.7. Novel Selectivity

The chemical reactions catalyzed by enzymes are limited (oxidation, hydrolysis, acylation, etc.), however, the range of molecules on which each class of enzyme oper-

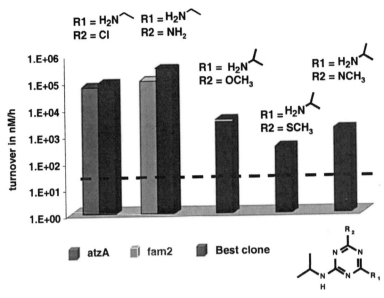

Fig. 5. Raillard et al. shuffled two triazine hydrolases and screened the resulting library for the ability to hydrolize a small set of triazine substrates. The substrates differered by the bulkiness of their ring substituents (R_1) and the nature of their leaving group (R_2). New enzymes (blue) were improved on the native substrates, and some had acquired the ability to function on substrates for which there was no detectable activity by the parental enzymes. The dotted line shows the threshold sensitivity of the high-throughput mass spectronomic assay used in the study.

ates is extremely diverse. The fact that naturally occurring enzyme homologs catalyze the same class of chemical reaction (go through a similar transition state) but function on distinct substrate molecules implies that family shuffling could lead to enzymes that function on new substrates for which there is undetectable activity from any of the parental enzymes. Gerlt and Babbitt's studies on the mechanistic conservation of enzyme superfamilies *(1)* provide further support for this theory. Raillard et al. tested this hypothesis by screening a library of family-shuffled atrazine hydrolases for activity on a panel of 15 triazine substrates that differed by leaving groups and ring substituents. The parental enzymes were active on a small set of the substrates, but there were members of the shuffled library that had acquired "new substrate specificity." These enzymes catalyzed the hydrolysis of substrates for which there was no detectable activity by the two parental enzymes (Fig. 5). This observation was the first demonstration that completely new substrate selectivity could be generated through DNA family shuffling. The study implies that it is not necessary to have initial detectable activity on a given substrate to initiate a directed evolution program. Thus, the massive screening of naturally occurring enzymes for one with detectable activity for a desired substrate is not necessary. One needs only to shuffle a family of related enzymes that catalyze the proper chemistry and screen for activity on the desired substrate.

3. Single Enzymes in Whole Cells

For a chapter dedicated to the discussion of the application of directed evolution to cell-culture engineering, the discussion has thus far focused solely on the improvement of individual enzymes for cell-free applications. This was to highlight the type of improvements that have been achieved through directed evolution. This section describes how these types of improvements can be applied to the improvement of whole cell systems, addresses the applications of improving single enzymes and complete pathways, and describes the expansion of DNA shuffling to whole-genome shuffling for the global improvement of whole-cell complex phenotypes.

3.1. DNA Shuffling as a Tool for Metabolic Engineering

Metabolic engineering can be broadly defined as the intentional manipulation of a cell by specific genetic modification. Although the goals are complex, the tools are relatively simple. Most efforts combine the modeling of a system, the identification of genetic limitations, and the correction of those limitations by targeted genetic modification. Common goals of metabolic engineering are improving the titer, yield, or productivity of a whole-cell catalyst or the expansion of the catalytic capacity—for example, to accept an alternative carbon source or to catalyze non-native chemical transformations. Genetic modifications include gene cloning, duplication, overexpression, deletion, mutagenesis, and selection for the loss of feedback inhibition, and often are supplemented with medium engineering. Each of these strategies has proven successful; however, there is still considerable room for improvement.

Targeted genetic modifications often do not result in the desired physiological changes, and can result in unexpected deleterious effects. This occurs because modern models greatly simplify biological systems, and the common methods of genetic manipulation diverge from natural biological strategies. DNA shuffling provides an additional tool in the engineer's toolbox that is based on the natural evolution algorithm for biological systems. Employing this technology, the ultimate goals of metabolic engineering described here can more routinely be achieved. For example, poor gene expression can result from regulation, RNA structure, or codon usage. These cannot be overcome by gene duplication, but rather require alteration of the gene sequence itself. Further, the activity of a gene product may be impaired simply as a result of the cellular environment in which it is expected to function. Commercial culture conditions seldom reflect a cell's natural environment. Thus, it is wrong to expect the physical and chemical intracellular environment to reflect the optimal functional state of the target gene. This is especially true when the target gene is heterologous, and cellular perturbations include variation in cellular proteins, metabolites, machinery, or chaperones, in addition to the stressful environment of a commercial bioreactor. As demonstrated in the cell-free systems described previously, one can direct the evolution of an enzyme to have improved expression, improved function in complex environments, to have improved selectivity for a single substrate present in the cell, to have enhanced catalytic activity, or to have completely novel substrate selectivity. Thus, in combination with a metabolic model, a targeted gene or metabolic pathway can be shuffled, and cells expressing the library can be screened for improved function. Completely new pathways can be constructed by using heterologous pathways that are adapted to a new host environment, or new pathways using evolved enzymes with completely new activities can be assembled. This approach greatly expands the possible genetic variations that can be tested in a metabolic engineering program and employs a strategy more likely to succeed, as it is the same strategy used in nature. **Subheading 3.2.**

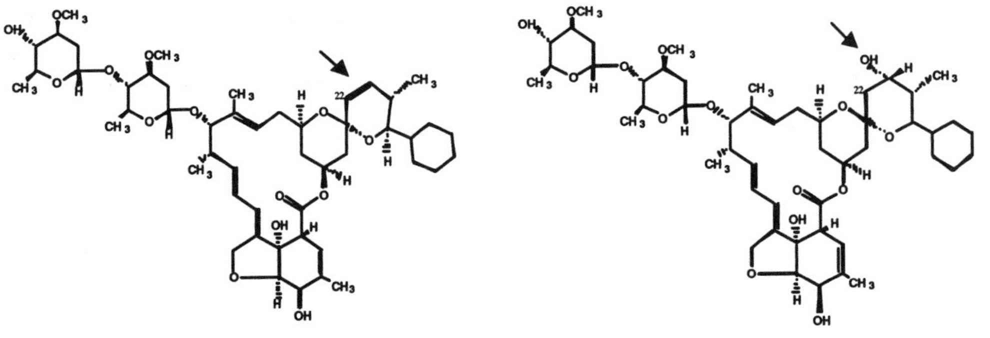

Fig. 6. Chemical structures of the CHC-B1 and CHC-B2 avermectin analogs.

describes two examples of how this approach has led to significant improvements in commercial fermentations.

3.2. Avermectin

Avermectin and its analogs are major commercial products for parasite control in the fields of animal health, agriculture, and human infection. The avermectins were discovered in 1979. The highly complex carbon chain backbone of avermectin is synthesized by the sequential activity of a multifunctional polyketide synthase (PKS) *(10–12)*. Mutants of *Streptomyces avermitilis* that are deficient in branched-chain-2-oxo acid dehydrogenase (*bkd*-deficient mutants) can only produce natural avermectins when fermentations are supplemented with the fatty acids S-(+)-2-methylbutyric acid and isobutyric acid. If supplemented with other fatty acids, novel avermectins are produced. Supplementation of fermentations with cyclohexanecarboxylic acid results in the production of the novel B1 avermectin, Doramectin (Fig. 6), which is sold commercially as Dectomax™. In the commercial production of Dectomax™, both the cyclohexyl-B1 (CHC-B1) and cyclohexyl-B2 (CHC-B2) forms are produced in fermentation. Although characterization of the genes in the avermectin gene cluster has clarified much of biosynthetic pathway *(10,12)*, the mechanism for determining the ratio of B2:B1 produced during fermentation remains unclear. The synthesis pathway branching point between the B1 and B2 compounds is probably an early event in the biosynthetic pathway *(3)*. A gene, *aveC*, which lies between two sets of PKS genes in the avermectin biosynthetic gene cluster, when mutated, resulted in the production of a single avermectin component, B2a *(11)*.

Among the related avermectins, the B1 type of avermectin is recognized as having the most effective antiparasitic activity and is therefore the most commercially desirable form. The production of a single B1 component would simplify production and purification of commercially important avermectins. Because directed evolution requires only a gene sequence, a link to the phenotype, and a functional assay, we opted to further evolve *aveC* by DNA shuffling and screening for variants producing increasing amounts of the CHC-B1 compound relative to the CHC-B2 compound. Libraries of the shuffled *aveC* gene were introduced into an *aveC* deletion strain of *S. avermitilis*, grown in a 96-well solid format and screened by mass spectroscopy for variants yielding an improved CHC-B2:CHC-B1 ratio. Positive clones were isolated

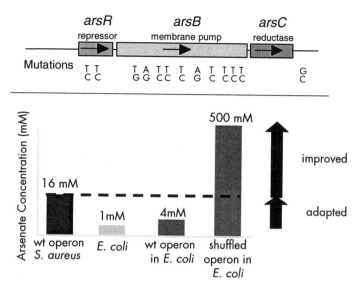

Fig. 7. Crameri et al. used DNA shuffling to improve the arsenate resistance operon from *Staphylococcus aureus* to confer arsenate resistance to *E. coli*. The operon contains three genes, a regulatory protein, a reductase, and an arsenite pump, and it confers to *S. aureus* resistance to 16 m*M* arsenate. When expressed in *E. coli*, the wt operon confers resistance to only 4 m*M*. After three rounds of shuffling and screening for improved resistance, a new operon containing 13 mutations was identified that conferred to *E. coli* resistance to 500 m*M* arsenate. Apparently, shuffling had both adapted the operon to the new environment of *E. coli* and improved its performance 30-fold over that observed in *S. aureus*.

and sequenced. The *aveC* mutations yielding the greatest improvement were introduced into the production strain chromosome, thereby improving the CHC-B2:CHC-B1 ratio obtained in fermentation by fourfold over wild-type levels.

3.3. Pathway Shuffling

The first demonstration of the directed evolution of a metabolic pathway was by Crameri et al. who showed that *E. coli*'s resistance to arsenate could be improved by evolution of the aresenate resistance operon from *Staphylococcus aureus (4)* (Fig. 7). The *ars*RBC operon, when introduced in *E. coli*, confers resistance up to only 4 m*M* arsenate (as opposed to 16 m*M* in *S. aureus*). After three rounds of shuffling, a clone was found that grew in 0.5 *M* arsenate. The improved clone was an enigma. Of the thirteen mutations, only three resulted in amino acid substitutions, and these were in the arsenite pump as opposed to the anticipated arsenate reductase *(4)*. In addition, the improved operon had integrated into the *E. coli* chromosome. The beauty of these findings is twofold. First, DNA shuffling provided the means to both adapt the pathway to the environment of the new host (4 m*M* to 16 m*M* resistance) and to further improve its function in *E. coli* to unprecedented levels (to 0.5 *M*, the solubility of arsenate). Second is the nature of the mutations. The fact that improvement required mutations in the pump as opposed to the reductase, which was hypothesized to be the limiting step, demonstrates the advantage of an evolutionary approach as opposed to a rational approach. If the reductase had been targeted, these improvements would probably not have been identified. The fact that an improvement in the pathway resulted from its integration into the genome is not surprising. However, it is surprising that integration can result simply from DNA shuffling, a very unexpected result of simple gene-sequence variation.

4. Genome Shuffling

Genome shuffling is a recent breakthrough in shuffling technology that combines the advantage of multi-parental crossing allowed by DNA shuffling with the recombination of entire genomes normally associated with conventional breeding. The technology was developed to catalyze the process of industrial strain improvement and to provide a means of rapidly evolving an entire organism in the absence of sophisticated models or genetic tools. Since most whole-cell phenotypes are complex by nature, it is difficult to model or identify which genes in the dynamic cellular machine may be limiting for a desired phenotype. By evolving an entire genome, no assumptions as to what is limiting are made, and an organism can improve through the evolution of its entire genome.

Genome shuffling is similar to classic strain improvement (CSI) in that it is a recursive cycle of genomic diversification and screening for improved strains (24). Its primary difference is that it is a sexual as opposed to an asexual process, and that populations of improved strains as opposed to individual strains are evolved. CSI is an asexual process of sequential random mutagenesis and screening (Fig. 2) (26). A promising microoganism is mutagenized to produce a diverse library of random mutants. Individuals from this population are then screened for improvement in a commercially relevant phenotype, such as titer, yield, or environmental tolerance, and the single best of the screened population is taken forward into additional cycles of mutagenesis and screening. CSI is inefficient, since each cycle produces mutations resulting in phenotypic improvements, other than the single best, which are discarded. The most resource- and time-intensive aspect of CSI is screening, so these discarded mutants are a waste of time and money. The shuffling of improved strains resulting from a single round of CSI theoretically should produce new strains of enhanced performance, as is observed from the shuffling of improved genes (Fig. 8). Zhang et al. tested this hypothesis (26). They first demonstrated that recursive genomic recombination within a population of bacteria efficiently generated a shuffled library of new strains. They then used this approach to shuffle a population of classically improved bacteria. Described here are the findings of these studies, as well as two examples of how this technology has been successfully applied to the rapid improvement of commercial microorganisms.

4.1. Recursive Recombination

The key to genome shuffling is providing a means to efficiently catalyze hyperrecombination within a population of related microorganisms. Bacteria generally are not considered sexual organisms; however, there are numerous mechanisms by which genetic material is exchanged between microbes in nature. These include natural competence, conjugation, transduction, and transposition. In addition, protoplast fusion, physical, and chemical transformation can enhance genetic exchange in the lab. Of all these methods, protoplast fusion is the most efficient, the most general, and requires the fewest genetic tools (9). For this reason, Zhang et al. explored protoplast fusion as a means of effecting genome shuffling in Streptomyces.

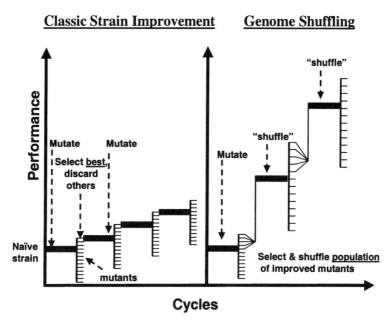

Fig. 8. Classic industrial strain improvement (CSI) is an asexual process of sequentially accumulating individual mutations. Selection of the "fittest" results in the capture of only a single mutant. It is slow because individuals within a population evolve alone as opposed to sharing information and evolving as a group. Genome shuffling allows the information within a population of classically improved strains to be shared. Mating within a selected population consolidates genetic information by providing a mechanism for the combination of useful mutations and the loss of deleterious mutations. Genome shuffling thus produces new populations containing individuals that are much more fit than their parents.

Hopwood, Baltz, and others, have reported that recombination between fused protoplasts of *Streptomyces* was extremely efficient, resulting in up to 50% recombination in regenerated fused protoplasts *(9)*. Hopwood and Wright further measured the efficiency of recombination between fused protoplasts of four multiply auxotrophic strains of *Streptomyces coelicolor (8)*. They found that the simple two marker progeny made up 3% of the regenerated population, and the complex three- and four-marker progeny were present at only 0.04%, and 0.00005% of the population. Since most strain-improvement programs require screening as opposed to selection (for example, for improved titer of a desired metabolite), efficient production of complex progeny was essential. This would decrease the number of individual colonies to be screened, and increase the likelihood of identifying strains that have accumulated multiple useful mutations.

Zhang et al. reasoned that improvement in the distribution of complex progeny could be achieved by the recursive fusion of a mixed protoplast population. Each pooled fusion would mimic each thermal cycle of a DNA-shuffling reaction. In the first fusion, strains with mutations from two parents would be generated. In the second fusion, these could recombine with each other to produce strains with three and four mutations, and so on. Thus, recursive fusions should result in the efficient shuffling of the population (Fig. 9). To test this idea, the same set of *S. coelicolor* strains used by Hopwood and Wright were shuffled by recursive protoplast fusion. Protoplasts from

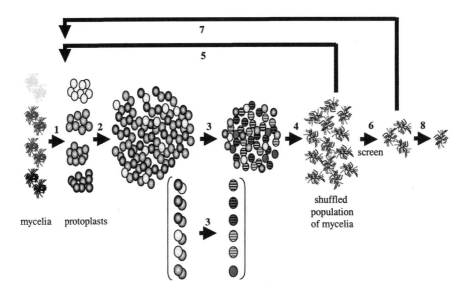

Fig. 9 . Genome shuffling by recursive protoplast fusion. Protoplasts are prepared *(1)* from a population of genetically distinct mycelial cultures. These are pooled *(2)* and recombined through fusion *(3)* and regeneration *(4)*. The majority of the regenerated population contains the original strains and recombinants from all pairwise combinations of those parents (parentheses). Mycelia from the regenerated population are then used to prepare a second population of protoplasts *(5)*, which are then fused as a pool. In subsequent rounds of pooled fusion, cells with DNA from two parents can breed with each other, generating strains with DNA from four parents, etc. Shuffled populations are screened *(6)* for desired phenotypes.

the four strains were mixed, fused, and regenerated under nonselective conditions. Spores from the regenerated protoplasts were pooled, resulting in the first fusion library (F1). F1 was grown as a population and used to prepare protoplasts, which were similarly fused and regenerated. This process was repeated four times, resulting in population F4. Consistent with Hopwood and Wright, a single fusion resulted in the efficient production of two-marker progeny (10%) and the poor production of three-marker (0.4%) and four-marker (0.00007%) progeny. However, recursive fusion efficiently produced the two (60%), three (17%), and four (2.5%) marker progeny, a 40–10^5-fold increase in complex progeny. This demonstrated that recursive protoplast fusion was successful in the shuffling of the four *S. coelicolor* strains, and could be used for the shuffling of a phenotypically selected population for the purposes of directed evolution.

4.2. Improved Tylosin Production from Streptomyces fradiae

Streptomyces fradiae is used for the commercial production of tylosin, a complex polyketide antibiotic sold for veterinary health and nutrition. The biosynthesis of tylosin requires a biosynthetic cluster of ~100 kb in addition to numerous genes distributed throughout the *S. fradiae* genome. Starting in 1961, Eli Lilly carried out an extensive strain-improvement program to improve tylosin production from *S. fradiae*. The program began with strain SF1, a stable derivative of the natural isolate obtained through natural selection. This strain was taken through 20 rounds of CSI for improved tylosin production, resulting in the commercial production strain SF21.

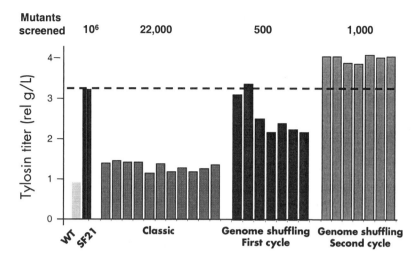

Fig. 10. Zhang et al. improved tylosin production from a strain of *Streptomyces fradiae*, SF1 (wt). A population of classically improved strains were shuffled by recursive protoplast fusion. 1000 members of the first genome shuffled library were screened in 96-well plates, and 7 were identified as having improved tylosin titers (1 cycle). This population was then further shuffled and from another 1000 screened, 7 were identified as having further improvements in tylosin titer. These new strains had tylosin titers significantly improved over SF1 as well as SF21, a commercial strain derived from SF1 through 20 rounds of CSI.

Zhang et al. tested their hypothesis that genome shuffling could catalyze the strain-improvement process by shuffling a population of classically improved strains derived from SF1. 22,000 mutants derived from SF1 via nitrosoguanidine mutagenesis were screened in 96-well plates for improved tylosin production. A total of 11 strains were identified and used as breeding stock for genome shuffling by recursive protoplast fusion. After two rounds of genome shuffling and screening of only 2000 mutants, the group identified new shuffled strains that produced tylosin at levels superior to SF21 in 96-well cultures (Fig. 10) and indistinguishable to SF21 in shake-flask cultures. A molecular and morphological comparison of two of the best performing strains, GS1 and GS2, to SF1 and SF21 demonstrated that GS1 and GS2 were morphologically similar to SF1, yet shared a functionally similar but structurally distinct mutation in a regulatory gene in the tylosin biosynthetic cluster. This study demonstrated that shuffling of the improved population resulting from one round of CSI could produce strains that performed as well as those resulting from 20 rounds of CSI. As a practical comparison, this represented the difference between 20 yr and over a million screens for the generation of SF21 as compared to 1 yr and less than 25,000 screens for the generation of GS1 and GS2.

4.3. Improved Acid Tolerance in Lactobacillus

Patnaik et al employed the same strategy of genome shuffling to improve the acid tolerance of a commercial strain of *Lactobacillus (18)*. *Lactobacilli* are of interest for the commercial production of lactic acid, a compound that has long been used as a food additive and has recently emerged as an important feedstock for the production of other chemicals such as polylactic acid (PLA), acetaldehyde, polypropylene glycol, acrylic acid, and 2,3-pentadione. Fermentation pH contributes to the economics of the lactic acid purification process. Although typical commercial *Lactobacillus* fermentations run at a minimum pH of 5.0 to 5.5, fermentation at or below the pK_a of lactic acid (~3.8) is desirable. At this low pH, significant product is in the free-acid form, and can be purified by direct organic extraction of the fermentation broth. At higher pH, lactate is the predominant form, and a more expensive purification is required. The improvement of microorganisms for improved growth and lactic production at low pH is a commercial goal.

Patnaik et al. took a similar approach to Zhang et al. *(26)* for evolving acid tolerance. Starting with a wild-type strain, LB-WT, which was potentially useful for commercial lactic acid production, the group classically produced a population of acid-tolerant strains. This population of strains was shuffled by recursive protoplast fusion, and the resulting population was enriched for new strains that had an improved ability to grow at low pH relative to LB-WT and the classically improved population. High-throughput screening of the enriched population identified strains that grew well at pH 3.8 and produced up to threefold more lactic acid at pH 4.0 than LB-WT. These results further confirm the value of genome shuffling for the rapid improvement of complex bacterial phenotypes that are of commercial importance.

5. Conclusions and Future Developments

DNA shuffling has proven to be essential for the rapid improvement of genes, proteins, enzymes, and pathways. Genome shuffling will likely prove just as essential for the rapid improvement of whole cells and organisms. As shuffling technology continues to improve, its applications will expand to even more complex systems. These will probably include: targeted pathways distributed throughout the genome of an organism, eukaryotic and tissue-cell culture, multi-celled organisms, and communities of microbes. The natural methods employed by shuffling technologies are the key to their success and the reason why these technologies continue to expand and affect diverse biological systems.

Acknowledgments

We would like to thank the diverse group of scientists at Maxygen who contributed to the research presented in this chapter.

References

1. Babbitt, P. C., Mrachko, G. T., Hasson, M. S., Huisman, G. W., Kolter, R., et al. (1995) A functionally diverse enzyme superfamily that abstracts the alpha protons of carboxylic acids. *Science* **267**, 1159–1161.
2. Baltz, R. H. (1986) Mutagenesis in *Streptomyces* spp., in *Manual of Industrial Microbiology and Biotechnology*. American Society for Microbiology, Washington, DC.
3. Chen, T. S. and Inamine, E. S. (1989) Studies on the biosynthesis of avermectins. *Archives of Biochemistry and Biophysics* **270**, 521–525.
4. Crameri, A., Dawes, G., Rodriguez, E., Jr., Silver, S., and Stemmer, W. P. (1997) Molecular evolution of an arsenate detoxification pathway by DNA shuffling. *Nat. Biotechnol.* **15**, 436–438.

5. Crameri, A., Raillard, S. A., Bermudez, E., and Stemmer, W. P. (1998). DNA shuffling of a family of genes from diverse species accelerates directed evolution. *Nature* **391**, 288–291.

6. Crameri, A., Whitehorn, E. A., Tate, E., and Stemmer, W. P. (1996). Improved green fluorescent protein by molecular evolution using DNA shuffling. *Nat. Biotechnol.* **14**, 315–319.

7. Giver, L. and Arnold, F. H. (1998). Combinatorial protein design by in vitro recombination. *Curr. Opin. Chem. Biol.* **2**, 335–338.

8. Hopwood, D. A. and Wright, H. M. (1978). Bacterial protoplast fusion: recombination in fused protoplasts of *Streptomyces coelicolor. Mol. Gen. Genet.* **162**, 307–317.

9. Hopwood, D. A. and Wright, H. M. (1979). Factors affecting recombinant frequency in protoplast fusions of *Streptomyces coelicolor. J. Gen. Microbiol.* **111**, 137–143.

10. Ikeda, H., Nonomiya, T., Usami, M., Ohta, T., and Omura, S. (1999). Organization of the biosynthetic gene cluster for the polyketide anthelmintic macrolide avermectin in *Streptomyces avermitilis. Proc. Natl. Acad. Sci. USA* **96**, 9509–9514.

11. Ikeda, H. and Omura, S.. (1995). Control of avermectin biosynthesis in *Streptomyces avermitilis* for the selective production of a useful component. *J. Antibiotics* **48**, 549–562.

12. MacNeil, D. J. (1995). Avermectins. Biotechnology 28: 421-42.

13. May, O., Nguyen, P. T., and Arnold, F. H. (2000). Inverting enantioselectivity by directed evolution of hydantoinase for improved production of L-methionine. *Nat. Biotechnol.* **18m** 317–320.

14. Minshull, J. and Stemmer, W. P. (1999). Protein evolution by molecular breeding. *Curr. Opin. Chem. Biol.* **3**, 284–290.

15. Moore, J. C. and Arnold, F. H. (1996). Directed evolution of a para-nitrobenzyl esterase for aqueous-organic solvents. *Nat. Biotechnol.* **14**, 458–467.

16. Ness, J. E., del Cardayre, S. B., Minshull, J., and Stemmer, W. P. (2000). Molecular breeding: the natural approach to protein design. *Adv. Protein Chem.* **55**, 261–292.

17. Ness, J. E., Welch, M., Giver, L., Bueno, M., Cherry, J. R., et al. (1999). DNA shuffling of subgenomic sequences of subtilisin. *Nat. Biotechnol.* **17**, 893–896.

18. Patnaik, R., Louie, S. K. P., and Del Cardayre, S. B. (2001). Genome shuffling of *Lactobacillus* for improved acid tolerance. *Nat. Biotechnol.* **38**, 707–712.

19. Powell, K. A., Ramer, S. W., del Cardayre, S. B., Stemmer, W. P. C., Tobin, M. B., et al. (2001). Directed Evolution and Biocatalysis. *Angewanndte Chemie International Edition* **40**, 3948–3959.

20. Raillard, S., Krebber, A., Chen, Y., Ness, J. E., Bermudez, E., et al. (2001). Novel enzyme activities and functional plasticity revealed by recombining highly homologous enzymes. *Chem. Biol.* **8**, 891–898.

21. Stemmer, W. P. (1994). DNA shuffling by random fragmentation and reassembly: in vitro recombination for molecular evolution. *Proc. Natl. Acad. Sci. USA* **91**, 10,747–10,751.

22. Stemmer, W. P. (1994). Rapid evolution of a protein in vitro by DNA shuffling. *Nature* **370**, 389–391.

23. Tobin, M. B., Gustafsson, C., and Huisman, G. W. (2000). DIrected Evolution: the 'rational' basis for 'irrational' design. *Current Opinion in Structural Biology* **10**, 421–427.

24. Vinci, V. A. and Byng, G. 1999. Strain improvement by non-recombinant methods, in *Manual of Industrial Microbiology and Biotechnology*, (Demain, A. L., Davies, J. E., eds.), ASM Press, Washington, DC.

25. Zhang, J. H., Dawes, G., and Stemmer, W. P. (1997). Directed evolution of a fucosidase from a galactosidase by DNA shuffling and screening. *Proc. Natl. Acad. Sci. USA* **94**, 4504–4509.

26. Zhang, Y. X., Perry, K., Vinci, V. A., Powell, K. A., Stemmer, W. P., and del Cardayre, S. B. (2001). Genome shuffling leads to rapid improvement in bacterial phenotype. *Nature* **415**, 644–646.

18

Cell Culture Preservation and Storage for Industrial Bioprocesses

James R. Moldenhauer

1. Introduction

The purpose of this chapter is to present the current understanding of issues and methods relating to the preservation and storage of cell cultures utilized for industrial bioprocesses. In order to serve as useful tools in biotechnology, cell cultures must be properly preserved and stored to ensure that their biological properties remain unchanged. Although other methods are mentioned, such as lyophilization, this chapter focuses on the use of ultra-cold temperatures to cryogenically preserve and store cells. Three major categories of cells are discussed—microbial, plant, and animal cells. This chapter provides an opportunity for the reader to review current technical, quality, and regulatory issues that influence the use of this broad biological spectrum of cells, which are currently used as tools of biotechnology. The focus of this chapter is the cryopreservation of these cell cultures—i.e., the maintenance of cell viability and biological properties through the application of freezing methodologies.

1.1. Use of Frozen Cells

The use of frozen cell cultures is of fundamental importance to industrial bioprocesses. Only through the application of proper laboratory techniques can cells be preserved for long periods of time to serve as useful substrates for the industrial production of biologically active molecules. These frozen cell cultures must provide a source of phenotypically and genotypically stable substrates for the manufacture of products with consistent properties over a long period of time that may span decades. Freezing of cells is the most commonly used method to ensure proper maintenance of stable cell cultures for manufacturing use. Properly frozen and preserved cell cultures form the basis of biological manufacturing. These cell cultures are cryogenically preserved to establish banks of cells for use in manufacturing processes. The goal of cryopreservation is to achieve metabolic stasis at the cellular level in order to prevent or suppress any genetic or biochemical drift that may occur during storage. Although a worthy goal would be to develop a common method to preserve all types of cells, this has not been achieved. Each type of cell possesses certain unique biological properties that must be retained throughout the preservation and storage pro-

From: *Handbook of Industrial Cell Culture: Mammalian, Microbial, and Plant Cells*
Edited by: V. A. Vinci and S. R. Parekh © Humana Press Inc., Totowa, NJ

cesses. However, despite the physiological differences between cell types, the basic techiques of cryopreservation are remarkably similar for such diverse organisms, thereby providing a surprisingly unified laboratory approach to the preparation of cryopreserved cell cultures.

1.2. Unifying Themes of Cryopreservation

One of the goals of this chapter is to bring to light the common themes and issues shared by this broad spectrum of industrially important cells. The solutions to many of the technical challenges that confront the laboratory practitioner are remarkably similar because of the common problems of preservation and storage shared by all three groups of cells. This chapter highlights the remarkable commonality among these three diverse biological systems with respect to the methods used to successfully preserve the integrity of cellular function during and after exposure to freezing conditions. This is a reflection of the fundamental characteristics shared by different types of cells (e.g., DNA, plasma membrane, and energy metabolism) that provide a unifying theme to cell biology. As a result, the basic principles learned from one type of cell have made it possible to apply and generalize to other cell types. It is hoped that the reader will gain an appreciation for the unity of laboratory strategies employed to successfully preserve and store such diverse cell cultures.

2. Principles of Cell Banking

2.1. Master Cell Bank

Cell cultures employed for laboratory and industrial use must be carefully preserved and catalogued, and properly stored and maintained. Typically, a three-tiered cell-banking system is used for biotechnology products (*see* Fig. 1). The first cell bank to be produced ultimately for use in manufacturing is the Master Cell Bank (MCB). The MCB (the second tier) is generally prepared from a well-studied research cell bank or pre-Master Cell Bank (*see* Fig. 1) generated in research and development laboratories by various technologies involving some method of transfection, transformation, or mutagenesis, and finally, selection of desirable cell clones. A series of tests is performed on this research cell bank prior to preparation of the MCB. These assays vary depending on cell type, but are conducted to confirm characteristics such as purity, phenotype, genotype, and product expression. The MCB is produced by filling vials with a homogenous cell suspension derived from a single harvest (e.g., passage) of cells propagated under controlled laboratory conditions using a well-defined medium. These vials are frozen in a medium designed for long-term preservation under cryogenic conditions.

The number of MCB vials are prepared must last for the entire commercial life of the product, typically about 200 cryogenically preserved vials (i.e., "cryovials"). The MCB vials—considered the "crown jewels" of the company—are subjected to the most rigorous set of tests to ensure that all applicable quality and regulatory standards are met. The MCB serves as the cornerstone of the global regulatory process, and is the foundation on which the company builds a product registration package for submission to regulatory agencies for review. The production process starts with creation of the MCB and ends with a final purified and biologically active product. Therefore, the product license and continued product livelihood are dependent on the proper preparation, preservation, and storage of the MCB.

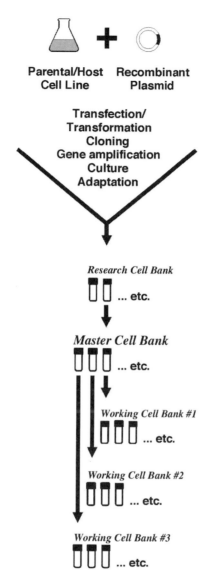

Fig. 1. Diagram of a three-tiered cell-banking process.

2.2. Working Cell Bank

The third tier of the cell-banking system involves the production of a Working Cell Bank (WCB) derived directly from a single cryovial or, if necessary, multiple (i.e., pooled) cryovials of the MCB. If cell banks are managed properly, an almost unlimited supply of WCB can be generated from a single MCB (*see* Fig. 1). The number of cryovials produced for a given WCB will depend on the manufacturing demands for a particular product. At Eli Lilly and Company, it ranges from a few hundred produced for mammalian cell banks to a few thousand produced for some microbial cell banks. The WCB serves as the first building block in the production process. It is the WCB that is used directly in the manufacturing scale-up process to achieve final production volume in the bioreactor. Since the production process starts with the WCB, success in manufacturing depends on its proper preparation, preservation, and storage. In this way, the WCB may constitute the most critical raw material to enter the production-process flow. Thus, the quality of this raw material must be rigorously controlled and meet the highest standards. As a result, both the MCB and WCB must undergo extensive post-bank testing for purity, absence of adventitious agents (e.g., bacteriophage, viruses, or mycoplasma), phenotypic properties, and genotypic stability in order to become fully qualified for manufacturing use.

3. Principles of Cryobiology

3.1. Cryopreservation in Nature

Cryobiology is the study of the effects of freezing temperatures on living organisms—microbes, plants, and animals. Freezing is usually lethal to living organisms. However, some animals in nature have evolved mechanisms to survive the effects of freezing. They possess properly adapted physiological properties to withstand the deleterious effects of freezing. One such example from nature is the wood frog, *Rana sylvatica (1)*. Wood frogs overwinter in leaf litter on the forest floor, where they may experience multiple freeze-thaw cycles during the winter months. They are able to endure freezing temperatures as low as –6°C to –8°C and durations of freezing of 2 wk or more. These frogs have adapted to such harsh climatic conditions through the use of glucose, as a natural cryoprotective agent to help counteract the osmotic shock and control changes in volume of tissue cells. Freezing triggers the rapid breakdown of liver glycogen reserves to produce large amounts of glucose, which is circulated to all organs of the frog's body. Other examples of natural cryopreservation include certain species of polar fish and overwintering insects that produce so-called "antifreeze" agents, which inhibit the growth of ice crystals by binding water molecules to prevent the buildup of ice in cells and tissues *(2)*. The larvae of *Bracon cephi*, a parasitic wasp, accumulate 25% glycerol in their hemolymph (i.e., body fluid) during the autumn. This adaptive process allows these insects to tolerate temperatures as low as –50°C *(2)* by placing them into a state of so-called "suspended animation," in which molecular activity has ceased or is minimized. Similarly, the adult northern carabid beetle, *Pterostichus brevicornis*, can withstand temperatures as low as –40°C during overwintering under natural conditions in its Alaskan habitat *(3)*. Studies of these winter beetles have confirmed this phenomenon in the laboratory *(3)*. These studies have also determined that the glycerol concentration in hemolymph samples varied from as low as 0.0 % in summertime to as high as 22.5 % in wintertime. Incidentally, summer beetles lacking glycerol could not tolerate even mild freezing down to –5°C in those laboratory studies.

3.2. Cryoprotectants

In the laboratory, attempts are made to mimic nature's ability to protect certain animals during times of low-temperature stresses. Using appropriate cryoprotectants at proper concentrations, laboratory practitioners can preserve cellular integrity and functions for long periods of time at extremely low temperatures. Since the 1949 report *(4)* that glycerol could protect spermatozoa during freezing and thawing, it has become a common practice to include one or more such cryoprotective chemicals in freezing media formulations. The effectiveness of glycerol, or any other cryoprotectant, is a direct result of its high solubility in water and its ability to maintain this property at very low temperatures. In this way, a cryoprotectant is able to induce a profound depression of the freezing point of the affected biological solution. It must easily permeate cell membranes, and be able to rapidly diffuse into and out of the cytosol. An effective cryoprotectant must also have a low toxicity for use at concentrations required to protect biological systems against the detrimental effects of freezing. Commonly used compounds that exhibit these properties are dimethylsulfoxide (DMSO), glycerol, 1, 2-ethanediol (ethylene glycol), and 1,2-propanediol (propylene glycol).

DMSO (Me_2SO) is one of the most commonly used cryoprotectants in the laboratory because of its universal ability to rapidly penetrate membranes of living cells and retain its fluidity at ultra-low temperatures as low as –60°C *(5)*. DMSO was first reported in the literature in 1959 *(6)* as an alternative to glycerol for the cryopreservation of human and bovine red cells. Not surprisingly, the cryoprotective effects of DMSO are influenced by its purity. Ultraviolet-spectroscopy-grade solution is recommended for use in cryopreservation *(7–9)*. Whereas cryoprotective agents are essential for the preservation of biological systems, they do not guarantee 100% survival of cells after freezing and thawing. In fact, the utility of cryoprotectants may be limited by their inherent toxicity at higher concentrations, which has been described as "cryoprotectant-associated freezing injury" *(10)*. Sufficiently high concentrations of cryoprotectants can suppress all ice formation in a biological system by a process known as vitrification, in which a vitreous or glassy state is induced during rapid cooling *(2)*. All commonly used cryprotectants will vitrify (i.e., form an amorphous solid) if used at sufficiently high concentrations. Unfortunately, the concentrations required for vitrification of cells, tissues, or organs can be toxic. Fahy *(10)* suggested that if there were no biological constraints on the concentration of cryoprotectants, then 100% survival of cells and tissues would be theoretically possible. For example, because of its intrinsic lack of toxicity, glycerol can be concentrated to 22.5 g % in the hemolymph of an Alaskan winter beetle and protect it during overwintering at temperatures as low as –40°C *(3)*.

The other category of cryoprotectants includes those nonpermeating solutes that cannot penetrate cell membranes, such as sucrose, lactose, trehalose, mannitol, sorbitol, hydroxyethyl starch, and polyvinyl pyrrolidone. These solutes are often used in combination with the permeating cryoprotectants such as DMSO as adjuncts in freezing media to increase the recovery of viable cells. The exact mechanism whereby these chemicals reduce freezing injury is not known. They appear to exert their effects through the alteration of membrane permeability induced by changes in extracellular osmotic forces *(5)*. The protective properties of these larger mol-wt substances may be a result of the ability to facilitate intracellular dehydration through diffusion of water out of the cell via osmotic forces and inhibition of ice-crystal formation.

3.3. Effects of Cooling and Warming Rates

3.3.1. Intracellular Freezing

Two critical factors in the survival and viability of organisms frozen in nature and in the recovery of cells frozen in the laboratory are the rates of cooling and warming. The challenge for the laboratory practitioner is to find the optimal cooling and warming rates for a particular cell type in order to prevent intracellular freezing and mitigate the osmotic stresses induced by high concentrations of solutes in the extracellular medium (i.e., hypertonicity). Since intracellular freezing is lethal, the prevention of intracellular ice formation is a universal goal of both natural and laboratory strategies for the cryopreservation of living cells. Although it is recognized that there is no universal freezing protocol that applies to all cells, the physical and chemical forces acting on all cells are the same, and as a result, finding an optimal cooling rate (i.e., a freezing rate) is critical to cell survival *(11)*. For example, in *Saccharomyces cerevisiae* cells, the incidence of intracellular ice formation increased dramatically when cooling rates were greater than −10°C/min *(11)*. Even in nature, the cooling rate is important. A slow cooling rate (<−1.0°C/h) is even important to the survival of the freeze-tolerant wood frog *Rana sylvatica (13)*. Similarly, the freeze-tolerant northern carabid beetle, *Pterostichus brevicornis*, could not tolerate rapid cooling (>200°C/h) induced in the laboratory *(3)*.

It has been established that the rate of change in temperature controls the diffusion of water across the cell membrane, and indirectly, the probability of intracellular freezing *(12)*. This rate of cooling controls the rate at which water is converted to ice, and therefore, controls the rate at which the concentration of the solutes change in the solution surrounding the cell. The resulting change in osmolarity (i.e., osmotic pressure) of the surrounding medium drives the rate at which water diffuses out of the cell during cooling and into the cell during warming. To prevent intracellular freezing, cells must be cooled slowly enough to allow for removal of about 90% of cytoplasmic water. Evidence has indicated that the residual water (approx 10%) is chemically bound within the cytoplasm and cannot freeze at any temperature *(12)*. Thus, the rate of freezing must be optimized for cell survival based on the physical principle that if cooling rates are too rapid, they will result in intracellular freezing and cell death. Intracellular freezing is a result of inadequate time for cytoplasmic water to diffuse out of the cell in response to the vapor pressure differential (i.e., osmotic pressure) between the extracellular environment and the supercooled cytosol. The optimum freezing rate for a particular cell type is achieved when a rate is found that is slow enough to prevent generation of intracellular ice (see Fig. 2c) and, at the same time, rapid enough to minimize exposure to the damaging hypertonic effects of the surrounding medium. These effects include reduction in cell volumes and stresses on cell membranes such as those of organelles (e.g., mitochondria or nuclei) *(14)*.

Ice crystals formed at very high rates of cooling are usually small (*see* Fig. 2a) or even undetectable using the electron microscope. However, these submicroscopic ice crystals are thermodynamically unstable in relation to larger crystals. Thus, they tend to aggregate during slow warming to form larger crystals in a process known as recrystallization. The rate of warming is the key factor in prevention of recrystallization of frozen water in cells during thawing. It has been shown that the survival of yeast cells is highly dependent on a rapid rate of warming following freezing to −196°C using a rapid cooling rate much faster than optimal *(12)*. As reviewed by Mazur *(12)*, there is a correlation between recrystallization of intracellular ice and cell death as observed in yeast, higher plant cells, ascites tumor cells, and hamster tissue cells. Therefore, by controlling the rates of freezing and thawing, the laboratory practitioner can successfully cryopreserve cells by reducing or eliminating the physical stresses on cells exposed to these critical temperature transitions.

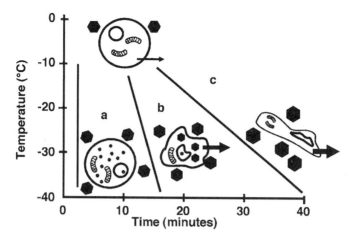

Fig. 2. Illustration showing the effects on mammalian cells of different cooling rates during freezing: (a) very rapid rate, e.g., –100°C/min; (b) rapid rate, e.g., >–10°C/min; (c) optimal rate, e.g., about –1°C/min. Size of arrow indicates extent of water diffusion out of the cell. Hexagons indicate the presence and relative size of ice crystals.

Additional factors for post-thaw cell viability (i.e., survivability) are the rates at which cryoprotectants are added before freezing and removed after thawing. Cryoprotectants are added and removed slowly to reduce the effects of cellular shrinkage and swelling, respectively, caused by osmotic fluctuations. Prior to freezing, cryoprotectants are added slowly to aid in acclimation of cells to the hypertonic environment created by these solutes. The inclusion in the freezing milieu of cryoprotectants such as DMSO and glycerol decreases the temperature at which ice will form within the cell, known as the intracellular ice nucleation temperature. This allows time for dehydration of cells to osmotic equilibrium before the intracellular freezing (ice nucleation) temperature is reached. Intracellular ice nucleation occurs when the cytoplasm is in a supercooled state (<0°C) and exposed to cooling rates too rapid to allow for sufficient osmotic movement of water out of the cytoplasm before the remaining cytosolic water freezes (*see* Fig. 2b).

3.3.2. Extracellular Freezing

Another important determinant in cell survival during freezing and thawing is the effect of extracellular ice formation. As extracellular ice forms, the concentration of solutes increases in the residual unfrozen milieu outside the cell. If the rate of cooling is not slow enough to allow time for equilibration of both the intra- and extracellular osmotic forces, then cells shrink as cytoplasmic water moves out through the cell membrane. This movement of water results in the increasing invagination of the cell membrane because of the reduction in membrane surface tension (i.e., decreasing cell volume and "turgor pressure"), with changes to the structure and function of phospholipid bilayers. Physical forces also contribute to the damage of cells by extracellular ice formation. The damage induced by extracellular ice formation is increased as the cell density increases. One explanation for this cellular damage *(12)* is that growth and expansion of ice crystals exert physical stresses on cells during freezing and cause deformation and disruption of cytoplasmic membranes.

3.4. Storage Temperature

Storage temperature is one of the most critical factors in maintaining the long-term viability of cryopreserved cells. At temperatures below the glass transition temperature (–139°C for pure water), no recrystallization of ice will occur and rates of chemical reactions and biophysical processes are too slow to affect cell survival *(15)*. Essentially, liquid water does not exist at temperatures below about –130°C. The only physical state that exists at these temperatures is crystalline or glassy with an associated viscosity high enough to prevent molecular diffusion. For this reason, laboratory practitioners universally recommend the use of temperatures below -130°C for long-term storage of frozen cell cultures. Additionally, at liquid nitrogen temperatures (e.g., –196°C) there is not enough thermal energy to drive chemical reactions. Therefore, at these temperatures, cell viability should be independent of the storage time. However, even at liquid nitrogen temperatures, photochemical insults can still occur and macromolecules, especially deoxyribonucleic acid (DNA), remain susceptible to background levels of ionizing radiation or cosmic rays *(16)*. At these ultra-low temperatures, the normal cellular DNA repair mechanisms are not functional, and any genetic damage induced will be accumulated throughout the period of storage and be expressed upon thawing and subsequent growth of cells. Interestingly, it has been calculated that with the background irradiation at existing levels, it would take approx 30,000 yr to accumulate the median lethal dose (approx 63% of exposed cells) for mammalian tissue-culture cells cryopreserved in liquid nitrogen at –196°C *(16)*. Conversely, many cells stored at temperatures above –80°C are not stable, and lose viability at rates ranging from several percent per h to several percent per yr, depending on the type of cell and conditions of cryopreservation *(12,14)*. Cellular instability at higher storage temperatures is most likely the result of to damage caused by the presence of residual intracellular liquid water and the progressive recrystallization of ice within the cytoplasm and organelles *(14)*.

4. General Methods of Cryopreservation

The following is a guide to common laboratory methods and practices applicable for long-term cryopreservation of the wide range of commercially important cell cultures covered in this chapter.

- Select a suitable cryogenic container or "cryovial" (i.e., vial or ampoule) that is designed to withstand the rigors of freezing to temperatures produced by the vapor phase of liquid nitrogen (below –130°C). This cryogenic container must either be pre-sterilized by the product vendor or be sterilizable by steam, gamma irradiation, or ethylene oxide gas. Cryovials should be pyrogen-free. It is preferable to use some type of screw-cap plastic cryovial (e.g., high-density polyethylene) to avoid the explosion hazards associated with the improper sealing of glass ampoules and after subsequent storage directly in liquid nitrogen. Pin-hole leaks in glass ampoules will allow liquid nitrogen to penetrate the ampoule, and upon thawing the rapid expansion of the liquid nitrogen to gas can cause ampoules to explode. If glass ampoules are used, they must be checked for leaks prior to freezing by immersion in 0.05% aqueous methylene blue at 4°C for about 30 min *(17)*. Any ampoules containing traces of the blue dye are discarded. Do not store plastic cryovials directly in liquid nitrogen, because most screw-cap seals are not designed to prevent the penetration of liquid nitrogen.

Because of these issues, cell banks at Eli Lilly and Company are stored in high-density polyethylene or polypropylene cryovials. Similarly, the American Type Culture Collection (ATCC) converted from glass to plastic ampoules for the liquid nitrogen storage of microbial cultures *(18)*. Interestingly, the ATCC reported that the contents of plastic ampoules took four times longer to thaw in a 35°C water bath than in glass ampoules, but the effect on viability of the thawed microorganisms was not studied. Additionally, the ATCC initiated an active program in the mid-1990s of substituting plastic for glass in its animal-cell-culture repository laboratories (personal observation). Another consideration is the use of newly manufactured plastic cryovials that are used within their expiration date. Notably, some plastics can deteriorate during storage at room temperatures and may exhibit post-thaw leakage after exposure to liquid nitrogen temperatures *(19)*.

- Harvest, transfer, and filling of cryovials with cell cultures should be conducted in a properly certified laminar airflow biological safety cabinet designed for the dual purpose of protecting both the operator and the cultures from exposure to extraneous microorganisms. This practice helps to maintain the purity of the culture and the health of the operator.

- Generally, most cell types should be harvested for preservation during late logarithmic-phase growth (or in early log phase for some plant cells) when cell viability is maximal.

- To properly identify each cryovial, use a suitable label or an indelible marking pen that can withstand the rigors of freezing and thawing at liquid nitrogen temperatures.

- All cryoprotectants (such as DMSO or glycerol) are sterilized by filtration prior to use and stored at room temperature. Cryoprotective freezing media containing glycerol or DMSO are used shortly after preparation (e.g., 24–48 h), stored and used at refrigerator temperatures.

- After the addition of cryprotectants and during filling of cryovials, cell suspensions should be kept cold (or as cool as possible) with the use of flexible frozen cold packs that can be wiped clean and disinfected (e.g., 70% isopropyl alcohol). These cold packs can be wrapped or placed against the sides of the vessel holding the cell suspension. To minimize the risk of contamination, avoid using wet ice in the laminar airflow biological safety cabinet.

- The cell suspension must be mixed or agitated during the filling process to help ensure the uniform distribution of cells into cryovials. This will help to minimize vial-to-vial variations within a given cell bank. This can be done manually by simply swirling the culture-dispensing vessel, through the use of a tissue-culture stir flask designed for gentle mixing of mammalian cells, or by more vigorous agitation with a stir bar for use with more robust cell types (e.g., bacterial cells).

- Most cell types have an optimal or limited range of cooling rates for maximum survival. The use of a programmable commercial instrument for controlled stepwise freezing of cells (i.e., "controlled-rate freezer") is recommended to ensure optimal freezing conditions for any cell bank used in manufacturing processes. This process of stepwise freezing, with an increased rate of cooling after –40°C to –60°C is reached, is used to minimize the size of ice crystals formed from any residual intracellular water that has not been removed osmotically during dehydration of cells in the first cooling step. These freezers are designed to produce accurate linear cooling and heating by controlled injection of atomized liquid nitrogen into a highly insulated chamber and the pulsing of a heater. A fan and chamber baffles are designed for uniform circulation of nitrogen gas and temperature control across the racks that hold cryovials. These freezers are equipped with thermocouples (e.g., platinum resistance thermometers) to provide precise monitoring of chamber and sample temperatures during freezing, and are programmable to allow the laboratory practitioner to optimize the conditions for a particular cell type and load or configuration of cryovials. Use of this equipment helps to ensure process reproducibility and lot-to-lot consistency when successive cell banks are prepared for manufacturing use. The laboratory practitioner must optimize the freezing program to ensure that controlled cooling rates are achieved during the critical fluctuations of temperatures induced by release of latent heat energy (i.e., "latent heat of fusion") as the water in cryovials freezes.

The latent heat of fusion (or crystallization) can be described as the release of energy in the form of heat (exothermic reaction) as ice crystals are formed from supercooled water molecules. As water molecules are cooled to the freezing point and bind to form an orderly crystalline structure (ice), the associated decrease in entropy results in the release of heat energy as molecules move from a higher to lower energy level. An increase in the cooling rate at this temperature is required to compensate for the transient increase in the temperature of the sample caused by the release of heat energy into the surrounding medium. Without control over the cooling rate, the release of this heat can cause a temperature spike (increase) of the supercooled liquid away from its freezing point and expose cells to a detrimental freeze/thaw cycle. The release of latent heat can significantly affect the cooling curve during freezing. In fact, temperature increases of up to 10°C have been observed in cryovials during this early stage of freezing *(20)*. The time-course of individual cryovials traversing the latent heat of fusion is dependent on the location of the sample within the freezing chamber, but typically occurs near –8°C (unpublished validation data). In order to account for positional variation of temperature, the freezing program must be developed with the most challenging (i.e., "worst case scenario") location within the chamber as a reference.

Fig. 3 is a copy of a temperature profile from an optimized program used to freeze mammalian cell lines at Eli Lilly and Company. This freezing program was developed through a series of pre-validation test runs using calibrated thermocouples interfaced with a temperature recording instrument (Validator® 2000, Kaye Instruments, Inc.). Eleven thermocouples were distributed throughout the cryovial load within the freezing chamber, and temperature data were gathered from these representative sample locations. The cryovials were allowed to equilibrate at 4°C for 20–30 min before starting the cooling program. This step also allowed time for cells to reach osmotic equilibrium with the surrounding medium containing a cryoprotectant such as DMSO. The goal of the freezing program validation was to achieve an average freezing rate of –1.0°C +/–0.1°C per min in all test cryovials immediately after traversing the latent heat of fusion (approx –8°C) to a temperature of approx –60°C. At this point in the program, the cooling rate was increased to –10°C/min until a final hold temperature of –110°C was reached. After completion of the cooling program at –110°C, cryovials would have been transferred to cryogenic storage in the vapor phase of liquid nitrogen.

- If a controlled-rate freezer is not available, then cryovials may be frozen manually by placing them in an insulated container within an ultra-cold mechanical freezer at –60°C or lower for at least several hours before transfer to the vapor phase of liquid nitrogen. Cryovials may be kept in the freezer for up to 24 h. This uncontrolled freezing does not provide a constant rate of cooling and does not compensate for the release of latent heat of fusion during freezing, but it may be used for cell cultures demonstrating an acceptable loss of viability upon thawing (e.g., <15%) when this method is employed.

- Whichever method of freezing is used, cryovials should always be stored in the vapor phase of liquid nitrogen, where temperatures colder than –130°C can be maintained. Whatever type of liquid nitrogen storage vessel (i.e., cryogen) is used, it should be continuously

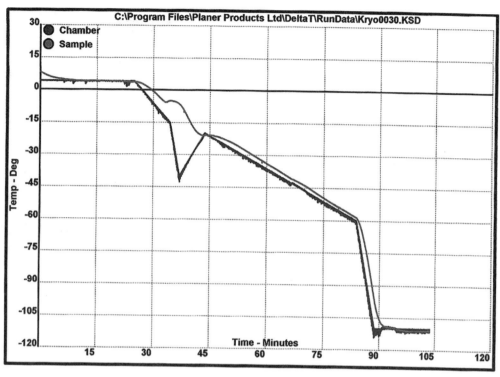

Fig. 3. Graph of a controlled-rate freezing run for a mammalian cell bank. The top line is the cryovial temperature and lower line is the chamber temperature. Notice the release of latent heat of fusion at about –8°C (i.e., slight "bump" in cooling curve) and need to compensate with a dramatic increase in the cooling rate of the chamber in order to maintain an optimal cooling rate in the cryovial of close to –1°C/min.

monitored either for temperature or level of liquid nitrogen (or both). Cryogens should be continuously alarmed to alert lab personnel to temperature excursions above –130°C in locations where cryovials are stored (except during transient opening of a vessel). Opening of cryogens should be minimized to prevent any repeated temperature fluctuations, which could reduce cell viability upon thawing. It is preferable to use cryogens that are designed for a low static rate of liquid nitrogen evaporation (*see* Fig. 4). Such high efficiency cryogens have a relatively small diameter offset neck opening (e.g., 12 in.) which reduces the rate of liquid nitrogen evaporation to approx 4 L per day. Other cryogens with larger diameter neck openings (e.g., 21 inches) may have static evaporation rates of approx 6 L per day. However, it should be noted that static evaporation rates can vary significantly with ambient temperature, barometric pressure, and altitude. Mechanical freezers capable of maintaining temperatures as low as –150°C should only be used if back-up power is available or if a liquid carbon dioxide back-up is installed to maintain ultra-cold temperatures in the event of a temporary power outage.

- Access to cryogens holding critical cell banks should be limited to only authorized laboratory personnel. Once again, as the "crown jewels" of the company, MCB and WCB must be adequately protected in a secured location.
- Generally, the best recovery of frozen cell cultures is achieved by removing the cryovials from liquid nitrogen storage and thawing them as quickly as possible in a water bath pre-

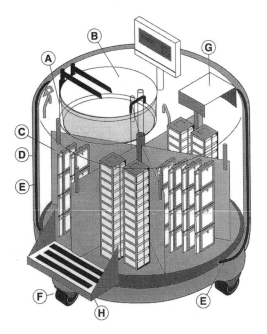

A   Offset neck design to
     maintain -150°C in vapor
     storage and provide low
     liquid nitrogen consumption
     with standard racks

B   Durable metal lid— designed
     for longer life

C   Rotating interior tray provides
     easy access to cryo-
     biological samples

D   Low Maintenance,
     all-stainless steel
     construction

E   Annular filling lines reduces
     frost and ice formation near
     lid

F   Super-tough, durable casters

G   Rack Stand

H   Step-up platform
     (XLC 1520HE, 1830HE and
     1830 2T)

Fig. 4. Drawing of a high-efficiency liquid nitrogen cryogen showing system of metal racks holding plastic boxes for storage of cryovials. Note the relatively small diameter of the neck opening and insulated lid (**B**) designed to reduce static evaporation of liquid nitrogen. (Reproduced by permission of MVE, Inc.)

warmed to 37–40°C. Cryovials should be gently agitated during thawing to facilitate uniform and rapid thawing of the frozen cell suspension. In this way, frozen cell cultures (1–2 mL volumes) can be thawed in less than 1 min. Cryovials should be removed from the water bath when a small sliver of ice remains in the liquid sample, and transferred immediately to a laminar airflow biological safety cabinet for inoculation of cells into an appropriate growth medium. Disinfection of the outside of the thawed cryovials with sterile 70% (v/v) isopropyl alcohol (available from vendors) prior to opening will help to reduce the risk of culture contamination. It is advisable to thaw cryovials in water that has been sterilized by autoclaving or filtration to minimize the risk of contamination.

• Generally, growth medium should be added slowly to freshly thawed cells to minimize the osmotic shock resulting from dilution of a cryoprotectant such as DMSO. Different cell types may be more or less affected by this osmotic perturbation. In any case, it is usually beneficial to introduce the first few milliliters of growth medium slowly with gentle agitation to the cell suspension or cell pellet, if centrifugation is used after thawing.

5. Microbial Cell Cultures

This section explores cryopreservation methods used with a wide range of microbial cells, from bacteria to multicellular fungi. These organisms exhibit a great diversity of morphology, genetics, biochemistry, and physiology, and would, *a priori*, pose a formidable challenge to the laboratory practitioner who is responsible for culture preservation. However, as indicated in **Subheading 4.**, all these groups share a number of common responses to the detrimental effects of freezing and thawing. This section presents a fairly unified set of laboratory techniques used to mitigate these damaging effects: use of common cryoprotectants, slow cooling, rapid warming, storage temperatures, and the condition of cells at the time of preservation.

5.1. Continuous Subculture

Probably the oldest and most traditional method of preserving microbial cultures has been the practice of serial transfer or continuous subculture on an agar medium stored at ambient temperatures on the laboratory benchtop or at selected temperatures in incubators. Agar slants were often covered with mineral oil to aid in the preservation of cultures. However, this practice is fraught with many problems. The most significant include:

- Genetic drift and associated change in phenotypes of sub-populations;
- Contamination with exogenous organisms either introduced by the lab technician or the environment;
- Cross-contamination between preserved strains;
- Loss of viability or death of cultures over time; and
- Lab bench or incubator space to store large numbers of agar slant cultures.

Cryopreservation techniques can either eliminate or minimize the problems associated with this traditional practice. For decades, cryopreservation has proven to be a relatively simple, reliable, and robust method for microbial cell culture preservation at Eli Lilly and Company. Of course, nothing can substitute for well-trained, conscientious laboratory practitioners who are skilled in the art of aseptic practice to maintain cell culture purity and viability!

5.2. Freeze-Drying

The role of lyophilization, or freeze-drying, of microorganisms—particularly viruses—has been firmly established over the years as a reliable means of preservation. During the freeze-drying process, water and other solvents are removed from a frozen aqueous sample by sublimation. Sublimation occurs when a frozen liquid changes directly to a gas without passing through the liquid phase *(21)*. The freeze-drying process consists of three steps, including pre-freezing of the product, primary drying to remove most of the water, and secondary drying to remove bound water. The result is a stable, dry product, which must be reconstituted with water before use. The use of freeze-drying for long-term preservation of more complex biological systems (such as animal cells) has not been successful. Even for microbial cells, the size and type of cell can affect recovery following freeze-drying. For example, it is easier to freeze-dry Gram-positive bacteria than Gram-negative cells *(21)*. Presumably, this phenomenon is the result of the differences in their cell-wall structures.

Freeze-drying has some undesirable side effects that are not observed with cryopreservation. These include the potential for genetic changes during freeze-drying caused by DNA-strand breakage and selection of mutants *(22)*, denaturation of sensitive proteins *(23)*, and decreased viability for many bacterial cell types *(24)*. Mutation has been reported to be induced in freeze-dried *Escherichia coli* during repair of damaged DNA after rehydration *(25)*.

Freeze-drying is labor-intensive, and requires more complicated equipment and protocols than cryopreservation. Use of freezing temperatures to preserve and store cultures is universally simpler and generally more robust for long-term maintenance of a broad range of biologically diverse cell types. For these reasons, only cryopreservation techniques are discussed in subsequent sections.

5.3. Bacteria

The metabolic state of bacterial cells harvested for cryopreservation is of major importance. To maximize recovery, cultures to be preserved should be grown under conditions of temperature, pH, and nutrient concentrations, that minimize lag phase after inoculation and support optimal growth of cells into late logarithmic phase or very early stationary phase (21,26,27). Generally, cells harvested from late log or early stationary growth phases are more resistant to freeze-thaw damage and provide the highest post-thaw recovery of viable cells. It has been demonstrated that aerated and shaken cultures are more resistant to freezing and thawing than cultures grown statically on agar (21). At Eli Lilly and Company, microbial cultures harvested for cell banking are typically propagated in shake flasks that are vented to ambient air and incubated in integrated shaker-incubators. One critical parameter to consider is the optimal volume of medium for a particular size and type of shake flask to provide adequate surface area for exchange of gases (i.e., aeration). Typically, growth conditions are chosen to achieve the highest viable cell density at the end of log-phase growth.

Broth cultures are harvested at the optimal time and concentrated by centrifugation. The resulting cell pellets are resuspended in freezing medium to a final concentration in the range of 10^8 to 10^{10} cells/mL (personal observations, 26). For maximal recovery, it is important to achieve a final cell suspension for freezing above the minimal concentration of 10^7 cells/mL (28). The cells are typically resuspended in freezing medium containing either 5–10% (v/v) DMSO or 5–10% (v/v) glycerol in addition to standard growth medium. A glycerol concentration exceeding 20% (v/v) in the freezing medium resulted in reduced post-thaw colony-forming units (CFU) of a recombinant E. coli production strain (W. Muth, unpublished data). Most recently, at Eli Lilly and Company, the freezing medium is composed primarily of a synthetic glycerin and a plant-derived peptone, both formulated to optimal concentrations.

The optimal cooling rate for most bacterial cell cultures is in the range of $-1°$ to $-5°$C/min down to about $-40°$C, followed by an increased rate of $-10°$ to $-30°$C/min down to a temperature of about $-100°$C or lower. The cryovials are then transferred for storage in the vapor phase of liquid nitrogen.

5.4. Actinomycetes

This group of microorganisms has been extensively studied for the production of useful metabolites, primarily antibiotics for treatment of infectious diseases. These organisms share cell-wall characteristics and multicellular structures such as filaments and conidia with both the bacteria and fungi, respectively. High-yielding cultures that produce valuable compounds, such as antibiotics or other secondary metabolites, must be properly preserved to retain these important biological properties for industrial production. The actinomycetes produce both aerial and submerged spores, and exhibit distinctive morphological features such as spore-chain structures, which can be useful in characterization and identification (29). These are a diverse group of organisms, which have been preserved with a variety of methods.

Cultures propagated for cryopreservation are harvested either from agar or broth media and resuspended in a freezing medium typically containing 10–20% (v/v) glycerol as a cryoprotectant. Alternatively, some cultures have been grown on agar, harvested as plugs in pre-sterilized plastic straws, and frozen directly in the vapor phase of liquid nitrogen or in a mechanical freezer at –70°C *(29)*. In that case, the agar medium provided cryoprotective or cryostabilizing properties to the frozen cultures.

At Eli Lilly and Company, the same methods used in the cryopreservation of bacterial cell cultures described here have been successfully applied to the actinomycetes. However, at least one production strain of *Streptomyces* has been cryopreserved successfully for many years (i.e., 2–3 decades) in the absence of glycerol or any other cryoprotective agent. It should be noted that this production strain was cryopreserved in conditioned medium which, like the agar mentioned here, apparently played a cryoprotective role in the process of freezing.

5.5. Fungi

The fungi comprise a large and diverse group of spore-forming organisms that have adapted to exploit a wide range of ecological niches. They have been utilized for the commercial production of enzymes, antibiotics, alcohols, and other industrial chemicals such as citric acid *(30)*. In general, it has been found that freezing and storage in liquid nitrogen is the preferred method of culture preservation. The advantage of cryopreservation it that is can be successfully used for both sporulating and non-sporulating cultures. In contrast, lyophilization has been useful primarily for sporulating cultures that will survive the combined freezing and drying process. In fact, not all types of spores will withstand the freeze-drying process *(31)*.

Generally, 10% (v/v) glycerol is the cryoprotectant of choice. At the International Mycological Institute (IMI, Egham, UK) over 4000 species belonging to 700 genera have been successfully preserved in media containing 10% (v/v) glycerol *(30)*. A proven method of cryopreservation at IMI has been controlled through cooling at a rate of –1°C /min to approx –35°C, followed by rapid uncontrolled freezing in a liquid nitrogen storage vessel. The use of either glycerol or DMSO as a cryoprotectant has been demonstrated to be effective. DMSO has been used successfully at concentrations of 5, 10, and 15% (v/v). A combination of 10% DMSO and 8% (v/v) glucose has also been used to preserve nine strains of Deuteromycetes *(29)*. Additionally, there is evidence to demonstrate that 10% (v/v) DMSO performed significantly better than 10% (v/v) glycerol for preservation of eight strains of mycelial cultures *(31)*. In fact, it was reported that the two strains surviving up to 6 yr in storage were preserved in DMSO, as compared to others preserved in glycerol. As seen with other cell cultures, slow cooling and rapid warming generally provide the highest recoveries of viable cells. It has been reported that pre-growth (i.e., conditioning) of cultures in the refrigerator at 4–7°C can improve post-thaw viabilities of some fungi *(30)*. This method is similar to the "cold hardening" or conditioning of plant cultures (or plant tissues) at reduced temperatures prior to freezing, and is discussed in **Subheading 6.**

5.6. Yeasts

The yeasts are a group of single-cell eukaryotic organisms that have been successfully manipulated to produce biotechnology products (e.g., recombinant proteins). A wide variety of strains have been preserved successfully by freeze-drying. However, some strains, including pseudomycelium-forming cultures, cannot be lyophilized. The National Collection of Yeast Cultures (NCYC, Norwich, UK) has found that post-thaw viabilities of their repository strains are much higher when cryopreserved than when freeze-dried *(32)*. Additionally, the ATCC has reported on the reduction in viability of two plasmid-bearing strains of *S. cerevisiae* by two to three orders of magnitude after lyophilization *(33)*. The NCYC *(32)* has found no significant loss of either viability or genetic stability of cultures cryopreserved for up to 10 yr. The standard method of cryopreservation at NCYC includes the use of 10% (v/v) glycerol as the cryoprotectant, uncontrolled cooling of filled polypropylene cryovials or straws by transfer to a –30°C methanol bath for 2 h, and final storage in liquid nitrogen. Apparently, this method using an uncontrolled cooling rate has proven successful for all NCYC repository strains.

Experiments conducted with two strains of *S. cerevisiae* at the NCYC using other cooling temperatures of –20°C and –40°C indicated that the primary freezing temperature did not affect the cell-culture viabilities. However, a study of plasmid-based gene expression in recombinant *S. cerevisiae* has indicated that cooling rates of greater than –8°/min resulted in a marked decrease in post-thaw cell viabilities, as well as permanent loss of plasmid DNA *(34)*. Results from another study of *S. cerevisiae* has demonstrated optimal cooling rates of between –3° and –10°C/min *(35)*. Using cryomicroscopy, they found that the incidence of intracellular ice formation increased markedly at cooling rates greater than –10°C/min.

Growth conditions prior to cryopreservation were found to have a profound effect on the response of *S. cerevisiae* cells to freezing and thawing *(35)*. They found that post-thaw viability was significantly higher when cell concentrations were greater than 5×10^8/mL. In addition, it was reported that cells obtained from early stationary-phase growth were more resistant to freezing damage.

At Eli Lilly and Company, methods used in the cryopreservation of a recombinant production culture (i.e., Working Cell Bank) of *Pichia pastoris* included the use of a filter-sterilized freezing medium comprised of a plant-derived peptone and 10% (v/v) synthetic glycerin. Cells were harvested from shake flasks in late log-phase growth and filled into plastic cryovials at cell concentrations in the range of 10^7 to 10^8 viable cells/mL. Cryovials were cooled at a slow rate and frozen in a controlled-rate freezer to approx –120°C and then transferred to the vapor phase of liquid nitrogen.

6. Plant Cell Cultures

The economic importance of plants cannot be overstated. Plants are sources of food, construction materials, fabrics, paper, and fuel. They are also important sources of pharmaceuticals, flavors, dyes, pigments, resins, enzymes, waxes, and agrochemicals. These plant-derived chemicals are referred to as secondary metabolites, meaning that these compounds are not essential to the survival of the plant, but fortunately can serve as useful products for human use or consumption. Some of the commonly known plant-derived pharmaceuticals and their natural sources include morphine (*Papaver somniferum*), digoxin (*Digitalis lanata*), theophylline (*Camellia sinensis*), vinblastine (*Catharanthus roseous*), quinine (*Cinchona sp.*), and codeine (*Papaver sp.*) *(36)*.

Plant tissue culture technology has found application in a number of important areas, including crop improvement, mass plant production, secondary metabolite production, plant physiology, biochemistry, molecular biology, and plant genetic conservation *(37)*. All these areas of research, development, and production require techniques for the preservation and maintenance of plant cell genotypes and phenotypes to ensure consistent and predictable production of cells, tissues, and whole plants. Cryopreservation and storage of plant cells and tissues are fundamental technologies underlying the current progress in plant biotechnology. Plant biotechnology is an emerging field that holds great promise for the use of transgenic plants in the production of recombinant products for human therapeutic use, with several products currently in clinical trials *(38)*.

A remarkable biological property unique to plant cells is known as totipotency. Unlike differentiated animal cells, many plant cells possess the inherent ability to generate the specialized cell types and structures that comprise an entire plant. Therefore, by appropriate laboratory manipulation of culture conditions, undifferentiated plant cells can be induced to form a variety of plant tissues, including roots, stems, and leaves. In fact, an entire plant can be regenerated from a single cultured cell *(39)*. This striking biological phenomenon has opened avenues for important progress in genetic manipulation of plants. One study demonstrated the successful application of cryopreservation methods for the regeneration of fertile maize (*Zea mays*) plants from protoplasts of elite maize inbreds *(40)*. Successful application of cryopreservation techniques to such research could have significant impact on genetic improvement of maize and other agricultural crops that are grown worldwide.

Although some success has been achieved in the freeze-drying of pollen, this preservation technique has not been successfully applied to other plant tissue *(37)*. The first report of successful cryopreservation and short-term storage of a plant cell suspension (i.e., flax cells) was published in 1968 *(41)*. That initial work opened the door to development and refinement of cryopreservation techniques over the last 15–20 yr for a variety of plant materials. In contrast to other cell types, the pre-freeze acclimation of cells and tissues (i.e., "cold-hardening") is one area of development in cryopreservation techniques that has been emphasized in plant-cell and tissue culture. The transient stage referred to as "pre-growth" provides an opportunity to increase the freeze tolerance of cells and tissues by adjusting culture conditions (e.g., reduced temperature, presence of cryoprotectants) prior to cryopreservation without inducing phenotypic or genotypic selection *(37)*. The phenomenon of "cold-hardening," or low-temperature acclimation, is more critical for cryopreservation of organized plant tissues such as shoot/meristem-tip cultures or callus cultures. Although such methods have also been developed to enhance post-thaw viability and regrowth of plant cell suspensions and protoplasts, it has been found that exposure to low, non-freezing temperatures induces changes in the proportions of all lipid components in plant-cell membranes, particularly in the composition of sterols and phospholipids *(42)*. Steponkus and Lynch *(42)* demonstrated that the alterations in lipid composition of cell membranes during exposure to low temperatures (e.g., 0° to –5°C) were correlated with increases in plasma membrane stability during freezing and associated freeze-tolerance. During pre-growth, in addition to low temperatures, the cells or tissues may be exposed to combinations of amino acids, cryoprotectants such as DMSO, and sugar alcohols such as mannitol or sorbitol to provide further pre-freeze conditioning *(37)*.

A number of cryoprotectants have been identified as effective in freezing plant cells and tissues, including DMSO, glycerol, sucrose, mannitol, trehalose, sorbitol, and amino acids such as proline. In general, combinations of these compounds have been empirically tested to optimize post-thaw viability and regrowth potential. An extensive list of cryopreservation methodologies for various plant materials has been presented by Owen *(43)*. An overview of this list reveals the use of DMSO (5–15% [v/v]), typically in combination with glycerol (5–15% [v/v]), and sucrose (0.5 to 1 *M*) as cryoprotectants. The reader is referred to Owen's list for specific laboratory applications.

As observed with other cell types, laboratory methods that involve slow, stepwise cooling rates during freezing followed by rapid warming during thawing, generally produce the highest post-thaw recovery of viable cells and plant materials. However, the development of optimal cryopreservation procedures for a given plant cell or tissue remains an empirical process. However, some generalizations can be made regarding successful methods. Generally, a cooling rate in the range of –0.25°C to –2.0°C/min, down to –30° to –40°C, held for 30–60 min, and finally rapid cooling to liquid nitrogen temperature, has proven to be effective for many species and different types of plant materials. Typically, a cooling rate of –1.0°C/min (to –40°C) is used for cell suspensions, protoplast cultures, callus cultures, and immature embryos *(20,37,43–45)*. These frozen plant materials are stored either in vapor phase or liquid phase of liquid nitrogen and rapid thawing of cryovials in a pre-warmed water bath at 35–40°C are proven methods for the successful regrowth of a majority of frozen plant cultures and materials.

A generally applicable procedure for cryopreservation of plant-cell materials is based on a model developed for cell suspension cultures *(37)*. This model protocol employs the following methods:

- use of cells harvested in early exponential growth phase (cell division at maximum rate) when cells are small and highly cytoplasmic as opposed to large, highly vacuolated cells found at later growth stages
- use of a cryprotectant solution containing 1 *M* DMSO, 1 *M* glycerol, and 2 *M* sucrose in standard culture medium
- pre-freeze acclimation of cell suspension in cryoprotectant solution at 4°C for 1 h
- cooling rate of –1°C/min to –35°C, hold for 40 min and then plunge into liquid nitrogen
- store directly in liquid nitrogen, or in vapor phase above liquid
- rapid warming of frozen ampoules in a 40°C water bath
- transfer of cells to a semisolid medium for recovery (2–4 wk) followed by inoculation into liquid medium and cultivation in shake flasks

For a detailed review of methods optimized for a variety plant materials including shoot tip (i.e., meristem), callus, protoplast, and embryo cultures, the reader is referred to Withers *(37)*. Additionally, for laboratory protocols specific for cryopreservation of protoplasts and cell suspensions, the reader is referred to Grout *(45)* and Schrijnemakers and Van Iren *(20)*, respectively.

It should be noted that such a model protocol as outlined here may not be universally successful. For particularly sensitive cell suspension cultures, a more directed empirical approach may be required to optimize cryopreservation conditions for maximum cell recovery and re-growth. For example, it has been found that the survival and regrowth of two-cell suspension lines, *Cinchona robusta* (quinine and derivatives) and

Tabaernaemontana divaricata were markedly increased when a cryoprotectant mixture or "cocktail" (proline/DMSO/glycerol) was modified to include sucrose as a substitute for proline *(46)*. What would appear to be a rather simple change had a significant impact on the viability and regrowth potential of these cell lines. The study demonstrated that larger mol-wt solutes (e.g., sucrose) that act as extracellular components in standard cryoprotectant mixtures could have a significant effect on the success of the cryopreservation method. Whereas these large mol-wt compounds exert their cryoprotective effects extracellularly during freezing and thawing, the contributions and mechanisms of action of these components in such a cryoprotectant cocktail are not precisely understood.

7. Mammalian Cell Cultures

Mammalian cell lines have become established as the true "work horses" for the industrial production of complex protein molecules that require post-translational modifications such as the proper glycosylation required for stability and biological activity in humans. Mammalian cells have been cultured in vitro since the beginning of the twentieth century. In 1955, a chemically defined basal medium, developed by Harry Eagle at the National Institutes of Health, opened the door to cell culture on a mass scale and ushered in the modern era of industrial cell culture. His formulation was comprised of key groups of nutrients including amino acids, vitamins, salts, and sugars. No longer were the development of cell lines and tissue cultures limited by the availability of crude extracts derived from chicken plasma, human serum, bovine embryos, and human placental-cord serum. The basal medium developed by Eagle remains one of the fundamental and universal medium formulations in use today, and is sold as Eagle's Minimum Essential Medium. Despite these early advances in the development of chemically defined media formulations, today's cell culturist still relies extensively on the use of fetal bovine serum (FBS) for nutritional supplementation of media, both for cell propagation, and in particular, for cryopreservation.

Not surprisingly, the methods employed for cryopreservation of mammalian cell cultures are a reflection of those already described for other cell types. The most commonly used cryoprotectant is DMSO. Typically, DMSO is used at a concentration of 7.5% (v/v) or 10% (v/v). However, it is has been used successfully at a concentration as low as 5% (v/v) for many cell lines (personal observations, *47*). Use of a lower concentration of DMSO has the benefit of quicker post-thaw removal of this toxic agent from cells upon dilution with a growth medium. The DMSO must be of the highest grade, and must either be provided as a sterile solution by the vendor or filter-sterilized by the user. Typically, DMSO manufactured and tested as "cell-culture grade" is purchased sterile from a vendor. The sterile DMSO can be purchased in 5- or 10-mL aliquots in sealed amber glass ampoules, which have an extended shelf-life. It must be stored at room temperature, as it will gel or solidify at refrigerator temperatures.

When preparing the final cryopreservation medium, DMSO should be added slowly to cold medium while mixing because of the heat generated by this exothermic reaction. Because of the inherent toxicity of DMSO, particularly at higher temperatures, the final cryopreservation medium must be cold (e.g., 2–8°C) and added quickly to cells. During the process of filling cryovials, the duration of exposure of cells to DMSO must be minimized to reduce the detrimental effects of increased osmolarity and pH of the cryopreservation medium. Interestingly, DMSO has both hydrophobic and hydrophilic properties, depending on temperature *(48)*. The toxic effects of DMSO at higher temperatures (e.g., room temperature) is are a result of the fact that it exhibits a hydrophobic character and preferentially binds to proteins, thus leading to denaturation. Conversely, at low temperatures (e.g., below 0°C), DMSO exhibits a hydrophilic quality and is preferentially excluded from the surface of proteins, thus leading to stabilization of the folded proteins.

Typically, mammalian cells are harvested for cryopreservation during mid-to-late logarithmic growth phase during active cell division, when mitotic indices are high. Cells that have entered a stationary or quiescent phase (i.e., G_0) should not be used. The effect of the cell cycle on recovery of cells has been investigated. One such study reported that Chinese hamster cells were most resistant to the stresses of freezing and thawing when harvested in the M and late S phases of the growth cycle, but least resistant in G_2 *(49)*. Another study reported similar results for HELA S3 cells, for which the highest recovery rate was found when cells were frozen in the mid-to-late S phase and lower in G_2 *(50)*. Most importantly, cultures with viability measurements of less than approx 90% should not be used for cell-banking purposes (personal observations, and *51*). However, others have recommended not freezing cell cultures with viabilities of less than 80% *(52)*. In either case, both cell viability and cell-cycle or growth phase are crucial factors in successful cryopreservation. Upon harvest, cells are centrifuged, and the resulting pellets are gently resuspended in cryopreservation medium using a larger bore pipet (e.g., 10 mL) to prepare a uniform single-cell suspension whenever possible. The viable cell density used for filling cryovials should fall into the range of 4×10^6/mL to 2×10^7/mL (personal observations, *51,52*).

Generally, the optimal cooling rate for mammalian cells is –1°C/min to a temperature of about –60°C, when the cooling rate is increased to –10° to –30°C/min down to a final hold temperature below –100°C before transfer to the vapor phase of liquid nitrogen. However, cooling rates in the ranges of –1° to –5°C/min *(19)* or –1° to –3° C/min *(51)* have been reported, and optimal rates are determined empirically. At the European Collection of Animal Cell Cultures (ECACC), a cooling rate of –3°C/min in a programmable freezer is used for the majority of cells *(51,52)*. The ECACC has observed post-thaw viabilities that typically exceed 85% using their controlled freezing process *(51)*. At Eli Lilly and Company, we have optimized and validated a controlled-rate freezing program in which the initial cooling rate is –1.0°C/min down to –60°C, followed by a more rapid cooling rate of –10°C/min down to a final holding temperature of –110°C before transfer to the liquid nitrogen vapor phase for cryogenic storage.

Typically, frozen cells are thawed as rapidly as possible in a warm water bath (e.g., 37–40°C) in order to effect a high warming rate. However, it was reported that the cooling rate had a markedly greater impact than warming rate on post-thaw recovery of a Chinese hamster ovary (CHO) cell line *(53)*. These authors reported that at the slowest cooling rate of –1.7°C/min (found to be optimal) the CHO cells were relatively unaffected by the warming rate. However, it should be noted that seemingly very high warming rates in the range of +140°C to +670°C/minute were tested. Additionally, CHO cell recovery was significantly affected by warming rates when the cells were frozen at the higher cooling rates of –10°C and –100°C/min.

Once thawed, cells should immediately be diluted with fresh growth medium at either ambient or refrigerator temperatures. It has been noted that dilution of freshly thawed cells with medium at room temperature may be less stressful and may facilitate cell recovery *(52,54)*. It is also recommended to slowly dilute the cryoprotectant from freshly thawed cells in order to reduce the osmotic shock of changes in cell volumes *(18,51,55)*. Slow dilution is particularly important if prewarmed medium is used, and may be accomplished by dropwise addition of fresh growth medium (first few milliliters) to the cells with gentle agitation. The use of prewarmed medium for post-thaw dilution has the major disadvantage of increasing the inherent toxicity of DMSO seen at higher temperatures. Alternatively, cold medium (e.g., 2–8°C) can be used to minimize the toxicity of DMSO exposure to cells after thawing (personal observations). As the temperature of the medium is decreased, it may be added to cells at a faster rate, presumably because of the reduction of membrane permeability at lower temperatures.

Post-thaw centrifugation of cells should be done at a minimal *g*-force (e.g.,100*g*) in order to reduce shear forces on cell membranes that have been altered by the presence of DMSO. However, the centrifugation *g*-force and time must be adequate to effectively pellet all (or most) of the intact cells, or a significant loss of viable cells may result (personal observations).

As for all other cell types, it is a universally recommended practice to store mammalian cells in the vapor phase of liquid nitrogen at temperatures lower than –130°C. One issue of particular importance for the storage of mammalian cell cultures in a liquid nitrogen environment is the potential for contamination with viruses. Transmission of two bovine viruses—bovine viral diarrhea virus (BVDV) and bovine herpesvirus-1 (BHV)—to frozen bovine embryos was demonstrated during storage directly in liquid nitrogen *(56)*. The liquid nitrogen was experimentally inoculated with viruses and unsealed containers of embryos were plunged and stored in the liquid phase. After 3–5 wk of storage, 21.3% of embryos tested positive for viral contamination. Conversely, all control embryos sealed in cryovials were free from contamination. This study illustrates the importance of maintaining cryovials in the vapor phase at all times during storage. As the reader may recall from **Subheading 4.**, plastic cryovials used for cryopreservation are not designed for storage in liquid and are prone to leakage if exposed to liquid nitrogen. This is particularly critical for storage in dewar vessels that are filled manually, where liquid nitrogen levels may not be monitored carefully and cryovials are allowed to be immersed in liquid nitrogen for varying lengths of time.

Because of the concern for contamination of mammalian cell lines with viruses, certain studies have eliminated the use of FBS or other serum sources (which may harbor viruses) in the formulations of cryopreservation media. Serum—particularly FBS—is used universally as a cryoprotective additive to provide a source of proteins and other factors to enhance the long-term cryostability of cells. The mode of action of serum during the freezing and thawing process is unknown, or at least has not been precisely described in the literature. However, FBS is well-recognized by laboratory practitioners for its anecdotal protective effects on cells during cryopreservation (personal observations, *57*). Although FBS is typically used in concentrations ranging from 5% (v/v) to 50% (v/v), it has been used at concentrations as high as 90% (v/v) for the development of hybridoma cell lines (i.e., early post-fusion cultures) and other particularly sensitive cell lines. Whereas there have been universal efforts in the development and use of serum-free and protein-free media for propagation of cells, by comparison, the development and use of serum-free cryopreservation media have received much less attention.

One early study in this area found that the addition of 0.1% (w/v) methylcellulose to freezing medium containing only 10% (v/v) DMSO and basal Minimal Essential Medium (MEM) increased by almost twofold the post-thaw viability of a mouse L cell line (L.P3) over the same freezing medium without methylcellulose (58). Additionally, Ohno et al. (58) reported the successful cryopreservation of HeLa cells and various hybridoma cell lines using serum-free freezing medium containing 0.1%(w/v) methylcellulose. Other investigators have reported the successful cryopreservation of a mouse hybridoma cell line and three transformed human lymphoblastoid cell lines using serum-free freezing medium containing 4.5 mg per mL of human serum albumin as a substitute for FBS (59). Another study reported comparable post-thaw viabilities and antibody titers of a hybridoma cell line frozen in medium containing 0.4 % (v/v) or 0.8% (v/v) bovine serum albumin (BSA) in place of 15% (v/v) fetal calf serum (60). Merten and Couveé (61) reported the successful replacement of 10% (v/v) fetal calf serum with either 0.1% (w/v) methylcellulose or 3.0% (w/v) polyvinyl pyrrolidone in freezing media for preservation of Vero and BHK-21 cell lines. It should be noted that in all these studies, the cell lines were propagated in serum-free media prior to freezing, and all cryopreservation media contained 5% (v/v) or 10% (v/v) DMSO. It should also be emphasized that replacement of FBS with human or BSA does not eliminate the risk of virus transmission to cell lines from these animal-sourced materials, unless they have been produced by recombinant methods using animal-source free processes.

Other cryoprotective compounds (of non-animal origin) used in cryopreservation media to enhance recovery of mammalian cells include trehalose and S-adenosylmethionine. It is currently understood that trehalose exerts its cryoprotective properties through stabilization of membranes and proteins both intracellularly and extracellularly during freezing and thawing. Trehalose, in combination with DMSO, has been introduced through the membranes of human pancreatic islet cells during the phase transition of membrane lipids, which increases membrane permeability (62). Another group has used a genetically engineered alpha-hemolysin to create pores in the cell membranes of 3T3 fibroblasts and human keratinocytes to induce uptake of trehalose into cells without the synergistic action of DMSO (63). The addition of S-adenosylmethionine to cryopreservation media containing DMSO proved to increase both viability and the metabolic activity of thawed rat hepatocyte cultures (64).

8. Facilities for Preparation and Storage of Cell Banks

Laboratory facilities used for the preparation and storage of cryopreserved cell banks should be designed and operated to preserve the purity (i.e., axenicity) of the cell cultures during propagation, harvest, and filling processes and to maintain the integrity of frozen cultures during long-term storage in liquid nitrogen vessels. At Eli Lilly and Company, cell-banking facilities are designed to provide a cleanroom-type environment through the use of HEPA-filtered air supplies, air-pressure differentials, air locks, gowning/degowning rooms, cleanable surfaces, and segregation of operations. The facilities are designed for unidirectional traffic flow of both personnel and equipment (see Fig. 5). The core cell-culture laboratories lead to the cryogenic storage areas through degowning rooms and airlocks.

At Eli Lilly and Company, each cryogenic storage room is equipped with automatic-fill liquid nitrogen cryogens for storage of cell banks in the vapor phase at approx −150°C or lower. Liquid nitrogen is delivered to the storage cryogens from an outside bulk storage tank through a vacuum-jacketed insulated piping system. The liquid nitrogen bulk tank is serviced by a vendor, and is continuously monitored for level and pressure. Each storage cryogen is continuously monitored and alarmed for temperature through a validated computerized system. Access to the storage cryogens is controlled both through the use of card readers and keypad password control.

Each cryogen has been validated for maintenance of required temperatures (i.e., <−130°C) throughout the vapor-phase storage compartment. This validation study was accomplished by the use of a calibrated temperature logging instrument (i.e., Validator® 2000, Kaye Instruments, Inc.) interfaced with thermocouples placed in representative locations within storage racks arranged to provide for maximum capacity. At Eli Lilly and Company, validation studies involved the use of 12 thermocouples distributed throughout the cryovial storage racks to test all positions from top to bottom within the vapor-phase storage areas of the cryogen (*see* Fig. 6). Those studies demonstrated the capability of the cryogens to maintain temperatures below −145°C at all locations and levels tested in the vapor-phase storage compartments even when lids were opened for 2 min (unpublished validation data). It is recommended that cryogens be validated to maintain the required vapor-phase temperature of <−130°C in all locations where cryovials are stored.

It is essential that liquid nitrogen storage vessels are located in a well-ventilated room to prevent depletion of oxygen by nitrogen gas. To safeguard laboratory personnel, oxygen monitors and alarms should be located within the room. For example, at the ECACC an oxygen monitor has been connected to an automatic ventilation system that operates when the oxygen level decreases to 18.5% (v/v) *(52)*. Similarly, at Eli Lilly and Company, each cryogenic storage room is equipped with two oxygen monitors and associated audible and visual alarms to alert personnel to reduced oxygen levels.

9. Current Quality/Regulatory Issues

The elimination of animal-sourced or animal-derived raw materials in cell banks is an emerging issue that impacts global licensing and marketing of biotechnology products generated through fermentation of cell cultures. Because of the continuing threat of transmissible spongiform encephalopathy (TSE)- and bovine spongiform encephalopathy (BSE)-related diseases, global regulatory agencies are closely scrutinizing the use of animal-sourced raw materials used either in the medium to propagate or preserve cell cultures used in manufacturing processes. At Eli Lilly and Company, we have been actively substituting all animal-derived medium components with animal-source medium (ASM)-free components, whenever possible. Unfortunately, in some cases—including the use of FBS for cryopreservation of mammalian cell cultures—there is currently no comparable substitute that will provide proven long-term stability to frozen cultures. This is becoming an active area of investigation for those in the biotechnology industry who use mammalian or insect cells in manufacturing processes. In other cases, we have successfully replaced animal-sourced peptones and glycerol with plant-derived or synthetic alternatives for both growth and cryopreservation, respectively, of recombinant *E. coli* cultures. Similarly, ASM have been removed from media used

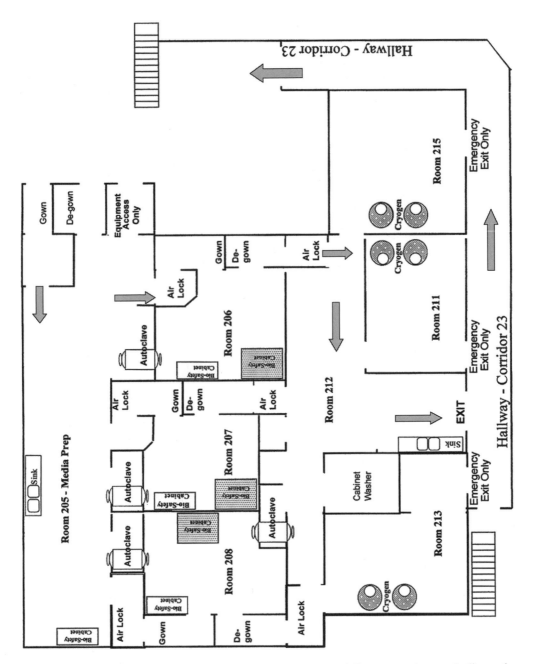

Fig. 5. Layout of cell-banking laboratories at Eli Lilly and Company. Arrows indicate the direction of personnel and equipment flow through the facility. Doors are not shown.

to propagate and freeze the cell-bank cultures of various antibiotic-producing microbial strains, including species of *Streptomyces* and *Amacolatopsis*. Additionally, ongoing efforts to replace FBS and serum derivatives in media used to cryopreserve mammalian cells (*see* **Subheading 7.**) is a result of the previously discussed global regulatory concerns.

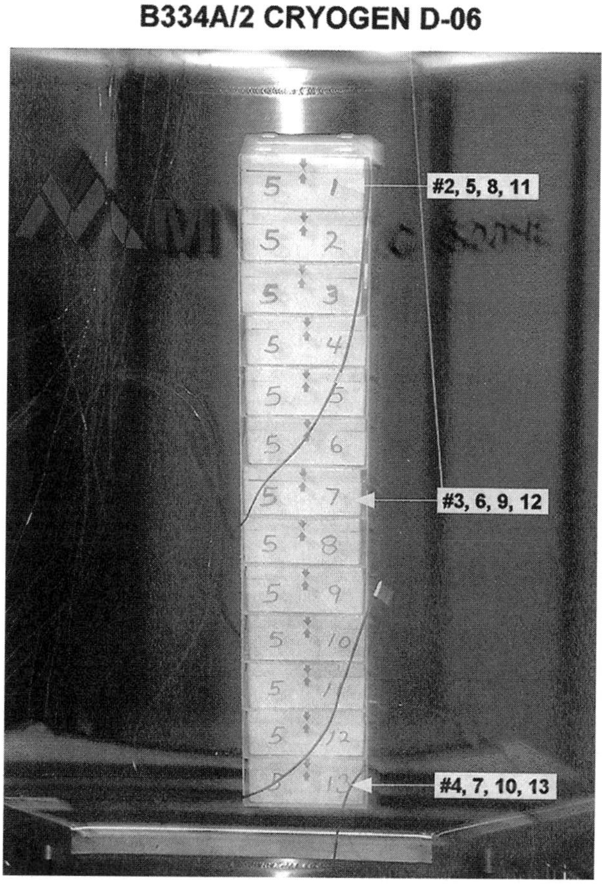

B334A/2 CRYOGEN D-06

Fig. 6. Digital photograph of a cryovial storage rack showing placement of thermocouples from top to bottom in the cryogens during validation studies. Four thermocouples were placed in cryovials within the top, middle, and bottom storage box (i.e., box #1, 7, and 13). The numbers with arrows indicate the thermocouple identification number.

The following list provides general guidance on quality and regulatory issues that influence the control and testing of cell banks for commercial use:

- The history of cells used to manufacture biotechnology products must be included in the product registration application package. The history of cells can identify potential risks and testing requirements associated with their use in a manufacturing process. Some commonly used host cells have already established safety profiles, and are more easily accepted by regulatory agencies. In any case, information should be gathered and filed on: 1) parentage/origin, including scientific references; 2) laboratory manipulations, including transfection/cloning records; 3) past testing results, including mycoplasma, microbial, and viral assays.
- All incoming cultures for banking should be obtained from a pre-approved or reputable source with appropriate documentation. Incoming cells should be segregated into a designated quarantine area until appropriate test results are confirmed (e.g., purity).

- Raw materials used in the preparation of cell banks can have a significant impact on the regulatory acceptance of a manufacturing process. One of the key areas of concern is the exposure of cells to animal-derived materials (e.g., trypsin, FBS) used during propagation and cryopreservation. Use of non-animal-sourced components (or defined protein-free media) is recommended. All FBS should be purchased from a vendor who can document sourcing from New Zealand herds. Additionally, the vendor must conduct appropriate testing to confirm absence of bovine viruses. As an added precaution, all FBS should be gamma-irradiated by a reputable vendor.

- The critical nature of cell-banking operations makes it essential to maintain accurate and detailed records of production, testing, and inventory transactions. All records should be filed to enable efficient retrieval for use in regulatory submissions, inspections, and quality audits. As an example, the following documents should be available: 1) manufacturing ticket or record; 2) pre-bank testing results; 3) certificates of analysis on raw materials; 4) environmental monitoring records; and 5) copies of relevant R&D notebook records, and 6) post-bank testing records.

- It is critical to conduct both pre-bank and post-bank testing on cells destined for manufacturing use. Pre-bank quality testing may include the following: 1) sterility (purity); 2) mycoplasma; 3) viral assays specific for the host-cell species; 4) bacteriophage; and 5) phenotyping. Post-bank testing may include all of these plus the following, 1) additional viral assays, both in vivo and in vitro, as required for a given cell type; 2) DNA sequencing; 3) plasmid copy number; and 4) restriction enzyme analysis.

- Stability of cell banks during storage should include testing for post-thaw viability (e.g., trypan blue or fluorescein diacetate staining), population doubling time, and consistent expression of product (e.g., yields or glycosylation patterns).

- Shipping and transport of critical cell banks (e.g., MCB, WCB) should be done in a vapor phase liquid nitrogen shipper designed and validated to maintain temperatures of <–130°C in the compartment holding cryovials. In this type of shipper, all liquid nitrogen is adsorbed into the liner of the vessel before shipment, thus creating a vapor phase environment and reducing the safety hazards of liquid nitrogen. *See* Fig. 7 for an example of a shipper designed for this purpose. It is also advisable to include some type of temperature logging/thermocouple device in the shipping container to record actual temperature conditions during transit. Data stored in these devices can be downloaded to a personal computer, and hard-copy printouts can be generated and filed for regulatory and quality review.

Interested readers are referred to key guidance documents *(65,66)* for additional information on current international regulatory and quality standards applied to biotechnology products derived from cell cultures.

10. Current and Emerging Trends

One important focus of investigation includes the continued search for animal-source-free (e.g., serum-free) cryopreservation medium formulations that are truly effective for long-term storage of cell banks—years to decades. Short-term studies have identified some potentially useful cryostabilizing agents such as methylcellulose, polyvinyl pyrrolidine, trehalose, and S-adenylsylmethionine, which appear to provide some measure of protection during freezing and thawing. However, long-term studies are needed to provide data on the efficacy of these and other agents, either alone or in combination for storage of cell cultures for the years and decades required to maintain the integrity of MCB and WCB used to support commercial manufacturing.

The successful development and availability of recombinant proteins such as the recombinant human serum albumin (Recombumin® 20%), now in clinical trials from

Aventis Behring, may prove to be effective replacements for FBS, which is universally used in cryopreservation media to stabilize frozen mammalian cell cultures. Unfortunately, unit costs are high for such products marketed for human therapeutic use, but use of small quantities required for cell banking should not be cost-prohibitive. Alternatively, the availability of a recombinant BSA product targeted for both research and commercial use could prove to be an excellent cost-effective substitute for FBS and bovine serum derivatives. In the meantime, to avoid TSE transmission potential in commercial products, any animal-derived materials used to propagate and cryopreserve MCB and WCB should be sourced from New Zealand or other countries that are documented to be BSE/TSE-free. Additionally, to further reduce the potential risk of virus transmission, these materials (e.g., FBS, BSA, or trypsin) should be gamma-irradiated by a qualified vendor. Alternatively, the use of plant-derived proteins and peptones may serve as effective substitutes for FBS and animal serum derivatives.

As mentioned in **Subheading 6.**, the introduction of technology for the plant-based production of biopharmaceuticals provides a promising avenue for avoidance of the regulatory and quality risks associated with use of animal-derived raw materials and use of animal cell lines. This alternative approach to the production of recombinant proteins has been developed using transgenic plants as so-called "in vivo bioreactors" *(38)*. It has been found that plants possess all the necessary cellular machinery for the post-translational modification and maturation of eukaryotic proteins, although differences do exist between mammalian and plant glycosylation patterns. This promising technology has generated several products that are now in clinical trials. Gruber and Theisen *(38)* predicted that within the next year or so, the first products will be licensed and will enter the marketplace.

As a result of the increasing scrutiny and inspection of commercial cell banks by global regulatory authorities, other key areas for further development and implementation to increase the degree of regulatory compliance include: 1) cell-bank computer-inventory databases; 2) equipment used to freeze, store, and ship cell banks; and, 3) state-of-the-art assays and procedures used to characterize the quality and genetic stability of cell banks. Recent inspections of biopharmaceutical companies by the Food and Drug Adminstration (FDA) have generated written observations (i.e., FDA inspection form "483") addressing all of these areas, particularly validation of equipment and databases (personal observations). Another area that has more recently gained the attention of regulatory authorities is the procedure or program used to monitor the stability of commercial cell banks during long-term cryogenic storage. This program or set of procedures should include measurements of post-thaw cell viability, product expression, and/or genetic stability at some predetermined frequency—for instance, as cryovials of MCB and WCB are thawed for use in clinical trial or commercial production *(65)*. Additionally, it is well-recognized by regulatory agencies that genetic stability is a critical component of the overall testing program required to fully qualify (i.e., certify) recombinant cell banks for commercial use. These tests, such as DNA sequence analysis and restriction enzyme mapping, are required to ensure genetic stability of the expression construct throughout the cell passages or generations (i.e., in vitro cell age) needed for scale-up through the end of fermentation, and even extended cell growth beyond the final bioreactor harvest. For more information on this issue, the reader is referred to a practical overview of the subject *(67)* and a key international regulatory document that provides guidance in the area *(68)*.

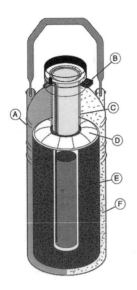

A Lightweight aluminum design reduces shipping costs

B All models come with locking tab for shipments

C Neck Tube– High strength neck tube reduces liquid nitrogen loss

D Advanced Chemical Vacuum Retention System provides superior vacuum performance for the life of the product

E Superior hydrophobic absorbent– repels moisture and humidity while maintaining a -150°C chamber environment

F Insulation — MVE's advanced insulation system provides maximum thermal performance

One model designed and approved to meet IATA and U.N. requirement for the shipment of infectious substances.

Fig. 7. Drawing of a vapor phase ("dry") liquid nitrogen shipping Dewar vessel. The heavily insulated protective case is not shown. Note the hydrophobic adsorbent liner (**E**). (Reproduced by permission of MVE, Inc.).

11. Conclusions

It is hoped that the reader has gained a deeper understanding and appreciation of the many similarities between laboratory methods employed in the successful cryopreservation of three very diverse categories of cells used in biotechnology—microbial, plant, and mammalian. As this chapter has illustrated, cryopreservation procedures are remarkably similar for very dissimilar organisms. As the literature and personal experiences suggest, all cells to be preserved by freezing share the same requirements for chemical cryoprotectants: slow rates of cooling, rapid rates of warming, optimal metabolic state (e.g., growth phase), cell concentrations, and storage at liquid nitrogen temperatures. However, it is obvious that all cells are not created equal. Thus, the development of an optimal cryopreservation protocol still requires an empirical approach to "fine tune" laboratory methods for a particular cell culture. It is hoped that this chapter provides useful guidance to the reader in their quest for that optimal process required to successfully cryopreserve cell cultures and create cell banks that can be used effectively for many years in industrial bioprocesses.

Acknowledgments

First, I want to thank Dr. Victor A. Vinci for providing me with the opportunity to contribute to this book. I thank my co-workers Robert Weeks, Dennis Genier, Renee Lawless-Justice, and Steven Mitchell for their technical assistance in completing the validation studies cited in this chapter. I also would like to thank Dr. William L. Muth for access to some unpublished data, as well as his expertise in reviewing and editing the manuscript. Finally, I am especially grateful to Dr. K. Roger Tsang for his invaluable mentorship over the years and the expert training he so unselfishly provided in the art and science of cell culture.

References

1. Devireddy, R. V., Barratt, P. R., Storey, K. B., and Bishof, J. C. (1999) Liver freezing response of the freeze-tolerant wood frog, *Rana sylvatica*, in the presence and absence of glucose; I. Experimental measurements. *Cryobiology* **38,** 310–326.
2. Pegg, D. E. (1994) Cryobiology: life in the deep freeze. *Biologist* **41,** 53–56.
3. Baust, J. G. and Miller, L. K. (1969) Mechanisms of freezing tolerance in an Alaskan insect. *Cryobiology* **6,** 258–259.
4. Polge, C., Smith, A. U., and Parkes, A. S. (1949) Revival of spermatozoa after vitrification and dehydration at low temperatures. *Nature* **164,** 666.
5. Meryman, H. T. (1971) Cryoprotective agents. *Cryobiology* **8,** 173–183.
6. Lovelock, J. E. and Bishop, M. W. H. (1959) Prevention of freezing damage to living cells by dimethyl sulphoxide. *Nature* **183,** 1394–1395.
7. Matthes, G. and Hackensellner, H. A. (1981) Correlations between purity of dimethyl sulfoxide and survival after freezing and thawing. *Cryo-Letters* **2,** 389–392.
8. Withers, L. A. (1990) Cryopreservation of plant cells, in *Methods in Molecular Biology, Vol. 6: Plant and Tissue Culture* (Pollard, J. W. and Walker, J. M., eds.), Humana Press, Inc., Totowa, NJ, pp. 39–48.
9. Grout, B. W. W. (1995) Cryopreservation of plant protoplasts, in *Methods in Molecular Biology, Vol. 38: Cryopreservation and Freeze-Drying Protocols* (Day, J. G. and McLellan, M. R., eds.), Humana Press, Inc., Totowa, NJ, pp. 91–101.
10. Fahy, G. M. (1986) The relevance of cryoprotectant 'toxicity' to cryobiology. *Cryobiology* **23,** 1–13.
11. Leibo, S. P. and Mazur, P. (1971) The role of cooling rates in low-temperature preservation. *Cryobiology* **8,** 447–452.
12. Mazur, P. (1984) Freezing of living cells: mechanisms and implications. *Am. J. Physiol.* **247,** C125–C142.
13. Costanzo, J. P., Lee, R. E. (Jr.), and Wright, M. F. (1992) Cooling rate influences cryoprotectant distribution and organ dehydration in freezing wood frogs. *J. Exp. Zool.* **261,** 373–378.
14. Mazur, P. (1970) Cryobiology: the freezing of biological systems. *Science* **168,** 939–949.
15. Grout, B. W. W. and Morris, G. J. (1987) Freezing and cellular organization, in *The Effects of Low Temperature on Biological Systems* (Grout, B. W. W. and Morris, G. J., eds.), Edward Arnold, London, UK, pp. 147–174.
16. Ashwood-Smith, M. J. and Friedmann, C. B. (1979) Lethal and chromosomal effects of freezing, thawing, storage time, and X-irradiation on mammalian cells preserved at –196° in dimethyl sulfoxide. *Cryobiology* **16,** 132–140.
17. Simione, F. P. and Brown, B. S., eds. (1991) Freezing methods, in *ATCC Preservation Methods: Freezing and Freeze-Drying, 2nd ed.*, American Type Culture Collection, Rockville, MD, pp. 5–46.
18. Simione, F. P., Jr., Daggett, P. M., McGrath, M. S., and Alexander, M. T. (1977) The use of plastic ampoules for freeze preservation of microorganisms. *Cryobiology* **14,** 500–502.
19. Doyle, A., Morris, C. B., and Armitage, W. J. (1988) Cryopreservation of animal cells, in *Advances in Biotechnological Processes, Vol. 7–Upstream Processes: Equipment and Techniques*, Alan R. Liss, Inc., pp. 1–17.
20. Schrijnemakers, E. W. M. and Van Iren, F. (1995) A two-step or equilibrium freezing procedure for the cryopreservation of plant cell suspensions, in *Methods in Molecular Biology, Vol. 38: Cryopreservation and Freeze-Drying Protocols* (Day, J. G. and McLellan M. R., eds.), Humana Press, Inc., Totowa, NJ, pp. 103–111.
21. Simione, F. P. and Brown, B. S., eds. (1991) Principles of freezing and freeze-drying, in *ATCC Preservation Methods: Freezing and Freeze-Drying, 2nd ed.*, 1991, pp. 1–4.

22. Heckly, R. J. (1978) Preservation of microorganisms, in *Advances in Applied Microbiology, Vol. 24*, pp. 1–53.
23. Carpenter, J. F., Crowe, L. M., and Crowe, J. H. (1987) Stabilization of phosphofructokinase with sugars during freeze-drying: characterization of enhanced protection in the presence of divalent cations. *Biochim. Biophys. Acta* **923,** 109–115.
24. MacKenzie, A. P. (1977) Comparative studies on the freeze-drying survival of various bacteria: Gram type, suspending media and freezing rate. *Dev. Biol. Stand.* **36,** 263–277.
25. Tanaka, Y., Yoh, M., Takeda, Y., and Miwatani, T. (1979) Induction of mutation in *Escherichia coli* by freeze-drying. *Appl. Environ. Microbiol.* **37,** 369–372.
26. Perry, S. F. (1995) Freeze-drying and cryopreservation of bacteria, in *Advances in Biotechnological Processes, Vol. 7–Upstream Processes: Equipment and Techniques*, Alan R. Liss, Inc., pp. 21–30.
27. Nakamura, L. K. (1996) Preservation and maintenance of eubacteria, in *Maintaining Cultures for Biotechnology and Industry* (Hunter-Cevera, J. C. and Belt, A., eds.), Academic Press, San Diego, CA, pp. 65–84.
28. Simione, F. P. and Brown, B. S., eds. (1991) Bacteria and bacteriophages, in *ATCC Preservation Methods: Freezing and Freeze-Drying, 2nd ed.*, American Type Culture Collection, Rockville, MD, pp. 14–16.
29. Dietz, A. and Currie, S. (1996) Actinomycetes, in *Maintaining Cultures for Biotechnology and Industry* (Hunter-Cevera, J. C. and Belt, A., eds.), Academic Press, San Diego, CA, pp. 85–99.
30. Smith, D. S. and Kolkowski, J. (1996). Fungi, in *Maintaining Cultures for Biotechnology and Industry* (Hunter-Cevera, J. C. and Belt, A., eds.), Academic Press, San Diego, CA, pp. 101–132.
31. Hwang, S.-W. (1976) Investigation of ultralow temperature for fungal cultures III: viability and growth rate of mycelial cultures following cryogenic storage. *Mycologia* **60,** 377–387.
32. Bond, C. J. (1995) Cryopreservation of yeast cultures, in *Methods in Molecular Biology, Vol. 38: Cryopreservation and Freeze-Drying Protocols* (Day, J. G. and McLellan M. R., eds.), Humana Press, Inc., Totowa, NJ, pp. 39–47.
33. Nierman, W. C. and Feldblyum, T. (1985) Cryopreservation of cultures that contain plasmids. *Dev. Ind. Microbiol.* **26,** 423–434.
34. Pearson, B. M., Jackman, P. J. H., Painting, K. A., and Morris, G. J. (1990) Stability of genetically manipulated yeasts under different crypreservation regimes. *Cryo Lett.* **11,** 205–210.
35. Morris, G. J., Coulson, G. E., and Clarke, K. J. (1988) Freezing injury in Saccharomyces cerevisiae: the effect of growth conditions. *Cryobiology* **25,** 471–482.
36. Fowler, M. W. and Scragg, A. H. (1988) Natural products from higher plants and plant cell culture, in *Plant Cell Biotechnology*, (Pais, M. S. S., et. al., eds.), Springer-Verlag, Berlin, Germany, pp. 165–177.
37. Withers, L. A. (1991) Maintenance of plant tissue cultures, in *Maintenance of Microorganisms, 2nd ed.* (Kirsop, B. E., and Doyle, A., eds.), Academic Press Ltd., London, UK, pp. 243–267.
38. Gruber, V. and Theisen, M. (2000) Transgenic plants in the production of therapeutic proteins. *Innov. Pharm. Technol.* **6,** 59–63.
39. Cooper, G. M. (2000) An overview of cells and cell research, in *The Cell, A Molecular Approach, 2nd ed.*, ASM Press, Washington, DC, pp. 3–39.
40. Shillito, R. D., Carswell, G. K., Johnson, C. M., DiMaio, J. J., and Harms, C. T. (1989) Regeneration of fertile plants from protoplasts of elite inbred maize. *Bio/Technology* **7,** 581–587.
41. Quantrano, R. S. (1968) Freeze-preservation of cultured flax cells utilizing DMSO. *Plant Physiol.* **43,** 2057.
42. Steponkus, P. L. and Lynch, D. V. (1989) Freeze/thaw-induced destabilization of the plasma membrane and the effects of cold acclimation. *J. Bioenerg. Biomembr.* **21,** 21–41.
43. Owen, H. R. (1996) Plant germplasm, in *Maintaining Cultures for Biotechnology and Industry* (Hunter-Cevera, J. C. and Belt, A., eds.), Academic Press, San Diego, CA, pp. 197–228.

44. Withers, L. A. (1987) The low temperature preservation of plant cell, tissue and organ cultures and seed for genetic conservation and improved agricultural practice, in *The Effects of Low Temperatures on Biological Systems* (Grout, B. W. W. and Morris, G. J., eds.), Edward Arnold, London, UK, pp. 389–409.

45. Grout, B. W. W. (1995) Cryopreservation of plant protoplasts, in *Methods in Molecular Biology, Vol. 38: Cryopreservation and Freeze-Drying Protocols* (Day, J. G. and McLellan M. R., eds.), Humana Press, Inc., Totowa, NJ, pp. 91–101.

46. McLellan, M. R., Schrijnemakers, E. W. M., and Van Iren, F. (1990) The responses of four cultured plant cell lines to freezing and thawing in the presence or absence of cryoprotectant mixtures. *Cryo Lett.* **11,** 189–204.

47. Shannon, J. E. and Macy, M. L. (1973) Freezing, storage, and recovery of cell stocks, in *Tissue Culture Methods and Applications* (Kruse, P. F. and Patterson, M. K., eds.), Academic Press, New York, NY, pp. 712–718.

48. Crowe, J. H., Carpenter, J. F., and Crowe, L. M. (1990) Are freezing and dehydration similar stress vectors? A comparison of modes of interaction of stabilizing solutes with biomolecules. *Cryobiology* **27,** 219–231.

49. Koch, G. J., Kruuv, J., and Bruckschwaiger, C. W. (1970) Survival of synchronized Chinese hamster cells following freezing in liquid nitrogen. *Exp. Cell Res.* **63,** pp. 476–477.

50. Terasima, T. and Yasukawa, M. (1977) Dependence of freeze-thaw damage on growth phase and cell cycle of cultured mammalian cells. *Cryobiology* **14,** 379–381.

51. Morris, C. B. (1995) Cryopreservation of animal and human cell lines, in *Methods in Molecular Biology, Vol. 38: Cryopreservation and Freeze-Drying Protocols* (Day, J. G. and McLellan M. R., eds.), Humana Press, Inc., Totowa, NJ, pp. 179–187.

52. Doyle, A. and Morris, C. B. (1991) Maintenance of animal cells, in *Maintenance of Microorganisms, 2nd ed.* (Kirsop, B. E., and Doyle, A., eds.), Academic Press Ltd., London, UK, pp. 227–241.

53. Harris, L. W. and Griffiths, J. B. (1977) Relative effects of cooling and warming rates on mammalian cells during the freeze-thaw cycle. *Cryobiology* **14,** 662–669.

54. Armitage, W. J. and Juss, B. K. (1996) The influence of cooling rate on survival of frozen cell differs in monolayers and in suspensions. *Cryo Lett.* **17,** 213–218.

55. Armitage, W. J. (1987) Cryopreservation of animal cells, in *Symposia of the Society for Experimental Biology, Number XXXXI, Temperature and Animal Cells* (Bowler, K. and Fuller, B. J., eds.), The Company of Biologists Limited, pp. 379–393.

56. Bielanski, A., Nadin-Davis, S., Sapp, T., and Lutze-Wallace, C. (2000) Viral contamination of embryos cryopreserved in liquid nitrogen. *Cryobiology* **40,** 110–116.

57. Doyle, A. and Morris, C. B. (1996) Cryopreservation, in *Cell & Tissue Culture: Laboratory Procedures* (Doyle, A., Griffiths, J. B., and Newell, D. G, eds.), John Wiley & Sons, West Sussex, UK, pp. 4C:1.1–4C:1.7.

58. Ohno, T., Kurita, K., Abe, S., Eimori, N., and Ikawa, Y. (1988) A simple freezing medium for serum-free cultured cells. *Cytotechnology* **1,** 257–260.

59. Hanak, J. A. J., Chung, R., Lewis, G., Faulkner, J., and Davis, J. M. (1992) Serum free cryopreservation of antibody producing cells, in *Animal Cell Technology: Developments, Processes, and Products* (Spier, R. E., Griffiths, J. B., and MacDonald, C., eds.), Butterworth-Heinemann, Oxford, Boston, MA, pp. 134–136.

60. Werz, W., Maucher, J., Reutter, B., Werner, R. G., and Berthold, W. (1992) Cryoprotection and cultivation of hybridoma cell lines under serum free condition, in *Animal Cell Technology: Developments, Processes, and Products* (Spier, R. E., Griffiths, J. B., and MacDonald, C., eds.), Butterworth-Heinemann, Oxford, Boston, MA, pp. 149–151.

61. Merten, O.-W., Petres, S., and Couve, E. (1995) A simple serum-free freezing medium for serum-free cultured cells. *Biologicals* **23,** 185–189.

62. Beattie, G. M., Crowe, J. H., Lopez, A. D., Cirulli, V., Ricordi, C., and Hayek, A. (1997) Trehalose: A cryoprotectant that enhances recovery and preserves function of human pancreatic islets after long-term storage. *Diabetes* **46,** 519–523.

63. Eroglu, A., Russo, M. J, Bieganski, R., Fowler, A., Cheley, S., Bayley, H., et al. (2000) Intracellular trehalose improves the survival of cryopreserved mammalian cells. *Nat. Biotechnol.* **18,** 163–167.

64. Vara, E., Arias-Diiaz, J., Villa, N., Hernandez, J., Garcia, C., Ortiz, P., et al. (1995) Beneficial effect of S-adenosylmethionine during both cold storage and cryopreservation of isolated hepatocytes. *Cryobiology* **32,** 422–427.

65. The European Agency for the Evaluation of Medicinal Products, Human Medicines Evaluation Unit (1997) ICH Topic Q 5 D, *Quality of biotechnological products: Derivation and characterization of cell substrates used for the production of biotechnological/biological products.*

66. Annex 2, Manufacture of biological medicinal products for human use, in *Medicines Control Agency, Rules and Guidance for Pharmaceutical Manufacturers and Distributors 1997*, The Stationary Office, pp. 107–114.

67. Kittle, J. D. and Pimental, B. J. (1997) Testing the genetic stability of recombinant DNA cell banks. *BioPharm* 48–51.

68. The European Agency for the Evaluation of Medicinal Products, Human Medicines Evaluation Unit (1996) ICH Topic Q 5 B, *Quality of biotechnological products: Analysis of the expression construct in cell lines used for production of r-DNA protein products.*

Index

Note:
    f: figure
    t: table